Nano-phytoremediation and Environmental Pollution

The book discusses nano-phytoremediation: the use of nanotechnology in combination with phytoremediation to restore polluted environs. The potentiality of plants in association with nanomaterials to effectively remediate polluted areas is elaborated meritoriously in this book. New strategies are necessary because anthropogenic actions represent a serious threat to life on Earth. This book has given enough space for a discussion of innovative and efficient technologies to restore damaged environs primarily focused on nano-phytoremediation. The first part of the book is dedicated to exploring organic and inorganic pollution and the threats they pose to living forms. The second part explores the joint use of plants and nanomaterials and the nano-phytoremediation of water and soil ecosystems. The book offers readers extensive knowledge on nano-phytoremediation as a feasible strategy to clean environmental pollution. The key features of the book are as follows:

- Nano-phytoremediation strategies to remediate soil and water ecosystems.
- Special chapters dedicated to different kinds of pollutants and methods of phytoremediation.
- Strategies to evaluate the success of nano-phytoremediation strategies, cost-effectiveness, and nano informatics for safe nanotechnology.

The book can be used as a primary or supplementary text in undergraduate, graduate, and post-graduate courses such as biotechnology, biochemistry, and environmental engineering. It is an interesting edition for instructors, researchers, and scientists working on environmental management and pollution control.

Nano-phytoremediation and Environmental Pollution

Strategies and Mechanisms

Edited by Fernanda Maria Policarpo Tonelli,
Rouf Ahmad Bhat, Gowhar Hamid Dar, and
Khalid Rehman Hakeem

CRC Press
Taylor & Francis Group
Boca Raton London

CRC Press is an imprint of the
Taylor & Francis Group, an **informa** business

Cover credit: © Fernanda Maria Policarpo Tonelli

First edition published 2025
by CRC Press
2385 NW Executive Center Drive, Suite 320, Boca Raton FL 33431

and by CRC Press
4 Park Square, Milton Park, Abingdon, Oxon, OX14 4RN

CRC Press is an imprint of Taylor & Francis Group, LLC

ISBN: 978-1-032-03023-4 (hbk)
ISBN: 978-1-032-03024-1 (pbk)
ISBN: 978-1-003-18629-8 (ebk)

DOI: 10.1201/9781003186298

Typeset in Times
by Apex CoVantage, LLC

Contents

Foreword

The advancement of knowledge in different fields of expertise has made possible the development of elegant solutions to the problems that humanity faces and has drawn human beings' attention to concepts necessary to a harmonic life on the Earth.

Therefore, diverse strategies to deal with environmental pollution, a consequence of human actions that seek development but ignore sustainability principles, are being proposed. Plants are an exciting tool in this sense as they promote the phytoremediation of contaminants. The nanotechnology field is also a relevant source of these strategies involving the use of nanomaterials.

The book "*Nano-phytoremediation and Environmental Pollution: Strategies and Mechanisms*" includes 15 chapters to allow readers access to a comprehensive view of the theme.

The first chapter entitled "Environmental Pollution Caused by Organic Pollutants, Their Harmful Impacts, and Treatment through a Microbiological Approach" deals with the type of pollution caused by organic contaminants, their harmful effects, and treatment with the help of different microbial species. Bioremediation with the help of indigenous microbial communities as a novel and effective solution for cleaning up the environment, specifically, organic pollutants, is also discussed in this chapter.

Authors from Morocco, China, and Iran have contributed Chapter 02, "Nanomaterial's Interesting Characteristics to Remediate Polluted Environs." The chapter focuses on the recent developments in nanomaterials and nanocomposites to remediate contaminated environs. Furthermore, it highlights the advantages and limitations of different nanomaterials in the cleanup of heavy metals, volatile organic compounds, and herbicides.

Chapter 03, written by authors from Brazil, discusses important classes of carbon-based nanomaterial and their derivatives for the remediation of environmental pollutants. The chapter addresses the main strategies using carbon-based nanomaterial to remediate soil, water, and air pollution.

Chapter 04, "Non-Carbon-Based Nanomaterials Used for Environmental Pollution Remediation," is contributed by Indian scientists. They discuss at length the non-carbon nanoparticles in nanoremediation. The physicochemical characteristics of these non-carbon nanoparticles to resolve environmental contaminants are the focus of this chapter.

Chapter 05, entitled "Green-Synthesis of Nanomaterials for Environmental Remediation," was written by researchers from India. They concentrate on nanomaterials' synthesis by using green techniques. Furthermore, the authors conclude that the green-synthesis of nanomaterials has emerged as an area that primarily considers environmental benefits.

Chapter 06, titled "Comparison between Traditional and Nanoremediation Technology with Special Reference to Soil and Heavy Metal Contamination," is comprehensively presented to compare traditional remediation technologies with nanotechnologies for dealing with a broad spectrum of environmental contaminants. A particular emphasis is put on heavy metal remediation techniques.

"Joint Use of Nanomaterials and Plants for the Remediation of Inorganic Pollutants" is the topic of Chapter 07 and was written by authors from Pakistan and Malaysia. This chapter explores the important strategies of phytoremediation assisted by nanotechnology to address inorganic pollutants. The chapter centers on the biosynthesis of nanoparticles from plants to remove inorganic pollutants from the environment.

Omics technology use for environmental remediation is the focus of Chapter 08. This chapter titled "Integration of Nano-Phytoremediation and Omics Technology for Sustainable Environmental Cleanup" explores the advances and challenges related to phytoremediation, particularly the essential potential of nanoparticles omics site-directed genome editing technology CRISPR/Cas9.

Authors from India write in Chapter 09, entitled "Water Nano-Phytoremediation," to address nano-phytoremediation strategies applied to polluted water. Current developments in nanotechnology to simplify the nano-phytoremediation of polluted water by using nanomaterials are examined in this chapter.

Researchers from Pakistan discuss in Chapter 10 "Nanotools-Coupled Phytoremediation: Auspicious Technology for the Detoxification of Contaminated Pedosphere Matrices." They concentrate on nanotool-enabled phytoremediation in the detoxification of the pedosphere zone. This chapter specifically investigates the potential of phytoremediation in environmental cleanup, the interplay of different mechanisms in phytoremediation and nano-phytoremediation, and the challenges and prospects associated with the nano-phytoremediation carried out with different types of nanoparticles and plants.

Chapter 11 is entitled "Iron Nanoparticles for nano-phytoremediation" and is written by authors from China, India, Iran, and Morocco; it reviews the potential of iron-based nanomaterials as unique and promising alternatives to traditional environmental restoration techniques. Specifically, it contemplates nano bioremediation using iron nanomaterial as an emerging environmentally friendly clean technology that requires further utilization.

Chapter 12, titled "Silver Nanoparticles for Nano-Phytoremediation: Recent Advancements and the Potential of Nano Silver Consolidated Phytoremediation in Ecospheric Decontamination," reviews phytoremediation assisted by silver nanoparticles and its use as green nanotechnology. The authors from Pakistan appraise this nanomaterial as an important tool to restore polluted environs. This chapter comprehensively elaborates the role of Ag nanoparticles in a detoxifying environment in collaboration with different types of hyper-accumulator plants.

Chapter 13, entitled "Magnetic Nanoparticles for Nano-Phytoremediation," is written by an author from Morocco who reviews the magnetic nanoparticle-assisted phytoremediation of polluted ecosystems. The chapter summarizes the current progress in our perception of the role of magnetic nanoparticles for nano-phytoremediation processes in pollutant removal from different ecosystems.

Chapter 14, "Sustainability Aspects of Nanoremediation and Nanophytoremediation," is written by Indian authors who give special attention to the sustainability issues surrounding nanoremediation techniques and nanomaterials. They argue that the ecological risks posed by the use of nanoparticles need to be assessed and monitored. In addition, the need for innovative and advanced greener ways to synthesize nanoparticles is the central theme of this chapter.

In Chapter 15, the authors discuss some emerging methods for the treatment of wastewater that are easy to implement and environmentally feasible. Furthermore, they emphasize the utilization of advanced oxidation processes in the complete mineralization of organic contaminants.

The book unites authors worldwide to offer a multidisciplinary point-of-view on nanotechnology and phytoremediation. Their capacity provides efficient strategies to remediate environmental pollution that are aligned with the idea of sustainable development. This book is unique and should be an idyllic source of scientific information to research scholars, faculty, and scientists globally.

Balal Yousaf, Ph.D./Eng.D.
1) Associate Professor
Environmental Science and Engineering,
University of Science and Technology of China, Hefei, Anhui, China
2) Leading Researcher/Principal Scientist
TUBITAK-2232 Program
Environmental Engineering Department,
Middle East Technical University, Ankara, Turkey

Editor Biography

Dr. Fernanda Maria Policarpo Tonelli (Ph.D.) is Assistant Professor at Federal University of São João del Rei, Divinópolis, MG, Brazil, where she teaches graduate and post-graduate students. She specializes in biotechnology/molecular biology and has been studying nanomaterials applications such as delivery and remediation as same as green synthesis protocols, and allelopathic potential of secondary metabolites from plants. The research projects have already resulted in 10 patent applications, three of them already analyzed and granted. She has published nine books, authored 15 scientific articles and more than 40 book chapters with international publishers. She has presented and participated in many national and international conferences and has helped in organizing various scientific events. Dr. Tonelli has also dedicated herself to the promotion of science and technology through co-funding a non-governmental organization. Dr. Tonelli is active in advocacy groups that support gender equality in science. Her efforts as a researcher have been recognized with various awards, including *For Women in Science Brazil* (awarded by L'Oréal Brasil, UNESCO-Brazil, and Brazilian Academy of Sciences) and *Under 30 Brazil* (Forbes) as well as various certificates of merit.

Dr. Rouf Ahmad Bhat (Ph.D.) pursued his doctorate at Sher-e-Kashmir University of Agricultural Sciences and Technology Kashmir (Division of Environmental Sciences), and is presently working in the Department of School Education, Government of Jammu and Kashmir. Dr. Bhat has been teaching graduate and postgraduate students of environmental sciences for the past three years. He is the author of more than 55 research articles (h-index 30; i-index 44; total citation 2500) and 40 book chapters, and has published more than 40 books with international publishers (Springer, Elsevier, CRC Press Taylor & Francis, Apple Academic Press, John Wiley, IGI Global). His specialization is in limnology, toxicology, phytochemistry, and phytoremediation. Dr. Bhat has presented and participated in numerous state, national, and international conferences, seminars, workshops, and symposia. In addition, he has worked as an associate environmental expert at the World Bank-funded Flood Recovery Project and as the environmental support staff in Asian Development Bank (ADB)-funded development projects. He has received many awards, appreciations, and recognition for his services to the science of water testing, air, and noise analysis. He has served as an editorial board member and a reviewer of reputed international journals. Dr. Bhat still writes and experiments with diverse capacities of plants for use in aquatic pollution remediation.

Dr. Gowhar Hamid Dar (Ph.D.) is currently Assistant Professor in Environmental Science, Sri Pratap College, Cluster University Srinagar, Department of Higher Education (J&K). He has a Ph.D. in Environmental Science with a specialization in Environmental Microbiology (fish microbiology, fish pathology, industrial microbiology, taxonomy, and limnology). He has been teaching postgraduate and graduate students for many years in the Post-Graduate Department of Environmental Science, Sri Pratap College, Cluster University Srinagar. He has written more than 70 research articles (h-index 16; i-index 22; total citation >950) in international and national journals of repute and more than 20 books with international publishers (Springer, Elsevier, CRC Press Taylor & Francis, Apple Academic Press, John Wiley, IGI Global). Moreover, he supervises a number of students for the completion of degrees. He has been working for the last several years on the isolation, identification and characterization of microbes, their pathogenic behaviour, and the impact of pollution on the development of diseases in fish fauna. He has received many awards and appreciations for his service to science and development. In addition, he is a member of various research and academic committees. Furthermore, Dr. Dar is Principal Investigator and Co-Principal Investigator for different R&D projects sanctioned by the Government of India and the Government of Jammu and Kashmir.

Dr. Khalid Rehman Hakeem (Ph.D., FRSB) is presently Professor at King Abdulaziz University, Jeddah, Saudi Arabia. After completing his Ph.D. in Botany, with a specialization in Plant Ecophysiology and Molecular Biology from Jamia Hamdard, New Delhi, India in 2011, he worked as Assistant Professor at the University of Kashmir, Srinagar for a short period. Later, he joined Universiti Putra Malaysia, Selangor, Malaysia and worked there as Post-Doctorate Fellow in 2012 and Fellow Researcher (Associate Prof.) from 2013–2016. He joined King Abdulaziz University in August 2016 and was recently promoted to Professor.

Dr. Hakeem has more than twelve years of teaching and research experience in plant eco-physiology, biotechnology and molecular biology, medicinal plant research, plant-microbe-soil interactions, and environmental studies. He is the recipient of several fellowships at both the national and international levels. He was recently elected Fellow of the Royal Society of Biology, London (FRSB). Professor Hakeem has served as Visiting Scientist at Fatih Universiti, Istanbul, Turkey, and at Jinan University, Guangzhou, China. Currently, he is involved in a number of international research projects with different government organizations.

So far, Dr. Hakeem has authored and edited more than 60 books with international publishers, including Springer Nature, Academic Press (Elsevier), CRC Press, etc. He also has to his credit 121 research publications in peer-reviewed international journals and 65 book chapters in edited volumes with international publishers.

At present, Dr. Hakeem is serving as an editorial board member and reviewer of several high-impact international scientific journals from Elsevier, Springer Nature, Taylor & Francis, Cambridge, and John Wiley Publishers.

Contributors

Khuram Shahzad Ahmad
Department of Environmental Sciences
Fatima Jinnah Women University
The Mall, Rawalpindi, Pakistan

Zubair Ahmad
Department of Botany
Aligarh Muslim University
Aligarh, UP, India

Ankita
Department of Anthropology
Punjab University
Chandigarh, Punjab

Sneha Bhandari
Department of Biotechnology
Kumaun University Nainital
Bhimtal Campus, Uttarakhand, India

Irshad Ahmad Bhat
Plant Tissue Culture Research Laboratory
Department of Botany
University of Kashmir
Hazratbal, Srinagar

Sarah Bouzroud
Laboratoire de Biotechnologie et Physiologie Végétales
Centre de biotechnologie végétale et microbienne biodiversité et environnement
Faculté des Sciences, Université Mohammed V de Rabat
Rabat, Morocco

Gul-e-Saba Chaudhry
Department of Biotechnology
Fatima Jinnah Women University
The Mall, Rawalpindi, Pakistan

Akshaya Prakash Chengatt
Department of Botany
St. Joseph's College
Devagiri, Calicut, Kerala, India

Sana Choudhary
Department of Botany
Aligarh Muslim University
Aligarh, UP, India

Xiaorong Fan
Key Laboratory of Plant Nutrition and Fertilization in Lower-Middle Reaches of the Yangtze River
Ministry of Agriculture, Nanjing Agricultural University
and
State Key Laboratory of Crop Genetics and Germplasm Enhancement,
Nanjing Agricultural University
Nanjing, China

Mudasir Fayaz
Plant Tissue Culture Research Laboratory
Department of Botany and Signal Transduction Laboratory
Department of Biotechnology
University of Kashmir Hazratbal
Srinagar, India

Khushboo Guleria
Research Scholar, Department of Zoology
School of Bioscience and Bioengineering
Lovely Professional University
Punjab, India

Asma Jabeen
Department of Environmental Sciences
Fatima Jinnah Women University
The Mall, Rawalpindi, Pakistan

Ekta B Jadhav
Government Institute of Forensic Science
Aurangabad, Maharashtra

Swapnali Jadhav
Department of Forensic Chemistry & Toxicology
Government Institute of Forensic Science
Aurangabad, Maharashtra, India

Shaan Bibi Jaffri
Department of Environmental Sciences
Fatima Jinnah Women University
The Mall, Rawalpindi, Pakistan

A. Judith
Department of Microbiology
The American College
Madurai, Tamil Nadu, India

Zahoor A. Kaloo
Department of Botany
University of Kashmir
Hazratbal, Srinagar, India

Rajeev Kumar
Department of Forensic Science
School of Basic and Applied Sciences
Galgotias University
Greater Noida, U.P., India

Neha Naaz
Department of Botany
Aligarh Muslim University
Aligarh, UP, India

Masarat Nabi
Department of Environmental Science/Centre of Research for Development (CORD)
University of Kashmir
Srinagar, J&K, India

Aqsa Nadeem Shah
Department of Environmental Sciences
Fatima Jinnah Women University
The Mall, Rawalpindi, Pakistan

Misbah Naz
State Key Laboratory of Crop Genetics and Germplasm Enhancement
Nanjing Agricultural University
Nanjing, China

Mohamed Nouri
Sultan Moulay Slimane University
Faculty of Polydisciplinary
Beni-Mellal, Morocco

Kapil Parihar
Department of Forensic Science
Vivekananda Global University
Jaipur, Rajasthan, India

Tanveer Bilal Pirzadah
University Centre for Research and Development (UCRD)
Chandigarh University
Punjab, India

Jasfeeda Qadir
Plant Tissue Culture Research Laboratory
Department of Botany, University of Kashmir
Hazratbal, Srinagar, India

Roof-Ul-Qadir
Plant Reproductive Biology
Genetic Diversity and Phytochemistry Research Laboratory
Department of Botany, University of Kashmir
Hazratbal, Srinagar, India

Muhammad Ammar Raza
College of Food Science and Biotechnology
Key Laboratory of Fruits and Vegetables Postharvest and Processing Technology Research of Zhejiang Province
Zhejiang Gongshang University
Hangzhou, China

Naila Safdar
Institute of Marine Biotechnology
Universiti Malaysia Terengganu
Kuala Terengganu, Malaysia

Mahipal Singh Sankhla
Department of Forensic Science
Institute of Sciences, SAGE University
Indore, M.P., India

Nair G Sarath
Plant Physiology and Biochemistry Division
Department of Botany
University of Calicut, C.U. Campus P.O.
Kerala, India

Mohammad Sarraf
Department of Horticulture Science
Shiraz Branch
Islamic Azad University
Shiraz, Iran

A. M. Shackira
Department of Botany
Sir Syed College, Taliparamba
Kannur, Kerala, India

Adil Shafi
Environmental Research Laboratory
Department of Chemistry
Aligarh Muslim University
Aligarh, UP, India

Gaurav Kumar Singh
Dr. APJ Abdul Kalam Institute of Forensic Science & Criminology
Bundelkhand University
Jhansi, U.P., India

Prashant Singh
School of Forensic Science
National Forensic Science University
Gandhinagar, Gujrat, India

Bhat Mohd Skinder
Department of Environmental Science/Centre of Research for Development (CORD)
University of Kashmir
Srinagar, J&K, India

Swaroop S Sonone
Government Institute of Forensic Science
Aurangabad, Maharashtra, India

J. Immanuel Suresh
Department of Microbiology
The American College
Madurai, Tamil Nadu, India

Fernanda Maria Policarpo Tonelli
Federal University of São João del Rei—Centro-Oeste Campus—Divinópolis-MG
Brazil

Flávia Cristina Policarpo Tonelli
Federal University of São João del Rei—Centro-Oeste Campus—Divinópolis-MG
Brazil

Ashutosh Tripathi
Institute of Sciences
SAGE University
Indore, M.P., India

Rohit Kumar Verma
Department of Forensic Science
University Institute of Applied Health Science
Chandigarh University
Gharuan, Punjab, India

Himanshu Yadav
Department of Forensic Science
Teerthanker Mahaveer University
Moradabad, U.P., India

Azra Yasmin
Department of Biotechnology
Fatima Jinnah Women University
The Mall, Rawalpindi, Pakistan

1 Environmental Pollution Caused by Organic Pollutants, Their Harmful Impacts, and Treatment through a Microbiological Approach

Sneha Bhandari

1.1 INTRODUCTION

Environmental pollution is a significant challenge and a serious issue of concern throughout the world. When any physicochemical contaminant or pollutant causes a disturbance and alteration of the environment, the phenomenon is termed environmental pollution. Human activity and natural calamities are two significant factors that lead to pollution. Industrialization, urbanization, deforestation, mining, excavation, and exploration have negatively influenced the environment and thus altered the ecological balance leading to pollution. This burden of environmental pollution is faced by both developed and developing nations, although developed nations have more awareness and strict guidelines and laws associated with environmental pollution. Despite worldwide awareness, the negative influence of environmental pollution can still be noticed because of long-term and continual impacts. Apart from the fundamental type of pollution related to air, water and soil, the world is also bearing the pollution associated with noise, radioactivity, plastic, and light. Environmental pollution directly affects the health of living organisms, not only humans but also non-human beings such as animals, plants, and others. It has been reported that environmental pollution results in a large number of diseases and premature deaths, and more than 9 million premature deaths have occurred because of air, soil, and water pollution. This death figure is three-fold more than the fatalities that occurred due to AIDS, tuberculosis, and malaria combined and is 15-fold greater than all wars and other types of violence that have occurred throughout the world. Moreover, soil and water pollution is a considerable crisis in the majority of regions of the world. This environmental pollution has led to several negative consequences such as vegetation loss (deforestation, overgrazing, and desertification) and soil fertility disruption, together with soil erosion and depletion, toxic acid rain, and species extinction (Bhandari and Nailwal 2020a).

Chemical pollutants can be fundamentally divided as organic and inorganic pollutants. Organic contaminants include polyaromatic hydrocarbons, phenols, pesticides, azo dyes, polychlorinated biphenyls, chlorinated phenols, herbicides, fertilizers, and endocrine-destroying chemicals, whereas inorganic contaminants include various heavy metals such as arsenic, cadmium, and lead. The recalcitrant, non-biodegradable nature of organic compounds has created an alarming situation, and their severity can be seen in the health and quality of living organisms. Industries such as textiles, pharmaceuticals, paper and pulp, chemical, sewage and water treatment plants, and petroleum and oil not only utilize these chemical compounds but also release them as industrial

DOI: 10.1201/9781003186298-1

effluents. For instance, phthalates are used as plasticizers for toys and personal care products. Organic chemical compounds that follow four characteristics *viz.*, persistence in the environment, capability of long-distance transport, bioaccumulation in the ecosystem, and potential of biological magnification, are termed persistent organic pollutants (POPs). These POPs pose serious health issues as they pass on from one living being to another from the food chain and food web, thus muddling the ecological balance. For example, dichlorodiphenyltrichloroethane (DDT) is one of the classic POPs that remain in the ecosystem for more than a decade, and it endangers not only humans but also other organisms. DDT and its related derivative compounds disturb the calcium metabolism in birds; thus, fragile eggshells are produced that are unable to hold the incubating bird, leading to the death of the developing embryos. The population of Brown Pelicans and Bald Eagles were greatly affected due to DDT. In humans, a higher percentage of DDT acts as mutagen and carcinogen and causes various symptoms and diseases such as seizures, neurotoxicity, tremors, vomiting, and cancer.

POPs contribute considerably to the environmental pollution of air, water, and soil. Organic pollutants decrease the fertility of the soil and cause reduced agricultural productivity. Industrial wastewater as an effluent is extensively released in water bodies (either in freshwater or marine water) and deteriorates the water quality and habiting living organisms. Higher levels of organic pollutants in water cause hypoxia (a reduction of dissolved oxygen), which threatens the lives of water creatures. In industries, a combustion of toxic chemicals and release of noxious gases into the atmosphere disrupt the atmospheric gas system, resulting in air pollution. Long-term exposure to polluted air leads to severe respiratory-associated diseases and disorders.

To address harmful organic pollutants, various strategic methods have been proposed. These include both biological and non-biological approaches. As physicochemical techniques are difficult to manage, a biological method involving microorganisms that degrade organic contaminants has evolved and opened new paths in environmental cleaning. Microorganisms are present in every part of the world and colonize even in very intense environments such as in caves, deep oceans, cold Antarctic areas, hot volcanic areas, and hot springs. When habitats change, this also affects the growth, development, and nutrition behavior in microorganisms and alters the metabolic activities within microbial communities (Bhandari and Nailwal 2020b). Microorganisms utilize energy sources according to the available nutrients present in their habitat. Therefore, in this chapter, we discuss the pollution caused by organic contaminants, their negative impacts on the health of living beings, and the microbiological approach for their treatment and removal from the environment.

1.2 ORGANIC POLLUTANTS

Organic chemical compounds are basically compounds that contain carbon-hydrogen bonding in their chemical structure. Organic compounds are in massive demand as manufacturing ingredients and are used in several industrial processes and applications, for instance, alcohols such as ethanol and isopropanol act as antiseptics and are utilized in the cosmetics industry and beverage industry. Organic compounds containing the carboxylic acid group are widely used in the pharmaceutical industry. Similarly, compounds such as toluene, benzene, naphthalene, carbon-tetrachloride, benzoic acid, benzaldehyde, and many more are used widely in numerous industries. These compounds have various industrial applications, but many of them act as POPs (Figure 1.1) and cause environmental pollution. Some of the major POPs such as chlordane, hexachlorobenzene, endrin, polychlorinated biphenyls (PCBs), dichlorodiphenyltricholorethane (DDT), polychlorinated dibenzofurans, hexabromodiphenyl ether (hexaBDE), heptabromodiphenyl ether (heptaBDE), pentachlorobenzene (PeCB), lindane, tetrabromodiphenyl ether (tetraBDE), perfluorooctaesulfonic acid (PFOS), and endosulfans are being used as crucial manufacturing reactants in various industries, but these organic compounds act as pollutants, pose a severe threat to the environment, and are thus a topic of global concern.

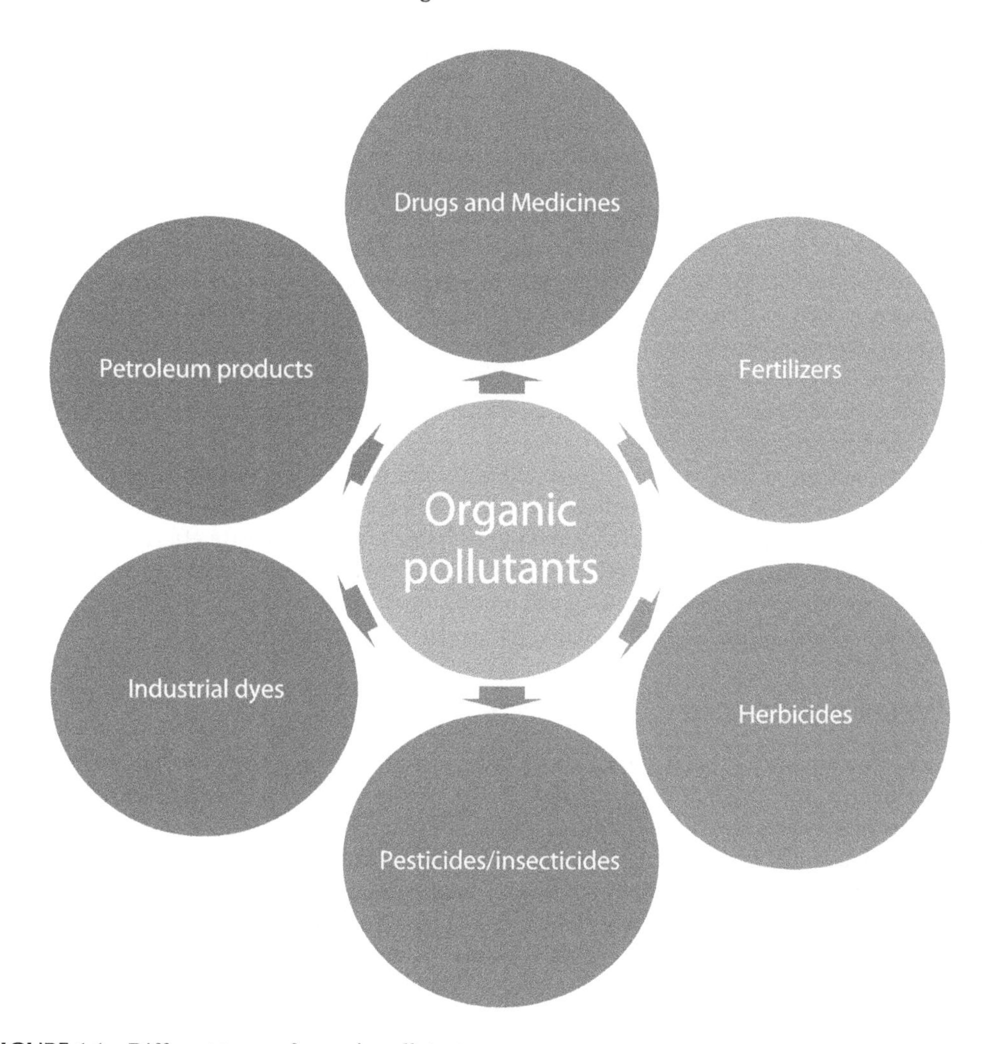

FIGURE 1.1 Different types of organic pollutants.

1.3 SOIL POLLUTION

Soil pollution occurs when any contaminant present in the soil degrades its quality and disturbs the natural soil composition. Organic pollutants include petroleum hydrocarbons, solvents, azo dyes, and polynuclear aromatic hydrocarbons, such as benzo-pyrene and naphthalene; apart from these chemicals, the extensive usage of chemical-based fertilizers, insecticides, pesticides, and herbicides are some factors that cause soil degradation. Heavy-duty vehicles transporting raw materials and other goods from place to place heightens the threat of oil spillage (which contains aromatic petroleum hydrocarbons and associated products), a common scenario on roads and land. It is difficult for these petroleum products to degrade from the soil, and extra endeavour is needed for their removal from the environment. The industrial discharge of both solid and liquid waste also consists of copious quantities of organic chemicals, which if untreated and directly discarded, causes both water and soil contamination. Agricultural wastes are composed of a higher concentration of chemical fertilizers, herbicides, and insecticides. Due to a lack of knowledge of fertilizers and other agro-chemicals, farmers use these chemicals in larger quantities than required in their fields. Not only does this result in the waste of these chemicals but also soil has to take up the burden of these chemicals. Moreover, it has been reported that the continual usage of fertilizer and pesticides has reduced both the structure of the soil and its fertility. Because of a higher accumulation of toxic organic compounds, crop productivity also decreases over time (Rahman and Zhang 2018). Furthermore, beneficial insects such as pollinators are highly affected by pesticide accumulation in the soil (Gullan and Cranston 2010). Fertilizers can help and support plant growth, but their excessive use causes only harm to the soil environment in terms of a disbalance in the nutrient and mineral biogeochemical cycles.

Direct exposure or contact with contaminated soil can act as a root cause of diseases both in humans and non-human species such as animals. Indirect effects may also arise when these organic pollutants leach and penetrate the groundwater, affecting its quality and making it unsafe for consumption. In addition, soil pollution changes the metabolism of soil-associated microorganisms and small arthropods colonizing in the soil (Prashar and Shah 2016). Consequences can be easily seen in the primary food chain, which affects the food chain and food web of higher trophic levels such as consumers and predator species.

Although there are strict laws and regulation for discarding industrial waste, it seems that these regulations are not being followed well. Even if the rules are followed, these rules are of no use because environmental pollution has increased so much that it needs extensive efforts and strategies to first clean up the environment. Therefore, environmental cleanup needs to have an amalgamation of geology, chemistry, computer modelling and processing skills, hydrology, and most importantly, a thorough knowledge of the environmental sciences and associated fields.

1.4 WATER POLLUTION

Water contamination is a crucial and agitating issue throughout the globe. When any toxic compounds or dangerous microorganisms pollute streams, rivers, lakes, canals, and oceans, deteriorating the water quality and making it unsafe for human consumption, the water is considered polluted. There are various sources that can contaminate and deteriorate water quality. Agricultural operations runoff (including animal excretory products, antibiotics, heavy metals, and salts), industrial effluents (from the pulp and paper industry, drug and pharmaceutical industry, mining and excavation, fossil fuel burning, and petrochemical industry), suburban and rural waste, underground storage, piping, and sewer leakages, hazardous chemical waste, radioactive resources, and landfill leakage are some of the common causes of pollutants for water pollution. Common organic pollutants of water are pharmaceutical drugs, aromatic petroleum hydrocarbon (PAH), microplastics, tannery waste, pigments, petroleum and other oil-based product spillage, and phthalates.

According to the survey of World Water Development 2017, barely 20% of produced wastewater is properly treated, while the remaining wastewater without any treatment is simply flushed into water bodies (Water 2016). The accumulation of organic pollutants has been increased tremendously in water bodies over recent years. Due to high toxicity, low water solubility, non-biodegradability, and a recalcitrant and semi-volatile nature, these organic contaminants have become a critical concern for the world (Abdeen and Mohammad 2013; Chen et al. 2011). During the transportation of goods and petroleum-based products, leakage and spillage occur quite frequently from ships into the oceans. This spillage not only plays a considerable role in polluting ocean water but also has become a grave issue for aquatic life. This is because due to the accumulation and non-biodegradable nature of organic pollutants, when water turbidity increases, it does not allow sunlight to pass through it, so the survival of photosynthetic organisms becomes difficult.

Consumption of contaminated water is a very basic reason for disease and disorder in humans. It has been reported that water contaminated by organic pollutants may lead to endocrine disruption, nephrotoxicity, cancer, hepatotoxicity, obesity, and reproductive system disorders (Nanseu-Njiki et al. 2010). Higher amounts of organic pollutants in water also lead to the death of aquatic animals due to asphyxiation and reduced fitness. Some marine organisms in which organic pollutants have been found are fishes (Moon et al. 2010; Wirnkor et al. 2020), *Prena viridis* (Bayen et al. 2004), barnacles (Karuppiah et al. 2004), and odontocete (Karups et al. 2004).

1.5 AIR POLLUTION

One of the most deadly types of pollution is air pollution, as living organisms share the common atmosphere so that any alteration in atmospheric gases that occurs in a particular region will also affect the gas composition of nearby places. Suspended solid and liquid particulates in the air are the main cause of air contamination. These suspended particles are gases, including nitrous oxide, carbon monoxide, ammonia, sulphur dioxide, and chlorofluorocarbon, biomolecules, and organic and inorganic particulate matter. Air pollution causes deleterious effects that involve diseases, disorders, allergies, and even the death of living beings. Volatile organic compounds (VOCs) are major air contaminants, which are generally classified as methane (MVOCs) and non-methane volatile organic compounds (NMVOCs). VOCs can be naturally synthesized by plants such as isoprenes, sesquiterpenes, methanol, and terpenes or they can be created from anthropogenic activities, including solvents used for paints and inks and biofuel such as bioethanol. It is speculated that VOCs are more consistently found indoors than outdoors; in fact, many routine chemicals including paints, varnish, cosmetics, and disinfectants can release organic pollutants.

Methane is regarded as one of the most potent greenhouse gases that causes global warming. Some VOCs, such as limonene and styrene, when they react with ozone products or nitrogen oxide, form new oxidation products and secondary aerosols that possess sensory irritation properties (Nielsen et al. 2007). Some of the key symptoms related to the exposure of VOCs are nose and throat discomfort, skin allergies, nausea, vomiting, a reduction in serum cholinesterase concentration, fatigue, and dizziness. Aromatic NMVOCs, such as toluene and benzene, are suspected to be carcinogens and are associated with leukaemia after long-term exposure. Chlorofluorocarbon, another important organic contaminant, is largely released from refrigerators and air conditioners into the stratosphere, where it comes into contact with other gases and damages the ozone layer. Earth is protected from harmful ultraviolet (UV) radiation that comes from the sun through the ozone, as large-scale air pollution has resulted in a thinning of the ozone layer in recent decades. Thinning of the ozone shield has allowed the penetration of UV radiation to the Earth's surface, resulting in various health-related concerns such as skin cancer, skin pigmentation, and eye-associated disorders. Peroxyacetyl nitrate, an organic nitrogen compound, is a component of photochemical smog that acts as a secondary air pollutant and is a potent lung and eye irritant. It is also estimated that numerous POPs attach to particulates and contaminate the air. Semi-volatile POPs (such as VHCHs, DDT, etc.) can vaporize from the soil, water, and foliage directly into the ecosystem. As these chemicals are non-degradable, they are therefore transported through the wind to distant places (Bacaloni et al. 2011).

1.6 NEGATIVE IMPACTS OF ORGANIC POLLUTANTS

Organic contaminants, major environmental pollutants, not only disturb the ecological sustenance but also cause severe negative impacts on the health of living organisms such as humans and non-human organisms, namely, microorganisms and plants. POPs also include many organo-metals, for example, methylmercury and lipophilic halogenated organics such as chlorinated pesticides, dioxins, perfluorinated compounds, polychlorinated biphenyls, and polybrominated flame retardants. Some of these chemicals are byproducts of combustion and different industries.

1.6.1 Impacts on Humans

A variety of health ailments have been identified based on exposure to these POPs such as cancer, a reduction in cognitive and neurobehavioral functions, immune system repression, and disturbances in thyroid function, and some of them even have the potential to enhance chronic diseases including hypertension, diabetes, and heart-related disorders (Carpenter 2011). Chlordane, one of the pesticides used in agriculture as a termite and insect repellent, accumulates in the lipids of humans and animals. This compound is highly toxic and causes several health issues such as diabetes, testicular cancer, prostate cancer, lymphoma, breast cancer, and obesity. Similarly, heptachlor, another organochlorine insecticide with no solubility in water, is a potent carcinogenic agent (Bandala et al. 2006). Oral exposure and inhalation of this compound for long periods but in a lower concentration may lead to neurological defects such as salivation, irritability, and negative blood effects (Ratola et al. 2003). Acute inhalation and exposure to this compound can also lead to gastrointestinal and nervous system issues.

Formaldehyde is a VOC that enters organisms upon inhalation and irritates the nose and lung linings; in addition, occupational asthma can be seen in some individuals. There are many other organic pollutants whose exposure is dangerous and sometimes lethal.

TABLE 1.1
Organic Pollutants and Their Negative Effects on Humans

Organic pollutant	Effects	References
Dichlorodiphenyltrichloroethane (DDT)	Carcinogenic; chronic exposure leads to reproductive inabilities and breast cancer; interferes with the thyroid function of developing babies in pregnant women, and chances are high in developing an autistic child	Chevrier et al. (2008), Brown et al. (2018), Thakur and Pathania (2020)
Aldrin	Neurotoxic effects; stimulates the central nervous system (CNS) leading to hyperexcitation and seizures, gastrointestinal disturbances, malaise, and lack of coordination; acute intoxication is lethal	Gupta (1975)
Chlordane	Inhalation of a higher concentration leads to headaches, irritation, dizziness, diarrhea, jaundice, nausea, tremors, slight involuntary muscular movements, testicular cancer, breast cancer, and obesity	Thakur and Pathania (2020)
Heptachlor	Negative impacts on immune, reproductive, renal, and hematopoietic systems, salivation, and irritability	Fendick et al. (1990), Snyder and Mulder (2001)
Polychlorinated biphenyls	Nail pigmentation, swelling of the eyelids, vomiting, pigmentation of the mucous membrane, fatigue, nausea, suppression of the immune system, thereby increasing the chances of autoimmune disorders, alteration of thyroid reproductive functions in both males and females, development of cardiovascular and liver diseases, and diabetes; exposed pregnant women give birth to a baby of low birth weight that is prone to various diseases throughout life	Carpenter (2006), Thakur and Pathania (2020)
Non phthalate plasticizers Bis(2-ethylhexyl) hexanedioate (DEHA), Acetyl Tri-n-Butyl Citrate (ATBC) and 1-isopropyl-2,2-dimethyltrimethyl ester or 2,2,4-trimethylpentanediol-1,3-diisobutyrate (TPIB)	These plasticizers interfere with the steroid ligand binding site on sex hormone binding globulin (SHBG); thus, they have the potential of causing the dysfunction of sex steroid homeostasis	Sheikh and Beg (2019)

(*Continued*)

TABLE 1.1 *(Continued)*
Organic Pollutants and Their Negative Effects on Humans

Organic pollutant	Effects	References
Hexachlorobenzene	Photosensitive skin lesions, porphyria turcica, colic, liver diseases, ulcers, hair loss, embryo fatality, negative effects on the thyroid and bones, and teratogenic effects; creates problems in infants as it can be transferred from the mother's milk	Chen et al. (2016), Jiang et al. (2018)
Hexachlorobutadiene	Disorders such as central nervous system depression, epithelial necrotizing nephritis, cyanosis, and fatty liver degeneration	Thakur and Pathania (2020)
Organophosphate esters (OPEs)	Loss of pregnancy as OPE could transfer from the mother to her developing baby through the placenta; negative consequences on cognitive and behavioural development in children; affects the timing of childbirth and causes premature delivery; also associated with obesity, allergic reactions, asthma, and neurodevelopmental issues	Doherty et al. (2019)
Short-chain chlorinated paraffins	Acts as a skin and eye irritant at low concentrations but upon ingestion, has negative impacts on the kidneys, liver, breast milk, and adipose tissues	Thakur and Pathania (2020)
Diethylhexyl phthalate (DEHP) and 17α-ethinylestradiol (EE2)	Phthalate organic compounds are endocrine-disrupting chemicals	
Methylene chloride	Carcinogenic in the livers and lungs of mice; related to hematopoietic cancer and associated cancers with multiple myeloma, and non-Hodgkin lymphoma; fatty liver, liver foci/areas of alteration, vacuole formation in hepatocytes, and necrosis	Schlosser et al. (2015)
Buta-1,3-diene	Carcinogenicity of multiple organs in mice, uncommon hemangiosarcomas of the heart, malignant lung tumors, partial regenerative anemia in the case of laboratory mice, and lymphatic and hematopoietic cancers in humans	Melnick and Huff (1992)
Toluene	Prolonged exposure leads to motor incoordination, cognitive impairment, dizziness, hallucinations, and illusions; acute inhalation also results in life-threatening conditions such as hypothermia, cardiac arrhythmias, hypoxia, or combined factors, muscle strength loss, cerebellar ataxia	Cruz et al. (2014)
Azo dyes	Carcinogenic agent to both animals and humans	Chung (2016)

1.6.2 Impacts on Animals

Organic pollutants have greatly affected several animal species, particularly marine and aquatic life. Marine water and freshwater bodies are the potential dumping reservoirs of industrial effluents that contain massive quantities of chemical and radioactive waste. Industrial discharges heavily comprise hydrocarbons, biphenyl and associated products, pesticides, plasticizers, oil, pharmaceuticals, dyes, greases, fertilizers, dioxins, and furans, which are collectively known as organic contaminants. The treatment of industrial waste is a highly laborious and complex process and requires financial credit and support, skilled labour, and enough time to become effluent-free from toxic compounds. However, when this treatment is not performed properly and is directly discharged to water bodies, water pollution *via* organic contaminants occurs. Organic pollutants cause various types of health issues in aquatic animals including neurotoxicity, genotoxicity, developmental issues, threats to larval and embryo survival, chromosomal aberrations, and anatomical and morphological issues. Dioxin (2,3,7,8-tetrachlorooxanthrene) exposure led to declined spermatozoa, with a simultaneous elevation in spermatogonia and a reduction in the thickness of the germinal epithelium in zebrafish (*Danio rerio*). Gene expression for the development of testis, spermatogenesis, hormone metabolism, steroidogenesis, and xenobiotic response was also changed. These genes are essential for the metabolism of lipids, cellular morphology, molecular transport, biochemistry of small molecules, and metabolism of vitamins and minerals. Overall, exposure to Tetrachlorodibenzodioxin caused transgenerational infertility in *Danio rerio* (Baker et al. 2016). Similarly, runoff of Cyhalofop-butyl (herbicide) from paddy fields exhibited toxic effects in zebrafish (*Danio rerio*). This herbicide elevated reactive oxygen species (ROS) accumulation and apoptosis near the heart area along with increased activity of caspase-9 and caspase-3 (Zhu et al. 2015). In the Canadian Arctic, the reproductive success of thick-billed murres and northern fulmars was greatly affected by dioxin-like polychlorinated biphenyls (PCBs) such as non-*ortho* PCBs (nPCBs), polychlorinated dibenzodioxins (PCDDs), and polychlorinated dibenzofurans (PCDFs) emission from industries (Braune and Mallory 2017). Polychlorinated dibenzo-p-dioxins/furans (PCDD/Fs) and dioxin-like polychlorinated biphenyls (DL-PCBs) accumulated inside the livers of *Caretta caretta* and *Chelonia mydas*, thereby leading to a threat to their survival (Lambiase et al. 2021).

Negative and toxic effects of organic contaminants can be seen in all kinds of living organisms, but for simplicity, we include all non-human organisms such as birds, aquatic species, and other mammals in one group, that is, animals.

TABLE 1.2
Organic Pollutants and Their Negative Effects on Animals

Organic pollutants	Effects	References
Benzo[*pqr*]tetraphene or (B[a]P) and Tetrachlorodibenzodioxin	Metabolic functions and immune system were impaired resulting in genotoxic damage in *Eisenia andrei (Earthworm)*	Sforzini et al. (2015)
Benzene	Haematopoietic disorders such as anemia, leucopenia, and bone marrow deficiency were exhibited with a reduction in the circulating blood cell count, aplastic anemia, thrombocytopenia, and acute myelogenous leukemia in both rodents and humans	Hayes et al. (2001)
Polybrominated diphenyl ethers (PBDEs) including 1,2,4-Tribromo-5-(2,4-dibromophenoxy)benzene, octaBDE, and 6,6'-Oxybis(1,2,3,4,5-pentabromobenzene)	Toxic to *D. magna*, shrimp, *Scenedesmus* quadricauda, *Carassius carassius*, chironomid larvae, mudsnai., Bufo gargarizans tadpoles, and *P. parva*	Lu et al. (2018)

(*Continued*)

TABLE 1.2 ***(Continued)***
Organic Pollutants and Their Negative Effects on Animals

Organic pollutants	Effects	References
Dioxins such as polychlorinated biphenyls (PCBs)	Males' secondary sexual characteristics were minimized, gene expression was altered, offspring care behavior patterns were reduced, and increased mortality of the embryo; thus, PCBs directly inhibit the paternal reproductive success and population size in male *Pimephales promelas* (fathead minnows)	Coulter et al. (2019)
Benzo(α)pyrene (BaP)	Deformities include elevated vacuolation in the cytoplasm, inflammatory cell infiltration, bulged nuclei, and non-uniform pigmentation, along with an induction of oxidative stress in zebrafish	Mai et al. (2021)
2,2-Bis(4-hydroxyphenyl)propane and Ethinylestradiol	Significant reduction in the fertilization rate in the offspring in the F2 generation and a decrease of embryo survival of the offspring in the F3 generation; thus, the exposure to 2,2-Bis(4-hydroxyphenyl)propane and Ethinylestradiol triggered reproductive impairment and reduced survival of embryos in subsequent generations in medaka fish (*Oryzias latipes*)	Bhandari et al. (2015)
Oxychlordane	Length of the telomere was reduced in *Rissa tridactyla* (adult black-legged kittiwakes)	Blévin et al. (2016)
2,3,7,8 tetrachlorooxanthrene	Elevated splenomegaly and leukemogenesis in *in vivo* pregnant mice	Ahrenhoerster et al. (2015)
Polyethylene microplastics (PE MPs)	Harmful effects on embryo's hatching rate and larvae survival in zebrafish (*Danio rerio*)	Malafaia et al. (2020)
Perfluorooctane sulfonate	Heartbeat, reproductive, biochemical, neurological, and antioxidant performance was reduced in *Daphnia magna* Disruption of wide ranges of the biological pathways specifically associated with lipid metabolism in the liver of chicken embryos	Liang et al. (2017), Jacobsen et al. (2018)
4,4'-Oxybis(1,3-dibromobenzene)	Negative impacts on the male reproductive system such as smaller testes, a reduction in sperm production due to the reduced testis size, and elevated levels of morphologically abnormal spermatozoa, along with an increased head size of spermatozoa; in short, the genes associated with immune reaction induction and spermatogenesis were suppressed. Protamine and transition protein gene expression in the testis of adult rats was decreased 4 times Alteration in the retinal metabolism and associated biological processes related to visual perception and eye morphogenesis, thereby affecting eye development in zebrafish larvae. Expression of the genes *pth1a* and collaborative cathepsin family was also altered, thereby disrupting the bone development mechanism Gene expression of *lingo1b*, *bcl-2*, *c-fos*, and *grin1b* in the brain of male zebrafish was elevated, which led to neurobehaviour and memory changes	Khalil et al. (2017), Xu et al. (2015), Zheng et al. (2017)

(*Continued*)

TABLE 1.2 *(Continued)*
Organic Pollutants and Their Negative Effects on Animals

Organic pollutants	Effects	References
Toxaphene	Acts as antiestrogen by inhibiting the vitellogenin synthesis in female fish (*Micropterus salmoides*); the immune system is suppressed as immune-related signaling cascade mechanisms such as lectin-like receptor and ITSM-Containing Receptor signaling, CD16/CD14 Proinflammatory Monocyte Activation, and CD38/CD3-JUN/FOS/NF-kB signaling in t-cell proliferation were suppressed; there were also changes in the androgen signaling pathways Induction of tumor formation was mediated by stimulating the constitutive androstane receptor (CAR) nuclear receptor in mouse liver	Wang et al. (2017b), Martyniuk et al. (2020)
Tetradioxin	Reduced growth rate, low development in terms of pericardial edema, yolk sac edema, tubular heart, yolk sac hemorrhage, eye deformities, including micro-opthalmia, and a reduction in swimming activity and endurance in *Acipenser fulvescens*	Tillitt et al. (2017)
Phthalates including bis(2-ethylhexyl) phthalate, Di-n-butyl phthalate Butyl phthalate and Diethyl 1,2-benzenedicarboxylate	Increase in reproduction output and inhibition of fatty acid uptake and catabolism led to an accumulation of the overall lipid content in and decreased body size and life span of *Daphnia magna*	Seyoum and Pradhan (2019)
Perfluorooctanesulfonic acid (PFOS)	Chronic reproductive toxicity in Northern Bobwhite Quail (*Colinus virginianus*) Gills and digestive glands of clam *Scrobicularia plana* were affected by generated neurotoxicity and oxidative stress. In chicken embryos, peroxisomal β-oxidation was elevated, whereas mitochondrial β-oxidation was reduced for lipid metabolism	Dennis et al. (2020), Islam et al. (2021)
Toxaphene and endrin	Affected the secretion of oxytocin, prostaglandin, testosterone, and oestradiol from uterine and ovarian cells; also inhibited myometrical contractions in cows	Wrobel and Mlynarczuk (2018)

1.6.3 Impacts on Plants

Plants take up nutrients in the form of minerals directly from the soil; thus, altered soil composition in terms of minerals and other growth factors would directly lead to changes in plant growth. Organic pollutants not only hamper the normal physiological processes carried out by plants such as photosynthesis, seed germination, and transpiration but also interfere with the metabolic pathways occurring in plants, for instance, the biosynthetic pathways of primary and secondary metabolites. Accumulation of pollutants in soil alters the soil pH and solubilizes the toxic salts of metals such as aluminium. Similarly, exposure to peroxyacetylnitrate, a secondary pollutant, leads to necrosis and chlorosis in leaves. It also impedes other plant physiological processes such as photosynthesis, respiration, carbon dioxide fixation, photorespiration, NADP reduction, and the synthesis of cellulose, carbohydrates, proteins, and enzymes involved in the processes of respiration and photosynthesis (Gheorghe and Ion 2011).

Traces or sometimes whole antibiotics, if present in the industrial effluents of pharmaceutical industries, cause severe negative influences on plant growth. According to reports, at low concentrations, tetracyclin has shown hormetic effects that induce plant early development and seed germination, but chronic exposure led to the suppression of plant performance and activity (Xie et al. 2011). Chlorophyll biosynthesis was inhibited in *Hordeum vulgare* when it was exposed to a higher concentration of streptomycin (Yaronskaya et al. 2007). Antibiotics such as enrofloxacin and sulfadimethoxine notably decreased plant growth (Migliore et al. 2003; Migliore et al. 1996), whereas ciprofloxacin (Aristilde et al. 2010) led to a reduction of both plant growth and photosynthesis in various plant species. Similarly, different organic pollutants have different influences on plant development and their performance.

TABLE 1.3
Organic Pollutants and Their Negative Effects on Plants

Organic pollutants	Effects	References
Peroxyactyl nitrate (PAN)	Highly phytotoxic even at low concentrations; results in declined stomatal conductance, elevated levels of reactive oxygen species (ROS), and intracellular H_2O_2 reduction in calcium chelators leading to plant cell death; also interferes with the production of macromoleucles and enzymes associated with photosynthesis and cellular respiration	Yukihiro et al. (2012)
Antibiotics such as teratcyclins, penicillins, cephalosporins, macrolides, flouroquinolones	Reduction in chlorophyll contents and foliage photosynthesis, stimulation of emission of lipooxygensase pathway products and monoterpenes, and enhancement of photoprotective carotenoid levels in wheat	Opriş et al. (2013)
Anthraquinone dyes (Lanasyn Blue and Optilan Blue, azo dyes (Lanasyn Red and Nylosan Red, Nylosan Dark Brown and Lanasyn Dark Brown and antibiotics such as penicillins (Amoxicillin, Ampicillin, and Penicillin G), cephalosporins (Ceftazidime and Ceftriaxone), tetracyclines (Tetracycline and Doxycycline), fluoroquinolones (Ciprofloxacin) and macrolides (Erythromycin)	Reduction of total flavonoid contents (flavonoids are crucial metabolites in plant defense) that provide protection against various biotic and abiotic stresses and enhance reproduction and protection of reproductive seeds and tissue, so a decreased concentration of flavonoids directly affects these activities	Copaciu et al. (2016)
Polychlorinated dibenzo-*p*-dioxins	Delayed flowering, decreased seed production with lower oil content and low viability, reduced levels of C18-unsaturated fatty acid, stimulation of the gene expression of certain genes such as *9-LOX* and *13-LOX* and the synthesis of their corresponding hydroperoxide such as 9- and 13-HPOD and 9- or 13-HPOT, originating from linoleic and linolenic acids, respectively, in *Arabidopsis thaliana*	Hanano et al. (2015)
Polychlorinated Biphenyls (PCBs) and their Hydroxylated Metabolites (OH-PCBs)	Numerous xenobiotic response genes such as cytochrome P-450 and glutathione S-transferases that participate in PCB metabolism were stimulated, which caused the inhibition of biomass growth, enhanced lipid eroxidation, and ROS	Subramanian et al. (2017)

(*Continued*)

TABLE 1.3 ***(Continued)***
Organic Pollutants and Their Negative Effects on Plants

Organic pollutants	Effects	References
Bisphenol A (BPA)	Exposure of higher concentrations of BPA inhibited seedling growth and reduced the levels of various plant hormones such as indole-3-acetic acid (IAA) zeatin (ZT), gibberellic acid (GA), and ethylene (ETH) in soybean Inhibition of root development, decrease of the stomatal size and chlorophyll concentration, leaf necrosis, a reduction in photosynthetic activity, and improvement in the phytoestrogen genistein level in mung bean (*Vigna radiata*)	Wang et al. (2015), Kim et al. (2018)
Pyriproxyfen	Reduction in photosynthetic levels in maize	Coskun et al. (2015)
Tricyclazole	Reduction in photosynthetic pigments in tomato (*Lycopersicon esculentum*)	Shanmugapriya et al. (2013)
2-(phosphonomethylamino)acetic acid and 1-Aminomethylphosphonic acid	Decreased photosynthetic rate and chlorophyll content in *Salix miyabeana* cultivar SX64 (willow species), induction of ROS accumulation, and lipid peroxidation due to the inhibition of antioxidant enzymes	Gomes et al. (2016)
Thiram	Increased levels of hydrogen peroxide and malondialdehyde (MDA) lead to oxidative stress, reduction in chlorophyll and carotenoid levels, upregulation of genes such as *GST1*, *GST2*, and *GST3* encoding for glutathione S-transferase and *P450* encoding for cytochrome P-450 monooxygenases in tomato (*Solanum lycopersicum* Mill.)	Yüzbaşıoğlu and Dalyan (2019)
Black dye	Chromosomal and nuclear aberrations including loss of genetic material, chromosomal stickiness, nuclear buds and binucleated cells, and cytotoxic effects led to the programmed death of cells in *Allium cepa*	de Campos Ventura-Camargo et al. (2016)
Petroleum	Reduction in stomatal conductivity and rate of transpiration and decreased levels of total chlorophyll in seedlings of *Amorpha fruticosa* resulting in a low rate of photosynthesis	Han et al. (2016)
Polyaromatic hydrocarbons (PAH)	Under phenanthrene (PHE)- and pyrene (PYR)-contamination, dehydrogenase activity was enhanced, whereas polyphenol oxidase activity was reduced in endophytic bacteria present in the root of salt plant *Spartina alterniflora* in which endophytic microbes play a crucial role in the phytormediation of phenanthrene and pyrene	Hong et al. (2015), Shen et al. (2017), Khpalwak et al. (2018)

(*Continued*)

TABLE 1.3 *(Continued)*
Organic Pollutants and Their Negative Effects on Plants

Organic pollutants	Effects	References
	Phenanthrene caused chlorosis due to increased degradation of chlorophyll and leaf moisture in wheat; phenanthrene treatment enhanced glutamate and porphyrins levels and activated gene encoding 5-Amino-4-oxopentanoic acid, uroporphyrinogen III, protoporphyrin IX, Mg-protoporphyrin IX and Monovinyl protochlorophyllide	
	Treatment of 1,2-Benzacenaphthene, phenanthrene, mannitol, and 2-Hydroxybenzoic acid to marigold (*Calendula officinalis*) led to the induction of oxidative stress due to increased levels of ROS, adverse effects on stomatal conductivity (G_s), photosynthesis at near saturating irradiance (A_{max}), concentration of internal carbon dioxide (C_i), leaf-water relations and photosynthetic pigments	
Perfluorooctanoic acid (PFOA)	Reduction in the chlorophyll pigment level and net photosynthetic rate, upregulation of phenol metabolism, and downregulation of native amino acids in cucumber leaves (*Cucumis sativus*)	Du et al. (2020)

1.7 MICROBIOLOGICAL APPROACH FOR REMOVING ORGANIC POLLUTANTS FROM THE ENVIRONMENT

Living organisms and the environment are massively affected by organic pollutants. Organic pollution results when larger amounts of organic contaminants originate from sewage waste, domestic waste, industrial effluents, metallurgical procedures, vehicle emissions, fossil fuel emissions, and petroleum hydrocarbons spillage and accumulate in the soil, water, and air. Various organic compounds have been characterized and categorized as POPs, which are persistent in nature, remain within living organisms for many years, and cause serious health issues. Several biological, chemical, and physical strategies have been formulated, designed, and utilized for removing these toxic organic contaminants. However, physical and chemical methods are quite tedious, are difficult to operate in rural regions, and require high-duty machinery, thereby increasing the cost of the treatment process, which results in a financial burden to industries and nations. Microbiological approaches involving different microbial species have captivated much attention and are becoming a popular strategy in the treatment of organic pollutants. Various microorganisms utilize organic pollutants as their energy source (if possible), and, thus, complex pollutants are degraded to simpler components with either low toxicity or no toxicity. If pollutants cannot be utilized as an energy source, then even microbial flora has the potential to degrade organic compounds *via* the 'bioremediation' and 'biotransformation' processes.

Pseudomonas putida, oil-eating bacteria, is a famous example of a genetically engineered microorganism that has been designed to degrade the complex petroleum hydrocarbon spillage in oceans. A foreign plasmid consisting of genes essentially required for the degradation of hydrocarbon was introduced in this bacterium. Wide varieties of hydrocarbon compounds including naphthalene, camphor, salicylate, and octane could be metabolized by *Pseudomonas putida*. In *Ralstonia eutropha* CH34 (a heavy metal-tolerant bacterial strain), the PCB degrading genes of *Achromobacter* sp. LBSIC1, *Ralstonia eutropha* A5, and *Achromobacter denitrificans* were introduced for metabolizing and degrading polychlorinated biphenyls (Joutey et al. 2013).

Apart from bacteria, algae, fungi, and yeast species are also involved in the 'bioremediation' process. Metabolic and enzymatic processes of microbial communities can biotransform these toxic organic compounds. The process of biotransformation is performed through two methods: growth and co-metabolism. In growth, complete degradation of organic contaminant takes place, as the organic compound is itself used as an individual source of carbon and energy by the microorganism. However, in the case of co-metabolism, apart from the organic pollutant, microbes use another growth substrate as a carbon and energy source, so a partial degradation of the organic compound occurs (Fritsche and Hofrichter 2000). The efficiency of the microbes in metabolizing organic contaminants depends upon various factors, for instance, the chemical nature and quantity of the pollutants, their availability to the microorganisms, and the physicochemical features of the environment (El Fantroussi and Agathos 2005). The microorganism involved in the process basically has to deal with biotic and environmental factors. The metabolic activity performed by the microbes is known as the biotic factor that is itself affected by the inhibition of elements of microbial enzymes and predation by other microbes such as bacteriophage and protozoa. Furthermore, a sufficient amount of nutrients and factors affecting enzyme action, for example, optimum oxygen, pH, and temperature, are equally important for organic pollutants to degrade microbes. Environmental factors including soil type and differences in the porosity of saturated and unsaturated regions of the aquifer matrix may impact the movement of the contaminants to the groundwater. In soil saturated with more water and in fine-grained sediments, the potential of the matrix to transmit gases such as oxygen, carbon dioxide, and methane is lowered. Thus, these factors can influence the type and rate of biodegradation occurring at the site.

Sulfamethoxazole (SMX), which is extremely important and extensively utilized in sulphonamide antibiotics, is becoming an issue as an environmental pollutant due to its recalcitrant nature. By adopting a variety of microbial communities, the degradation of SMX has become possible. Bacterial species including *Rhodococcus rhodochrous*, *Bacillus subtilis*, *Pseudomonas aeruginosa*, *Alcaligenes faecalis*, and *Pycnoporus sanguineus* metabolize SMX when provided with an additional carbon source in the culture medium. Some other bacteria such as *Acinetobacter* sp. and *Microbacterium* sp. can totally metabolize the SMX, whereas species such as *Achromobacter denitrificans* strain PR1 and *Rhodococcus equi* require an additional carbon source that significantly enhances the rate of SMX degradation (Reis et al. 2014; Nguyen et al. 2017). Fungal species, namely, *Pleurotus ostreatus*, *Pleurotus pulmonarius*, and *Tramete* can also degrade SMX (de Araujo et al. 2017). Similarly, oleophilic microorganisms are employed for 'bioremediation' at oil spillage sites. Microbes such as *Trichoderma harzianum*, *Cunninghamella elegans*, and *Aspergillus niger* can metabolize petroleum hydrocarbon products, that is, naphthalene, phenanthrene, and n-haxadecane, respectively (Mollea et al. 2005; Romero et al. 1998; Volke-Sepulveda et al. 2003).

Bacterial species such as *Pseudomonas fluorescens*, *Ochrobactrum anthropi*, *Pseudomonas aeruginosa*, and *Agrobacterium radiobacter* were isolated from the PCB-contaminated sediments from the Strážsky canal and the Zemplínska šírava water reservoir. The *bphA1* gene reported in the bacterial species encoded the enzyme biphenyldioxygenase, which catalyzed the first reaction in PCB degradation (Dercová et al. 2008). Similarly, bacterial strains *Achromobacter* sp. NP03, *Ochrobactrum* sp. NP04, *Lysinibacillus* sp. NP05, and *Pseudomonas* sp. NP06 were identified from the sediment samples of the Brisbane River, and Coombabah Lake, Gold Coast and a soil sample from the Brisbane City area, Australia, contaminated with PCB. These strains were able to degrade PCB, with *Lysinibacillus* sp. NP05 exhibiting the highest efficiency among all other microbial species (Pathiraja et al. 2019).

It was reported that *Streptomyces alanosinicus*, *Streptoverticillium album*, *Nocardia farcinia*, *Streptomyces atratus*, *Nocardia vaccini*, *Nocardia amarae*, and *Micromonospora chalcea* can efficiently grow and metabolize pesticides (Jayabarath et al. 2010). Atrazine was successfully degraded by white-rot fungi (Elgueta et al. 2016) and green microalgae *Chlamydomonas mexicana* (Kabra et al. 2014), and in fact, the latter exhibited a degradation rate of 14–36%. Species of genera *Pseudomonas* can effectively degrade pesticides such as endosulfan hexachlorocyclohexane, aldrin, parathion (Verma et al. 2014; Upadhyay and Dutt 2017), and many more; similarly, *Bacillus* sp. are efficient at degrading organic pollutants, for example, chlorpyrifos, dieldrin, monocrotophos, DDT, and PAH (Verma et al. 2014). Likewise, small green algae can degrade phorate and parathion (Tang et al. 2018), whereas some genera of diatoms can metabolize patoran and DDT (Shehata et al. 1997). Pyridine, an important organic compound used in several applications such as paints, medicine, dyes, insecticides, and herbicides, has been reported to be effectively degraded by *Paracoccus* sp. NJUST30 with the production of major intermediates including piperidin-2-ol, 2,4-dihydroxy-2H-pyridine-3-one, 4-formylamino-butyric acid, 2-carbonyl-succinic acid, and dihydropyridine alcohol or (1,2-dihydro-pyridin-2-ol) (Wang et al. 2018c).

A moderate thermophile *Paenibacillus napthalenovorans* strain 4B1 discovered from the dioxin-contaminated soil in Vietnam was capable of degrading dibenzofuran (DF), taking it as an exclusive source of carbon and energy. The rate of degradation was comparatively higher than the reference *Paenibacillus* strain YK5, which proved that the *Paenibacillus napthalenovorans* strain 4B1 produced a *dbf* gene cluster (*dbfA1A2BC*) (by encoding the alpha subunit of DF dioxygenase enzyme) and this expression level was four times higher than the one from reference strain (Thi et al. 2019).

1.8 INVOLVEMENT OF GENES AND ENZYMES IN THE 'BIOREMEDIATION PROCESS'

Microorganisms can utilize organic compounds as their energy and carbon source and *via* different metabolic pathways, degrade these complex compounds into simpler forms. To achieve this challenge of biodegradation, various microbial genes are involved in this process. In simpler terms, genes encode enzymes that help in the catalysis and metabolism of these toxic organic compounds. The large group of dioxins and their derived compounds are chiefly present in the industrial effluents as POPs. Research shows that various microbial species have the potential to degrade and metabolize dioxins and related compounds. For example, *Burkholderia cenocapacia* 869T2 (an endophytic bacterium) has been reported to be effective in degrading the very toxic dioxin compounds 2,3,7,8-Tetrachlorooxanthrene by utilizing gene encoding L-2-haloacid dehalogenase (2-HAD) in the dioxin dehalogenation process (Nguyen et al. 2021). Bacterial species such as *Cupriavidus necator* JMP134 and *Pseudomonas* strains remove the side chain of 2, 4-D by α-ketoglutarate-dependent 2,4-D dioxygenase (*tfdA*) and form 2,4-dichlorophenol (2,4-DCP) and use 2,4-DCP hydroxylase (*tfdB*) hydroxylate 2,4-DCP to form dichlorocatechol. Chlorocatechol 1,2-dioxygenase (*tfdC*) is involved in the *ortho* or *meta* cleavage of dichlorocatechol by forming 2,4-dichloro-*cis,cis*-muconate, which, in turn, is cleaved by chloromuconate cycloisomerase (*tfdD*) to form 2-chlorodienelactone. Chlorodienelactone hydrolase (*tfdE*) converts 2-chlorodienelactone to 2-chloromaleylacetate, and ultimately, the catalysis of 2-chloromaleylacetate to 3-oxoadepate via maleylacetate is performed by chloromaleylacetate reductase and maleylacetate reductase (*tfdF*), respectively. The final products are thus involved in the Kreb cycle within the microbes (Kumar et al. 2016). β-propeller phytase-like sequences (BPP-like sequences) are found in bacterial species colonizing in land, water, and plants and are actively involved in the degradation of dioxins and related compounds. *Sphingomonas wittichii* RW-1 exhibits the putative phytase gene (*physw*) that encodes the novel enzyme β-propeller phytases to degrade polychlorinated dibenzo-p-dioxins and dibenzofurans (Sanangelantoni et al. 2018).

TABLE 1.4
Degradation of Halogenated Aromatic Hydrocarbons by Different Microbial Communities

Microorganisms	Halogenated aromatic hydrocarbons	References
Aquincola tertiaricarbonis L10, *Denitratisoma oestradiolicum* AcBE2–1, *Xylophilus ampelinus* BPIC 48, and *Geothrix fermentans* H5	PCBs and PBDEs	Song et al. (2015)
Ochrobactrum anthropi, *Pseudomonas veronii*, *Pseudomonas stutzeri*, and *Alcaligenes xylosoxidans*	PCB congeners [1-chloro-2-(2-chlorophenyl) benzene, 1-chloro-4-(4-chlorophenyl) benzene, 1,4-dichloro-2-(2-chlorophenyl) benzene, 2,4-dichloro-1-(4-chlorophenyl) benzene, 1,4-dichloro-2-(2,5-dichlorophenyl) benzene, 1,2,4-trichloro-5-(2,5-dichlorophenyl)benzene, 21,2,3,4-tetrachloro-5-(4-chlorophenyl)benzene, 1,2,3,4-tetrachloro-5-(4-chlorophenyl) benzene, 1,2,3,4-tetrachloro-5-(2,4-dichlorophenyl)benzene, 1,2,4-trichloro-5-(2,4,5-trichlorophenyl)benzene, 1,2,3,4-tetrachloro-5-(2,4,5-trichlorophenyl) benzene, and 1,2,4,5-tetrachloro-3-(2,4,5-trichlorophenyl)benzene]	Murínová et al. (2014)
Achromobacter xylosoxidans, *Stenotrophomonas maltophilia*, and *Ochrobactrum anthropi*	PCB	Horváthová et al. (2019)
Serratia sp. SSA1 and *Bacillus* sp. SS2	Dichlorinated dibenzo-*p*-dioxin and dichlorinated dibenzofuran	Saibu et al. (2020)
Thermoascus crustaceus, *Doratomyces purpureofuscus*, *Myceliophthora thermophila*, *Doratomyces nanus*, *Phoma eupyrena*, and *Doratomyces verrucisporus*	PCB congeners [1,4-dichloro-2-(2-chlorophenyl)benzene, 2,4-dichloro-1-(4-chlorophenyl)benzene, 1,4-dichloro-2-(2,5-dichlorophenyl)benzene, 1,2,4-trichloro-5-(2,5-dichlorophenyl) benzene, 1,2,4-trichloro-5-(3,4-dichlorophenyl)benzene1,2,3,4-tetrachloro-5-(2,4-dichlorophenyl)benzene, 1,2,4-trichloro-5-(2,4,5-trichlorophenyl) benzene, 1,2,3,4-tetrachloro-5-(2,4,5-trichlorophenyl)benzene, and 1,2,4,5-tetrachloro-3-(2,4,5-trichlorophenyl) benzene]	Mouhamadou et al. (2013)
Sinorhizobium meliloti	2,4, 4'—trichlorobiphenyl	Tu et al. (2011)
Rhodococcus biphenylivorans strain TG9T	PCB	Ye et al. (2020)
Clostridia, Alphaproteobacteria, Actinobacteria, Bacilli, and *Gammaproteobacteria as main classes with Ralstonia* sp. and *Burkholderia* sp., and *Pseudomonas* sp. as the main species	Dioxins	Tran et al. (2020)
Rhodococcus ruber and *Rhodococcus pyridinivorans*	low-chlorinated PCB congeners, Benzoic acid, 3-chlorobenzoic acid, and PCB congeners	Wang et al. (2018b)

(*Continued*)

TABLE 1.4 *(Continued)*
Degradation of Halogenated Aromatic Hydrocarbons by Different Microbial Communities

Microorganisms	Halogenated aromatic hydrocarbons	References
Debaryomyces maramus and CW36 *Pseudomonas* sp. MO2A	2,4-dichloro-1-(4-chlorophenyl)benzene, 1,4-dichloro-2-(2,5-dichlorophenyl)benzene, 1,2,4-trichloro-5-(2,5-dichlorophenyl) benzene, 1,2,4-trichloro-5-(2,4,5-trichlorophenyl)benzene, and 1,2,3,4-tetrachloro-5-(2,4,5-trichlorophenyl) benzene	Chen et al. (2015)
Anabaena PD-1	PCBs	Zhang et al. (2015)
Dehalococcoides mccartyi	PCB congeners including hepta-, hexa-, and penta-chlorobiphenyls (CBs), tri-CBs, and tetra-CBs	Matturro et al. (2016)
Mycolicibacterium frederiksbergense IN53, *Rhodococcus erythropolis* IN129, and *Rhodococcus* sp. IN306	PCB congeners and total petroleum hydrocarbons (TPH)	Steliga et al. (2020)
Castellaniella sp. strain SPC4	1,2-dichloro-4-(3,4-dichlorophenyl)benzene	Su et al. (2019)
Paraburkholderia xenovorans LB400	PCB congeners including 3,3',5,5'-tetrachlorobiphenyl	Bako et al. (2021)
Bacillus sp. GZT	2,4,6-Tribromophenol (TBP)	Xiong et al. (2017)
Mesorhizobium sp. ZY1	3,3′,4,4′-TCB	Teng et al. (2016)
White-rot fungi, including *Irpex lacteus* and *Pleurotus ostreatus*	PCB	Stella et al. (2017)

TABLE 1.5
Degradation of Herbicides by Different Microbial Communities

Microorganisms	Herbicides	References
Comamonas odontotermitis P2	Glyphosate (GLYP)	Firdous et al. (2017)
Rhodococcus sp. T1	Clodinafop propargyl	Hou et al. (2011)
Micrococcus sp., *Arthrobacter* spp., *Aspergillus* sp., *Bacillus* spp., *Streptomyces* sp., *Pseudomonas* sp., and *Mycobacterium* sp.	Oxyfluorfen	Mohamed et al. (2011)
Thauera sp. strain DKT	(2,4-Dichlorophenoxy)acetic acid	Ha (2018)
Bacillus badius ABP6 strain	atrazine	Khatoon and Rai (2020)
Cupriavidus oxalaticus Strain X32	Phenoxyalkanoic acid herbicide	Xiang et al. (2020)
Sphingomonas sp. Ndbn-20 and *Sphingobium* sp. Ndbn-10	Dicamba (chlorinated benzoic acid)	Yao et al. (2015)
Lysinibacillus sphaericus	GLYP	Pérez Rodríguez et al. (2019)
Pseudomonas sp. and *Achromobacter* sp.	atrazine	Fernandes et al. (2018)
Pseudomonas sp. strain B2	Clodinafop propargyl	Singh (2013)
Cupriavidus gilardii T-1	2,4-Dichlorophenoxyacetic acid (2,4-D), 2-(4-chloro-2-methylphenoxy)acetic acid), 6-methylheptyl 2-(4-chloro-2-methylphenoxy)acetate, and Dichlorprop	Wu et al. (2017)

(Continued)

TABLE 1.5 ***(Continued)***
Degradation of Herbicides by Different Microbial Communities

Microorganisms	Herbicides	References
Rhizorhabdus dicambivorans Ndbn-20	Dicamba	Li et al. (2020)
Cryptococcus laurentii	atrazine	Evy et al. (2012)
Sphingobium sp. Strain SMB	Linuron	Zhang et al. (2020b)
Achromobacter sp. LZ35	2,4-dichlorophenoxyacetic acid (2,4-D) and 2-methyl-4-chlorophenoxy acetic acid (MCPA)	Xia et al. (2017)
Providencia rettgeri GDB 1	GLYP	Xu et al. (2019)
Rhodocyclaceae, *Desulfitobacterium*, and *Desulfuromonas*	atrazine and hexachlorobenzene (HCB)	Wang et al. (2018a)
Sphingopyxis sp. DBS4	Dichlorprop	Zhang et al. (2020a)
Cunninghamella sp. PFCM-1 and *Aspergillus tamari-*	paraquat	Wongputtisin et al. (2020)
Novosphingobium strain DY4	2-(2,4-dichlorophenoxy)acetic acid	Dai et al. (2015)
Penicillium oxalicum MET-F-1	2-chloro-*N*-(2-ethyl-6-methylphenyl)-*N*-(1-methoxypropan-2-yl)acetamide	Chang et al. (2020)
Burkholderia vietnamiensis strain AQ5–12	GLYP	Manogaran et al. (2018)
Bacillus licheniformis ATLJ-5 and *Bacillus megaterium* ATLJ-11	atrazine	Zhu et al. (2019)
Delftia sp.	Hedonal and 4-chloro-*o*-toyloxyacetic acid	González et al. (2017)
Bacillus subtilis Bs-15	GLYP	Yu et al. (2015)
Klebsiella sp. A1 and *Comamonas* sp. A2	atrazine	Yang et al. (2010)
Aspergillus oryzae A-F02	GLYP	Fu et al. (2017)
Rhizorhabdus dicambivorans sp. nov. Ndbn-20T	Dicamba	Yao et al. (2016)
Pseudomonas geniculata PQ01	paraquat	Wu et al. (2020)
Citricoccus sp. strain TT3	atrazine	Yang et al. (2018)
Salinicoccus spp.	GLYP	Sharifi et al. (2015)

Paracetamol is extensively employed as an antipyretic and analgesic drug but is a potential organic contaminant. Bacterial genera *Acinetobacter, Bacillus, Sphingomonas,* and *Pseudomonas* can metabolize paracetamol. Bacterial strain *Pseudomonas moorei* KB4 among all isolated bacteria was reported to be the highest degrader of paracetamol. Deaminase, acyl amidohydrolase, and hydroquinone 1,2-dioxygenase were found to be the major enzymes associated with the degradation of paracetamol by *Pseudomonas moorei* KB4 (Żur et al. 2018). The enzyme 2-hydroxy-6-oxo-6-phenylhexa-2,4-dienoate hydrolase participates in the aromatic and biphenyl compound breakdown. Members of this enzyme family participate in the hydrolysis of C-C bonding in breaking down the pathway of biphenyls (Horsman et al. 2006). For metabolizing naphthalene sulphonates and anthracene, enzyme 2-hydroxychromene-2-carboxylate isomerase has been identified from *Pseudomonas testosteroni* A3 and *Sphingomonas yanoikuyae* B1 (Kuhm et al. 1993; Kim et al. 1997).

Aromatic amines are a crucial ingredient of dyes, polymers, and other industrially derived products. Aromatic amines enter the soil ecosystem due to the usage of pesticides for the control of insects and pests in soybean, corn, and other crops. Enzyme 4-hydroxyphenylacetate 3-monooxygenase is actively involved in the degradation of aromatic ester 4-hydroxyphenylacetate (an intermediate

TABLE 1.6
Degradation of Pesticides/Insecticides by Different Microbial Communities

Microorganisms	Pesticide	References
Klebsiella pneumoniae BPBA052 and *Acinetobacter junii* LH-1–1	[(*S*)-cyano-(3-phenoxyphenyl)methyl] (1*R*,3*R*)-3-(2,2-dibromoethenyl)-2,2-dimethylcyclopropane-1-carboxylate and its major metabolite 3-Carboxydiphenyl ether	Tang et al. (2020)
Stenotrophomonas acidaminiphila BJ1	Chlorothalonil (or 2,4,5,6-tetrachloroisophthalonitrile)	Zhang et al. (2019)
Ochrobactrum sp. CPD-03	Diethoxy-sulfanylidene-(3,5,6-trichloropyridin-2-yl)oxy-λ5-phosphane and 3,5,6-trichloro-1*H*-pyridin-2-one	Nayak et al. (2019)
Aspergillus oryzae M-4 and *Bacillus licheniformis* B-1	Beta-cypermethrin (β-CY) and its associated metabolite 3-Carboxydiphenyl ether	Zhao et al. (2018)
Pseudomonas azotoformanss strain ACP1, *Pseudomonas putida* ACP3, and *Pseudomonas aeruginosa* strain ACP2	Acephate (or N-(Methoxy-methyl sulfanyl phosphoryl)-acetamide)	Singh et al. (2020)
Bacillus sp. strain Neni-12	Coumaphos (or 3-chloro-7-diethoxyphosphinothioyloxy-4-methyl-2-chromenone)	Gusmanizar (2019)
Pseudomonas aeruginosa, Enterobacter cloacae, and *Candida tropicalis*	2-[2-(1-chlorocyclopropyl)-3-(2-chlorophenyl)-2-hydroxypropyl]-1,2-dihydro-3H-1,2,4-triazole-3-thione	Shi et al. (2020)
Candida tropicalis	Methyl 2-(*N*-(2-methoxyacetyl)-2,6-dimethylanilino)propanoate	Derbalah et al. (2020)
Pseudomonas sp. CB2	Bifenthrin (BF) and chlorpyrifos (CP)	Zhang et al. (2018)
Sulfitobacter sp. S-1, *Sphingobium* sp. P-1, *Marinicauda* sp. M-1, *Erythrobacter* sp. E-17, *Stakelama* sp. P-2, and *Oceanicaulis* sp. B-1	Chlorothalonil (or 2,4,5,6-tetrachloroisophthalonitrile)	Hu et al. (2020)
Bacillus thuringiensis strain SG4	Alpha-cypermethrin	Bhatt et al. (2020)
Genera such as *Pseudomonas, Ochrobactrum*, and *Comamonas*	Epoxiconazole (EPO) and fludioxonil (FLU)	Alexandrino et al. (2020)
Enterobacter sp.	Carbofuran	Ekram et al. (2020)
Sphingomonas sp. NJUST37	8-methyl-[1,2,4]triazolo[3,4-b][1,3] benzothiazole	Wu et al. (2018)
Pseudomonas pseudoalcaligenes ADI-03, *Lysinibacillus fusiformis* ADI-, *Pseudomonas* species ADI-04, 01*Pseudomonas pseudoalcaligenes* ADI-06, and atypical VP negative *Bacillus cereus* ADI-10	Acephate (or O,S-dimethyl acetylphosphoramidothioate)	Mohan and Naveena (2015)

compound of amino acid, styrene, and lignin) (Mohamed et al. 2002). The enzyme Acetyl-CoA acetyltransferase plays an active role in the breakdown of benzoate through CoA ligation and participates in the degradation pathways such as aromatic amino acid metabolism. Gram-negative bacterium *Azoarcus evansii* and gram-positive bacterium *Bacillus stearothermophilus* are known to degrade benzoate in aerobic conditions (Zaar et al. 2001). The dioxygenation of carbazole (the *N*-heterocyclic aromatic compound present in the petroleum products) has been reported in phyla Proteobacteria

TABLE 1.7
Degradation of Industrial Dyes by Different Microbial Communities

Microorganisms	Industrial dyes	References
Streptomyces chrestomyceticus S20	Malachite green	Vignesh et al. (2020)
Cedecea davisae	Crystal violet	Cao et al. (2019)
Bacillus stratosphericus SCA1007	Sodium 4-[(4-dimethylamino) phenylazo]benzenesulfonate	Akansha et al. (2019)
Acinetobacter spp.	Carbol fuchsin	Mhatre (2021)
Trichoderma lixii F21	Alizarin Red S and Quinizarine Green SS	Adnan et al. (2017)
Green algae *Scenedesmus obliquuss*, *Chlorella* sp., *Chlorella vulgaris*, *Scenedesmus officinalis*, *Haematococcus* sp., and *Scenedesmus quadricauda* and blue green algae *Arthospira maxima*	Congo red	Mahalakshmi et al. (2015)
Bacillus cereus AZ27, *Alcaligenes faecalis* AZ26, and *Bacillus* sp. AZ28	Novacron Super Black G (NSB-G)	Hossen et al. (2019)
Saccharomyces cerevisiae ATCC 9763	Disodium;4-hydroxy-3-[(4-sulfonatonaphthalen-1-yl)diazenyl] naphthalene-1-sulfonate	Kiayi et al. (2019)
Myrothecium verrucaria ITCC-8447	5-methyl-2-[[5-(4-methyl-2-sulfoanilino)-9,10-dioxoanthracen-1-yl]amino]benzenesulfonic acid and Alizarin cyanine green (ACG)	Agrawal and Verma (2020)
Brevibacillus laterosporus	Methyl 3-[[4-(2-chloro-4-nitrophenyl) diazenylphenyl]-(2-cyanoethyl) amino]propanoate	Kurade et al. (2016)
Saccharomyces cerevisiae	Alizarin Red S	Ramavandi et al. (2019)
Aspergillus flavus	Malachite green	Barapatre et al. (2017)
Pseudomonas stutzeri MN1 and *Acinetobacter baumannii* MN3	Disodium;4-amino-3-[[4-[4-[(1-amino-4-sulfonatonaphthalen-2-yl)diazenyl] phenyl]phenyl]diazenyl]naphthalene-1-sulfonate and gentian violet	Kuppusamy et al. (2017)
Aeromonas hydrophila	Crystal violet	Bharagava et al. (2018)
Serratia liquefaciens	Azure-B	Haq and Raj (2018)
Candidatus Methanoperedens and *Pseudoxanthomonas sp.*	Methyl orange (MO)	Fu et al. (2019)
Aspergillus niger	Congo red	Asses et al. (2018)
Oscillatoria limnetica and *Hydrocoleum oligotrichum*	Basic fuchsin and methyl red	Abou-El-Souod and El-Sheekh (2016)
Phoma tropica MRCH 1–3 and *Dichotomomyces cejpii* MRCH 1–2	Azo dye including Congo red, methyl red, and reactive dyes	Krishnamoorthy et al. (2018)
Azospirillum brasilense strain SR80	Remazol brilliant blue and malachite green	Kupryashina et al. (2020)
Novel thermophilic microflora	Direct Black G textile dye	Chen et al. (2018)
Shewanella oneidensis MR-1	[9-(2-carboxyphenyl)-6-(diethylamino)xanthen-3-ylidene]-diethylazanium; chloride	Xiao et al. (2019)
Stropharia sp. ITCC-8422	Alizarin Cyanine Green (ACG)	Agrawal and Verma (2019)
Halopiger aswanensis strain ABC_IITR	Malachite green	Chauhan and Choudhury (2021)
Pseudoalteromonas sp CF10–13	Naphthol Green B (NGB)	Cheng et al. (2019)
Ceriporia lacerata. ZJSY	Congo red	Wang et al. (2017a)
Bacillus fermus (Kx898362)	Direct Blue-14 dye	Sandesh et al. (2019)

TABLE 1.8
Degradation of Petroleum Hydrocarbon and Its Derivatives by Different Microbial Communities

Microorganisms	Petroleum hydrocarbon and derived products	References
Pseudomonas aeruginosa NCIM 5514	Petroleum hydrocarbon oil	Varjani and Upasani (2021)
Bacillus amyloliquefaciens 6A	Kerosene	Shahzadi et al. (2019)
Bacillus cereus and *Pseudomonas putida*	1,3,5-Cyclohexatriene, Methyl benzene, Phenylethane, and Dimethylbenzene	Handayani et al. (2019)
Acinetobacter sp. K-6	Diesel and *n*-alkanes (C18, C20, and C22)	Chaudhary et al. (2020)
Talaromyces sp. and *Acinetobacter baumannii*	Crude oil	Zhang et al. (2021)
Sporosarcina globispora and *Bacillus cereus*	Poly (1-methylethylene)	Helen et al. (2017)
Paraburkholderia aromaticivorans BN5	Naphthalene, 1,3,5-Cyclohexatriene, Methyl benzene, Phenylethane, 1,2-Dimethylbenzene, 1,3-Dimethylbenzene, 1,4-Dimethylbenzene, and aliphatic hydrocarbons	Lee et al. (2019)
Halomonas sp. BR04 and *Cryptococcus* sp. MR22	Phenanthrene and anthracene	Al Farraj et al. (2020)
Bacillus velezensis KLP2016	Engine oil	Meena et al. (2021)
Acinetobacter venetianus strain RAG-1 (RAG-1)	Crude oil	Liu et al. (2021)
Genera *Talaromyces, Penicillium*, and *Aspergillus*, along with Proteobacteria, Actinobacteria, Bacteroidetes, and Firmicutes	Aviation kerosene	Shapiro et al. (2021)
Candidatus argoarchaeum, Candidatus syntrophoarchaeum, and *Candidatus methanoliparia*	Non-methane alkanes	Laso-Pérez et al. (2019)
Enterobacter aerogenes	Heavy oil	Rastib et al. (2019)
Stenotrophomonas pavanii IRB19	Total petroleum hydrocarbon (TPH)	Behera et al. (2020)
Rhodococcus sp. strain AQ5–07	Diesel	Roslee et al. (2020)
Genera such as *Georgfuchsia, Sulfuritalea* (all Betaproteobacteria), *Rhodoferax, Pelotomaculum* (Firmicutes), and *Azoarcus*	1,3,5-Cyclohexatriene, Methyl benzene, Phenylethane, and Dimethylbenzene	Sperfeld et al. (2018)
Ochrobactrum sp. MB-2	Napthalene and pyrene	Wang et al. (2019b)
Chelatococcus daeguensis HB-4	Crude oil	Ke et al. (2019)

and Actinobacteria. For the breakdown of carbazole, enzyme 2-hydroxy-6-oxo-6-phenylhexa-2,4-dienoate hydrolase (HOPD hydrolase) is actively involved (Salam et al. 2017). Different gene/enzymes including Phenylacetate-CoA oxygenase (PaaG subunit), Phenylacetate-CoA oxygenase (PaaH subunit), Phenylacetate-CoA oxygenase/reductase (PaaK subunit), Phenylacetic acid degradation protein PaaD (thioesterase), and transcripts related to the phenylacetic acid degradation proteins PaaE, PaaN, and ketothiolase, and ring-opening aldehyde dehydrogenase (EC 1.2.1.3) have been discovered for phenylacetyl-CoA catabolic pathway in metatranscriptome samples (Singh et al. 2018).

TABLE 1.9
Degradation of Drugs and Pharmaceuticals by Different Microbial Communities

Microorganisms	Drugs and pharmaceuticals	References
Achromobacter sp. L3	Sulfamethoxazole (SMX)	Liang and Hu (2021)
Raoultella sp. KDF8	Diclofenac and codeine analgesics	Palyzová et al. (2018)
Cupriavidus sp., *Sphingomonas* sp., *Delftia* sp., *Acinetobacter* sp., and *Methylobacterium* sp.	Carbamazepine (CBZ)	González-Benítez et al. (2020)
Dyella sp. WW1	Triclosan	Wang et al. (2018d)
Bacillus thuringiensis B1(2015b)	Naproxen	Dzionek et al. (2020)
Rhodococcus, *Streptomyces*, *Pseudomonas*, *Sphingomonas*, *Methylobacillus*, and *Stenotrophomonas*	Carbamazepine (CBZ), triclocarban (TCC), and triclosan (TCS)	Thelusmond et al. (2019)
Proteobacteria	Enrofloxacin (ENR) and oxytetracycline (OXY)	Harrabi et al. (2019)
Bacterial taxa, Latescibacteria, Actinobacteria, Bacteroidetes, and Proteobacteria Acidobacteria, Gemmatimonadetes, genera *Pseudomonas* and *Novosphingobium*	2-[4-(1-Hydroxy-2-methylpropyl)phenyl] propionic acid, 2-[4-(2-hydroxy-2-methylpropyl)phenyl]propanoic acid, 2-[4-(3-hydroxy-2-methylpropyl)phenyl] propanoic acid, and 3-[4-(1-carboxyethyl) phenyl]-2-methylpropanoic acid	Rutere et al. (2020)
Sphingobacterium thalpophilum P3d	Nitrofurantoin	Pacholak et al. (2019)
Aminobacter strain	*N*, *N*-dimethylbiguanide and its transformation product diaminomethylideneurea	Poursat et al. (2019)
Shewanella, *Bacillus*, and *Pseudomonas*	Tetracycline	Wang et al. (2019a)
Bacillus drentensis strain S1	*N*-acetyl-para-aminophenol or acetaminophen (APAP, paracetamol)	Chopra and Kumar (2020)
Ganoderma lucidum, *Pleurotus ostreatus*, and *Trametes versicolor*	1-[2-(dimethylamino)-1-(4-methoxyphenyl)ethyl]cyclohexan-1-ol and 4-[(1*S*)-2-(dimethylamino)-1-(1-hydroxycyclohexyl)ethyl]phenol	Llorca et al. (2019)
Micrococcus yunnanensis KGP04	Ibuprofen	Sharma et al. (2019)
Enterococcus, *Ochrobactrum*, *Achromobacter*, *Lactococcus*, and *Bacillus*	Ciprofloxacin	Feng et al. (2019)
Genera *Pseudomonas*, *Flavobacterium*, *Dokdonella*, and *Methylophilus*	Paracetamol	Palma et al. (2018)
Penicillium oxalicum B4	5-chloro-2-(2,4-dichlorophenoxy)phenol	Tian et al. (2018)
Rhodotorula mucilaginosa (strain LMB-D), *Bacillus subtilis* (strain LMB-A), and *Penicillium oxalicum* (strain LMB-E)	Lincomycin	Li et al. (2021)

*Lad*A gene encoding a novel alkane monooxygenase enzyme was reported in the thermophilic strain NG80–2 of *Geobacillus thermodenitrificans* and can catalyze the first reaction for the degradation pathway of n-alkanes (Feng et al. 2007). The expression of the *alk*B gene was induced when different species of *Geobacillus* (*G. thermoleovorans* strain 27, *Geobacillus caldoxylosilyticus* 17, *Geobacillus pallidus* 2, *Geobacillus toebii* 1, and *Geobacillus* sp. 3) were used for the degradation of n-hexadecane (Marchant et al. 2006). It was later determined that three strains, *G. toebii* B-1024, *Geobacillus* sp. 1017, and *Aeribacillus pallidus* 8m3, with the potential of degrading C10–C30 n-alkanes, showed elevated expression of *alk*B and *lad*A gene encoding the enzymes rubredoxin-dependent alkane monooxygenase and flavin-dependent alkane monooxygenase, respectively. The *alk*B gene was expressed in all three strains, whereas the expression of the *lad*A gene was limited to *G. toebii*B-1024 and *Geobacillus*sp. 1017 (Tourova et al. 2016). Similarly, a thermophilic bacterial strain *Sulfolobus solfataricus*P2 identified from a volcanic site, when sequenced, revealed the presence of genes encoding enzymes by participating in monocyclic aromatic hydrocarbon (MAH) degradation, including mono- and dioxygenases (Chae et al. 2007).

1.9 CONCLUSION

Due to globalization and high-speed industrialization, the world has evolved a great deal, and some have become comfortable with a gracious lifestyle. However, because of industrial development and other human activities, the ecological balance has been disturbed by the continuous occurrence of environmental pollution. Organic pollutants are one of the major contaminates that pollutes the environment including the soil, water, and air. POPs due to their recalcitrant and toxic nature cause severe health issues in both humans and other animals. Organic pollutants are also toxic to the physiological processes of plants. These pollutants are mutagenic, carcinogenic, hepatotoxic, nephrotoxic, and endocrine-disrupting for both humans and other animals. In the case of plants, these pollutants affect the photosynthetic and transpiration rates, along with stomatal conductance and the biosynthetic and metabolic pathways. Various strategies have been applied for removing these noxious chemicals from the environment, but the problem with these techniques is their high cost. Moreover, physical and chemical processes to remove organic pollutants cannot be applied to remote and rural areas. Therefore, in the past few decades, scientists have been applying microbiological and biotechnological techniques to remove noxious chemicals. The biological method involving microorganisms and genetically modified microbes is one of the safest and easiest approaches with less financial burden than conventional methods. Microbes are actively involved in various environmental cleaning processes such as 'bioremediation', 'bioaugmentation', 'bioventing', 'biotransformation', 'mineralization', and 'composting'. Bacteria and fungus are the most common species involved in the process of biodegrading complex organic pollutants. These microbial groups utilize organic contaminants as their energy and carbon sources and, thus, by different metabolic pathways, convert them into simpler compounds that are either less or not as toxic. The breakdown of organic contaminants through 'bioremediation' is an eco-friendly strategy that can be easily employed at contaminated sites. The large microbial diversity offers considerable potential in the biotransformation of toxic organic compounds to simpler compounds with less or no toxicity. The degradation and metabolic pathways of different organic pollutants in various microbial communities are still unknown. By adopting a metagenomics approach, a molecular basis for the degradation pathways of organic compounds could be achieved. In fact, the 'bioremediation' process could be made more efficient if certain constraints such as the bioavailability of organic compounds in the environment, functional capability within a limited range of physical and chemical conditions, and deactivation of enzymes associated with the bioremediation process at high threshold concentrations of recalcitrant compounds were addressed in better ways.

REFERENCES

Abdeen, Z., and Mohammad, S. G. (2013). Study of the adsorption efficiency of an eco-friendly carbohydrate polymer for contaminated aqueous solution by organophosphorus pesticide. *Open Journal of Organic Polymer Materials, 2014*.

Abou-El-Souod, G. W., and El-Sheekh, M. M. (2016). Biodegradation of basic fuchsin and methyl red. *Environmental Engineering and Management Journal, 15*(2), 279–286.

Adnan, L. A., Sathishkumar, P., Yusoff, A. R. M., Hadibarata, T., and Ameen, F. (2017). Rapid bioremediation of Alizarin Red S and Quinizarine Green SS dyes using *Trichoderma lixii* F21 mediated by biosorption and enzymatic processes. *Bioprocess and Biosystems Engineering, 40*(1), 85–97.

Agrawal, K., and Verma, P. (2019). Biodegradation of synthetic dye Alizarin Cyanine Green by yellow laccase producing strain *Stropharia* sp. ITCC-8422. *Biocatalysis and Agricultural Biotechnology, 21*, 101291.

Agrawal, K., and Verma, P. (2020). Myco-valorization approach using entrapped *Myrothecium verrucaria* ITCC-8447 on synthetic and natural support via column bioreactor for the detoxification and degradation of anthraquinone dyes. *International Biodeterioration & Biodegradation, 153*, 105052.

Ahrenhoerster, L. S., Leuthner, T. C., Tate, E. R., Lakatos, P. A., and Laiosa, M. D. (2015). Developmental exposure to 2, 3, 7, 8 tetrachlorodibenzo-p-dioxin attenuates later-life Notch1-mediated T cell development and leukemogenesis. *Toxicology and Applied Pharmacology, 283*(2), 99–108.

Akansha, K., Chakraborty, D., and Sachan, S. G. (2019). Decolorization and degradation of methyl orange by *Bacillus stratosphericus* SCA1007. *Biocatalysis and Agricultural Biotechnology, 18*, 101044.

Alexandrino, D. A., Mucha, A. P., Almeida, C. M. R., and Carvalho, M. F. (2020). Microbial degradation of two highly persistent fluorinated fungicides-epoxiconazole and fludioxonil. *Journal of Hazardous Materials, 394*, 122545.

Al Farraj, D. A., Hadibarata, T., Yuniarto, A., Alkufeidy, R. M., Alshammari, M. K., and Syafiuddin, A. (2020). Exploring the potential of halotolerant bacteria for biodegradation of polycyclic aromatic hydrocarbon. *Bioprocess and Biosystems Engineering, 43*(12), 2305–2314.

Aristilde, L., Melis, A., and Sposito, G. (2010). Inhibition of photosynthesis by a fluoroquinolone antibiotic. *Environmental Science & Technology, 44*(4), 1444–1450.

Asses, N., Ayed, L., Hkiri, N., and Hamdi, M. (2018). Congo red decolorization and detoxification by *Aspergillus niger*: Removal mechanisms and dye degradation pathway. *BioMed Research International, 2018*.

Bacaloni, A., Insogna, S., and Zoccolillo, L. (2011). Remote zones air quality: Persistent organic pollutants: Sources, sampling and analysis. *Air Quality Monitoring, Assessment and Management, 223*.

Baker, B. B., Yee, J. S., Meyer, D. N., Yang, D., and Baker, T. R. (2016). Histological and transcriptomic changes in male zebrafish testes due to early life exposure to low level 2, 3, 7, 8-tetrachlorodibenzo-p-dioxin. *Zebrafish, 13*(5), 413–423.

Bako, C. M., Mattes, T. E., Marek, R. F., Hornbuckle, K. C., and Schnoor, J. L. (2021). Biodegradation of PCB congeners by *Paraburkholderia xenovorans* LB400 in presence and absence of sediment during lab bioreactor experiments. *Environmental Pollution, 271*, 116364.

Bandala, E. R., Andres-Octaviano, J., Pastrana, P., and Torres, L. G. (2006). Removal of aldrin, dieldrin, heptachlor, and heptachlor epoxide using activated carbon and/or *Pseudomonas fluorescens* free cell cultures. *Journal of Environmental Science and Health Part B, 41*(5), 553–569.

Barapatre, A., Aadil, K. R., and Jha, H. (2017). Biodegradation of malachite green by the ligninolytic fungus *Aspergillus flavus*. *CLEAN—Soil, Air, Water, 45*(4), 1600045.

Bayen, S., Lee, H. K., and Obbard, J. P. (2004). Determination of polybrominated diphenyl ethers in marine biological tissues using microwave-assisted extraction. *Journal of Chromatography A, 1035*(2), 291–294.

Behera, I. D., Basak, G., Kumar, R. R., Sen, R., and Meikap, B. C. (2020). Treatment of petroleum refinery sludge by petroleum degrading bacterium *Stenotrophomonas pavanii* IRB19 as an efficient novel technology. *Journal of Environmental Science and Health, Part A*, 1–13.

Bhandari, R. K., vom Saal, F. S., and Tillitt, D. E. (2015). R transgenerational effects from early developmental exposures to bisphenol A or 17α-ethinylestradiol in medaka, *Oryzias latipes*. *Scientific Reports, 9303*.

Bhandari, S., and Nailwal, T. K. (2020a). Role of brassinosteroids in mitigating abiotic stresses in plants. *Biologia*, 1–28.

Bhandari, S., and Nailwal, T. K. (2020b). Exploration of microbial communities of Indian hot springs and their potential biotechnological applications. In *Recent Advancements in Microbial Diversity* (pp. 251–288). Academic Press.

Bharagava, R. N., Mani, S., Mulla, S. I., and Saratale, G. D. (2018). Degradation and decolourization potential of an ligninolytic enzyme producing *Aeromonas hydrophila* for crystal violet dye and its phytotoxicity evaluation. *Ecotoxicology and Environmental Safety*, *156*, 166–175.

Bhatt, P., Huang, Y., Zhang, W., Sharma, A., and Chen, S. (2020). Enhanced cypermethrin degradation kinetics and metabolic pathway in *Bacillus thuringiensis* strain SG4. *Microorganisms*, *8*(2), 223.

Blévin, P., Angelier, F., Tartu, S., Ruault, S., Bustamante, P., Herzke, D., . . . and Chastel, O. (2016). Exposure to oxychlordane is associated with shorter telomeres in arctic breeding kittiwakes. *Science of the Total Environment*, *563*, 125–130.

Braune, B. M., and Mallory, M. L. (2017). Declining trends of polychlorinated dibenzo-p-dioxins, dibenzofurans and non-ortho PCBs in Canadian Arctic seabirds. *Environmental Pollution*, *220*, 557–566.

Brown, A. S., Cheslack-Postava, K., Rantakokko, P., Kiviranta, H., Hinkka-Yli-Salomäki, S., McKeague, I. W., . . . and Sourander, A. (2018). Association of maternal insecticide levels with autism in offspring from a national birth cohort. *American Journal of Psychiatry*, *175*(11), 1094–1101.

Cao, D. J., Wang, J. J., Zhang, Q., Wen, Y. Z., Dong, B., Liu, R. J., . . . and Geng, G. (2019). Biodegradation of triphenylmethane dye crystal violet by *Cedecea davisae*. *Spectrochimica Acta Part A: Molecular and Biomolecular Spectroscopy*, *210*, 9–13.

Carpenter, D. O. (2006). Polychlorinated biphenyls (PCBs): Routes of exposure and effects on human health. *Reviews on Environmental Health*, *21*(1), 1.

Carpenter, D. O. (2011). Health effects of persistent organic pollutants: The challenge for the Pacific Basin and for the world. *Reviews on Environmental Health*, *26*(1), 61–69.

Chae, J. C., Kim, E., Bini, E., and Zylstra, G. J. (2007). Comparative analysis of the catechol 2, 3-dioxygenase gene locus in thermoacidophilic archaeon *Sulfolobus solfataricus* strain 98/2. *Biochemical and Biophysical Research Communications*, *357*(3), 815–819.

Chang, X., Liang, J., Sun, Y., Zhao, L., Zhou, B., Li, X., and Li, Y. (2020). Isolation, degradation performance and field application of the metolachlor-degrading fungus *Penicillium oxalicum* MET-F-1. *Applied Sciences*, *10*(23), 8556.

Chaudhary, D. K., Bajagain, R., Jeong, S. W., and Kim, J. (2020). Biodegradation of diesel oil and n-alkanes (C 18, C 20, and C 22) by a novel strain *Acinetobacter* sp. K-6 in unsaturated soil. *Environmental Engineering Research*, *25*(3), 290–298.

Chauhan, A. K., and Choudhury, B. (2021). Synthetic dyes degradation using lignolytic enzymes produced from *Halopiger aswanensis* strain ABC_IITR by Solid State Fermentation. *Chemosphere*, 129671.

Chen, B., Yuan, M., and Liu, H. (2011). Removal of polycyclic aromatic hydrocarbons from aqueous solution using plant residue materials as a biosorbent. *Journal of Hazardous Materials*, *188*(1–3), 436–442.

Chen, F., Hao, S., Qu, J., Ma, J., and Zhang, S. (2015). Enhanced biodegradation of polychlorinated biphenyls by defined bacteria-yeast consortium. *Annals of Microbiology*, *65*(4), 1847–1854.

Chen, J., Zhong, D., Hou, H., Li, C., Yang, J., Zhou, H., . . . and Wang, L. (2016). Ferrite as an effective catalyst for HCB removal in soil: Characterization and catalytic performance. *Chemical Engineering Journal*, *294*, 246–253.

Chen, Y., Feng, L., Li, H., Wang, Y., Chen, G., and Zhang, Q. (2018). Biodegradation and detoxification of Direct Black G textile dye by a newly isolated thermophilic microflora. *Bioresource Technology*, *250*, 650–657.

Cheng, S., Li, N., Jiang, L., Li, Y., Xu, B., and Zhou, W. (2019). Biodegradation of metal complex Naphthol Green B and formation of iron–sulfur nanoparticles by marine bacterium *Pseudoalteromonas* sp CF10-13. *Bioresource Technology*, *273*, 49–55.

Chevrier, J., Eskenazi, B., Holland, N., Bradman, A., and Barr, D. B. (2008). Effects of exposure to polychlorinated biphenyls and organochlorine pesticides on thyroid function during pregnancy. *American Journal of Epidemiology*, *168*(3), 298–310.

Chopra, S., and Kumar, D. (2020). Characterization, optimization and kinetics study of acetaminophen degradation by *Bacillus drentensis* strain S1 and waste water degradation analysis. *Bioresources and Bioprocessing*, *7*(1), 1–18.

Chung, K. T. (2016). Azo dyes and human health: A review. *Journal of Environmental Science and Health, Part C*, *34*(4), 233–261.

Copaciu, F., Opriş, O., Niinemets, Ü., and Copolovici, L. (2016). Toxic influence of key organic soil pollutants on the total flavonoid content in wheat leaves. *Water, Air, & Soil Pollution*, *227*(6), 1–10.

Coskun, Y., Kilic, S., and Duran, R. E. (2015). The effects of the insecticide pyriproxyfen on germination, development and growth responses of maize seedlings. *Fresenius Environmental Bulletin*, *24*(1b), 278–284.

Coulter, D. P., Huff Hartz, K. E., Sepúlveda, M. S., Godfrey, A., Garvey, J. E., and Lydy, M. J. (2019). Lifelong exposure to dioxin-like PCBs alters paternal offspring care behavior and reduces male fish reproductive success. *Environmental Science & Technology*, *53*(19), 11507–11514.

Cruz, S. L., Rivera-García, M. T., and Woodward, J. J. (2014). Review of toluene action: Clinical evidence, animal studies and molecular targets. *Journal of Drug and Alcohol Research*, *3*.

Dai, Y., Li, N., Zhao, Q., and Xie, S. (2015). Bioremediation using *Novosphingobium* strain DY4 for 2, 4-dichlorophenoxyacetic acid-contaminated soil and impact on microbial community structure. *Biodegradation*, *26*(2), 161–170.

de Araujo, C. A. V., Maciel, G. M., Rodrigues, E. A., Silva, L. L., Oliveira, R. F., Brugnari, T., . . . and de Souza, C. G. M. (2017). Simultaneous removal of the antimicrobial activity and toxicity of sulfamethoxazole and trimethoprim by white rot fungi. *Water, Air, & Soil Pollution*, *228*(9), 1–12.

de Campos Ventura-Camargo, B., de Angelis, D. D. F., and Marin-Morales, M. A. (2016). Assessment of the cytotoxic, genotoxic and mutagenic effects of the commercial black dye in *Allium cepa* cells before and after bacterial biodegradation treatment. *Chemosphere*, *161*, 325–332.

Dennis, N. M., Karnjanapiboonwong, A., Subbiah, S., Rewerts, J. N., Field, J. A., McCarthy, C., . . . and Anderson, T. A. (2020). Chronic reproductive toxicity of perfluorooctane sulfonic acid and a simple mixture of perfluorooctane sulfonic acid and perfluorohexane sulfonic acid to northern bobwhite quail (*Colinus virginianus*). *Environmental Toxicology and Chemistry*, *39*(5), 1101–1111.

Derbalah, A. S. H., El-Banna, A., and Allah, M. S. (2020). Efficiency of *Candida tropicalis* for potential degradation of metalaxyl in the aqueous media. *Current Microbiology*, *77*(10), 2991–2999.

Dercová, K., Čičmanová, J., Lovecká, P., Demnerová, K., Macková, M., Hucko, P., and Kušnír, P. (2008). Isolation and identification of PCB-degrading microorganisms from contaminated sediments. *International Biodeterioration & Biodegradation*, *62*(3), 219–225.

Doherty, B. T., Hammel, S. C., Daniels, J. L., Stapleton, H. M., and Hoffman, K. (2019). Organophosphate esters: Are these flame retardants and plasticizers affecting Children's health? *Current Environmental Health Reports*, *6*(4), 201–213.

Du, W., Liu, X., Zhao, L., Xu, Y., Yin, Y., Wu, J., . . . and Guo, H. (2020). Response of cucumber (*Cucumis sativus*) to perfluorooctanoic acid in photosynthesis and metabolomics. *Science of the Total Environment*, *724*, 138257.

Dzionek, A., Wojcieszyńska, D., Adamczyk-Habrajska, M., and Guzik, U. (2020). Enhanced degradation of naproxen by immobilization of *Bacillus thuringiensis* B1 (2015b) on loofah sponge. *Molecules*, *25*(4), 872.

Ekram, M. A. E., Sarker, I., Rahi, M. S., Rahman, M. A., Saha, A. K., and Reza, M. A. (2020). Efficacy of soil-borne *Enterobacter* sp. for carbofuran degradation: HPLC quantitation of degradation rate. *Journal of Basic Microbiology*, *60*(5), 390–399.

El Fantroussi, S., and Agathos, S. N. (2005). Is bioaugmentation a feasible strategy for pollutant removal and site remediation? *Current Opinion in Microbiology*, *8*(3), 268–275.

Elgueta, S., Santos, C., Lima, N., and Diez, M. C. (2016). Immobilization of the white-rot fungus *Anthracophyllum* discolor to degrade the herbicide atrazine. *AMB Express*, *6*(1), 1–11.

Evy, A. A. M., Lakshmi, V., and Nilanjana, D. (2012). Biodegradation of atrazine by *Cryptococcus laurentii* isolated from contaminated agricultural soil. *Journal of Microbiology and Biotechnology Research*, *2*(3), 450–457.

Fendick, E. A., Mather-Mihaich, E., Houck, K. A., Clair, M. S., Faust, J. B., Rockwell, C. H., and Owens, M. (1990). Ecological toxicology and human health effects of heptachlor. *Reviews of Environmental Contamination and Toxicology*, 61–142.

Feng, L., Wang, W., Cheng, J., Ren, Y., Zhao, G., Gao, C., . . . and Wang, L. (2007). Genome and proteome of long-chain alkane degrading *Geobacillus thermodenitrificans* NG80–2 isolated from a deep-subsurface oil reservoir. *Proceedings of the National Academy of Sciences*, *104*(13), 5602–5607.

Feng, N. X., Yu, J., Xiang, L., Yu, L. Y., Zhao, H. M., Mo, C. H., . . . and Li, Q. X. (2019). Co-metabolic degradation of the antibiotic ciprofloxacin by the enriched bacterial consortium XG and its bacterial community composition. *Science of the Total Environment*, *665*, 41–51.

Fernandes, A. F. T., Braz, V. S., Bauermeister, A., Paschoal, J. A. R., Lopes, N. P., and Stehling, E. G. (2018). Degradation of atrazine by *Pseudomonas* sp. and *Achromobacter* sp. isolated from Brazilian agricultural soil. *International Biodeterioration & Biodegradation*, *130*, 17–22.

Firdous, S., Iqbal, S., and Anwar, S. (2017). Optimization and modeling of glyphosate biodegradation by a novel *Comamonas odontotermitis* P2 through response surface methodology. *Pedosphere*, *30*(5), 618–627.

Fritsche, W., and Hofrichter, M. (2000). Aerobic degradation by microorganisms. *Biotechnology: Environmental Processes II*, *11*, 144–167.

Fu, G. M., Chen, Y., Li, R. Y., Yuan, X. Q., Liu, C. M., Li, B., and Wan, Y. (2017). Pathway and rate-limiting step of glyphosate degradation by *Aspergillus oryzae* A-F02. *Preparative Biochemistry and Biotechnology*, *47*(8), 782–788.

Fu, L., Bai, Y. N., Lu, Y. Z., Ding, J., Zhou, D., and Zeng, R. J. (2019). Degradation of organic pollutants by anaerobic methane-oxidizing microorganisms using methyl orange as example. *Journal of Hazardous Materials*, *364*, 264–271.

Gheorghe, I. F., and Ion, B. (2011). The effects of air pollutants on vegetation and the role of vegetation in reducing atmospheric pollution. In *The Impact of Air Pollution on Health, Economy, Environment and Agricultural Sources* (pp. 241–280). IntechOpen.

Gomes, M. P., Le Manac'h, S. G., Maccario, S., Labrecque, M., Lucotte, M., and Juneau, P. (2016). Differential effects of glyphosate and aminomethylphosphonic acid (AMPA) on photosynthesis and chlorophyll metabolism in willow plants. *Pesticide Biochemistry and Physiology*, *130*, 65–70.

González, A. J., Fortunato, M. S., Gallego, A., and Korol, S. E. (2017). Simultaneous biodegradation and detoxification of the herbicides 2, 4-dichlorophenoxyacetic acid and 4-chloro-2-methylphenoxyacetic acid in a continuous biofilm reactor. *Water, Air, & Soil Pollution*, *228*(8), 1–7.

González-Benítez, N., Molina, M. C., and Arrayás, M. (2020). Empirical evidence and mathematical modelling of carbamazepine degradative kinetics by a wood-rotting microbial consortium. *Waste and Biomass Valorization*, 1–9.

Gullan, P. J., and Cranston, P. S. (2010). *The Insects: An Outline of Entomology*. Willey.

Gupta, P. C. (1975). Neurotoxicity of chronic chlorinated hydrocarbon insecticide poisoning—a clinical and electroencephalographic study in man. *The Indian Journal of Medical Research*, *63*(4), 601–606.

Gusmanizar, N. (2019). Isolation and characterization of a molybdenum-reducing and coumaphos-degrading *Bacillus* sp. strain Neni-12 in soils from West Sumatera, Indonesia. *Journal of Environmental Microbiology and Toxicology*, *7*(2), 20–25.

Ha, D. D. (2018). Anaerobic degradation of 2, 4-dichlorophenoxyacetic acid by *Thauera* sp. DKT. *Biodegradation*, *29*(5), 499–510.

Han, G., Cui, B. X., Zhang, X. X., and Li, K. R. (2016). The effects of petroleum-contaminated soil on photosynthesis of *Amorpha fruticosa* seedlings. *International Journal of Environmental Science and Technology*, *13*(10), 2383–2392.

Hanano, A., Almousally, I., Shaban, M., Moursel, N., Shahadeh, A., and Alhajji, E. (2015). Differential tissue accumulation of 2, 3, 7, 8-Tetrachlorinated dibenzo-p-dioxin in *Arabidopsis thaliana* affects plant chronology, lipid metabolism and seed yield. *BMC Plant Biology*, *15*(1), 1–13.

Handayani, S., Safitri, R., Surono, W., Astika, H., Damayanti, R., and Agung, M. (2019, August). Biodegradation of BTEX by indigenous microorganisms isolated from UCG project area, South Sumatra. In *IOP Conference Series: Earth and Environmental Science* (Vol. 308, No. 1, p. 012017). IOP Publishing.

Haq, I., and Raj, A. (2018). Biodegradation of Azure-B dye by *Serratia liquefaciens* and its validation by phytotoxicity, genotoxicity and cytotoxicity studies. *Chemosphere*, *196*, 58–68.

Harrabi, M., Alexandrino, D. A., Aloulou, F., Elleuch, B., Liu, B., Jia, Z., . . . and Carvalho, M. F. (2019). Biodegradation of oxytetracycline and enrofloxacin by autochthonous microbial communities from estuarine sediments. *Science of the Total Environment*, *648*, 962–972.

Hayes, R. B., Songnian, Y., Dosemeci, M., and Linet, M. (2001). Benzene and lymphohematopoietic malignancies in humans. *American Journal of Industrial Medicine*, *40*(2), 117–126.

Helen, A. S., Uche, E. C., and Hamid, F. S. (2017). Screening for polypropylene degradation potential of bacteria isolated from mangrove ecosystems in Peninsular Malaysia. *International Journal of Bioscience, Biochemistry and Bioinformatics*, *7*(4), 245–251.

Hong, Y., Liao, D., Chen, J., Khan, S., Su, J., and Li, H. (2015). A comprehensive study of the impact of polycyclic aromatic hydrocarbons (PAHs) contamination on salt marsh plants *Spartina alterniflora*: Implication for plant-microbe interactions in phytoremediation. *Environmental Science and Pollution Research*, *22*(9), 7071–7081.

Horsman, G. P., Ke, J., Dai, S., Seah, S. Y., Bolin, J. T., and Eltis, L. D. (2006). Kinetic and structural insight into the mechanism of BphD, a C– C Bond hydrolase from the biphenyl degradation pathway. *Biochemistry*, *45*(37), 11071–11086.

Horváthová, H., Lászlová, K., and Dercová, K. (2019). Bioremediation vs. nanoremediation: Degradation of polychlorinated biphenyls (PCBS) using integrated remediation approaches. *Water, Air, & Soil Pollution*, *230*(8), 1–11.

Hossen, M. Z., Hussain, M. E., Hakim, A., Islam, K., Uddin, M. N., and Azad, A. K. (2019). Biodegradation of reactive textile dye Novacron Super Black G by free cells of newly isolated *Alcaligenes faecalis* AZ26 and *Bacillus* spp obtained from textile effluents. *Heliyon*, *5*(7), e02068.

Hou, Y., Tao, J., Shen, W., Liu, J., Li, J., Li, Y., . . . and Cui, Z. (2011). Isolation of the fenoxaprop-ethyl (FE)-degrading bacterium *Rhodococcus* sp. T1, and cloning of FE hydrolase gene feh. *FEMS Microbiology Letters*, *323*(2), 196–203.

Hu, S., Cheng, X., Liu, G., Lu, Y., Qiao, W., Chen, K., and Jiang, J. (2020). Degradation of chlorothalonil via thiolation and nitrile hydration by marine strains isolated from the surface seawater of the Northwestern Pacific. *International Biodeterioration & Biodegradation*, *154*, 105049.

Islam, N., da Fonseca, T. G., Vilke, J., Gonçalves, J. M., Pedro, P., Keiter, S., . . . and Bebianno, M. J. (2021). Perfluorooctane sulfonic acid (PFOS) adsorbed to polyethylene microplastics: Accumulation and ecotoxicological effects in the clam *Scrobicularia plana*. *Marine Environmental Research*, 105249.

Jacobsen, A. V., Nordén, M., Engwall, M., and Scherbak, N. (2018). Effects of perfluorooctane sulfonate on genes controlling hepatic fatty acid metabolism in livers of chicken embryos. *Environmental Science and Pollution Research*, *25*(23), 23074–23081.

Jayabarath, J., Musfira, S. A., Giridhar, R., and Arulmurugan, R. (2010). Biodegradation of carbofuran pesticide by saline soil actinomycetes. *International Journal of Biotechnology and Biochemistry*, *6*(2), 187–193.

Jiang, Y., Shang, Y., Yu, S., and Liu, J. (2018). Dechlorination of hexachlorobenzene in contaminated soils using a nanometallic Al/CaO dispersion mixture: Optimization through response surface methodology. *International Journal of Environmental Research and Public Health*, *15*(5), 872.

Joutey, N. T., Bahafid, W., Sayel, H., and El Ghachtouli, N. (2013). Biodegradation: Involved microorganisms and genetically engineered microorganisms. *Biodegradation-Life of Science*, 289–320.

Kabra, A. N., Ji, M. K., Choi, J., Kim, J. R., Govindwar, S. P., and Jeon, B. H. (2014). Toxicity of atrazine and its bioaccumulation and biodegradation in a green microalga, *Chlamydomonas mexicana*. *Environmental Science and Pollution Research*, *21*(21), 12270–12278.

Karuppiah, S., Subramanian, A., and Obbard, J. P. (2004). The barnacle, *Xenobalanus globicipitis* (Cirripedia, Coronulidae), attached to the bottle-nosed dolphin, *Tursiops truncatus* (Mammalia, Cetacea) on the southeastern coast of India. *Crustaceana*, *77*(7), 879–894.

Karups, S., Annamalai, S., and Obbard, J. P. (2004). Organochlorine residues in marine mammals from the southeast coast of India. *Chemosphere*, *60*, 891–897.

Ke, C. Y., Lu, G. M., Wei, Y. L., Sun, W. J., Hui, J. F., Zheng, X. Y., . . . and Zhang, X. L. (2019). Biodegradation of crude oil by *Chelatococcus daeguensis* HB-4 and its potential for microbial enhanced oil recovery (MEOR) in heavy oil reservoirs. *Bioresource Technology*, *287*, 121442.

Khalil, A., Parker, M., Brown, S. E., Cevik, S. E., Guo, L. W., Jensen, J., . . . and Suvorov, A. (2017). Perinatal exposure to 2, 2′, 4′ 4′– Tetrabromodiphenyl ether induces testicular toxicity in adult rats. *Toxicology*, *389*, 21–30.

Khatoon, H., and Rai, J. P. N. (2020). Optimization studies on biodegradation of atrazine by *Bacillus badius* ABP6 strain using response surface methodology. *Biotechnology Reports*, *26*, e00459.

Khpalwak, W., Abdel-Dayem, S. M., and Sakugawa, H. (2018). Individual and combined effects of fluoranthene, phenanthrene, mannitol and sulfuric acid on marigold (*Calendula officinalis*). *Ecotoxicology and Environmental Safety*, *148*, 834–841.

Kiayi, Z., Lotfabad, T. B., Heidarinasab, A., and Shahcheraghi, F. (2019). Microbial degradation of azo dye carmoisine in aqueous medium using *Saccharomyces cerevisiae* ATCC 9763. *Journal of Hazardous Materials*, *373*, 608–619.

Kim, D., Kwak, J. I., and An, Y. J. (2018). Effects of bisphenol A in soil on growth, photosynthesis activity, and genistein levels in crop plants (*Vigna radiata*). *Chemosphere*, *209*, 875–882.

Kim, E., Zylstra, G. J., Freeman, J. P., Heinze, T. M., Deck, J., and Cerniglia, C. E. (1997). Evidence for the role of 2-hydroxychromene-2-carboxylate isomerase in the degradation of anthracene by *Sphingomonas yanoikuyae* B1. *FEMS Microbiology Letters*, *153*(2), 479–484.

Krishnamoorthy, R., Jose, P. A., Ranjith, M., Anandham, R., Suganya, K., Prabhakaran, J., . . . and Kumutha, K. (2018). Decolourisation and degradation of azo dyes by mixed fungal culture consisted of *Dichotomomyces cejpii* MRCH 1–2 and Phoma tropica MRCH 1–3. *Journal of Environmental Chemical Engineering*, *6*(1), 588–595.

Kuhm, A. E., Knackmuss, H. J., and Stolz, A. (1993). 2-Hydroxychromene-2-carboxylate isomerase from bacteria that degrade naphthalenesulfonates. *Biodegradation*, *4*(3), 155–162.

Kumar, A., Trefault, N., and Olaniran, A. O. (2016). Microbial degradation of 2, 4-dichlorophenoxyacetic acid: Insight into the enzymes and catabolic genes involved, their regulation and biotechnological implications. *Critical Reviews in Microbiology*, *42*(2), 194–208.

Kuppusamy, S., Sethurajan, M., Kadarkarai, M., and Aruliah, R. (2017). Biodecolourization of textile dyes by novel, indigenous *Pseudomonas stutzeri* MN1 and Acinetobacter baumannii MN3. *Journal of Environmental Chemical Engineering*, *5*(1), 716–724.

Kupryashina, M. A., Ponomareva, E. G., and Nikitina, V. E. (2020). Ability of bacteria of the genus *Azospirillum* to decolorize synthetic dyes. *Microbiology*, *89*(4), 451–458.

Kurade, M. B., Waghmode, T. R., Khandare, R. V., Jeon, B. H., and Govindwar, S. P. (2016). Biodegradation and detoxification of textile dye Disperse Red 54 by *Brevibacillus laterosporus* and determination of its metabolic fate. *Journal of Bioscience and Bioengineering*, *121*(4), 442–449.

Lambiase, S., Serpe, F. P., Pilia, M., Fiorito, F., Iaccarino, D., Gallo, P., and Esposito, M. (2021). Polychlorinated organic pollutants (PCDD/Fs and DL-PCBs) in loggerhead (*Caretta caretta*) and green (*Chelonia mydas*) turtles from Central-Southern Tyrrhenian Sea. *Chemosphere*, *263*, 128226.

Laso-Pérez, R., Hahn, C., van Vliet, D. M., Tegetmeyer, H. E., Schubotz, F., Smit, N. T., . . . and Wegener, G. (2019). Anaerobic degradation of non-methane alkanes by "*Candidatus Methanoliparia*" in hydrocarbon seeps of the Gulf of Mexico. *MBio*, *10*(4).

Lee, Y., Lee, Y., and Jeon, C. O. (2019). Biodegradation of naphthalene, BTEX, and aliphatic hydrocarbons by *Paraburkholderia aromaticivorans* BN5 isolated from petroleum-contaminated soil. *Scientific Reports*, *9*(1), 1–13.

Li, N., Peng, Q., Yao, L., He, Q., Qiu, J., Cao, H., . . . and Hui, F. (2020). Roles of the gentisate 1, 2-dioxygenases DsmD and GtdA in the catabolism of the herbicide dicamba in *Rhizorhabdus dicambivorans* Ndbn-20. *Journal of Agricultural and Food Chemistry*, *68*(35), 9287–9298.

Li, Y., Fu, L., Li, X., Wang, Y., Wei, Y., Tang, J., and Liu, H. (2021). Novel strains with superior degrading efficiency for lincomycin manufacturing biowaste. *Ecotoxicology and Environmental Safety*, *209*, 111802.

Liang, D. H., and Hu, Y. (2021). Application of a heavy metal-resistant *Achromobacter* sp. for the simultaneous immobilization of cadmium and degradation of sulfamethoxazole from wastewater. *Journal of Hazardous Materials*, *402*, 124032.

Liang, R., He, J., Shi, Y., Li, Z., Sarvajayakesavalu, S., Baninla, Y., . . . and Lu, Y. (2017). Effects of Perfluorooctane sulfonate on immobilization, heartbeat, reproductive and biochemical performance of *Daphnia magna*. *Chemosphere*, *168*, 1613–1618.

Liu, J., Zhao, B., Lan, Y., and Ma, T. (2021). Enhanced degradation of different crude oils by defined engineered consortia of *Acinetobacter venetianus* RAG-1 mutants based on their alkane metabolism. *Bioresource Technology*, 124787.

Llorca, M., Castellet-Rovira, F., Farré, M. J., Jaén-Gil, A., Martínez-Alonso, M., Rodríguez-Mozaz, S., . . . and Barceló, D. (2019). Fungal biodegradation of the N-nitrosodimethylamine precursors venlafaxine and O-desmethylvenlafaxine in water. *Environmental Pollution*, *246*, 346–356.

Lu, C., Yang, S., Yan, Z., Ling, J., Jiao, L., He, H., . . . and Fan, J. (2018). Deriving aquatic life criteria for PBDEs in China and comparison of species sensitivity distribution with TBBPA and HBCD. *Science of the Total Environment*, *640*, 1279–1285.

Mahalakshmi, S., Lakshmi, D., and Menaga, U. (2015). Biodegradation of different concentration of dye (Congo red dye) by using green and blue green algae. *International Journal of Environmental Research*, *9*(2), 735–744.

Mai, Y., Peng, S., Li, H., Gao, Y., and Lai, Z. (2021). NOD-like receptor signaling pathway activation: A potential mechanism underlying negative effects of benzo (α) pyrene on zebrafish. *Comparative Biochemistry and Physiology Part C: Toxicology & Pharmacology*, *240*, 108935.

Malafaia, G., de Souza, A. M., Pereira, A. C., Gonçalves, S., da Costa Araújo, A. P., Ribeiro, R. X., and Rocha, T. L. (2020). Developmental toxicity in zebrafish exposed to polyethylene microplastics under static and semi-static aquatic systems. *Science of The Total Environment*, *700*, 134867.

Manogaran, M., Shukor, M. Y., Yasid, N. A., Khalil, K. A., and Ahmad, S. A. (2018). Optimisation of culture composition for glyphosate degradation by *Burkholderia vietnamiensis* strain AQ5-12. *3 Biotech*, *8*(2), 1–13.

Marchant, R., Sharkey, F. H., Banat, I. M., Rahman, T. J., and Perfumo, A. (2006). The degradation of n-hexadecane in soil by thermophilic *Geobacilli. FEMS Microbiology Ecology*, *56*(1), 44–54.

Martyniuk, C. J., Mehinto, A. C., Colli-Dula, R. C., Kroll, K. J., Doperalski, N. J., Barber, D. S., and Denslow, N. D. (2020). Transcriptome and physiological effects of toxaphene on the liver-gonad reproductive axis in male and female largemouth bass (*Micropterus salmoides*). *Comparative Biochemistry and Physiology Part D: Genomics and Proteomics*, *36*, 100746.

Matturro, B., Ubaldi, C., Grenni, P., Caracciolo, A. B., and Rossetti, S. (2016). Polychlorinated biphenyl (PCB) anaerobic degradation in marine sediments: Microcosm study and role of autochthonous microbial communities. *Environmental Science and Pollution Research*, *23*(13), 12613–12623.

Meena, K. R., Dhiman, R., Singh, K., Kumar, S., Sharma, A., Kanwar, S. S., . . . and Mandal, A. K. (2021). Purification and identification of a surfactin biosurfactant and engine oil degradation by *Bacillus velezensis* KLP2016. *Microbial Cell Factories*, *20*(1), 1–12.

Melnick, R. L., and Huff, J. (1992). 1, 3-Butadiene: Toxicity and carcinogenicity in laboratory animals and in humans. In *Reviews of Environmental Contamination and Toxicology* (pp. 111–144). Springer.

Mhatre, K. J. (2021). Microbial decolorization of carbol fuchsin dye by *Acinetobactor spp*. *IJRAR-International Journal of Research and Analytical Reviews (IJRAR)*, *8*(1), 26–34.

Migliore, L., Brambilla, G., Casoria, P., Civitareale, C., Cozzolino, S., and Gaudio, L. (1996). Effect of sulphadimethoxine contamination on barley (*Hordeum distichum* L., Poaceae, Liliposida). *Agriculture, Ecosystems & Environment*, *60*(2–3), 121–128.

Migliore, L., Cozzolino, S., and Fiori, M. (2003). Phytotoxicity to and uptake of enrofloxacin in crop plants. *Chemosphere*, *52*(7), 1233–1244.

Mohamed, A. T., El-Hussein, A. A., El-Siddig, M. A., and Osman, A. G. (2011). Degradation of oxyfluorfen herbicide by soil microorganisms biodegradation of herbicides. *Biotechnology*, *10*(3), 274–279.

Mohamed, M., Ismail, W., Heider, J., and Fuchs, G. (2002). Aerobic metabolism of phenylacetic acids in *Azoarcus evansii*. *Archives of Microbiology*, *178*(3), 180–192.

Mohan, N., and Naveena, L. (2015). Isolation and determination of efficacy of acephate degrading bacteria from agricultural soil. *Journal of Environmental Science, Toxicology and Food Technology*, *9*(3), 10–20.

Mollea, C., Bosco, F., and Ruggeri, B. (2005). Fungal biodegradation of naphthalene: Microcosms studies. *Chemosphere*, *60*(5), 636–643.

Moon, H. B., Kim, H. S., Choi, M., and Choi, H. G. (2010). Intake and potential health risk of polycyclic aromatic hydrocarbons associated with seafood consumption in Korea from 2005 to 2007. *Archives of Environmental Contamination and Toxicology*, *58*(1), 214–221.

Mouhamadou, B., Faure, M., Sage, L., Marçais, J., Souard, F., and Geremia, R. A. (2013). Potential of autochthonous fungal strains isolated from contaminated soils for degradation of polychlorinated biphenyls. *Fungal Biology*, *117*(4), 268–274.

Murínová, S., Dercová, K., and Dudášová, H. (2014). Degradation of polychlorinated biphenyls (PCBs) by four bacterial isolates obtained from the PCB-contaminated soil and PCB-contaminated sediment. *International Biodeterioration & Biodegradation*, *91*, 52–59.

Nanseu-Njiki, C. P., Dedzo, G. K., and Ngameni, E. (2010). Study of the removal of paraquat from aqueous solution by biosorption onto Ayous (*Triplochiton schleroxylon*) sawdust. *Journal of Hazardous Materials*, *179*(1–3), 63–71.

Nayak, T., Panda, A. N., Adhya, T. K., Das, B., and Raina, V. (2019). Biodegradation of Chlorpyrifos and 3, 5, 6-trichloro-2-pyridinol (TCP) by *Ochrobactrum* sp. CPD-03: Insights from genome analysis on organophosphorus pesticides degradation, chemotaxis and PGPR activity. *bioRxiv*. https://doi.org/10.1101/2019.12.12.866210

Nguyen, B. A. T., Hsieh, J. L., Lo, S. C., Wang, S. Y., Hung, C. H., Huang, E., . . . and Huang, C. C. (2021). Biodegradation of dioxins by *Burkholderia cenocepacia* strain 869T2: Role of 2-haloacid dehalogenase. *Journal of Hazardous Materials*, *401*, 123347.

Nguyen, P. Y., Carvalho, G., Reis, A. C., Nunes, O. C., Reis, M. A. M., and Oehmen, A. (2017). Impact of biogenic substrates on sulfamethoxazole biodegradation kinetics by *Achromobacter denitrificans* strain PR1. *Biodegradation*, *28*(2), 205–217.

Nielsen, G. D., Larsen, S. T., Olsen, O., Løvik, M., Poulsen, L. K., Glue, C., and Wolkoff, P. (2007). Do indoor chemicals promote development of airway allergy? *Indoor Air*, *17*(3), 236–255.

Opriş, O., Copaciu, F., Soran, M. L., Ristoiu, D., Niinemets, Ü., and Copolovici, L. (2013). Influence of nine antibiotics on key secondary metabolites and physiological characteristics in *Triticum aestivum*: Leaf volatiles as a promising new tool to assess toxicity. *Ecotoxicology and Environmental Safety*, *87*, 70–79.

Pacholak, A., Smułek, W., Zgoła-Grześkowiak, A., and Kaczorek, E. (2019). Nitrofurantoin—Microbial degradation and interactions with environmental bacterial strains. *International Journal of Environmental Research and Public Health, 16*(9), 1526.

Palma, T. L., Donaldben, M. N., Costa, M. C., and Carlier, J. D. (2018). Putative role of *Flavobacterium*, *Dokdonella* and *Methylophilus* strains in paracetamol biodegradation. *Water, Air, & Soil Pollution, 229*(6), 1–23.

Palyzová, A., Zahradník, J., Marešová, H., Sokolová, L., Kyslíková, E., Grulich, M., . . . and Kyslík, P. (2018). Potential of the strain *Raoultella* sp. KDF8 for removal of analgesics. *Folia Microbiologica, 63*(3), 273–282.

Pathiraja, G., Egodawatta, P., Goonetilleke, A., and Te'o, V. S. J. (2019). Solubilization and degradation of polychlorinated biphenyls (PCBs) by naturally occurring facultative anaerobic bacteria. *Science of the Total Environment, 651*, 2197–2207.

Pérez Rodríguez, M., Melo, C., Jiménez, E., and Dussán, J. (2019). Glyphosate bioremediation through the sarcosine oxidase pathway mediated by *Lysinibacillus sphaericus* in soils cultivated with potatoes. *Agriculture, 9*(10), 217.

Poursat, B. A., van Spanning, R. J., Braster, M., Helmus, R., de Voogt, P., and Parsons, J. R. (2019). Biodegradation of metformin and its transformation product, guanylurea, by natural and exposed microbial communities. *Ecotoxicology and Environmental Safety, 182*, 109414.

Prashar, P., and Shah, S. (2016). Impact of fertilizers and pesticides on soil microflora in agriculture. In *Sustainable Agriculture Reviews* (pp. 331–361). Springer.

Rahman, K. M., and Zhang, D. (2018). Effects of fertilizer broadcasting on the excessive use of inorganic fertilizers and environmental sustainability. *Sustainability, 10*(3), 759.

Ramavandi, B., Najafpoor, A. A., Alidadi, H., and Bonyadi, Z. (2019). Alizarin red-S removal from aqueous solutions using *Saccharomyces cerevisiae*: Kinetic and equilibrium study. *Desalin Water Treat, 144*, 286–291.

Rastib, A., Memariani, M., and Riahi, M. A. (2019). Investigation of *Enterobacter aerogenes* effects on heavy oil from biological degradation aspects by GC* GC technique. *International Journal of Petrochemical Science & Engineering, 4*(2), 47–52.

Ratola, N., Botelho, C., and Alves, A. (2003). The use of pine bark as a natural adsorbent for persistent organic pollutants—study of lindane and heptachlor adsorption. *Journal of Chemical Technology & Biotechnology: International Research in Process, Environmental & Clean Technology, 78*(2–3), 347–351.

Reis, P. J., Reis, A. C., Ricken, B., Kolvenbach, B. A., Manaia, C. M., Corvini, P. F., and Nunes, O. C. (2014). Biodegradation of sulfamethoxazole and other sulfonamides by *Achromobacter denitrificans* PR1. *Journal of Hazardous Materials, 280*, 741–749.

Romero, M. C., Cazau, M. C., Giorgieri, S., and Arambarri, A. M. (1998). Phenanthrene degradation by microorganisms isolated from a contaminated stream. *Environmental Pollution, 101*(3), 355–359.

Roslee, A. F. A., Zakaria, N. N., Convey, P., Zulkharnain, A., Lee, G. L. Y., Gomez-Fuentes, C., and Ahmad, S. A. (2020). Statistical optimisation of growth conditions and diesel degradation by the Antarctic bacterium, *Rhodococcus* sp. strain AQ5–07. *Extremophiles, 24*(2), 277–291.

Rutere, C., Knoop, K., Posselt, M., Ho, A., and Horn, M. A. (2020). Ibuprofen degradation and associated bacterial communities in hyporheic zone sediments. *Microorganisms, 8*(8), 1245.

Saibu, S., Adebusoye, S. A., Oyetibo, G. O., and Rodrigues, D. F. (2020). Aerobic degradation of dichlorinated dibenzo-p-dioxin and dichlorinated dibenzofuran by bacteria strains obtained from tropical contaminated soil. *Biodegradation, 31*(1), 123–137.

Salam, L. B., Ilori, M. O., and Amund, O. O. (2017). Properties, environmental fate and biodegradation of carbazole. *3 Biotech, 7*(2), 1–14.

Sanangelantoni, A. M., Malatrasi, M., Trivelloni, E., Visioli, G., and Agrimonti, C. (2018). A novel β-propeller phytase from the dioxin-degrading bacterium *Sphingomonas wittichii* RW-1. *Applied Microbiology and Biotechnology, 102*(19), 8351–8358.

Sandesh, K., Kumar, G., Chidananda, B., and Ujwal, P. (2019). Optimization of Direct Blue-14 dye degradation by *Bacillus fermus* (Kx898362) an alkaliphilic plant endophyte and assessment of degraded metabolite toxicity. *Journal of Hazardous Materials, 364*, 742–751.

Schlosser, P. M., Bale, A. S., Gibbons, C. F., Wilkins, A., and Cooper, G. S. (2015). Human health effects of dichloromethane: Key findings and scientific issues. *Environmental Health Perspectives, 123*(2), 114–119.

Seyoum, A., and Pradhan, A. (2019). Effect of phthalates on development, reproduction, fat metabolism and lifespan in *Daphnia magna*. *Science of the Total Environment*, *654*, 969–977.

Sforzini, S., Moore, M. N., Boeri, M., Bencivenga, M., and Viarengo, A. (2015). Effects of PAHs and dioxins on the earthworm *Eisenia andrei*: A multivariate approach for biomarker interpretation. *Environmental Pollution*, *196*, 60–71.

Shahzadi, S., Khan, Z., Rehman, A., Nisar, M. A., Hussain, S. Z., and Asma, S. T. (2019). Isolation and characterization of *Bacillus amyloliquefaciens* 6A: A novel kerosene oil degrading bacterium. *Environmental Technology & Innovation*, *14*, 100359.

Shanmugapriya, A. K., Sivakumar, T., and Panneerselvam, R. (2013). Difenoconazole and Tricyclazole induced changes in photosynthetic pigments of *Lycopersicon esculentum*. L. *International Journal of Agriculture and Food Science*, *3*, 72–75.

Shapiro, T., Chekanov, K., Alexandrova, A., Dolnikova, G., Ivanova, E., and Lobakova, E. (2021). Revealing of non-cultivable bacteria associated with the mycelium of fungi in the kerosene-degrading community isolated from the contaminated jet fuel. *Journal of Fungi*, *7*(1), 43.

Sharifi, Y., Pourbabaei, A. A., Javadi, A., Abdolmohammad, M. H., Saffari, M., and Morovvati, A. (2015). Biodegradation of glyphosate herbicide by *Salinicoccus* spp isolated from Qom Hoze-soltan lake, Iran. *Environmental Health Engineering and Management Journal*, *2*(1), 31–36.

Sharma, K., Kaushik, G., Thotakura, N., Raza, K., Sharma, N., and Nimesh, S. (2019). Fate of ibuprofen under optimized batch biodegradation experiments using *Micrococcus yunnanensis* isolated from pharmaceutical sludge. *International Journal of Environmental Science and Technology*, *16*(12), 8315–8328.

Shehata, S. A., El-Dib, M. A., and Abou Waly, H. F. (1997). Effect of certain herbicides on the growth of freshwater algae. *Water, Air, and Soil Pollution*, *100*(1), 1–12.

Sheikh, I. A., and Beg, M. A. (2019). Structural characterization of potential endocrine disrupting activity of alternate plasticizers di-(2-ethylhexyl) adipate (DEHA), acetyl tributyl citrate (ATBC) and 2, 2, 4-trimethyl 1, 3-pentanediol diisobutyrate (TPIB) with human sex hormone-binding globulin. *Reproductive Toxicology*, *83*, 46–53.

Shen, Y., Li, J., Gu, R., Yue, L., Zhan, X., and Xing, B. (2017). Phenanthrene-triggered Chlorosis is caused by elevated Chlorophyll degradation and leaf moisture. *Environmental Pollution*, *220*, 1311–1321.

Shi, Y., Ye, Z., Hu, P., Wei, D., Gao, Q., Zhao, Z., . . . and Cao, H. (2020). Removal of prothioconazole using screened microorganisms and identification of biodegradation products via UPLC-QqTOF-MS. *Ecotoxicology and Environmental Safety*, *206*, 111203.

Singh, B. (2013). Degradation of clodinafop propargyl by Pseudomonas sp. strain B2. *Bulletin of Environmental Contamination and Toxicology*, *91*(6), 730–733.

Singh, D. P., Prabha, R., Gupta, V. K., and Verma, M. K. (2018). Metatranscriptome analysis deciphers multifunctional genes and enzymes linked with the degradation of aromatic compounds and pesticides in the wheat rhizosphere. *Frontiers in Microbiology*, *9*, 1331.

Singh, S., Kumar, V., Singla, S., Sharma, M., Singh, D. P., Prasad, R., . . . and Singh, J. (2020). Kinetic study of the biodegradation of acephate by indigenous soil bacterial isolates in the presence of humic acid and metal ions. *Biomolecules*, *10*(3), 433.

Snyder, M. J., and Mulder, E. P. (2001). Environmental endocrine disruption in decapod crustacean larvae: Hormone titers, cytochrome P450, and stress protein responses to heptachlor exposure. *Aquatic Toxicology*, *55*(3–4), 177–190.

Song, M., Luo, C., Li, F., Jiang, L., Wang, Y., Zhang, D., and Zhang, G. (2015). Anaerobic degradation of Polychlorinated Biphenyls (PCBs) and Polychlorinated Biphenyls Ethers (PBDEs), and microbial community dynamics of electronic waste-contaminated soil. *Science of the Total Environment*, *502*, 426–433.

Sperfeld, M., Rauschenbach, C., Diekert, G., and Studenik, S. (2018). Microbial community of a gasworks aquifer and identification of nitrate-reducing Azoarcus and Georgfuchsia as key players in BTEX degradation. *Water Research*, *132*, 146–157.

Steliga, T., Wojtowicz, K., Kapusta, P., and Brzeszcz, J. (2020). Assessment of biodegradation efficiency of polychlorinated biphenyls (PCBs) and petroleum hydrocarbons (TPH) in soil using three individual bacterial strains and their mixed culture. *Molecules*, *25*(3), 709.

Stella, T., Covino, S., Čvančarová, M., Filipová, A., Petruccioli, M., D'Annibale, A., and Cajthaml, T. (2017). Bioremediation of long-term PCB-contaminated soil by white-rot fungi. *Journal of Hazardous Materials*, *324*, 701–710.

Su, X., Li, S., Cai, J., Xiao, Y., Tao, L., Hashmi, M. Z., . . . and Sun, F. (2019). Aerobic degradation of 3, 3′, 4, 4′-tetrachlorobiphenyl by a resuscitated strain *Castellaniella* sp. SPC4: Kinetics model and pathway for biodegradation. *Science of the Total Environment, 688*, 917–925.

Subramanian, S., Schnoor, J. L., and Van Aken, B. (2017). Effects of polychlorinated biphenyls (PCBs) and their hydroxylated metabolites (OH-PCBs) on *Arabidopsis thaliana. Environmental Science & Technology, 51*(12), 7263–7270.

Tang, J., Hu, Q., Lei, D., Wu, M., Zeng, C., and Zhang, Q. (2020). Characterization of deltamethrin degradation and metabolic pathway by co-culture of *Acinetobacter junii* LH-1-1 and *Klebsiella pneumoniae* BPBA052. *AMB Express, 10*(1), 1–11.

Tang, W., Ji, H., and Hou, X. (2018). Research progress of microbial degradation of organophosphorus pesticides. *Progress in Applied Microbiology, 1*, 29–35.

Teng, Y., Li, X., Chen, T., Zhang, M., Wang, X., Li, Z., and Luo, Y. (2016). Isolation of the PCB-degrading bacteria *Mesorhizobium* sp. ZY1 and its combined remediation with *Astragalus sinicus* L. for contaminated soil. *International Journal of Phytoremediation, 18*(2), 141–149.

Thakur, M., and Pathania, D. (2020). Environmental fate of organic pollutants and effect on human health. In *Abatement of Environmental Pollutants* (pp. 245–262). Elsevier.

Thelusmond, J. R., Strathmann, T. J., and Cupples, A. M. (2019). Carbamazepine, triclocarban and triclosan biodegradation and the phylotypes and functional genes associated with xenobiotic degradation in four agricultural soils. *Science of the Total Environment, 657*, 1138–1149.

Thi, T. V. N., Shintani, M., Moriuchi, R., Dohra, H., Loc, N. H., and Kimbara, K. (2019). Isolation and characterization of a moderate thermophilic *Paenibacillus naphthalenovorans* strain 4B1 capable of degrading dibenzofuran from dioxin-contaminated soil in Vietnam. *Journal of Bioscience and Bioengineering, 128*(5), 571–577.

Tian, H., Ma, Y. J., Li, W. Y., and Wang, J. W. (2018). Efficient degradation of triclosan by an endophytic fungus *Penicillium oxalicum* B4. *Environmental Science and Pollution Research, 25*(9), 8963–8975.

Tillitt, D. E., Buckler, J. A., Nicks, D. K., Candrl, J. S., Claunch, R. A., Gale, R. W., . . . and Baker, M. (2017). Sensitivity of lake sturgeon (*Acipenser fulvescens*) early life stages to 2, 3, 7, 8-tetrachlorodibenzo-P-dioxin and 3, 3′, 4, 4′, 5-pentachlorobiphenyl. *Environmental Toxicology and Chemistry, 36*(4), 988–998.

Tourova, T. P., Sokolova, D. S., Semenova, E. M., Shumkova, E. S., Korshunova, A. V., Babich, T. L., . . . and Nazina, T. N. (2016). Detection of n-alkane biodegradation genes alkB and ladA in thermophilic hydrocarbon-oxidizing bacteria of the genera *Aeribacillus* and *Geobacillus. Microbiology, 85*(6), 693–707.

Tran, H. T., Lin, C., Hoang, H. G., Nguyen, M. T., Kaewlaoyoong, A., Cheruiyot, N. K., . . . and Vu, C. T. (2020). Biodegradation of dioxin-contaminated soil via composting: Identification and phylogenetic relationship of bacterial communities. *Environmental Technology & Innovation, 19*, 101023.

Tu, C., Teng, Y., Luo, Y., Li, X., Sun, X., Li, Z., . . . and Christie, P. (2011). Potential for biodegradation of polychlorinated biphenyls (PCBs) by *Sinorhizobium meliloti. Journal of Hazardous Materials, 186*(2–3), 1438–1444.

Upadhyay, L. S., and Dutt, A. (2017). Microbial detoxification of residual organophosphate pesticides in agricultural practices. In *Microbial Biotechnology* (pp. 225–242). Springer.

Varjani, S., and Upasani, V. N. (2021). Bioaugmentation of Pseudomonas aeruginosa NCIM 5514—A novel oily waste degrader for treatment of petroleum hydrocarbons. *Bioresource Technology, 319*, 124240.

Verma, J. P., Jaiswal, D. K., and Sagar, R. (2014). Pesticide relevance and their microbial degradation: A-state-of-art. *Reviews in Environmental Science and Bio/Technology, 13*(4), 429–466.

Vignesh, A., Manigundan, K., Santhoshkumar, J., Shanmugasundaram, T., Gopikrishnan, V., Radhakrishnan, M., . . . and Balagurunathan, R. (2020). Microbial degradation, spectral analysis and toxicological assessment of malachite green by *Streptomyces chrestomyceticus* S20. *Bioprocess and Biosystems Engineering, 43*(8), 1457–1468.

Volke-Sepulveda, T. L., Gutiérrez-Rojas, M., and Favela-Torres, E. (2003). Biodegradation of hexadecane in liquid and solid-state fermentations by *Aspergillus niger. Bioresource Technology, 87*(1), 81–86.

Wang, H., Cao, X., Li, L., Fang, Z., and Li, X. (2018a). Augmenting atrazine and hexachlorobenzene degradation under different soil redox conditions in a bioelectrochemistry system and an analysis of the relevant microorganisms. *Ecotoxicology and Environmental Safety, 147*, 735–741.

Wang, H., Hu, J., Xu, K., Tang, X., Xu, X., and Shen, C. (2018b). Biodegradation and chemotaxis of polychlorinated biphenyls, biphenyls, and their metabolites by *Rhodococcus* spp. *Biodegradation*, *29*(1), 1–10.

Wang, J., Jiang, X., Liu, X., Sun, X., Han, W., Li, J., . . . and Shen, J. (2018c). Microbial degradation mechanism of pyridine by *Paracoccus* sp. NJUST30 newly isolated from aerobic granules. *Chemical Engineering Journal*, *344*, 86–94.

Wang, N., Chu, Y., Zhao, Z., and Xu, X. (2017a). Decolorization and degradation of Congo red by a newly isolated white rot fungus, *Ceriporia lacerata*, from decayed mulberry branches. *International Biodeterioration & Biodegradation*, *117*, 236–244.

Wang, Q., Li, X., Yang, Q., Chen, Y., and Du, B. (2019a). Evolution of microbial community and drug resistance during enrichment of tetracycline-degrading bacteria. *Ecotoxicology and Environmental Safety*, *171*, 746–752.

Wang, S., Wang, L., Hua, W., Zhou, M., Wang, Q., Zhou, Q., and Huang, X. (2015). Effects of bisphenol A, an environmental endocrine disruptor, on the endogenous hormones of plants. *Environmental Science and Pollution Research*, *22*(22), 17653–17662.

Wang, S., Yin, Y., and Wang, J. (2018d). Microbial degradation of triclosan by a novel strain of *Dyella* sp. *Applied Microbiology and Biotechnology*, *102*(4), 1997–2006.

Wang, Z., Li, X., Wu, Q., Lamb IV, J. C., and Klaunig, J. E. (2017b). Toxaphene-induced mouse liver tumorigenesis is mediated by the constitutive androstane receptor. *Journal of Applied Toxicology*, *37*(8), 967–975.

Wang, Z., Wang, W., Li, Y., and Yang, Q. (2019b). Co-metabolic degradation of naphthalene and pyrene by acclimated strain and competitive inhibition kinetics. *Journal of Environmental Science and Health, Part B*, *54*(6), 505–513.

Water, U. (2016). *WWAP (United Nations World Water Assessment Programme)* (pp. 1–148). The United Nations World Water Development Report.

Wirnkor, V. A., Iheanyichukwu, O. A., Christian Ebere, E., Ngozi, V. E., Kingsley, O. U., Chizoruo, I. F., and Amaka, A. P. (2020). Petroleum hydrocarbons and heavy metals risk of consuming fish species from oguta lake, Imo State, Nigeria. *Journal of Chemical Health Risks*, *11*(1), 1–15.

Wongputtisin, P., Supo, C., Suwannarach, N., Honda, Y., Nakazawa, T., Kumla, J., . . . and Khanongnuch, C. (2020). Filamentous fungi with high paraquat-degrading activity isolated from contaminated agricultural soils in northern Thailand. *Letters in Applied Microbiology*, *72*(4), 467–475.

Wrobel, M. H., and Mlynarczuk, J. (2018). Chlorinated insecticides (toxaphene and endrin) affect oxytocin, testosterone, oestradiol and prostaglandin secretion from ovarian and uterine cells as well as myometrial contractions in cow in vitro. *Chemosphere*, *198*, 432–441.

Wu, C., Wu, X., Chen, S., and Wu, D. (2020). A newly discovered humic-reducing bacterium, *Pseudomonas geniculata* PQ01, isolated from paddy soil promotes paraquat anaerobic transformation. *Frontiers in Microbiology*, *11*, 2003.

Wu, H., Shen, J., Jiang, X., Liu, X., Sun, X., Li, J., . . . and Wang, L. (2018). Bioaugmentation potential of a newly isolated strain *Sphingomonas* sp. NJUST37 for the treatment of wastewater containing highly toxic and recalcitrant tricyclazole. *Bioresource Technology*, *264*, 98–105.

Wu, X., Wang, W., Liu, J., Pan, D., Tu, X., Lv, P., . . . and Hua, R. (2017). Rapid biodegradation of the herbicide 2, 4-dichlorophenoxyacetic acid by *Cupriavidus gilardii* T-1. *Journal of Agricultural and Food Chemistry*, *65*(18), 3711–3720.

Xia, Z. Y., Zhang, L., Zhao, Y., Yan, X., Li, S. P., Gu, T., and Jiang, J. D. (2017). Biodegradation of the herbicide 2, 4-dichlorophenoxyacetic acid by a new isolated strain of *Achromobacter* sp. LZ35. *Current Microbiology*, *74*(2), 193–202.

Xiang, S., Lin, R., Shang, H., Xu, Y., Zhang, Z., Wu, X., and Zong, F. (2020). Efficient degradation of phenoxyalkanoic acid herbicides by the alkali-tolerant *Cupriavidus oxalaticus* strain X32. *Journal of Agricultural and Food Chemistry*, *68*(12), 3786–3795.

Xiao, X., Ma, X. L., Liu, Z. Y., Li, W. W., Yuan, H., Ma, X. B., . . . and Yu, H. Q. (2019). Degradation of rhodamine B in a novel bio-photoelectric reductive system composed of *Shewanella oneidensis* MR-1 and Ag3PO4. *Environment International*, *126*, 560–567.

Xie, X., Zhou, Q., Lin, D., Guo, J., and Bao, Y. (2011). Toxic effect of tetracycline exposure on growth, antioxidative and genetic indices of wheat (*Triticum aestivum* L.). *Environmental Science and Pollution Research*, *18*(4), 566–575.

Xiong, J., Li, G., and An, T. (2017). The microbial degradation of 2, 4, 6-tribromophenol (TBP) in water/sediments interface: Investigating bioaugmentation using *Bacillus* sp. GZT. *Science of the Total Environment*, *575*, 573–580.

Xu, B., Sun, Q. J., Lan, J. C. W., Chen, W. M., Hsueh, C. C., and Chen, B. Y. (2019). Exploring the glyphosate-degrading characteristics of a newly isolated, highly adapted indigenous bacterial strain, *Providencia rettgeri* GDB 1. *Journal of Bioscience and Bioengineering*, *128*(1), 80–87.

Xu, T., Zhao, J., Yin, D., Zhao, Q., and Dong, B. (2015). High-throughput RNA sequencing reveals the effects of 2, 2′, 4, 4′-tetrabromodiphenyl ether on retina and bone development of zebrafish larvae. *BMC Genomics*, *16*(1), 1–12.

Yang, C., Li, Y., Zhang, K., Wang, X., Ma, C., Tang, H., and Xu, P. (2010). Atrazine degradation by a simple consortium of *Klebsiella* sp. A1 and *Comamonas* sp. A2 in nitrogen enriched medium. *Biodegradation*, *21*(1), 97–105.

Yang, X., Wei, H., Zhu, C., and Geng, B. (2018). Biodegradation of atrazine by the novel *Citricoccus* sp. strain TT3. *Ecotoxicology and Environmental Safety*, *147*, 144–150.

Yao, L., Jia, X., Zhao, J., Cao, Q., Xie, X., Yu, L., . . . and Tao, Q. (2015). Degradation of the herbicide dicamba by two *Sphingomonads* via different O-demethylation mechanisms. *International Biodeterioration & Biodegradation*, *104*, 324–332.

Yao, L., Zhang, J. J., Yu, L. L., Chen, Q., Zhu, J. C., He, J., and Ding, D. R. (2016). *Rhizorhabdus dicambivorans* sp. nov., a dicamba-degrading bacterium isolated from compost. *International Journal of Systematic and Evolutionary Microbiology*, *66*(9), 3317–3323.

Yaronskaya, E. B., Gritskevich, E. R., Trukhanovets, N. L., and Averina, N. G. (2007). Effect of kinetin on early stages of chlorophyll biosynthesis in streptomycin-treated barley seedlings. *Russian Journal of Plant Physiology*, *54*(3), 388–395.

Ye, Z., Li, H., Jia, Y., Fan, J., Wan, J., Guo, L., . . . and Shen, C. (2020). Supplementing resuscitation-promoting factor (Rpf) enhanced biodegradation of polychlorinated biphenyls (PCBs) by *Rhodococcus biphenylivorans* strain TG9T. *Environmental Pollution*, *263*, 114488.

Yu, X. M., Yu, T., Yin, G. H., Dong, Q. L., An, M., Wang, H. R., and Ai, C. X. (2015). Glyphosate biodegradation and potential soil bioremediation by *Bacillus subtilis* strain Bs-15. *Genetics and Molecular Research*, *14*(4), 14717–14730.

Yukihiro, M., Hiramatsu, T., Bouteau, F., Kadono, T., and Kawano, T. (2012). Peroxyacetyl nitrate-induced oxidative and calcium signaling events leading to cell death in ozone-sensitive tobacco cell-line. *Plant Signaling & Behavior*, *7*(1), 113–120.

Yüzbaşıoğlu, E., and Dalyan, E. (2019). Salicylic acid alleviates thiram toxicity by modulating antioxidant enzyme capacity and pesticide detoxification systems in the tomato (*Solanum lycopersicum* Mill.). *Plant Physiology and Biochemistry*, *135*, 322–330.

Zaar, A., Eisenreich, W., Bacher, A., and Fuchs, G. (2001). A novel pathway of aerobic benzoate catabolism in the bacteria *Azoarcus evansii* and *Bacillus stearothermophilus*. *Journal of Biological Chemistry*, *276*(27), 24997–25004.

Zhang, H., Jiang, X., Lu, L., and Xiao, W. (2015). Biodegradation of polychlorinated biphenyls (PCBs) by the novel identified cyanobacterium *Anabaena* PD-1. *PLoS ONE*, *10*(7), e0131450.

Zhang, L., Hang, P., Zhou, X. Y., Qiao, W. J., and Jiang, J. D. (2020a). Enantioselective catabolism of the two enantiomers of the phenoxyalkanoic acid herbicide dichlorprop by *Sphingopyxis* sp. DBS4. *Journal of Agricultural and Food Chemistry*, *68*(26), 6967–6976.

Zhang, L., Hu, Q., Liu, B., Li, F., and Jiang, J. D. (2020b). Characterization of a linuron-specific amidohydrolase from the newly isolated bacterium *Sphingobium* sp. Strain SMB. *Journal of Agricultural and Food Chemistry*, *68*(15), 4335–4345.

Zhang, Q., Li, S., Ma, C., Wu, N., Li, C., and Yang, X. (2018). Simultaneous biodegradation of bifenthrin and chlorpyrifos by *Pseudomonas* sp. CB2. *Journal of Environmental Science and Health, Part B*, *53*(5), 304–312.

Zhang, Q., Liu, H., Saleem, M., and Wang, C. (2019). Biotransformation of chlorothalonil by strain *Stenotrophomonas acidaminiphila* BJ1 isolated from farmland soil. *Royal Society Open Science*, *6*(11), 190562.

Zhang, X., Kong, D., Liu, X., Xie, H., Lou, X., and Zeng, C. (2021). Combined microbial degradation of crude oil under alkaline conditions by *Acinetobacter baumannii* and *Talaromyces* sp. *Chemosphere*, 129666.

Zhao, J., Li, X., Chi, Y., Jia, D., and Yao, K. (2018). Characterization of co-culturing microorganisms for simultaneous degradation of β-cypermethrin and 3-phenoxybenzoic acid. *Fresenius Environmental Bulletin*, *27*(6), 4249–4257.

Zheng, S., Liu, C., Huang, Y., Bao, M., Huang, Y., and Wu, K. (2017). Effects of 2, 2′, 4, 4′-tetrabromodiphenyl ether on neurobehavior and memory change and bcl-2, c-fos, grin1b and lingo1b gene expression in male zebrafish (Danio rerio). *Toxicology and Applied Pharmacology*, *333*, 10–16.

Zhu, J., Fu, L., Jin, C., Meng, Z., and Yang, N. (2019). Study on the isolation of two atrazine-degrading bacteria and the development of a microbial agent. *Microorganisms*, *7*(3), 80.

Zhu, L., Mu, X., Wang, K., Chai, T., Yang, Y., Qiu, L., and Wang, C. (2015). Cyhalofop-butyl has the potential to induce developmental toxicity, oxidative stress and apoptosis in early life stage of zebrafish (*Danio rerio*). *Environmental Pollution*, *203*, 40–49.

Żur, J., Wojcieszyńska, D., Hupert-Kocurek, K., Marchlewicz, A., and Guzik, U. (2018). Paracetamol—toxicity and microbial utilization. *Pseudomonas moorei* KB4 as a case study for exploring degradation pathway. *Chemosphere*, *206*, 192–202.

2 Nanomaterial's Interesting Characteristics to Remediate Polluted Environs

Misbah Naz, Muhammad Ammar Raza, Sarah Bouzroud, A. M. Shackira, Mohammad Sarraf, and Xiaorong Fan

2.1 INTRODUCTION

Although various studies have underlined the promising applications of nanotechnology in many fields, concerns about their use for the remediation of environmental compartments have not been resolved (Macoubrie 2006). Similarly, although many studies have indeed proven nanotechnology's efficiency in a laboratory environment, more studies are still needed to fully determine how nanotechnology can have a significant impact on the remediation of environmental contaminants in practical situations (for example, the effects of pollution in industrial processes and water, soil and air restoration) (Zhang et al. 2021). Similarly, although the mechanisms for applying different nanotechnologies are well known, the treatment of these materials after they are used for contaminant neutralization or elimination has not been fully studied (Zhou et al. 2020). Even though some of these materials are known to be recyclable, it seems that their efficiency decreases at some point, which ultimately limits their use. As a result, it is critical to conduct research to determine what happens to these materials after they repair and are released into the environment to avoid the risk of them becoming environmental threats. To access all of the possibilities of nano(scale)materials (NMs) in environmental applications, these challenges must be overcome (Roco, Mirkin, and Hersam 2011). Nanotechnology has a wide range of strategies to address environmental pollution. Nanotechnology has received a great deal of attention over recent decades due to their high physical characteristics (Lee, Park, and Lee 2020). NMs have increased reactivity because of their higher surface area-to-volume ratio and are therefore more effective than larger materials of the same type. NMs can also be functionalized or associated with other functional groups to target specific molecular contamination for effective repair because of their unique surface chemistry (Zhang, Mou, and Jiang 2020). In addition, deliberate changes to NMs' physical properties (such as the size, morphology, porosity, and chemical composition) can confer additional benefits. These characteristics modify the efficacy of the material for environmental cleanup (Manzano and Vallet-Regí 2020). Compared with traditional methods to solve environmental pollution, the abundant surface modification chemistry and adjustable physical parameters of NMs provide obvious advantages (Khine and Stenzel 2020). Through expansion, the development of hybrid or composite materials by combining different materials with desired characteristics may be more effective, selective and certain than using a single nanoplatform. For example, when compared to using nanoparticles (NPs) alone, grafting NPs to the frame can improve the stability of the material. To a certain extent, the use of specific chemicals that can target certain pollutants to functionalize materials can improve the selectivity and efficacy of the materials. The use of NMs for environmental remediation is discussed in this chapter (Guerra et al. 2018). In the field of environmental remediation, NMs can efficiently eliminate biological and other pollutants. Detecting and removing contaminated gases (SO2, CO, NOx, etc.), chemicals (heavy metals, etc.), organic pollutants (aliphatic and aromatic hydrocarbons) and certain biological

DOI: 10.1201/9781003186298-2

compounds, nanoparticles, tubes, metal wires, fibres, and other composite materials with polymers are among the many shapes/forms of NM activity, and they act as adsorbents and catalysts (Khin et al. 2012). NMs outperform other conventional technologies in terms of remediation due to their surface area-to-volume ratio and associated high reactivity. This chapter focuses on the development of new NMs and methods for the treatment of toxic metal ions, organic and inorganic solutes, radionuclides, bacteria and viruses that pollute consumption and industrial wastewater (Wang et al. 2019). Furthermore, the most recent developments in the pollutant treatment and monitoring of polymer nanocomposites are debated. Additionally, imminent research trends and prospects are briefly discussed, because engineered nanoparticles (ENPs) have been designed for industrial and human commodities, by-products of waste and chemical industrial reactions are common, and incidental nanoparticles can also be found in the environment (Qu et al. 2013).

Nanotechnology is currently being used to remediate contaminated soil, and it has great promise in terms of improving the performance of traditional remediation methods. When nanotechnology is released into the ecosystem, however, it causes environmental issues that affect human and environmental health. The goal of this chapter is to highlight the benefits and risks of using nanotechnology to remediate contaminated soil oil (Madhura et al. 2019). Soil nanoremediation has been found to have not only some benefits but also some drawbacks in terms of the final disposal of nanoparticles, NMs, or nanodevices (Wu et al. 2019). The most recent applications of these carbon NMs have been linked to heavy metal treatment in polluted water and environmental remediation. The use of carbon NMs (fullerene carbon nanotubes, graphene, graphene oxide and activated carbon) for environmental and special decontamination of heavy metals from pollutants is discussed (Zou et al. 2016). Heavy metal-polluted water has been successfully purified using carbon NMs. The reason for their widespread use is due to their intriguing properties, such as their high surface area, ease of recycling, and ease of adsorption, as inorganic acid solutions and regenerated materials can be recycled while still retaining adsorption capacity (Guerra et al. 2018). In addition to these characteristics, carbon NMs can also be easily manufactured in conjunction with other NMs and easily adapted to form a multifunctional nanoadsorbent material. Moreover, the biocompatibility of carbon-based materials with organisms and the environment is extremely high (Wu et al. 2019). Factors such as the pH, interaction time and adsorbent category have a significant impact on metal ion adsorption. According to this review, carbon NMs have excellent physical and chemical properties and a high potential for environmental remediation (Abbas et al. 2016). Here, structures up to 1 μm will be referred to as NMs; NPs belong to this group of materials and present size most commonly in the range of 1 to 100 nm (Buzea, Pacheco, and Robbie 2007).

2.2 ENVIRONMENTAL APPLICATIONS OF NANOTECHNOLOGY

The wide distribution of pollutants in the air, water and soil provides us with a novel technology that can detect, screen and remove pollutants irrespective of the medium. As a result, researchers and environmentalists have turned their attention to nanotechnology, which is part of an emerging field of purification with broad prospects for improving the quality of the existing environment. Nanotechnology includes NPs with a size range of 1–100 nm and has several unique features that are lacking in batch forms. The most important features are the increased surface area and the high reactivity of the nanoparticles, which are critical to the decontamination process. Therefore, nanoparticles can be effectively exploited to prevent and detect pollution and to decontaminate the pollutant screen (Yunus et al. 2012). In the following section, the role of NPs in the decontamination of air, water and soil is detailed.

2.2.1 NP Application for the Decontamination of Air

Air pollution is currently a critical concern as it is increasing every day, especially in cities all over the world. The major pollutants of air are chloroflurocarbons (CFCs), carbon monoxide (CO), sulfur dioxide, nitrogen oxides, toxic metal ions (e.g., arsenic (As), chromium (Cr), lead (Pb), cadmium (Cd), mercury (Hg), zinc (Zn), etc.), hydrocarbons and organic compounds (e.g., dioxins). Most of these pollutants are released into the air through various anthropogenic activities including the combustion of fossil fuel, industrial activities, and smelting (Yunus et al. 2012). Nanotechnology can be effectively employed to clean polluted air through a sustainable approach. NP-like carbon nanotubes (CNTs) and gold NPs have been the most exploited NPs in recent years for the successful remediation of polluted air (Yunus et al. 2012).

For instance, CNTs are found to be exceptionally beneficial for the removal of dioxin, CO_2 and NO from polluted air as suggested by Long and Yang (2001). This increased activity is due to the curved surface of CNTs that provides a stronger force of interaction between the pollutant and the NP (Bhushan 2017). In addition to CNTs, gold NPs, especially manganese oxide coated with gold nanoparticles, are excellent absorbers for volatile molecules since they are highly porous in nature and have a large surface area. Due to the presence of free radicals on the surface, the adsorbed molecules are effectively degraded, and gold NPs play an important part in the formation of free radicals apart from the NPs stated earlier, such as zeolite, silica-based NPs and activated carbon, which are also employed in the successful remediation of air pollutants (Yunus et al. 2012).

2.2.2 NP Application for the Decontamination of Soil

Decontamination and detoxification (phytoremediation) of pollutants in the soil with plants is a very mature strategy that has been applied since ancient times. However, in recent years, following the development of nanotechnology, the use of NPs in the phytoremediation process is also robust and has been widely accepted because of its high repair efficiency. Several NPs are employed for nano-phytoremediation, and most of them are excellent for removing/degrading the pollutant in situ. Some of them include CNTs, fullerenes, metallic NPs, iron NPs, grapheme oxide NPs and silver and gold NPs (Baragaño et al. 2020).

Quantum dots (QDs) have been shown to bind metals such as copper (Cu) and Pb in soil, which reduces their availability, and the cell walls provide additional protection by preventing the entry of these molecules (Worms et al. 2012). In addition, zero-valent iron (ZVI)-treated *Panicum maximum* showed a higher potential for the breakdown of Trinitrotoluene (TNT). It is therefore used for the remediation of TNT from contaminated soil; similarly, treatment with nano zero-valent iron (NZVI) in *Panicum maximum* and *Helianthus annuus* detoxifies TNT contaminants in soil. Moreover, plants treated with NZVI showed significantly higher polychlorinated biphenyls (PCB) detoxification efficiency than untreated plants (Jiamjitrpanich et al. 2012). Recently, Baragaño et al. (2020) described that two NPs, namely, graphene oxide nanoparticles (nGOx) and NZVI, are efficient in the decontamination of toxic metal ions in soil, which may include As, Pb, Zn, Cd and Cu. These types of nano-phytoaccumulation may be used to improve the texture of degraded soil.

2.2.3 NP Application for the Decontamination of Water

The pollution of water bodies must be considered a serious issue since the pollutants from water bodies may easily enter the food chain through bio magnification. The major source of pollutants in water bodies are oil spillage, fertilizers, pesticides, herbicides, and waste disposal (Krantzberg et al. 2010). A number of NPs are shown to be effective for the successful recovery of contaminated water including the NPs of ZVI, biopolymers, single-enzyme NPs, CNTs, zeolite, and self-assembled monolayers on mesoporous supports (Sharma et al. 2009).

ZVI is found to have reactivity towards a range of pollutants including NO_3^-, Cu, Cr, and chlorinated hydrocarbons (Miehr et al. 2004). Another widely used NP for water purification is TiO_2 NP mainly because of its low toxicity, high photoconductivity and photostability; it is also cost effective (Yunus et al. 2012). Generally, the antimicrobial properties of silver NP are well known, and it is commonly used to remove microbes such as bacteria from contaminated water. Polyacrylonitrile

(PAN) nanofibre and polyvinyl alcohol (PVA) with silver NP have the potential to reduce the microbial level up to 91% without leaching the silver particles into the water, thus safeguarding the quality of drinking water (Du Plessis 2011). Accordingly, antimicrobial nanotechnology is now emerging as a promising branch of nanotechnology to decontaminate and thereby ensuring the quality of water bodies. In addition to silver NP, a number of NP-like TiO_2, zinc oxide (ZnO), fullerol, CNTs, chitosan and peptides are also efficient in their antimicrobial activity and can act as potential agents of remediation (Li et al. 2008)(Figure 2.1).

2.3 THE PROSPECTS OF BIOSYNTHETIC NMS TO REPAIR ORGANIC AND INORGANIC ENVIRONMENTAL POLLUTANTS

Nanotechnology involves synthesis and extremely small NMs used in the fields of chemistry, biology, physics, materials science and engineering. These uniquely structured NMs have different chemical, physical, electrical and mechanical properties. NMs have structures up to 1 μm (1000 nm), and particles in the sub-micron range are called NPs (Buzea, Pacheco, and Robbie 2007), which usually range from 1 to 100 nm. Such nanoscale materials are potentially beneficial for economic and ecological applications. As NMs have a wide variety of semiconductors, examples of industrial applications include storage, display, optical and photonic technologies, energy and biotechnology and health care, and they produce the maximum commodities, which are NMs (Meena et al. 2014). They are also used to store, produce and convert electricity, improve farm efficiency, treat and remedy water, and diagnose and detect systems for diseases; they are used in medicines, food handling and storage systems, air pollution and remediation, buildings and medical systems and as a vector for the control of pests (Murty et al. 2013).

However, nanotechnology increasingly employs various techniques to aid in the detection of pollution, for example, fluorescence, electrochemical or optical detection, pollution prevention, and the long-term handling and cleaning of hazardous waste sites, as time and tools allow (Karn, Kuiken, and Otto 2009, Ghasemzadeh et al. 2014). These technologies are proving to be viable alternatives to existing in-situ organic and inorganic pollution remediation methods. The impact of nanotechnology is the focus of current research, which emphasizes preventing pollution and removing various contaminated land, sediments, solid waste and water and air pollutants (Lowry 2007). NM-based

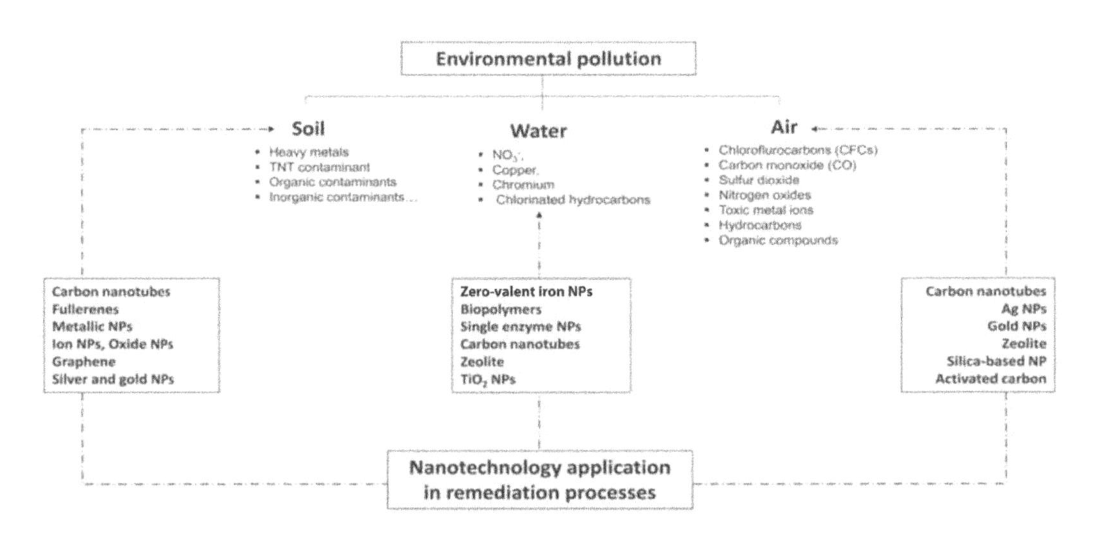

FIGURE 2.1 Nanoparticle application in air, soil, and water remediation in the environment.

separation membranes have also been used to produce drinking water as part of a low-energy power supply process (Gehrke, Geiser, and Somborn-Schulz 2015).

The plan for the use of nanotechnology-mediated multi-functional processes is to provide high-performance and reasonable wastewater action solutions without relying on large-scale infrastructure (Qu, Alvarez, and Li 2013). Wastewater treatment based on nanotechnologies not only overcomes the major drawbacks of current treatment technologies and adds new treatment capabilities but also expands the supply of non-traditional water sources and therefore provides a cost-effective, safe and green method for various applications (Drexler and Minsky 1990; Ball 1996; Ando et al. 2012; Mehndiratta et al. 2013). The reduction of the source, the degradation of pollutants and the sensing of pollutants are the three steps in pollution control. Through different techniques, such as surface adsorption and surface interaction of metal ions with ohydroxy groups through permeable reactive barriers, several chemically synthesized NMs/NPs can remedy toxic metals and organic pollutants (Tungittiplakorn, Cohen, and Lion 2005; Cundy, Hopkinson, and Whitby 2008; Liu et al. 2011; Kharisov et al. 2012; Torabian et al. 2014). All the possible uses of biosynthetic NMs/NPs for sustainable environmental applications are still unknown; however, the prospects for biosynthetic NMs and their mechanisms to remediate organic and inorganic environmental pollutants are discussed in this chapter.

2.3.1 Biosensing Mechanisms and the Remediation of Environmental Contaminants

In the past ten years, NMs based on environmental protection technology have been continuously developed for environmental restoration. Today, a variety of microorganisms has synthesized many inorganic and organic NMs/NPs with excellent chemical and physical properties, and they have been used in various applications in innovative and environmentally friendly technologies (Figure 2.2). Among several possible mechanisms, NMs/NPs follow three main methods, namely, adsorption, conversion and catalysis, to remedy environmental pollutants. Biological inorganic and organic NMs/NPs have so far reduced toxic metal and organic compound absorption, processing, photocatalysis, and catalytic reduction. The different mechanisms of biosynthetic NMs for the environmental remediation of organic and inorganic pollutants are highlighted here.

TABLE 2.1
The Application of Nanotechnology in Resisting Mechanical Stress in a Clean Environment

Nanotechnology	Uses against mechanical stress	References
Nanomaterials	They can replace ecologically problematic chemicals in certain applications	(Villaseñor and Ríos 2018)
Titanate nanofibers	NMs for the purification of radioactive waste in water are used as absorbents to remove radioactive ions from water	(Shen et al. 2020)
Ceramic nanofibers	They exclude various pollutants and make the water and environment clean	(Malwal and Gopinath 2016)
Ion concentration polarization	Conducts remediation, detection and pollution prevention	(Li and Anand 2016)
Filtration	Carbon dioxide capture	(Pires et al. 2012)
Hydrogen production from artificial-sunlight photosynthesis	Renewable energy technologies, such as wind, solar, geothermal and hydroelectric power generation split water into hydrogen and oxygen	(Pace 2005)
Adsorption, membrane process, photocatalysis and disinfection process	Environmental applications of nanotechnology for the remediation of air, soil and water through cleaning	(Ibrahim et al. 2016)

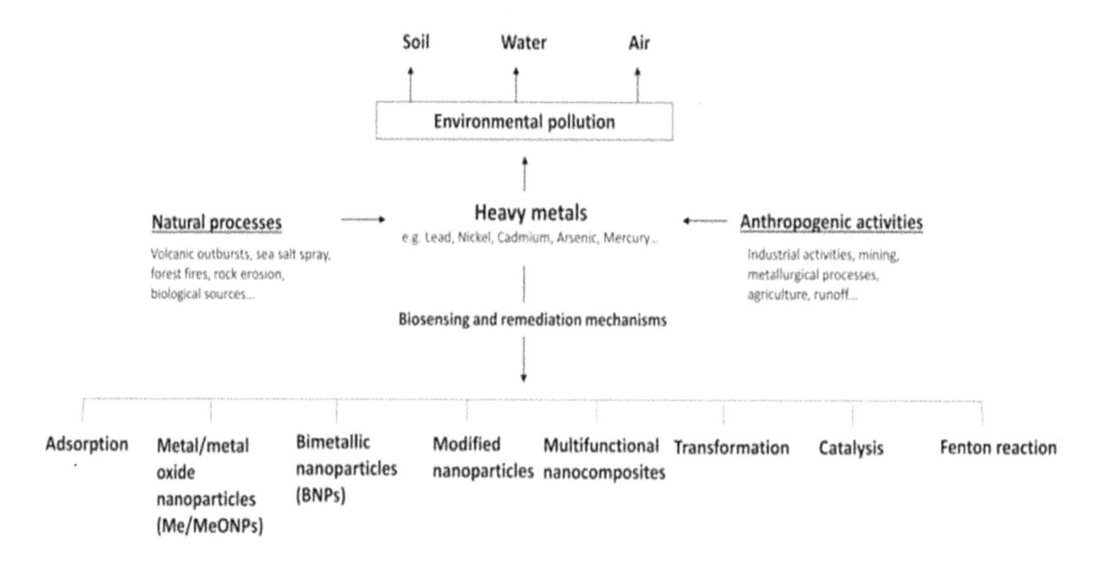

FIGURE 2.2 Biosensing mechanism and remediation of environmental pollutants.

TABLE 2.2
Elimination of the Organic and Inorganic Pollutants That Are Produced Biologically by NMs

NMs	Plants/Crop	Contamination	Approach of remediation	References
α-Fe2O3	Aloe vera	As (v)	Adsorption	(Jerin et al. 2019)
CuO-NPs	Citrus limon	CrIJVI)	-	(Nemati, Sarkheil, and Johari 2019)
CdS-NPs	*Pseudomonas aeruginosa* JP-11	CdIJII	-	(Chakraborty et al. 2018)
Zv-INP	Grape marc, black tea and vine leaves	Ibuprofen	-	(Das et al. 2018)
Pd-NPs	Fruit extract	Nitroarenes	Catalysis	(Das et al. 2018)
Ag-NPs	Tea polyphenols	Phenolic compound	-	(Das et al. 2018; Wang et al. 2020)
SnO2-NPs	Microwave	Organics (nitrophenol)	-	(Matussin et al. 2020)
CuO-NPS	Plant (*Thymus vulgaris*)	Organics	-	(Das et al. 2018)
Ag-NPs	Fungus (*Penicillium oxalicum*)	Organics (methylene blue)	Catalytic reduction	(Wang et al. 2020)
Biomatrixed AuNPs	Fungus (*Flammulina velutipes*)	Organics (methylene blue, nitrophenol)	Catalytic reduction	(Narayanan, Park, and Han 2015)
Au and Ag-NPs	Fungus (*Gordonia amicalis* HS-11)	Organics (hexadecane)	-	(Wang et al. 2020)
Ca-Alginate NPs	Honey	CrIJVI	Sorption	(Soylak, Ozalp, and Uzcan 2020; Das et al. 2018)
NiO-NPs	*Microbacterium* sp.	Nickel(II)	Transformation	(Vishnu and Dhandapani 2021)
Co3O4-NPs	*Bacillus subtilis*	Lithium	Transformation	(Kavitha et al. 2017)
Ag-NPs	Spinach	Organics	Electrocatalysis	(Singh et al. 2019; Wang et al. 2020)
Cu-NPs	Plant (*Citrus grandis*)	Organics (methylene blue)	Photocatalysis	(Westerhoff, Kiser, and Hristovski 2013)

2.3.1.1 Adsorption

The combination of pollutants on the surface of the adsorbent is called adsorption. This process is mediated by the interaction of exothermic surfaces and ions. Enthalpy is the amount of heat generated by the adsorbent during the adsorption of one mole of the adsorbate. Enthalpy adsorption is continually specified as a negative value because the liberty of movement of the adsorbate is restricted during the adsorption process, which leads to a decrease in entropy.

Adsorption occurs spontaneously under constant temperature and pressure, further reducing Gibbs free energy (Ruthven 1984). A variety of adsorption methods tend to effectively eliminate environmental pollutants. Due to their environmental protection and cost-effective characteristics, there is great demand for biologically produced nanoscale adsorbents to remove organic and inorganic pollutants (Rizwan et al. 2014). Many organisms are used as factories to produce nanoscale adsorbents from prokaryotes to eukaryotes. All of them have components that can be used currently to make nanoscale adsorbents, and many biosynthetic nano-sorbents are employed to capture pollutants from contaminated sites.

2.3.1.2 Nanoparticles of Metal/Metal Oxide (Me/MeONPs)

Heavy metal adsorbents are nanoparticles of metal/metal oxide (Me/MeONP). Recent research has shown that magnetic nanostructures can also be used to remove organic pollutants, and magnetic NPs are crucial in contemporary Me/MeONPs because they can treat vast quantities of wastewater while still allowing for rapid magnetic separation (Hu, Wang, and Pan 2010). Many studies have shown that among other low-cost adsorbents, NP iron oxide has the highest adsorption capacity. A study executed by Lunge, Singh, and Sinha (2014) found that in Fe3O4 NPs in the range of 188.69 and 153.8 mg g-1, As (III) has a maximum capacity for adsorption, and As (v) ions have been determined to be much higher than those in conventional methods. This shows that the rapid removal and recovery of metal ions from wastewater discharge through a biologically produced Fe3O4 nanosorbent tea extract is actually economical.

2.3.2 Bimetallic Nanoparticles (BNPs)

Biological systems such as bacteria, yeast at room temperature, fungi and plants biologically produce BNPs inside or outside the cell, which provides a biocompatible method for the synthesis of NMs. Compared with uncoated iron metal, the reactivity of nanoscale iron particles enhances the deposition of discontinuous precious metal layers, thereby increasing the reaction speed (Chang, Lian, and Zhu 2011). Accordingly, in the field of remediation, the characteristics of BNPs make them useful in removing environmental pollutants. The use of biosynthetic BNPs is a low-cost, advanced green method to increase the rate of decontamination of organic and inorganic pollutants in the environment. *Spirulina Platensis*, a single-cell protein, has been used to synthesize bimetallic Au/Ag NPs (Chang, Lian, and Zhu 2011). Plants have also been studied as non-toxic and environmentally friendly reducing agents for metal ions in the synthesis of bimetallic nanoparticles, which is different from the reduction of chemicals. Some reports on BNP biosynthesis have been published that indicate that this process has been used to produce materials capable of removing environmental pollutants (Litter et al. 2014). However, further research is needed in this field concerning green BNP synthesis being used to eliminate environmental pollutants.

2.3.3 Modified Nanoparticles

Bimetallic NPs and Me/MeONP have been synthesized for environmental remediation in recent years. These BNPs and Me/MeONP usually have high reactivity, surface energy and magnetism, which makes it easy to oxidize the air dissolved in water. In addition, due to their high concentration in wastewater, they may compete with other interference molecules at the adsorption site.

This limits the effectiveness of the nano-adsorbent. To prevent NM aggregation, the surface of NMs must be modified to make it stable in the system for a longer period of time, as organic and inorganic materials (e.g., polymers, biomolecules, silica and metals) may change the surface of NMs/NPs. Organic molecules (including small organic molecules, i.e., surfactants, polymers and biomolecules) and inorganic layer coatings are examples of this (such as silica, metal or non-metal elements, metal oxides or metals). The main stability strategy of the sulfide is as follows. The mechanism of the surface-modified NM/NP adsorption of wastewater pollutants includes surface bonding, selective magnetic adsorption, electrostatic interaction and modified ligand bonding. The biosynthesis of Me/MeONPs was noted by Li et al. (2011) and Jeevanandam, Chan, and Danquah (2016). Nonetheless, the modification of these NMs/NPs has been investigated by employing biodegradable material to remove organic and inorganic pollutants (Cho et al. 2015; Martínez-Cabanas et al. 2016). Biodegradable materials can be utilized to add an additional surface of biologically produced NMs/NPs that can be modified by binding sites for the enhanced removal of heavy metals and other contaminants from wastewater. Organic materials have been added to the surface of Me/MeONPs to create additional binding sites for Cd adsorption (II)(Singh et al. 2014). Organic (tricholorehylene) and inorganic (chromium) pollutants have been simultaneously removed from hydrogel green-synthesized agarose-Fe nanoparticles (Luo et al. 2016).

2.3.4 Multifunctional Nano-Composite

NMs/NPs are currently being explored because of their multi-functional actions, creating smart materials with two or more separate actions. Nanotechnology-based products have been developed recently to perform a variety of functions. For example, to perform multiple tasks simultaneously in the study, Ag NP was biosynthesized using vegetable A extract to produce NP-embedded polyvinyl alcohol film. Ag, Cu and zero-valent iron (ZVI) NP hybrid clusters have been patented and can be used to treat various industrial wastewaters in tannery, textile and pharmaceutical waste.

In water inclosing contaminated metals such as Cu (II), Cd (II) and Pb(II), the efficacy of chitosan/silver nanoparticle/Cu nanoparticle/CNT multifunctional nanocomposite has been studied. In just 10 minutes, these multifunctional nanocomposites almost completely removed metal ions from the solution. It is also possible to regenerate this multifunctional nanocomposite.

2.3.5 Transformation

Environmental human activities are leading to the accretion of new pollutants of organic and inorganic contaminants in the environment, which can be purified from contaminated sites by converting them through oxidation or reduction; these reactions may lead to the formation of metals that may reduce their toxicity. In addition, site change and/or toxicity result from the presence of electron transfer reactions in a stable oxidation state. Heavy metal conversions such as Cd, Cr, Ni, Zn and Pb have been extensively studied using ZVI NP in several experimental and field studies to reduce Cr (VI) by NZVI particles (Singh, Misra, and Singh 2012). NZVI particles act as an electron donor to reduce Cr (VI); during the reduction of Cr, ferrous iron is released from NZVI particles (VI) and molecular hydrogen, atomic active hydrogen and Fe (II)-containing solid minerals such as NZVI corrosion products, all of which contribute to the reduction process. NZVI particles effectively removed 85% of Zn2+ from an aqueous solution in the presence of dissolved oxygen (DO) in a separate study. Due to DO corrosion, the surface of NZVI particles was coated with an iron (oxy) hydroxide shell, resulting in increased Zn2+ adsorption and co-precipitation (Liang et al. 2014). Numerous organic chemical pollutants have also been used to reduce NZVI particles. Chlorinated pesticides, organophosphates, nitroamines, nitroaromatics and chlorinated solvents may all contribute to organic chemical pollution (Liang et al. 2014). The metal oxidation states of ions and NMs offer considerable possibilities for redox cycling. For environmental remediation, the term transformation refers to the elimination of organic and inorganic chemical pollutants and their transportation and toxicity.

2.3.5.1 Catalysis

Catalytic technologies are a prerequisite for present and future energy, chemical processes and environmental sectors (Centi et al. 2002). Catalytic technology is used to convert crude oil, coal and natural gas into fuels and chemical raw materials, to produce various petrochemical and chemical products and to control CO, hydrocarbon and NO emissions (Centi, Quadrelli, and Perathoner 2013). For efficient catalysts, high-quality products made from low-cost, environmentally friendly raw materials are required. Active catalysts must increase the number of sites not only to expand their surface area but also to reduce the size of catalytic particles (Chaturvedi, Dave, and Shah 2012). The prospects for biological pathways for NM/NP synthesis therefore indicate that the two functions required for active catalysis are both low-cost and environmentally friendly. Most of the benchtop research and field applications for wastewater treatment materials are now concentrating their efforts on magnetic nanoscale materials (Westerhoff et al. 2016). Magnetic iron oxide NM/NP appears to have the ability to treat a large amount of wastewater and is suitable for magnetic separation, making it one of the most promising heavy metal treatment materials (Neyaz, Siddiqui, and Nair 2014). Photocatalysis is an acceleration of a photoreaction in the presence of a catalyst and is suitable for inorganic and organic photo degradations. First, in the photocatalysis process, visible light or ultraviolet radiation sensitizes a semiconductor catalyst, causing it to form electron and hole pairs (Navarro Yerga et al. 2009). If the incident light energy is equal to or greater than the band gap energy of the semiconductor, then the electrons in the valence band are elated and move up to the semiconductor conduction band, thus generating holes in the valence band (Umar and Aziz 2013). When these pores divide water molecules into hydrogen and hydroxyl radicals, organic pollutants undergo oxidative degradation (Gratzel 2001). Redox reactions are caused by the reaction of conductive band electrons with DO, which produces superoxide anions.

2.3.5.2 Fenton Reaction

A Fenton reaction is a progressed oxidation process (AOP) that has been advocated as a way to boost the efficiency of removing organic and inorganic pollutants from specific agents (Cuerda-Correa, Alexandre-Franco, and Fernández-González 2020). For different industries containing phenol in the wastewater (such as pharmaceutical industries and their wastewater treatment plants) and other organic pollutants, this response is one of the alternate methods for wastewater treatment. Since iron oxides are rich in content, easy to separate and low in cost, they are often used as heterogeneous catalysts in Fenton oxidation systems (Gadipelly et al. 2014). The heterogeneous Fenton reaction produces a hydroxyl radical (OH) and an ion of hydroxide (OH) when hydrogen peroxide (H2O2) oxidizes Fe2+ ion to $Fe3^{+.}$ Another H2O2 molecule further catalyses the dissolution of Fe3+ to form Fe2+ and generates a hydroperoxyl radical (OOH) and a proton (He et al. 2016). These free radicals are powerful oxidants associated with the oxidation process of organic pollutants. Li and Zhang (2016) noted that hematite NP, with the help of white rot fungi, promoted the degradation of bisphenol A caused by a Fenton-like reaction. The removal rate of nitrate in wastewater is evaluated by using green synthetic iron nanoparticles through a Fenton reaction (Wang et al. 2014). The Fenton reaction, along with the fluidized-bed Fenton process (FBF), has recently been thoroughly reviewed (Garcia-Segura et al. 2016). This is one of the most promising methods for biologically producing iron nanoparticles to reduce the burden on organic and inorganic environments (Tisa, Raman, and Daud 2014).

2.3.6 Contamination Characteristics and Potential Environmental Implications

2.3.6.1 Environmental Contamination

The environment is often contaminated or polluted. There are different forms of pollution, and pollutants may cause harmful effects on the environment. According to the literature, pollution is defined as the direct or indirect introduction of human-released substances or energy into the environment, resulting in harmful effects such as damage to life resources, human health, environmental activities and the environment, including the quality of use of the environment and the reduction of facilities. Environmental pollution is an unfavorable and undesirable change. It has an impact on the physical, chemical and biological properties of air, water and soil. These will, in turn, have an injurious effect on organisms. Pollution may originate from natural or foreign chemical or energy substances (Wong 2012).

2.3.6.2 Types of Pollutants

Environmental pollutants are still a global problem and a challenge that mankind has faced in recent decades. Various types of pollutants exist, such as inorganic and organic pollutants (Amin, Alazba, and Manzoor 2014).

2.3.6.3 Heavy Metals

Heavy metals are defined in the literature as natural elements with atomic weights (AWs) and densities five times higher than water (Bánfalvi 2011). Amongst all contaminants, substantial metals have received considerable attention due to their enormous toxicity. Heavy metals in natural water usually exist in the form of trace elements, but even at very low concentrations, they are highly toxic.

Even in small amounts, metals including arsenic, lead, cadmium (Cd), nickel (Ni), mercury (Hg), chromium (Cr), cobalt (Co), zinc (Zn) and selenium (Si) are very toxic. Currently, the rise in heavy metal content in our resources is a major issue, particularly because many industries discharge metal-containing waste that has not been treated (Förstner, Mader, and Salomons 1995). Heavy metals become poisonous materials when they are not metabolized by the human body and are stored in soft tissues. They often impact people in agriculture, and one study (Singh et al. 2011) indicates that the collective route for adult exposure is industrial exposure. Ingestion is the most common way that children come into contact. The environment and its resources have been polluted by natural and human activities, and the pollutants released exceed the environment's capacity (Herawati et al. 2000; He, Yang, and Stoffella 2005).

2.3.6.4 Sources of Heavy Metals

Heavy metals could be the result of both natural and man-made processes and may eventually enter various environmental zones (soil, water, air and their interfaces).

2.3.6.5 Natural Processes

A number of studies have identified natural sources of several heavy metals. Under certain environmental conditions, natural emissions of heavy metals will occur (Masindi and Muedi 2018). These emissions can be caused by volcanic eruptions, sea salt spray, forest fires, rock weathering, biological sources and wind-borne soil particles, which are just a few of the sources of these emissions. Metal may be released from its endemic areas into altered ecological sections because of natural weathering. Hydroxides, oxides, sulfides, sulfates, phosphates, silicates and organic compounds are all examples of heavy metals (Tchounwou et al. 2012). The most common heavy metals are the following: Pb, Ni, Cr, Cd, As, Hg, Zn and Cu. Even when trace amounts of these heavy metals can be identified, they may still cause serious health problems for humans and other mammals (Herawati et al. 2000).

2.3.6.6 Anthropogenic Processes

Pollution is also released into various settings as a consequence of agriculture, industrial activities, mining, metallurgical methods and runoff. For some metals, the amount of metal released by man-made heavy metal processes is higher than that released by natural processes. Naturally discharged metals in wind-blown dust mainly come from the industrial sector. Automobile exhaust, smelting, pesticide production and fossil fuel combustion also lead to the release of large amounts of Pb, As,

Cu, Zn, Ni, vanadium, Hg, Si and tin, which ultimately increase the heavy metal pollution in the environment. Therefore, human activities amplify environmental pollution due to daily production that meets the needs of the world's population (He, Yang, and Stoffella 2005).

2.3.6.7 Categories of Nanomaterials

NPs are mainly divided into various categories according to their shape, scope and chemical properties (Khan, Saeed, and Khan 2019). According to these physical and chemical characteristics, five categories can be distinguished, namely, carbon-based nanoparticles, metal NPs, polymer NPs, semiconductor NPs and lipid-based NPs.

2.3.7 Carbon-Based Nanoparticles

Carbon-based nanoparticles are made entirely of carbon. Since their discovery, they have attracted the scientific community (Zhang et al. 2013). They are formed by a cloud of weakly delocalized electrons along the z-axis and a layer of sp2-bonded carbon atoms in which each atom is connected to the other three carbon atoms in the x–y plane (Scida et al. 2011). This configuration has many advantages, such as high electrical conductivity, π-plasmon resonance and the capacity to form charging-transfer complexes in contact with the donor groups of electrons (Barnes et al. 2007; Valcárcel et al. 2008). This configuration also allows the occurrence of robust van der Waals forces obstructing in a significant way the dispersion and solubility of this group of nanoparticles (Scida et al. 2011). Nanoparticles based on carbon are formed of fullerenes, CNTs, graphene and its derivatives, graphene oxide, nanodiamonds and carbon-based quantum dots (Patel, Singh, and Kim 2019).

2.3.7.1 Fullerenes

Fullerenes are synthesized by evaporating, in an inert helium atmosphere, graphite electrodes (Stevenson et al. 1999). They can be synthesized in different ways, such as discharge, electron beam ablation and sputtering (Smalley 1997; Ross et al. 2009; Churilov et al. 2013). However, the high synthesis cost and low yield of existing production technologies considerably limit the use of these carbon-based nanoparticles. Fullerenes have high electrochemical stability due to their special characteristics, small size, specific shape and neat structure, so they can be used in energy conversion

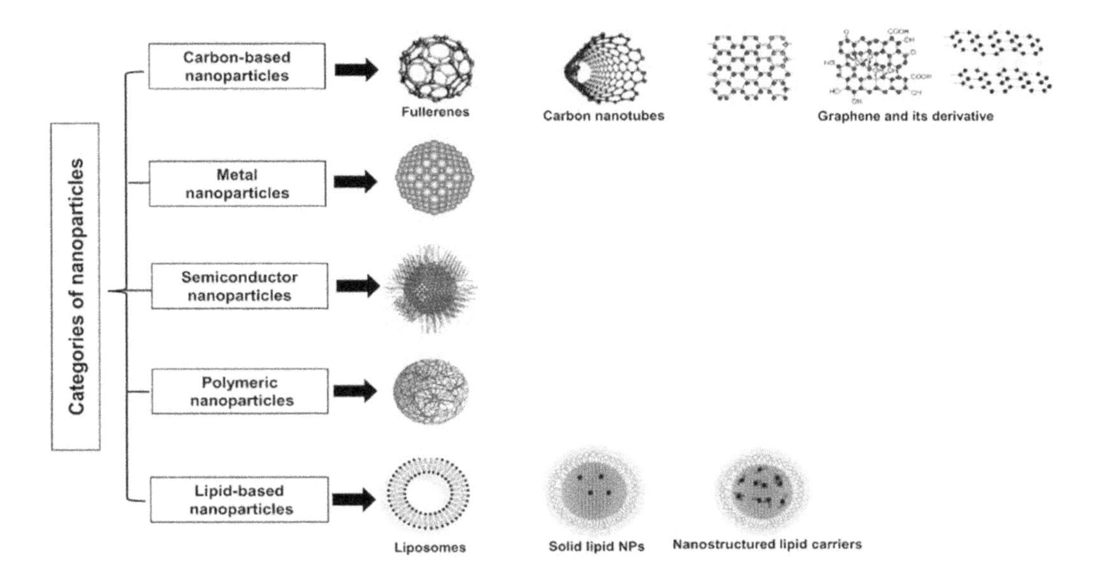

FIGURE 2.3 Categories of nanoparticles according to their shape and chemical properties.

systems (Kamat 2006). This finding marked the start of the era of carbon NMs, which are also called synthetic carbon allotropes (SCAs). Fullerenes have received significant consideration since their discovery due to their unique photophysical and photochemical properties. They are a 0D form of graphitic carbon and resemble an irregular sheet of graphenium curled up into a sphere by adding pentagons to its structure. They can be presented in different shapes and sizes ranging from 30 to 300 carbon atoms. The majority of fullerenes (e.g., C60) have a spheroidal shape, but there are also some with oblong shapes similar to a rugby ball (e.g., C70). The smallest stable and most prominent fullerene is the Buckminsterfullerene (C60) that is spheroid in shape and formed by 60 structurally equivalent sp2 hybridized carbon atoms with 12 pentagons and 20 hexagons or 60 C atoms consisting of 12-five member rings and 20-six member rings (Kumar and Kumbhat 2016; Kour et al. 2020). With 0.7 nm in diameter, the C sites are equivalent, and the double and single bond lengths are 0.14 nm and 0.146 nm, respectively.

2.3.7.2 Carbon Nanotubes

Carbon nanotube (CNT) is derived from carbon fiber and fullerene. The molecule has 60 carbon atoms and is cylindrical in shape (Polizu et al. 2006). CNTs can be classified into two separate categories depending on the number of carbon layers present in these nanostructures (Anzar et al. 2020). We distinguish single wall (SW) CNTs, which comprise graphene layers with a diameter between 0.4 and 2 nm (usually in the form of hexagonal bundles), from multi-walled CNTs containing two or more CNTs. The composition of CNTs (MWCNT) includes many cylinders, and each cylinder is composed of graphene sheets. Their diameter is between 1 and 3 nm (He et al. 2013).

The arc discharge method, laser ablation method and chemical vapour deposition method are three different ways to produce CNTs (Anzar et al. 2020). CNTs display a wide range of chemical and physical properties, mainly a high tensile strength, ultra-light weight, special electronic structures and high chemical and thermal stability. Due to these outstanding properties, carbon NMs are the most widely applied nanostructures and are utilized for a number of applications, including their use as biomolecules, drugs and drug supply to target organs, biosensor diagnostics and analysis (Che, Cagin, and Goddard III 2000).

2.3.7.3 Graphene and Its Derivatives

Graphene discovery was the result of the work of Andre Geim and Konstantin Novoselov in 2004 (Lee et al. 2019). Graphene has drawn the scientific community's attention since its finding, prompting extensive research into the NM's potential applications (Novoselov et al. 2004). A single-atom-thick flat sheet of sp2—bonded carbon atoms perfectly arranged in a honeycomb lattice—makes up grapheme (Zhu et al. 2016; Paul et al. 2019). The carbon atoms forming this two-dimensional structure are bonded at a length of 0.142 nm (Paul et al. 2019). A graphene sheet's thickness is approximately 0.34 nm, and its size can range from a few nanometers to several centimetres depending on the synthesis method (Mahmoudi, Wang, and Hahn 2018). In addition to its particular structure, Each carbon atom is covalently bound together in the same plane in graphene, which is made of pure carbon, and the monolayer graphene sheets are bound together by van der Waals force (Lee et al. 2019).

Qualified as "supermaterial" or "miracle material," graphene and its derivatives are studied in various fields due to the presence of an aromatic ring, free π-electron and functional reactive groups (Paul et al. 2019). It can be achieved using three unique techniques, namely, exfoliation, chemical vapour deposition and chemical-based techniques (Tahriri et al. 2019). Graphene is widely known for its high electrical conductivity, light weight (0.77 mg/m2), exceptional properties and high opacity (only 2.3% of the white light is absorbed by graphene) (Nair et al. 2008; Chauhan, Yadav, and Sehrawat 2020). These great properties offer a large number of applications in mechanical engineering, electrical engineering and micro-electronics (Bharech and Kumar 2015).

2.3.8 Metal Nanoparticles

Metal nanoparticles are engineering NMs that have been generally used in the production of commercial, industrial and environmental controls (Zhang et al. 2018). Due to their smaller size and higher specific surface area, Fe, Co and Ni metal nanoparticles have rare magnetic, optical, electrochemical and chemical catalytic properties, which have piqued the interest of the scientific community (Khalid et al. 2017). Sunscreens, cosmetics and toothpaste are just a few examples of customer products, and in industrial catalysis, titanium dioxide nanoparticles (nTiO2) are added (Zhang et al. 2018). More than 400 products contain silver nanoparticles (Ag NPs), including detergents, food packaging, cosmetics and textiles, while ceria nanoparticles (nCeO2) are part of the composition of diesel fuel additives and polishing agents (Ma et al. 2015; Vance et al. 2015).

Metal nanoparticles have long been used to remove and repair pollutants (organic pollutants, heavy metals and radionuclides) and pathogens in the environment. They can absorb toxic substances and convert them into non-toxic substances. They ensure the degradation and absorption of pollutants through different mechanisms (including redox reactions, ion exchange, surface complexation, surface processes, adsorption and electrostatic interactions), (Trujillo-Reyes, Peralta-Videa, and Gardea-Torresdey 2014). Iron NPs are said to be the first nanoparticles to be used for environmental cleanup (Chauhan, Yadav, and Sehrawat 2020). Nanoscale zero-valent iron (NZVI) has also been used to remove As (III), immobilize Cr (VI) and Pb (II) from aqueous solution and reduce chromium to Cr (III) and Pb to Pb (0) (Ponder, Darab, and Mallouk 2000; Kanel et al. 2005). Organic contaminants such as atrazine and molinate can be suppressed with nano-sized zero-valent ions (Chauhan, Yadav, and Sehrawat 2020). Zero-valent iron particles immobilising in the silica microsphere can also be used to remediate soils from polybrominated diphenyl ethers, which are a class of environmental contaminants (Xie et al. 2016). This extraordinary ability to neutralize hazardous environmental contamination is partially attributed to their potent reductant nature (Kanel et al. 2012).

2.3.9 Semiconductor Nanoparticles

Semiconductor nanoparticles are crystalline clusters of a few hundred to a few thousand atoms (Chen and Dobson 2012). They are made from a wide range of different compounds (Suresh 2013). Their size is typically comprised between 1–100 nm (Smith and Nie 2010). They are classified into three categories of nanocrystal semiconductor, specifically, II-VI, III-V and IV-VI, based on the periodic table groups forming these elements. For instance, ZnO, zinc sulfide and cadmium sulfide are II-VI semiconductors, and group IV-VI materials include silicon and germanium (Sahu 2019). Semiconductor nanoparticles show in most cases the behaviour of quantum containment in a size ranged between 1 and 20 nm (Suresh 2013). Semiconductor nanoparticles display attractive physical and chemical properties that include high chemical and photo bleaching stability, surface functionality, constant absorption bands and a slight and intensive emission spectra (Sahu 2019). Because of these outstanding properties, they have become the research focus for the last two decades, and several applications of these nanoparticles are used in a variety of disciplines, mainly solid state physics, chemistry, material sciences, biological sciences, medical sciences, and the engineering and interdisciplinary fields (Smith and Nie 2010).

2.3.10 Polymeric Nanoparticles

Polymeric nanoparticles (PNPs) are solid or colloidal particles with a tiny size ranging from 10 to 1000 nanometers (Nasir, Kausar, and Younus 2015). The term polymeric nanoparticle is attributed to any type of polymer nanoparticles, especially to nanocapsules and nanospheres (Rao and Geckeler 2011). Nanocapsules are defined as vesicular systems that act as a reservoir. They consist of a cavity containing a liquid core in which substances can be trapped in the shell of a solid material (El-Say and El-Sawy 2017). In contrast, nanospheres are the particles of a spherical matrix whole mass is solid. As a result, the molecules can be adsorbed on the surface of the sphere or encapsulated inside the particle (Vauthier and Couvreur 2000).

The main component of polymeric nanoparticles is the polymer, which can be natural or synthetic depending on the polymer source of origin. Many materials need to be reassembled to create a polymeric nanoparticle (El-Say and El-Sawy 2017). Several methods can be applied to produce polymeric nanoparticles, depending on the requirements of their application (Nasir, Kausar, and Younus 2015). The different preparation methods that have been developed can be divided into two separate group: those that rely on monomer polymerization and those using predetermined polymers (Allouche 2013). Evaporation of solvents, salting-out, supercritical fluid technology and dialysis procedures are the main techniques used for polymeric nanoparticle formulation from pre-existing polymers (Rao and Geckeler 2011).

2.3.11 Lipid-Based Nanoparticles

Lipid-based nanoparticles are colloidal carriers with a lipid matrix that is solid at body temperature (Shah et al. 2015). Lipid nanoparticles (LNPs) have piqued scholarly interest in recent decades due to their unique and tremendous properties that allow them to be used in different domains (Mutyam Pallerla and Prabhakar 2013). LNPs can be divided into three categories according to their structure and composition: liposomes, solid lipids and nanostructured lipids (Shah et al. 2015).

2.3.11.1 Liposomes

Liposomes are circular vesicles covering the internal aqueous cavity of the lipid bilayer membrane (Naseri, Valizadeh, and Zakeri-Milani 2015). The diameter of the liposomes is between 400 nm and 2.5 µm (Malam, Loizidou, and Seifalian 2009). Both synthetic and natural phospholipids can form the liposome bilayer. Therefore, these nanoparticles are mainly defined by their physical and chemical properties and the properties of bilayer phospholipids (Bawarski et al. 2008). Liposomes can be synthesized using a wide range of procedures such as the extrusion technique, probe or bath sonication and microfluidization (Panahi et al. 2017).

Since their discovery in 1965, they are considered the basic models of lipid-based formulations widely used for bioactive component encapsulation and drug delivery (Mukherjee, Ray, and Thakur 2009). They offer encouraging advantages as a pharmaceutical carrier by ensuring good protection of drugs against enzyme degradation and having less toxicity and great biocompatibility, biodegradability and non-immunogenicity (Kumar et al. 2010). Liposomes are widely used as effective DNA supply systems, siRNA and antisense oligonucleotides (asODN) (Jayaraman et al. 2012; Kawabata et al. 2012). In addition to their use as a drug and DNA carrier, liposomes are also employed in the food industry to prevent the degradation of key enzymes, sugars and vitamins for great nutritional value (Panahi et al. 2017). They have great properties, but the use of liposomes can present some limitations due to their low encapsulation efficiency and low storage stability (Bamrungsap et al. 2012).

2.3.11.2 Solid Lipid Nanoparticles (SLNs)

Nanoparticles of solid lipids, also referred to as lipospheres, are colloidal particles between 50 and 1000 nm in size (Malam, Loizidou, and Seifalian 2009). They are formed from a range of physiological biodegradable lipids that remain constantly in a solid state despite temperature variation (García-Pinel et al. 2019). The synthesis of SLNs is accomplished by replacing the liquid lipid (oil) in an oil-in-water emulsion with a solid lipid, such as mono-, di- and triglycerides, fatty acids, waxes and combinations thereof (Malam, Loizidou, and Seifalian 2009). Several substances, such as phospholipids, Poloxamers and Polysorbates, are added to the SLN formulation to provide the steric stabilisation of nanoparticles (Geszke-Moritz and Moritz 2016). The majority of SLNs can be obtained by high shear, hot or cold homogenization, by ultrasonication or by solvent emulsification/evaporation (Mukherjee, Ray, and Thakur 2009).

SLNs emerged in the early 1990s to overcome the weakness of other conventional lipid carriers by providing biocompatibility and storage stability and avoiding drug degradation (Geszke-Moritz and Moritz 2016). They display attractive features such as a small size, giant surface zone, low toxicity and greater cellular uptake than other carriers, which justify their use in the pharmaceutical and biomedical fields as carrier systems, thus substituting for liposomes (Yadav, Khatak, and Sara 2013; Lingayat, Zarekar, and Shendge 2017). Despite SLNs' great advantages, they show many weaknesses that can considerably limit their use. The limitations include that they present low space for drug encapsulation, which ultimately leads to poor loading capacity, unwanted interaction among the different components and a moderate elevation in the water content of the dispersions (Mishra et al. 2018).

2.3.11.3 Nanostructured Lipid Carriers (NLC)

Nanostructured lipids are modified SLNs in which the lipid phase is established by a mixture of solid and liquid phase lipids at ambient temperature (Tamjidi et al. 2013). They can be synthesized utilizing the same procedures used for the formation of SLNs (Yoon, Park, and Yoon 2013). Three forms of structure can be distinguished. Solid and liquid lipids are mixed in the imperfect type (Naseri, Valizadeh, and Zakeri-Milani 2015). The second form matches the non-crystalline matrix, that is, it is a formless type with no crystalline structure (Jaiswal, Gidwani, and Vyas 2016). The third or multiple type is identical to w/o/w emulsions and is characterized by an enormous drug solubility in liquid lipids compared with solids (Müller, Staufenbiel, and Keck 2014).

NLCs are the next generation of SLNs that were introduced at the end of the 1990s to overcome the drawbacks of SLNs (Müller, Staufenbiel, and Keck 2014). Indeed, NLCs have a higher loading capacity and a lower water content in the dispersion; more importantly, with NLCs, drug expulsion is minimized compared with SLNs (Naseri, Valizadeh, and Zakeri-Milani 2015).

2.4 CONCLUSION

To understand the full implications of NMs in eco-friendly applications, several challenges must be overcome. However, nanotechnology proposes many approaches that can be used to solve environmental pollution. Although the detailed mechanism of the interaction among NMs' interesting characteristics to remediate pollution and soil components are outside the scope of this work and should be clarified in future revisions, here, we determine the possibility of Me/MeONPs as an addition or additional nanoremediation technology that is directed at heavy metal adsorbents. Based on the advantages and limits of each material, we provide here an overall view of some NMs that have been used in environmental remediation.

REFERENCES

Abbas, Aamir, Adnan M Al-Amer, Tahar Laoui, Mohammed J Al-Marri, Mustafa S Nasser, Majeda Khraisheh, and Muataz Ali Atieh. 2016. "Heavy metal removal from aqueous solution by advanced carbon nanotubes: critical review of adsorption applications." *Separation and Purification Technology* 157:141–161.

Allouche, Joachim. 2013. "Synthesis of organic and bioorganic nanoparticles: an overview of the preparation methods." In *Nanomaterials: A Danger or a Promise?*, 27–74. Springer.

Amin, MT, AA Alazba, and Umair Manzoor. 2014. "A review of removal of pollutants from water/wastewater using different types of nanomaterials." *Advances in Materials Science and Engineering* 2014.

Ando, Tsuneya, Yasuhiko Arakawa, Kazuhito Furuya, Susumu Komiyama, and Hisao Nakashima. 2012. *Mesoscopic Physics and Electronics*. Springer Science & Business Media.

Anzar, Nigar, Rahil Hasan, Manshi Tyagi, Neelam Yadav, and Jagriti Narang. 2020. "Carbon nanotube: a review on synthesis, properties and plethora of applications in the field of biomedical science." *Sensors International* 1:100003.

Ball, Philip. 1996. *Designing the Molecular World: Chemistry at the Frontier*. Vol. 117. Princeton University Press.

Bamrungsap, Suwussa, Zilong Zhao, Tao Chen, Lin Wang, Chunmei Li, Ting Fu, and Weihong Tan. 2012. "Nanotechnology in therapeutics: a focus on nanoparticles as a drug delivery system." *Nanomedicine* 7 (8):1253–1271.

Bánfalvi, Gáspár. 2011. *Cellular Effects of Heavy Metals*. Springer.

Baragaño, Diego, Rubén Forján, Lorena Welte, and José Luis R Gallego. 2020. "Nanoremediation of as and metals polluted soils by means of graphene oxide nanoparticles." *Scientific Reports* 10 (1):1–10.

Barnes, TM, J Van de Lagemaat, D Levi, Gary Rumbles, TJ Coutts, CL Weeks, DA Britz, I Levitsky, J Peltola, and P Glatkowski. 2007. "Optical characterization of highly conductive single-wall carbon-nanotube transparent electrodes." *Physical Review B* 75 (23):235410.

Bawarski, Willie E, Elena Chidlowsky, Dhruba J Bharali, and Shaker A Mousa. 2008. "Emerging nanopharmaceuticals." *Nanomedicine: Nanotechnology, Biology and Medicine* 4 (4):273–282.

Bharech, Somnath, and Richa Kumar. 2015. "A review on the properties and applications of graphene." *Journal of Materials Science and Mechanical Engineering* 2 (10):70.

Bhushan, Bharat. 2017. *Springer Handbook of Nanotechnology*. Springer.

Buzea, Cristina, Ivan I Pacheco, and Kevin Robbie. 2007. "Nanomaterials and nanoparticles: sources and toxicity." *Biointerphases* 2 (4):MR17–MR71.

Centi, Gabriele, P Ciambelli, S Perathoner, and P Russo. 2002. "Environmental catalysis: trends and outlook." *Catalysis Today* 75 (1–4):3–15.

Centi, Gabriele, Elsje Alessandra Quadrelli, and Siglinda Perathoner. 2013. "Catalysis for CO 2 conversion: a key technology for rapid introduction of renewable energy in the value chain of chemical industries." *Energy & Environmental Science* 6 (6):1711–1731.

Chakraborty, Jaya, Sagarika Mallick, Ritu Raj, and Surajit Das. 2018. "Functionalization of extracellular polymers of Pseudomonas aeruginosa N6P6 for synthesis of CdS nanoparticles and cadmium bioadsorption." *Journal of Polymers and the Environment* 26 (7):3097–3108.

Chang, Chun, Fei Lian, and Lingyan Zhu. 2011. "Simultaneous adsorption and degradation of γ-HCH by nZVI/Cu bimetallic nanoparticles with activated carbon support." *Environmental Pollution* 159 (10):2507–2514.

Chaturvedi, Shalini, Pragnesh N Dave, and NK Shah. 2012. "Applications of nano-catalyst in new era." *Journal of Saudi Chemical Society* 16 (3):307–325.

Chauhan, R, Hari OS Yadav, and N Sehrawat. 2020. "Nanobioremediation: a new and a versatile tool for sustainable environmental clean up-Overview." *Journal of Materials and Environmental Sciences* 11 (4):564–573.

Che, Jianwei, Tahir Cagin, and William A Goddard III. 2000. "Thermal conductivity of carbon nanotubes." *Nanotechnology* 11 (2):65.

Chen, Xianfeng, and Peter J Dobson. 2012. "Synthesis of semiconductor nanoparticles." In *Nanoparticles in Biology and Medicine*, 103–123. Springer.

Cho, Eunae, Muhammad Nazir Tahir, Jae Min Choi, Hwanhee Kim, Jae-Hyuk Yu, and Seunho Jung. 2015. "Novel magnetic nanoparticles coated by benzene-and β-cyclodextrin-bearing dextran, and the sorption of polycyclic aromatic hydrocarbon." *Carbohydrate Polymers* 133:221–228.

Churilov, Grigory N, Wolfgang Krätschmer, Irina V Osipova, Gariy A Glushenko, Natalia G Vnukova, Andrey L Kolonenko, and Aleksander I Dudnik. 2013. "Synthesis of fullerenes in a high-frequency arc plasma under elevated helium pressure." *Carbon* 62:389–392.

Cuerda-Correa, Eduardo Manuel, María F Alexandre-Franco, and Carmen Fernández-González. 2020. "Advanced oxidation processes for the removal of antibiotics from water. An overview." *Water* 12 (1):102.

Cundy, Andrew B, Laurence Hopkinson, and Raymond LD Whitby. 2008. "Use of iron-based technologies in contaminated land and groundwater remediation: a review." *Science of the Total Environment* 400 (1–3):42–51.

Das, Surajit, Jaya Chakraborty, Shreosi Chatterjee, and Himanshu Kumar. 2018. "Prospects of biosynthesized nanomaterials for the remediation of organic and inorganic environmental contaminants." *Environmental Science: Nano* 5 (12):2784–2808.

Drexler, KE, and M Minsky. 1990. *Engines of Creation*. Fourth Estate, 171–190.

Du Plessis, Danielle Marguerite. 2011. *Fabrication and Characterization of Anti-Microbial and Biofouling Resistant Nanofibers with Silver Nanoparticles and Immobilized Enzymes for Application in Water Filtration*. University of Stellenbosch.

El-Say, Khalid M, and Hossam S El-Sawy. 2017. "Polymeric nanoparticles: promising platform for drug delivery." *International Journal of Pharmaceutics* 528 (1–2):675–691.

Förstner, Ulrich, Pavel Mader, and Willem Salomons. 1995. *Heavy Metals: Problems and Solutions*. Springer.

Gadipelly, Chandrakanth, Antía Pérez-González, Ganapati D Yadav, Inmaculada Ortiz, Raquel Ibáñez, Virendra K Rathod, and Kumudini V Marathe. 2014. "Pharmaceutical industry wastewater: review of the technologies for water treatment and reuse." *Industrial & Engineering Chemistry Research* 53 (29):11571–11592.

García-Pinel, Beatriz, Cristina Porras-Alcalá, Alicia Ortega-Rodríguez, Francisco Sarabia, Jose Prados, Consolación Melguizo, and Juan M López-Romero. 2019. "Lipid-based nanoparticles: application and recent advances in cancer treatment." *Nanomaterials* 9 (4):638.

Garcia-Segura, Sergi, Luzvisminda M Bellotindos, Yao-Hui Huang, Enric Brillas, and Ming-Chun Lu. 2016. "Fluidized-bed Fenton process as alternative wastewater treatment technology—A review." *Journal of the Taiwan Institute of Chemical Engineers* 67:211–225.

Gehrke, I, A Geiser, and A Somborn-Schulz. 2015. "Innovations in nanotechnology for water treatment." *Nanotechnology, Science and Applications* 8:1–17.

Geszke-Moritz, Małgorzata, and Michał Moritz. 2016. "Solid lipid nanoparticles as attractive drug vehicles: composition, properties and therapeutic strategies." *Materials Science and Engineering: C* 68:982–994.

Ghasemzadeh, Gholamreza, Mahdiye Momenpour, Fakhriye Omidi, Mohammad R Hosseini, Monireh Ahani, and Abolfazl Barzegari. 2014. "Applications of nanomaterials in water treatment and environmental remediation." *Frontiers of Environmental Science & Engineering* 8 (4):471–482.

Gratzel, Michael. 2001. "Photoelectrochemical cells." *Nature* 414 (6861):338–345.

Guerra, Fernanda D, Mohamed F Attia, Daniel C Whitehead, and Frank Alexis. 2018. "Nanotechnology for environmental remediation: materials and applications." *Molecules* 23 (7):1760.

He, Hua, Lien Ai Pham-Huy, Pierre Dramou, Deli Xiao, Pengli Zuo, and Chuong Pham-Huy. 2013. "Carbon nanotubes: applications in pharmacy and medicine." *BioMed Research International* 2013.

He, Jie, Xiaofang Yang, Bin Men, and Dongsheng Wang. 2016. "Interfacial mechanisms of heterogeneous Fenton reactions catalyzed by iron-based materials: a review." *Journal of Environmental Sciences* 39:97–109.

He, Zhenli L, Xiaoe E Yang, and Peter J Stoffella. 2005. "Trace elements in agroecosystems and impacts on the environment." *Journal of Trace Elements in Medicine and Biology* 19 (2–3):125–140.

Herawati, Netti, S Suzuki, K Hayashi, IF Rivai, and H Koyama. 2000. "Cadmium, copper, and zinc levels in rice and soil of Japan, Indonesia, and China by soil type." *Bulletin of Environmental Contamination and Toxicology* 64 (1):33–39.

Hu, Haibo, Zhenghua Wang, and Ling Pan. 2010. "Synthesis of monodisperse Fe3O4@ silica core—shell microspheres and their application for removal of heavy metal ions from water." *Journal of Alloys and Compounds* 492 (1–2):656–661.

Ibrahim, Rusul Khaleel, Maan Hayyan, Mohammed Abdulhakim AlSaadi, Adeeb Hayyan, and Shaliza Ibrahim. 2016. "Environmental application of nanotechnology: air, soil, and water." *Environmental Science and Pollution Research* 23 (14):13754–13788.

Jaiswal, Piyush, Bina Gidwani, and Amber Vyas. 2016. "Nanostructured lipid carriers and their current application in targeted drug delivery." *Artificial Cells, Nanomedicine, and Biotechnology* 44 (1):27–40.

Jayaraman, Muthusamy, Steven M Ansell, Barbara L Mui, Ying K Tam, Jianxin Chen, Xinyao Du, David Butler, Laxman Eltepu, Shigeo Matsuda, and Jayaprakash K Narayanannair. 2012. "Maximizing the potency of siRNA lipid nanoparticles for hepatic gene silencing in vivo." *Angewandte Chemie* 124 (34):8657–8661.

Jeevanandam, Jaison, Yen San Chan, and Michael K Danquah. 2016. "Biosynthesis of metal and metal oxide nanoparticles." *ChemBioEng Reviews* 3 (2):55–67.

Jerin, VM, R Remya, Mariyam Thomas, and Jaya T Varkey. 2019. "Investigation on the removal of toxic chromium ion from waste water using Fe2O3 nanoparticles." *Materials Today: Proceedings* 9:27–31.

Jiamjitrpanich, Waraporn, Preeda Parkpian, Chongrak Polprasert, and Rachain Kosanlavit. 2012. "Enhanced phytoremediation efficiency of TNT-contaminated soil by nanoscale zero valent iron." *2nd International Conference on Environment and Industrial Innovation IPCBEE*. Available from: https://www.researchgate.net/publication/229090834_Enhanced_Phytoremediation_Efficiency_of_TNT-Contaminated_Soil_by_Nanoscale_Zero_Valent_Iron#:~:text=Overall%2C%20the%20highest%20removal%20efficiency%20of%20nano-phytoremediation%20was,with%20the%20complete%20TNT%20remediation%20by%20day%2060

Kamat, Prashant. 2006. "Carbon nanomaterials: building blocks in energy conversion devices." *Interface* 15 (1):45–47.

Kanel, Sushil Raj, Bruce Manning, Laurent Charlet, and Heechul Choi. 2005. "Removal of arsenic (III) from groundwater by nanoscale zero-valent iron." *Environmental Science & Technology* 39 (5):1291–1298.

Kanel, Sushil R, Chunming Su, Upendra Patel, and Abinash Agrawal. 2012. "Use of metal nanoparticles in environmental cleanup." *Science and Applications*:271.

Karn, Barbara, Todd Kuiken, and Martha Otto. 2009. "Nanotechnology and in situ remediation: a review of the benefits and potential risks." *Environmental Health Perspectives* 117 (12):1813–1831.

Kavitha, Thangavelu, Sajjad Haider, Tahseen Kamal, and Mazhar Ul-Islam. 2017. "Thermal decomposition of metal complex precursor as route to the synthesis of Co3O4 nanoparticles: antibacterial activity and mechanism." *Journal of Alloys and Compounds* 704:296–302.

Kawabata, Atsushi, Abdulgader Baoum, Naomi Ohta, Stephanie Jacquez, Gwi-Moon Seo, Cory Berkland, and Masaaki Tamura. 2012. "Intratracheal administration of a nanoparticle-based therapy with the angiotensin II type 2 receptor gene attenuates lung cancer growth." *Cancer Research* 72 (8):2057–2067.

Khalid, Sana, Muhammad Shahid, Nabeel Khan Niazi, Behzad Murtaza, Irshad Bibi, and Camille Dumat. 2017. "A comparison of technologies for remediation of heavy metal contaminated soils." *Journal of Geochemical Exploration* 182:247–268.

Khan, Ibrahim, Khalid Saeed, and Idrees Khan. 2019. "Nanoparticles: properties, applications and toxicities." *Arabian Journal of Chemistry* 12 (7):908–931.

Kharisov, Boris I, HV Rasika Dias, Oxana V Kharissova, Victor Manuel Jiménez-Pérez, Betsabee Olvera Perez, and Blanca Muñoz Flores. 2012. "Iron-containing nanomaterials: synthesis, properties, and environmental applications." *RSC Advances* 2 (25):9325–9358.

Khin, Mya Mya, A Sreekumaran Nair, V Jagadeesh Babu, Rajendiran Murugan, and Seeram Ramakrishna. 2012. "A review on nanomaterials for environmental remediation." *Energy & Environmental Science* 5 (8):8075–8109.

Khine, Yee Yee, and Martina H Stenzel. 2020. "Surface modified cellulose nanomaterials: a source of non-spherical nanoparticles for drug delivery." *Materials Horizons* 7 (7):1727–1758.

Kour, Ravinder, Sandeep Arya, Sheng-Joue Young, Vinay Gupta, Pankaj Bandhoria, and Ajit Khosla. 2020. "Recent advances in carbon nanomaterials as electrochemical biosensors." *Journal of the Electrochemical Society* 167 (3):037555.

Krantzberg, Gail, Aysegul Tanik, José Simão Antunes do Carmo, Antonius Indarto, Alpaslan Ekdal, Melike Gurel, Elif pehlivanoglu Mantas, Zongping Wang, Guanyun Wang, and ChunLu Zhao. 2010. *Advances in Water Quality Control*. Scientific Research Publishing, Inc. USA.

Kumar, Ajay, Shital Badde, Ravindra Kamble, and Varsha B Pokharkar. 2010. "Development and characterization of liposomal drug delivery system for nimesulide." *International Journal of Pharmacy and Pharmaceutical Sciences* 2 (4):87–89.

Kumar, Narendra, and Sunita Kumbhat. 2016. *Essentials in Nanoscience and Nanotechnology*. Vol. 486. Wiley Online Library.

Lee, Haesung A, Eunsook Park, and Haeshin Lee. 2020. "Polydopamine and its derivative surface chemistry in material science: a focused review for studies at KAIST." *Advanced Materials* 32 (35):1907505.

Lee, Xin Jiat, Billie Yan Zhang Hiew, Kar Chiew Lai, Lai Yee Lee, Suyin Gan, Suchithra Thangalazhy-Gopakumar, and Sean Rigby. 2019. "Review on graphene and its derivatives: synthesis methods and potential industrial implementation." *Journal of the Taiwan Institute of Chemical Engineers* 98:163–180.

Li, Min, and Robbyn K Anand. 2016. "Recent advancements in ion concentration polarization." *Analyst* 141 (12):3496–3510.

Li, Mingzhu, and Chengdong Zhang. 2016. "γ-Fe2O3 nanoparticle-facilitated bisphenol a degradation by white rot fungus." *Science Bulletin* 61 (6):468–472.

Li, Qilin, Shaily Mahendra, Delina Y Lyon, Lena Brunet, Michael V Liga, Dong Li, and Pedro JJ Alvarez. 2008. "Antimicrobial nanomaterials for water disinfection and microbial control: potential applications and implications." *Water Research* 42 (18):4591–4602.

Li, X, H Xu, ZS Chen, and G Chen. 2011. "Biosynthesis of nanoparticles by microorganisms and their applications." *Journal of Nanomaterials* 2011:1–16.

Liang, Wen, Chaomeng Dai, Xuefei Zhou, and Yalei Zhang. 2014. "Application of zero-valent iron nanoparticles for the removal of aqueous zinc ions under various experimental conditions." *PLoS ONE* 9 (1):e85686.

Lingayat, Vishal J, Nilesh S Zarekar, and Rajan S Shendge. 2017. "Solid lipid nanoparticles: a review." *Nanoscience and Nanotechnology Research* 2:67–72.

Litter, Marta I, José Luis Cortina, António MA Fiúza, Aurora Futuro, and T Christos. 2014. "In-situ technologies for groundwater treatment: the case of arsenic." In *In-Situ Remediation of Arsenic-Contaminated Sites*, 1–33. Taylor & Francis Group.

Liu, Yanyan, Gaoxing Su, Bin Zhang, Guibin Jiang, and Bing Yan. 2011. "Nanoparticle-based strategies for detection and remediation of environmental pollutants." *Analyst* 136 (5):872–877.

Long, Richard Q, and Ralph T Yang. 2001. "Carbon nanotubes as superior sorbent for dioxin removal." *Journal of the American Chemical Society* 123 (9):2058–2059.

Lowry, Gregory V. 2007. *Nanomaterials for Groundwater Remediation*. McGraw-Hill.

Lunge, Sneha, Shripal Singh, and Amalendu Sinha. 2014. "Magnetic iron oxide (Fe3O4) nanoparticles from tea waste for arsenic removal." *Journal of Magnetism and Magnetic Materials* 356:21–31.

Luo, Fang, Zuliang Chen, Mallavarapu Megharaj, and Ravendra Naidu. 2016. "Simultaneous removal of trichloroethylene and hexavalent chromium by green synthesized agarose-Fe nanoparticles hydrogel." *Chemical Engineering Journal* 294:290–297.

Ma, Yuhui, Peng Zhang, Zhiyong Zhang, Xiao He, Junzhe Zhang, Yayun Ding, Jing Zhang, Lirong Zheng, Zhi Guo, and Lijuan Zhang. 2015. "Where does the transformation of precipitated ceria nanoparticles in hydroponic plants take place?" *Environmental Science & Technology* 49 (17):10667–10674.

Macoubrie, Jane. 2006. "Nanotechnology: public concerns, reasoning and trust in government." *Public Understanding of Science* 15 (2):221–241.

Madhura, Lavanya, Shalini Singh, Suvardhan Kanchi, Myalowenkosi Sabela, and Krishna Bisetty. 2019. "Nanotechnology-based water quality management for wastewater treatment." *Environmental Chemistry Letters* 17 (1):65–121.

Mahmoudi, Tahmineh, Yousheng Wang, and Yoon-Bong Hahn. 2018. "Graphene and its derivatives for solar cells application." *Nano Energy* 47:51–65.

Malam, Yogeshkumar, Marilena Loizidou, and Alexander M Seifalian. 2009. "Liposomes and nanoparticles: nanosized vehicles for drug delivery in cancer." *Trends in Pharmacological Sciences* 30 (11):592–599.

Malwal, Deepika, and P Gopinath. 2016. "Fabrication and applications of ceramic nanofibers in water remediation: a review." *Critical Reviews in Environmental Science and Technology* 46 (5):500–534.

Manzano, Miguel, and María Vallet-Regí. 2020. "Mesoporous silica nanoparticles for drug delivery." *Advanced Functional Materials* 30 (2):1902634.

Martínez-Cabanas, María, Marta López-García, José L Barriada, Roberto Herrero, and Manuel E Sastre de Vicente. 2016. "Green synthesis of iron oxide nanoparticles. Development of magnetic hybrid materials for efficient As (V) removal." *Chemical Engineering Journal* 301:83–91.

Masindi, Vhahangwele, and Khathutshelo L Muedi. 2018. "Environmental contamination by heavy metals." *Heavy Metals* 10:115–132.

Matussin, Shaidatul, Mohammad Hilni Harunsani, Ai Ling Tan, and Mohammad Mansoob Khan. 2020. "Plant-extract-mediated SnO2 nanoparticles: synthesis and applications." *ACS Sustainable Chemistry & Engineering* 8 (8):3040–3054.

Meena, Jagan Singh, Simon Min Sze, Umesh Chand, and Tseung-Yuen Tseng. 2014. "Overview of emerging nonvolatile memory technologies." *Nanoscale Research Letters* 9 (1):1–33.

Mehndiratta, Poorva, Arushi Jain, Sudha Srivastava, and Nidhi Gupta. 2013. "Environmental pollution and nanotechnology." *Environment and Pollution* 2 (2):49.

Miehr, Rosemarie, Paul G Tratnyek, Joel Z Bandstra, Michelle M Scherer, Michael J Alowitz, and Eric J Bylaska. 2004. "Diversity of contaminant reduction reactions by zerovalent iron: role of the reductate." *Environmental Science & Technology* 38 (1):139–147.

Mishra, Vijay, Kuldeep K Bansal, Asit Verma, Nishika Yadav, Sourav Thakur, Kalvatala Sudhakar, and Jessica M Rosenholm. 2018. "Solid lipid nanoparticles: emerging colloidal nano drug delivery systems." *Pharmaceutics* 10 (4):191.

Mukherjee, S, S Ray, and RS Thakur. 2009. "Solid lipid nanoparticles: a modern formulation approach in drug delivery system." *Indian Journal of Pharmaceutical Sciences* 71 (4):349.

Müller, Rainer H, Sven Staufenbiel, and CM Keck. 2014. "Lipid Nanoparticles (SLN, NLC) for innovative consumer care & household products." *Household and Personal Care Today* 9:18–24.

Murty, Budaraju S, P Shankar, Baldev Raj, BB Rath, and James Murday. 2013. *Textbook of Nanoscience and Nanotechnology*. Springer Science & Business Media.

Mutyam Pallerla, Swathi, and Bala Prabhakar. 2013. "A review on solid lipid nanoparticles." *International Journal of Pharmaceutical Sciences Review and Research* 20 (2):196–206.

Nair, Rahul Raveendran, Peter Blake, Alexander N Grigorenko, Konstantin S Novoselov, Tim J Booth, Tobias Stauber, Nuno MR Peres, and Andre K Geim. 2008. "Fine structure constant defines visual transparency of graphene." *Science* 320 (5881):1308–1308.

Narayanan, Kannan Badri, Hyun Ho Park, and Sung Soo Han. 2015. "Synthesis and characterization of biomatrixed-gold nanoparticles by the mushroom Flammulina velutipes and its heterogeneous catalytic potential." *Chemosphere* 141:169–175.

Naseri, Neda, Hadi Valizadeh, and Parvin Zakeri-Milani. 2015. "Solid lipid nanoparticles and nanostructured lipid carriers: structure, preparation and application." *Advanced Pharmaceutical Bulletin* 5 (3):305.

Nasir, Amara, Ayesha Kausar, and Ayesha Younus. 2015. "A review on preparation, properties and applications of polymeric nanoparticle-based materials." *Polymer-Plastics Technology and Engineering* 54 (4):325–341.

Navarro Yerga, Rufino M, M Consuelo Alvarez Galvan, F Del Valle, Jose A Villoria de la Mano, and Jose LG Fierro. 2009. "Water splitting on semiconductor catalysts under visible-light irradiation." *ChemSusChem: Chemistry & Sustainability Energy & Materials* 2 (6):471–485.

Nemati, Tayebeh, Mehrdad Sarkheil, and Seyed Ali Johari. 2019. "Trophic transfer of CuO nanoparticles from brine shrimp (Artemia salina) nauplii to convict cichlid (Amatitlania nigrofasciata) larvae: uptake, accumulation and elimination." *Environmental Science and Pollution Research* 26 (10):9610–9618.

Neyaz, Nikhat, Weqar Ahmad Siddiqui, and Kishore K Nair. 2014. "Application of surface functionalized iron oxide nanomaterials as a nanosorbents in extraction of toxic heavy metals from ground water: a review." *International Journal of Environmental Sciences* 4 (4):472.

Novoselov, Kostya S, Andre K Geim, Sergei V Morozov, Dingde Jiang, Yanshui Zhang, Sergey V Dubonos, Irina V Grigorieva, and Alexandr A Firsov. 2004. "Electric field effect in atomically thin carbon films." *Science* 306 (5696):666–669.

Pace, Ron J. 2005. "An integrated artificial photosynthesis model." *Artificial Photosynthesis*:13–34.

Panahi, Yunes, Masoud Farshbaf, Majid Mohammadhosseini, Mozhdeh Mirahadi, Rovshan Khalilov, Siamak Saghfi, and Abolfazl Akbarzadeh. 2017. "Recent advances on liposomal nanoparticles: synthesis, characterization and biomedical applications." *Artificial Cells, Nanomedicine, and Biotechnology* 45 (4):788–799.

Patel, Kapil D, Rajendra K Singh, and Hae-Won Kim. 2019. "Carbon-based nanomaterials as an emerging platform for theranostics." *Materials Horizons* 6 (3):434–469.

Paul, Gayatri, Harish Hirani, Tapas Kuila, and NC Murmu. 2019. "Nanolubricants dispersed with graphene and its derivatives: an assessment and review of the tribological performance." *Nanoscale* 11 (8):3458–3483.

Pires, JCM, MCM Alvim-Ferraz, FG Martins, and M Simões. 2012. "Carbon dioxide capture from flue gases using microalgae: engineering aspects and biorefinery concept." *Renewable and Sustainable Energy Reviews* 16 (5):3043–3053.

Polizu, Stefania, Oumarou Savadogo, Philippe Poulin, and L'Hocine Yahia. 2006. "Applications of carbon nanotubes-based biomaterials in biomedical nanotechnology." *Journal of Nanoscience and Nanotechnology* 6 (7):1883–1904.

Ponder, Sherman M, John G Darab, and Thomas E Mallouk. 2000. "Remediation of Cr (VI) and Pb (II) aqueous solutions using supported, nanoscale zero-valent iron." *Environmental Science & Technology* 34 (12):2564–2569.

Qu, Xiaolei, Pedro JJ Alvarez, and Qilin Li. 2013. "Applications of nanotechnology in water and wastewater treatment." *Water Research* 47 (12):3931–3946.

Qu, Xiaolei, Jonathon Brame, Qilin Li, and Pedro JJ Alvarez. 2013. "Nanotechnology for a safe and sustainable water supply: enabling integrated water treatment and reuse." *Accounts of Chemical Research* 46 (3):834–843.

Rao, J Prasad, and Kurt E Geckeler. 2011. "Polymer nanoparticles: preparation techniques and size-control parameters." *Progress in Polymer Science* 36 (7):887–913.

Rizwan, Md, Man Singh, Chanchal K Mitra, and Roshan K Morve. 2014. "Ecofriendly application of nanomaterials: nanobioremediation." *Journal of Nanoparticles* 2014.

Roco, Mihail C, Chad A Mirkin, and Mark C Hersam. 2011. *Nanotechnology Research Directions for Societal Needs in 2020: Retrospective and Outlook*. Vol. 1. Springer Science & Business Media.

Ross, Russel B, Claudia M Cardona, Dirk M Guldi, Shankara Gayathri Sankaranarayanan, Matthew O Reese, Nikos Kopidakis, Jeff Peet, Bright Walker, Guillermo C Bazan, and Edward Van Keuren. 2009. "Endohedral fullerenes for organic photovoltaic devices." *Nature Materials* 8 (3):208–212.

Ruthven, Douglas M. 1984. *Principles of Adsorption and Adsorption Processes*. John Wiley & Sons.

Sahu, Manas Kumar. 2019. "Semiconductor nanoparticles theory and applications." *International Journal of Applied Engineering Research* 14 (2):491–494.

Scida, Karen, Patricia W Stege, Gabrielle Haby, Germán A Messina, and Carlos D García. 2011. "Recent applications of carbon-based nanomaterials in analytical chemistry: critical review." *Analytica Chimica Acta* 691 (1–2):6–17.

Shah, Rohan, Daniel Eldridge, Enzo Palombo, and Ian Harding. 2015. *Lipid Nanoparticles: Production, Characterization and Stability*. Springer.

Sharma, Yogesh C, Varsha Srivastava, VK Singh, SN Kaul, and CH Weng. 2009. "Nano-adsorbents for the removal of metallic pollutants from water and wastewater." *Environmental Technology* 30 (6):583–609.

Shen, Shuling, Cheng Wang, Minquan Sun, Mengmeng Jia, Zhihong Tang, and Junhe Yang. 2020. "Free-standing sodium titanate ultralong nanotube membrane with oil-water separation, self-cleaning, and photocatalysis properties." *Nanoscale Research Letters* 15 (1):22.

Singh, Dharmveer, Ravindra Kumar Gautam, Rajendra Kumar, Brajesh Kumar Shukla, Vijay Shankar, and Vijay Krishna. 2014. "Citric acid coated magnetic nanoparticles: synthesis, characterization and application in removal of Cd (II) ions from aqueous solution." *Journal of Water Process Engineering* 4:233–241.

Singh, Jagpreet, Vanish Kumar, Sukhwinder Singh Jolly, Ki-Hyun Kim, Mohit Rawat, Deepak Kukkar, and Yiu Fai Tsang. 2019. "Biogenic synthesis of silver nanoparticles and its photocatalytic applications for removal of organic pollutants in water." *Journal of Industrial and Engineering Chemistry* 80:247–257.

Singh, Reena, Neetu Gautam, Anurag Mishra, and Rajiv Gupta. 2011. "Heavy metals and living systems: an overview." *Indian Journal of Pharmacology* 43 (3):246.

Singh, Ritu, Virendra Misra, and Rana Pratap Singh. 2012. "Removal of hexavalent chromium from contaminated ground water using zero-valent iron nanoparticles." *Environmental Monitoring and Assessment* 184 (6):3643–3651.

Smalley, Richard E. 1997. "Discovering the fullerenes." *Reviews of Modern Physics* 69 (3):723.

Smith, Andrew M, and Shuming Nie. 2010. "Semiconductor nanocrystals: structure, properties, and band gap engineering." *Accounts of Chemical Research* 43 (2):190–200.

Soylak, Mustafa, Ozgur Ozalp, and Furkan Uzcan. 2020. "Magnetic nanomaterials for the removal, separation and preconcentration of organic and inorganic pollutants at trace levels and their practical applications: a review." *Trends in Environmental Analytical Chemistry*:e00109.

Stevenson, S, G Rice, T Glass, K Harich, F Cromer, MR Jordan, J Craft, E Hadju, R Bible, and MM Olmstead. 1999. "Small-bandgap endohedral metallofullerenes in high yield and purity." *Nature* 401 (6748):55–57.

Suresh, Sagadevan. 2013. "Semiconductor nanomaterials, methods and applications: a review." *Journal of Nanoscience and Nanotechnology* 3 (3):62–74.

Tahriri, M, M Del Monico, A Moghanian, M Tavakkoli Yaraki, R Torres, A Yadegari, and L Tayebi. 2019. "Graphene and its derivatives: opportunities and challenges in dentistry." *Materials Science and Engineering: C* 102:171–185.

Tamjidi, Fardin, Mohammad Shahedi, Jaleh Varshosaz, and Ali Nasirpour. 2013. "Nanostructured lipid carriers (NLC): a potential delivery system for bioactive food molecules." *Innovative Food Science & Emerging Technologies* 19:29–43.

Tchounwou, Paul B, Clement G Yedjou, Anita K Patlolla, and Dwayne J Sutton. 2012. "Heavy metal toxicity and the environment." *Molecular, Clinical and Environmental Toxicology*:133–164.

Tisa, Farhana, Abdul Aziz Abdul Raman, and Wan Mohd Ashri Wan Daud. 2014. "Applicability of fluidized bed reactor in recalcitrant compound degradation through advanced oxidation processes: a review." *Journal of Environmental Management* 146:260–275.

Torabian, Azadeh, Homayoun Ahmad Panahi, Gholam Reza Nabi Bid Hendi, and Naser Mehrdadi. 2014. "Synthesis, modification and graft polymerization of magnetic nano particles for PAH removal in contaminated water." *Journal of Environmental Health Science and Engineering* 12 (1):1–10.

Trujillo-Reyes, J, JR Peralta-Videa, and JL Gardea-Torresdey. 2014. "Supported and unsupported nanomaterials for water and soil remediation: are they a useful solution for worldwide pollution?" *Journal of Hazardous Materials* 280:487–503.

Tungittiplakorn, Warapong, Claude Cohen, and Leonard W Lion. 2005. "Engineered polymeric nanoparticles for bioremediation of hydrophobic contaminants." *Environmental Science & Technology* 39 (5):1354–1358.

Umar, Muhammad, and Hamidi Abdul Aziz. 2013. "Photocatalytic degradation of organic pollutants in water." *Organic Pollutants-Monitoring, Risk and Treatment* 8:196–197.

Valcárcel, Miguel, Soledad Cárdenas, Bartolomé M Simonet, Yolanda Moliner-Martínez, and Rafael Lucena. 2008. "Carbon nanostructures as sorbent materials in analytical processes." *TrAC Trends in Analytical Chemistry* 27 (1):34–43.

Vance, Marina E, Todd Kuiken, Eric P Vejerano, Sean P McGinnis, Michael F Hochella Jr, David Rejeski, and Matthew S Hull. 2015. "Nanotechnology in the real world: redeveloping the nanomaterial consumer products inventory." *Beilstein Journal of Nanotechnology* 6 (1):1769–1780.

Vauthier, C, and P Couvreur. 2000. "Development of nanoparticles made of polysaccharides as novel drug carrier systems." In *Handbook of Pharmaceutical Controlled Release Technology*. Wise DL (Ed.), 13–429. Marcel Dekker.

Villaseñor, M Jesús, and Ángel Ríos. 2018. "Nanomaterials for water cleaning and desalination, energy production, disinfection, agriculture and green chemistry." *Environmental Chemistry Letters* 16 (1):11–34.

Vishnu, Dhanya, and Balaji Dhandapani. 2021. "A review on the synergetic effect of plant extracts on nanomaterials for the removal of metals in industrial effluents." *Current Analytical Chemistry* 17 (2):260–271.

Wang, Ting, Xiaoying Jin, Zuliang Chen, Mallavarapu Megharaj, and Ravendra Naidu. 2014. "Green synthesis of Fe nanoparticles using eucalyptus leaf extracts for treatment of eutrophic wastewater." *Science of the Total Environment* 466:210–213.

Wang, Yinan, David O'Connor, Zhengtao Shen, Irene MC Lo, Daniel CW Tsang, Simo Pehkonen, Shengyan Pu, and Deyi Hou. 2019. "Green synthesis of nanoparticles for the remediation of contaminated waters and soils: constituents, synthesizing methods, and influencing factors." *Journal of Cleaner Production* 226:540–549.

Wang, Yuting, Liyan He, Yuanshuai Li, Lingyun Jing, Jianzhi Wang, and Xiaoli Li. 2020. "Ag NPs supported on the magnetic Al-MOF/PDA as nanocatalyst for the removal of organic pollutants in water." *Journal of Alloys and Compounds* 828:154340.

Westerhoff, Paul K, Pedro Alvarez, Qilin Li, Jorge Gardea-Torresdey, and Julie Zimmerman. 2016. "Overcoming implementation barriers for nanotechnology in drinking water treatment." *Environmental Science: Nano* 3 (6):1241–1253.

Westerhoff, Paul K, Mehlika A Kiser, and Kiril Hristovski. 2013. "Nanomaterial removal and transformation during biological wastewater treatment." *Environmental Engineering Science* 30 (3):109–117.

Wong, Ming Hung. 2012. *Environmental Contamination: Health Risks and Ecological Restoration*. CRC Press.

Worms, Isabelle AM, Jonathan Boltzman, Miguel Garcia, and Vera I Slaveykova. 2012. "Cell-wall-dependent effect of carboxyl-CdSe/ZnS quantum dots on lead and copper availability to green microalgae." *Environmental Pollution* 167:27–33.

Wu, Yihan, Hongwei Pang, Yue Liu, Xiangxue Wang, Shujun Yu, Dong Fu, Jianrong Chen, and Xiangke Wang. 2019. "Environmental remediation of heavy metal ions by novel-nanomaterials: a review." *Environmental Pollution* 246:608–620.

Xie, Yingying, Wen Cheng, Pokeung Eric Tsang, and Zhanqiang Fang. 2016. "Remediation and phytotoxicity of decabromodiphenyl ether contaminated soil by zero valent iron nanoparticles immobilized in mesoporous silica microspheres." *Journal of Environmental Management* 166:478–483.

Yadav, Neha, Sunil Khatak, and UV Singh Sara. 2013. "Solid lipid nanoparticles-a review." *International Journal of Applied Pharmaceutics* 5 (2):8–18.

Yoon, Goo, Jin Woo Park, and In-Soo Yoon. 2013. "Solid lipid nanoparticles (SLNs) and nanostructured lipid carriers (NLCs): recent advances in drug delivery." *Journal of Pharmaceutical Investigation* 43 (5):353–362.

Yunus, Ian Sofian, Harwin, Adi Kurniawan, Dendy Adityawarman, and Antonius Indarto. 2012. "Nanotechnologies in water and air pollution treatment." *Environmental Technology Reviews* 1 (1):136–148.

Zhang, Bo-Tao, Xiaoxia Zheng, Hai-Fang Li, and Jin-Ming Lin. 2013. "Application of carbon-based nanomaterials in sample preparation: a review." *Analytica Chimica Acta* 784:1–17.

Zhang, Jiangjiang, Lei Mou, and Xingyu Jiang. 2020. "Surface chemistry of gold nanoparticles for health-related applications." *Chemical Science* 11 (4):923–936.

Zhang, Jie, Wenli Guo, Qingqing Li, Zhe Wang, and Sijin Liu. 2018. "The effects and the potential mechanism of environmental transformation of metal nanoparticles on their toxicity in organisms." *Environmental Science: Nano* 5 (11):2482–2499.

Zhang, Yuenan, Yujie Zhang, Ozioma Udochukwu Akakuru, Xiawei Xu, and Aiguo Wu. 2021. "Research progress and mechanism of nanomaterials-mediated in-situ remediation of cadmium-contaminated soil: a critical review." *Journal of Environmental Sciences* 104:351–364.

Zhou, Jiarong, Ashley V Kroll, Maya Holay, Ronnie H Fang, and Liangfang Zhang. 2020. "Biomimetic nanotechnology toward personalized vaccines." *Advanced Materials* 32 (13):1901255.

Zhu, Xiaohua, Yang Liu, Pei Li, Zhou Nie, and Jinghong Li. 2016. "Applications of graphene and its derivatives in intracellular biosensing and bioimaging." *Analyst* 141 (15):4541–4553.

Zou, Yidong, Xiangxue Wang, Ayub Khan, Pengyi Wang, Yunhai Liu, Ahmed Alsaedi, Tasawar Hayat, and Xiangke Wang. 2016. "Environmental remediation and application of nanoscale zero-valent iron and its composites for the removal of heavy metal ions: a review." *Environmental Science & Technology* 50 (14):7290–7304.

3 Carbon-Based Nanomaterials to Environs' Remediation

Flávia Cristina Policarpo Tonelli and
Fernanda Maria Policarpo Tonelli

3.1 INTRODUCTION

Environmental pollution is a problem that threatens the lives of countless species, is a major cause of mortality and morbidity and, although quite present today, is not recent. One of the main causes of this is human actions through industrialization, urbanization and mining exploration. This problem plagues not only developed countries but also underdeveloped countries and countries that are undergoing intense development. However, these countries have different ways of dealing with the problem. In general, developed countries have stricter laws and awareness of the environmental situation (Ukaogo et al., 2020).

Pollutants can be divided into two main groups, namely, organic and inorganic, and are spread over water, air and soil. Organic pollutants include pesticides and dyes, and inorganic pollutants encompass radionuclides and heavy metals (Tonelli and Tonelli, 2020). Most of these pollutants are persistent and accumulate in plant and animal species. Through the food chain, pollutants are passed from species to species. Therefore, humans can be in contact with pollution not just through direct exposure. Humans, when eating, for example, can also ingest pollutants and have their health impaired. This can intoxicate them, damage their DNA or even lead to death (Kazemzadeh and Zali, 2023; Venkatraman et al., 2024).

Regarding water pollution, special care must be taken since water is an essential substance for living beings whose quantity is limited and stocks are being depleted on Earth. Therefore, strategies to purify water become even more necessary (Ying et al., 2017). Two sources of water pollution deserve to be highlighted: landfills and wastewater. Through the composition of wastewater, it is possible to obtain some information about the way of life of a given population. Organic pollutants are the main contaminants found when analyzing domestic wastewater. In general, there is a large amount of estrogenic hormones and drugs in wastewater from large urban centers. Estrogen and endocrine disrupting products, such as polychlorinated biphenyls, can be present in polluted waters, which can cause living systems to be unbalanced (Yeung et al., 2011; Fan et al., 2020). The presence of drugs in water is mainly due to the ease with which patients obtain them, even without a medical prescription, and their incorrect disposal. These residues can contaminate animals, extend throughout the food chain and impact human health, even at low casualties, since they remain present even after water treatment (Mezzelani et al., 2018).

In addition to domestic sewage, another important source of water pollution is industrial sewage contamination. In it, it is possible to find pollutants (products and by-products) that vary according to the type of industry from which water is discharged, such as veterinary products, personal hygiene products, drugs, heavy metals, dyes and food additives. Organic pollutants and phosphorus from chemically improved primary industries are not usually removed from the water after treatment; then, they are released into the environment and modify the chemical composition of a given ecosystem. Dyes are also great environmental polluters. Azo dyes, for example, have great carcinogenic and mutagenic potential, and may be present in polluted water (Alharbi et al., 2018; Guo et al., 2019).

 DOI: 10.1201/9781003186298-3

In addition to water pollution, air pollution is another extremely important factor concerning the destruction of the environment. With the increase in industrial activities, population growth and, consequently, consumption patterns, there is an increase in the concentration of toxic gases in the air, such as SO_2, NO, CO_2 and O_3, in addition to suspended particles and other compounds. The origin of most of these pollutants relates to the production of energy by burning fossil fuels. These compounds are extremely harmful to humans and can cause respiratory problems and severe intoxication. In addition, through them, there may be an intensification of the greenhouse effect, which leads to an increase in Earth's temperatures, unbalancing ecosystems and threatening species. Climate change can also lead to the production of two allergens that become available in the air, specifically, pollen and mold, due not only to the lengthening of the pollen season and its greater production but also to the increase in humidity by floods and extreme climates.

Air pollutants can be deadly and present severe health risks. During the combustion of gas, gasoline and coal or incineration, they are released. Some of the most common pollutants are mercury, benzene, dioxin and lead. Mercury causes damage to the central nervous system. Polycyclic aromatic hydrocarbons, such as benzene, in large quantities, can lead to problems with the liver and blood and irritation of the lungs and eyes, in addition to being carcinogenic. Dioxins can damage the nervous, endocrine, reproductive and immune systems. Lead in higher concentrations can affect the nervous system and kidneys, especially in children. In small amounts, it can affect their learning and IQ.

Another environmental aspect directly affected is the soil. According to the Food and Agriculture Organization of the United Nations (FAO), because of population growth, which leads to greater food production, and incorrect soil planting and burning, soil degradation is already occurring, resulting in a lack of nutrients (micro and macronutrients) in and scarcity of soil (FAO, 2024). In addition to these issues, there continues to be great concern with the contamination of the soil by pharmaceutical products, toxic metals, such as Cd, Pb and Ni and pesticides that compromise the quality and safety of the food produced.

Nanoscale materials have been explored as important tools for the efficient remediation of contaminated environs. In this chapter, they receive attention, and some of the main uses of carbon-based nanomaterials to promote the remediation of water, soil and air are reviewed.

3.1.1 Nanomaterials

The term "nanotechnology" was created in 1974 by Norio Taniguchi and refers to nanoscale materials that have been produced and used since the 1950's (Taniguchi, 1974). Many scientists have been searching different protocols to chemically modify and to even synthesize them to perform a specific task (Zhang et al., 2008). The field of nanotechnology has been in a development process that offers low-cost strategies to improve not only environmental remediation but also sensors, drug delivery and gene delivery, among other applications. Especially green synthesis has been explored to produce green-fabricated nanomaterials (NMs) as eco-friendly tools (Nasrollahzadeh et al., 2021).

NMs are classified based on their morphology/shape. In this way, they are mainly classified as nanoparticles (NPs—zero-dimensional), nanowires (NWs—one-dimensional), nanotubes (NTs—one-dimensional), nanofibers (NFs—one-dimensional) and nanomembranes (NMBs—two-dimensional); there are also some 3D structures that can be synthesized and a one-dimensional variation for some nanostructures such as gold that can be organized into rods (Figure 3.1). According to their composition, they can be classified as organic NMs based on organic molecules or inorganic NMs based on inorganic molecules (Lu and Astruc, 2020).

To be considered a nanomaterial, at least one dimension of the material has to present a size ranging from 1 to 100 nanometers. It must also demonstrate high stability (chemical, thermal and physical), catalytic activity and chemical reactivity (Xu et al., 2019). Nanomaterials usually have

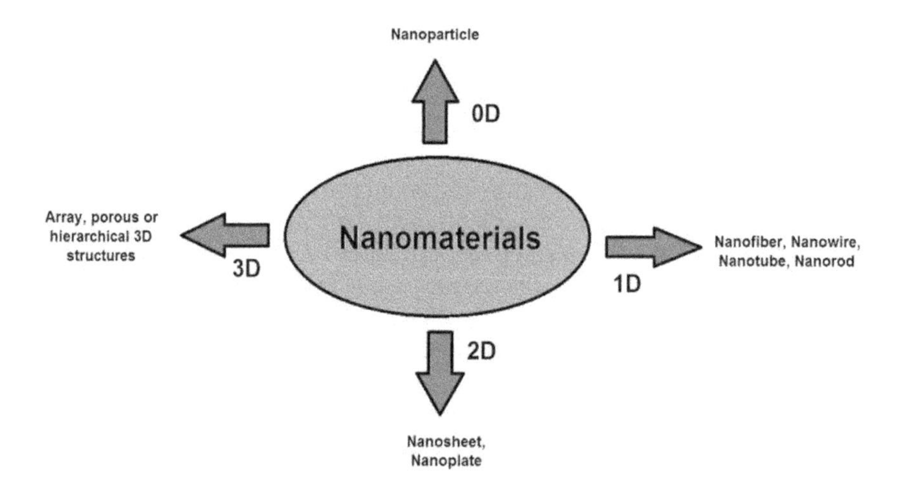

FIGURE 3.1 Nanomaterial classification based on morphology/shape.

the property of aggregating when hydrated and being sedimented at different rates, which depends on the charge of their surface and the charge magnitude. Regarding nanomaterials' surfaces, they can be made by natural organic matter (NOM), other organic molecules or mono- or divalent cations (Keller et al, 2010).

Due to all these characteristics, NMs are versatile in their functions and can be used in processes such as imaging/diagnosis, vaccine development, tissue engineering and the treatment of health problems through drug and gene delivery. Scientists can chemically modify nanomaterials' surface and functionalize them to perform specific desirable tasks, which makes them acquire specific characteristics that optimize them for the specific role to be performed (Tonelli et al., 2015, 2016, 2020).

Safety for living beings is the main aspect regarding the extensive use of nanomaterials for the remediation of water, soil and air. The use of these materials and the final product of this remediation must not, under any circumstances, pose a high-risk threat to fauna and flora of the ecosystem that will undergo nanoremediation. Thus, a concern and a search for safety and biocompatibility must always be maintained so that this strategy can be used on large scales (Pulizzi and Sun, 2018; Tonelli and Tonelli, 2023). One of the strategies for the remediation of polluted water, for example, is the use of nanostructures (Khan and Malik, 2019) and nanosensors (Xie et al., 2019) to detect and remove contaminants. Different NMBs and nano-absorbents are in the development stage for large-scale industrialization.

The chemical constitution of nanomaterials can be classified into carbon- (CB) and non-carbon-based (NCB) structures. Examples are graphene (CB), graphene oxide (CB), carbon nanotubes (CB), carbon quantum dots (CB), fullerenes (CB) (Figure 3.2), silver quantum dots (NCB) and magnetic NPs (NCB) (Trivedi et al., 2020).

The life of animal and plant species, including human life, is directly dependent on water, air and soil, and because these elements have been undergoing intense degradation, new safety and low-cost remediation strategies are increasingly required to promote efficient environmental pollution remediation; nanomaterials can be designed to perform this role with excellent results (Ghadimi et al., 2020).

3.2 CARBON-BASED NANOMATERIALS TO REMEDIATE POLLUTED WATER

The carbon element has some properties that make it extremely useful in the production of nanomaterials. First, it has natural (graphite and diamond) and synthetic (carbon-based nanomaterials)

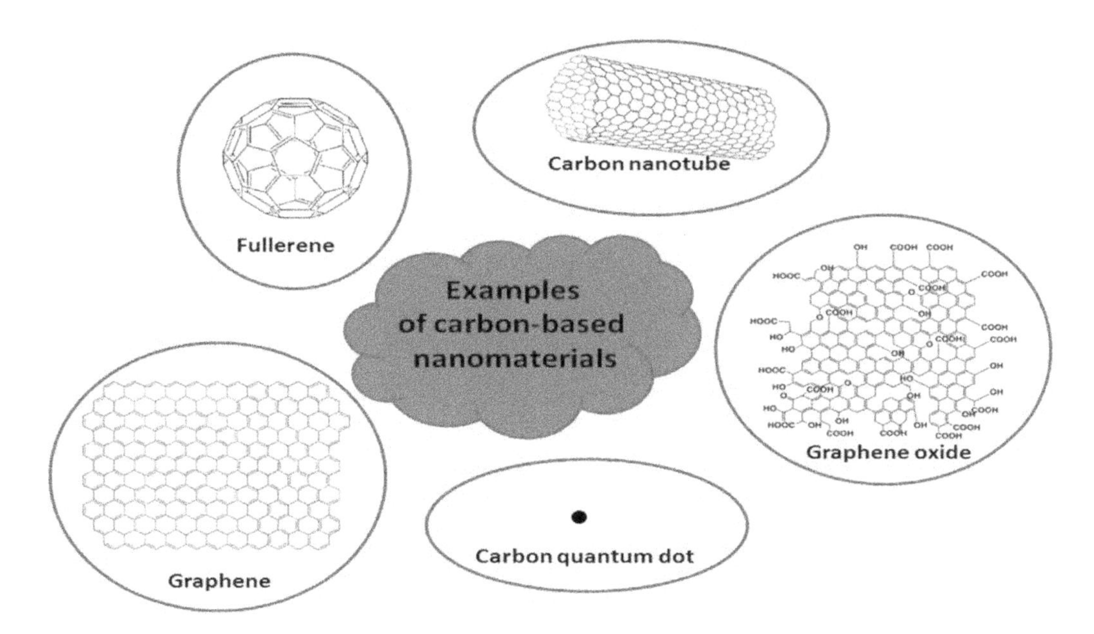

FIGURE 3.2 Examples of carbon-based nanomaterials.

allotropes. Second, it can hybridize sp, sp^2 and sp^3 and can thus bind to several other elements, and even to another carbon atom, to form the structures of interest (Rauti et al., 2019).

Several carbon-based nanomaterials are used to remediate polluted water (**Table 3.1**), as highlighted by the following: amino-functionalized C_{60} immobilized on 3-(2-succinic anhydride) propyl functionalized silica to remediate Cimetidine and Ranitidine; C_{60} modified by Ag_3PO_4 to remediate Dye acid red 18; C_{60} modified graphite phase carbonitride to remediate Methylene blue and phenol; carbon dots impregnated in defect-rich g-C_3N_4 to remediate hexavalent chromium, ofloxacin, bisphenol A and ciprofloxacin; nitrogen-doped carbon quantum dots hybridized with g-C_3N_4 to remediate Methylene; graphene associated with iron oxide to remediate Methylene blue, rhodamine B, acid orange 7 and phenol; reduced graphene oxide functionalized and associated with MnO_2 to remediate Methylene blue and arsenite; carbon nanotubes (CNTs) grafted-antifouling layer of polyacryloyl hydrazide in a poly(vinylidene fluoride) membrane to remediate micro pollutants and heavy metals; Ag-embedded C_{60} to remediate 4-nitrophenol and orange G dye; fullerene coating nano titanium dioxide and Polyhydroxy fullerene to remediate the herbicide Mesotrione; anatase-TiO_2 NPs modified by fullerene to remediate Methylene blue; carbon dots decorating hollow carbon nitride nanospheres to remediate naproxen; carbon dots attached to tungsten oxide to remediate Cd^{2+} and crystal violet; graphene oxide in magnetic aerogel to remediate Congo red, Methylene blue, Cu^{2+}, Pb^{2+}, Cd^{2+} and Cr^{3+}; a graphene oxide platform containing a nickel-benzene dicarboxylate nano-sized nickel metal organic framework and a CNT platform containing a nickel-benzene dicarboxylate nano-sized nickel metal organic framework to remediate Methylene blue; chitosan-CNT-supporting palladium NPs to remediate 2-nitroaniline, 4-nitrophenol, 4-nitro-*o*-phenylenediamine and 2,4-dinitrophenol, methyl orange, Congo red, methyl red and Methylene blue dyes in water (Tonelli et al., 2020).

Carbon dots have a small size, high catalytic activity, superconductivity, good dispersibility, high crystallization and strong fluorescence, and they are considered materials with zero dimension (Semeniuk et al., 2019). They can easily be functionalized through various packages. They are useful in the remediation of polluted water in a chemical sensor/bioimaging field (Rani et al., 2020). Water contaminated with naproxen can, for example, be remedied by a metal-free

TABLE 3.1
The Main Approaches, Proposed in 2020, Applying Carbon-Based Nanomaterial to Remediate Polluted Water

Nanomaterial	Chemical modification	Pollutant(s)	Reference
Graphene	Binding to iron oxide	Methylene blue, rhodamine B, acid orange 7 and phenol	(Hammad et al., 2020)
Graphene oxide	Magnetic aerogel	Congo red, Methylene blue, Cu^{2+}, Pb^{2+}, Cd^{2+} and Cr^{3+}	(Xiong et al., 2020)
	Nickel-benzene dicarboxylate nano-sized nickel metal organic framework	Methylene blue	(Ahsan et al., 2020)
Reduced graphene oxide	Associated with MnO_2	Methylene blue and arsenite	(Tara et al., 2020)
Fullerene	Ag-embedded	4-nitrophenol and orange G dye	(Liu R. et al., 2020)
Carbon dot	Impregnated in defect-rich g-C_3N_4	Hexavalent chromium, ofloxacin, bisphenol A and ciprofloxacin	(Liu H. et al., 2020)
	Nitrogen-doped and hybridized with g-C_3N_4	Methylene	(Seng et al., 2020)
	Binding to tungsten oxide	Cd^{2+} and crystal violet	(Smrithi et al., 2020)
Carbon nitride nanospheres	Decorated by carbon dots	Naproxen	(Wu Y. et al., 2020)
Carbon nanotube	Platform containing nickel-benzene dicarboxylate nano-sized nickel metal organic framework		
	Polyacryloyl hydrazide in poly(vinylidene fluoride) membrane	Micro pollutants and heavy metals	(Chen Z. et al., 2020)
	Chitosan and palladium nanoparticles	2-nitroaniline, 4-nitrophenol, 4-nitro-*o*-phenylenediamine and 2,4-dinitrophenol, methyl orange, Congo red, methyl red and Methylene blue dyes	(Sargin et al., 2020)

photocatalytic nanoreactor composed of hollow carbon nitride nanospheres decorated with carbon dots, with an efficiency of 10 mg/L in five minutes of natural solar irradiation (Wu Y. et al., 2020). Regarding mixed pollution involving the inorganic contaminant hexavalent chromium and the organic ones ofloxacin, ciprofloxacin and bisphenol A, remediation can be carried out using g-C3N4 rich in defects impregnated with N-doped carbon quantum dots (Liu H. et al., 2020). Water polluted by Methylene blue can be remedied with nitrogen-doped quantum carbon dots that have been hybridized with g-C3N4. This hybridization leads to a 2.6-fold increase in efficiency compared to non-hybrid g-C3N4 (Seng et al., 2020).

Carbon-based materials obtained by a green synthesis protocol can also be used in water remediation. For the removal of Cd^{2+} and degradation of violet crystal, it is possible to use a nanosystem composed of tungsten oxide and *Trichosanthes cucumerina* shells (Smrithi et al., 2020). TiO_2 can be converted into a material with greater photocatalytic capacity through sustainable nitrogen-doped quantum carbon dots derived from Bombyx mori silk fibroin (Wang et al., 2020).

Graphene, another type of carbon-based nanomaterial, contains some interesting characteristics, such as high thermal conductivity and mechanical and optical properties. It is a two-dimensional nanomaterial composed of hexagonal rings of carbon atoms sp^2 (Novoselov et al., 2004). The remediation of water can be accomplished through graphene, including connecting it to a biochar (Tiwari et al., 2020). It is thus possible to form a graphene nanosystem containing biochar

(Fang et al., 2020). In addition, graphene can be linked to iron oxide to degrade waste pollutants, such as rhodamine B, Methylene blue, orange acid 7 and phenol, which leads to an increase in its activity. Another advantage of the system is the stability in different systems, from pH 3 to 9, allowing it to be applied to water treatment (Hammad et al., 2020).

Graphene oxide can be generated by the oxidation of graphene flakes followed by exfoliation. Generating this structure is interesting, since it presents less toxicity and greater compatibility than graphene and does not need surfactant to be dispersed in water (Kiew et al., 2016)

Water polluted with Methylene blue has been remedied by the adsorption of the compound in graphene oxide bonded to the nickel-benzene dicarboxylate platform, an organic nickel metal nanostructure (Ahsan et al., 2020). CNTs can also be used. Another way to remedy this and arsenide contamination is performed with reduced graphene oxide functionalized with functional groups of black cumin seeds and associated with MnO_2 particles (Tara et al., 2020).

Graphene oxide can also be used to make an airgel. The oxide is self-gelling with Fe_3O_4 and forms a magnetic airgel. When in use, it has the capacity to adsorb metal ions such as Cr^{3+}, Pb^{2+}, Cd^{2+} and Cu^{2+} and dyes such as Methylene blue and Congo red (Xiong et al., 2020).

C_{60} belongs to the allotropic family of fullerene and is formed by the sp^2 hybridization of carbon atoms arranged in rings of five or six members to form a hollow sphere. Its discovery was recognized in 1996 by the Nobel Prize in Chemistry (Biglova and Mustafina, 2019). The distance between the edges of two hexagons are 1.38 Å long (double bonds), and at each end of the pentagon and hexagon, they are 1.45 Å long (Hirsch, 1995). It is possible to functionalize fullerene to use it in the remediation of polluted water. The poly (N-vinylpyrrolidone) compound, for example, can be encapsulated and have greater efficiency than a hydroxylated version to generate reactive oxygen species that can destroy microorganisms sensitive to superoxide and singlet oxygen and remedy pollutants (Brunet et al., 2009). To degrade Methylene blue in water, fullerene photocatalytic activity can modify graphite phase carbonitride. It is also possible to degrade phenol in this way (Bai et al., 2014).

Methylene blue can also be degraded by fullerene and anatase-TiO_2, as modified by C_{60}, to produce an increase in its photocatalytic capacity (Qi et al., 2016).

To remedy pharmaceutical pollutants in water, such as ranitidine and cimetidine, 3-(2-succinic anhydride) propyl-silica functionalized in C_{60} amino-functionalized (photoactive structure) can be immobilized (Lee et al., 2010).

The orange dye G can be photodegraded by the incorporation of C_{60} in Ag, which leads to an increase in the catalytic reduction of 4-nitrophenol (Liu R. et al., 2020).

To remedy water contaminated with the herbicide mesotrione, it is possible to coat other nanomaterials such as titanium nano dioxide with polyhydroxy fullerene and fullerene (Djordjevic et al., 2018).

3.3 CARBON-BASED NANOMATERIALS TO REMEDIATE POLLUTED SOIL

Carbon-based nanomaterials have been configured as a good soil remediation alternative without prejudice to crops that depend on this soil (Lei et al., 2019). Carbon-based nanomaterials are easily absorbed by plant species and can influence their root growth and increase it and their development. Other possible actions are facilitated by the germination of seeds and increasing vegetal biomass. Precisely, due to the ease of plant species in absorbing nanomaterials, nanofertilizers have been researched. Due to carbon-based nanomaterials' nanostructured information, porous structure, high specific surface area and crystallinity, nutrients can be stored in the nanostructure and be delivered more easily. In addition, these nanomaterials allow the slow release of active compounds for the development of plant culture (Achari and Kowshik, 2018).

Carbon-based nanomaterials are very effective in remedying contaminated soil, and they immobilize contaminants. Antibiotics, paracetamol and chlorinated hydrocarbons can be immobilized by CNTs (Yan et al., 2017). Lead and cadmium can be adsorbed by graphene (GN) and graphene oxide (GO). These materials can also be used for the transport and slow release of micronutrients in plants due to their adapted groups on the surface and high mechanical resistance. Stabilities can also be increased by these nanomaterials (Kabiri et al., 2017).

Soil improvement, which leads to increased crop knowledge, can be conducted through two other carbon-based nanomaterials: biochar-based fertilizer (BBF) and biochar (BC). These are nanostructured materials with a mixture of nanographic and amorphous carbon. They are inspired by the properties of the "Black Lands of Indians" (BLI) found in the Amazon region with great resilience and fertility (Glaser, 2007). The use of these materials leads to the improvement of soil physical and active properties and crop yield, nutrient retention and contaminant adsorption (Chew et al., 2020).

The organochlorine pesticide dichlorophenyl trichloroethane (DDT) can be immobilized by multi-walled carbon nanotubes (MWCNTs). The same is true for naphthalene, pyrene, fluorene and phenanthrene. However, some factors may interfere with the efficiency of this immobilization, such as the soil composition and NT concentration. Lead, copper, nickel, mercury, lead and antimony can be immobilized on the ground by MWCNTs. However, the immobilization capacity of the nanomaterial also depends on the characteristics of the metal (Vithanage et al., 2017).

In addition, MWCNTs can be used as an adjunct in the adsorption of other metals. MWCNT participates, for example, with citric acid in the reduction of Cr^{6+} by citric acid in oxisol. The 1,2-dichlorobenzene compound has been electrokinetically remedied in clay using a CNT barrier (Yuan et al., 2009).

Biochar can be used to remedy soil contaminated by heavy metals through several vehicles. One is the ion exchange between the metallic cation and cations associated with the biochar (Mg^{2+} and Ca^{2+}) from mineral oxides and complexed humic matter. Another mechanism is the complexation of phosphates, carbonates and sulfates in which heavy metals form poorly soluble salts (Lei et al., 2019).

Cation-exchange capacity (CEC) is more present in animal-derived than plant-derived biochar due to the high Ca^{2+} content. Thus, the ion exchange mechanism is dominant in biochars of animal origin. This can be used to immobilize ions from bovine and pig carcasses, such as Cu^{2+} and Cd^{2+}. Pb^{2+} immobilization can be accomplished by using biochar. Notably, the metal adsorption mechanism depends on its own characteristics and the structure of the biochar (Lei et al., 2019).

Polycyclic aromatic hydrocarbons, agrochemicals, phenols, benzene, derivatives, plasticizers and medicines (for example, used to treat disorders of the endocrine system and antibiotics) in the soil can be adsorbed by biochar due to its great porosity, aromatic structure and large specific surface (Dai et al., 2019). It can also interact with several other aromatic and hydrophobic organic compounds (Jung et al., 2013). However, such interaction depends on the structure of the compound to be adsorbed and can occur in different ways such as through hydrogen bonding, complexation and electrostatic attraction.

17β-estradiol, estrone, estriol, bisphenol A and 17α-ethinylestradiol are adsorbed on the biochar of *Eucalyptus globulus* wood with hydrogen bonds. This type of interaction, however, depends on the pH change (Ahmed et al., 2018).

A factor that can limit adsorption by biochar is an electrostatic repulsion between groups charged with the same signal in the biochart and from the ion to an adsorbed one. This can occur, for example, in the adsorption of sulfonic metazachlor and imazamox (Petter et al., 2019).

Graphene, graphene oxide and CNT are not commonly used in the remediation of contaminated soil. However, they are widely used for the remediation of water and air by sequestering toxic metal ions, organic compounds and rare earth metal ions (Wang et al., 2013). As they are non-porous adsorbent materials, contaminants adsorb only on their external surface. However, this does not limit their use for soil remediation. They can interact with different contaminants and be covalently modified to make such interactions. Graphene oxide, for example, has groups with oxygen, such as carbonyl, hydroxyl and carboxyl, which favours electrostatic and hydrophobic interaction, complexation or the establishment of hydrogen bonds (Zhao et al., 2019).

One of the reasons why graphene oxide or graphene is not so used in soil remediation is the blocking of adsorptive sites by interactions with soil particles (Zhao et al., 2019). A possible solution is the use of 3D composites containing graphene oxide that can retain solid particles on the external surface (Zhao et al., 2019).

Engineered carbonaceous nanomaterials are materials created by mixing carbon-based nanomaterials with other materials, or modifying them, to improve some characteristic that favors their use to remedy the soil. In general, they have a reduction in aggregation capacity, a higher adsorption density of active sites on the surface and an increase in the specific surface area. Thus, new strategies are being adopted to expand the use of these nanomaterials in soil remediation, surpassing obstacles (Zhou et al., 2013).

3.4 CARBON-BASED NANOMATERIALS TO REMEDIATE POLLUTED AIR

Carbon-based nanomaterials can also be used in the remediation of polluted air, which can be performed by the isolation and filtering of nano-filters, nanocatalysis available in degradation and adsorption by nanomaterials (Mohamed, 2017).

The presence of pollutants in the air can be detected when the CNT in filter, for example, contacts the air with the pollutant; this leads to a change in the conductance of the nanomaterial. Carbon-based nanomaterials have also been used in filters for filtering carbon dioxide needed by plants. This is due to the large specific area and the great adsorption capacity of CNTs (Ong et al., 2010). Carbon dioxide filtration membranes from plants where they are emitted can also be built (Lin et al., 2008).

CNTs build high-energy binding sites, which gives them good adsorptive capacity to remove difficult air pollutants (PAHs). Prior to the use of CNT, it must be treated chemically, physically or physical-chemically, which will allow access to the internal structure of the tube as a passage for pollutants and the removal of impurities. For physical treatment, heat and an oxidizing agent, such as oxygen, can be applied. Chemical treatment can use acids, such as sulfuric acid. It is also possible to vary the time and temperature at which the process occurs (Gangupomu et al., 2014).

CNTs can be further modified with many amine groups to increase the absorption of carbon dioxide in a temperature range from 20°C to 100°C (Su et al., 2009). In addition to the absorption of carbon dioxide (CO_2), CNTs are used to remove NO_x, a mixture of NO and NO_2. For this, the air containing NO and O_2 passes through the NT where the NO is oxidized to NO_2 and is then adsorbed on the surface of the nitrate species (Zhang et al., 2012). Isopropyl vapor, in contrast, can be chemically adsorbed on CNT on the surface of functional groups and physically adsorbed by van Der Waals forces. Volatile organic compounds can be adsorbed onto CNTs deposited on quartz filters. Carbon monoxide can be adsorbed by physisorption or chemisorption on Si-doped and boron-doped single-walled CNTs (Azama et al., 2017).

Fullerene is sphere-shaped and can have varying numbers of carbon, ranging among 20, 60, 70, 82, 100 and even 960, and is organized into 20 hexagons and 12 pentagons (Inagaki and Kang, 2014). Its electron acceptors are efficient. They can undergo chemical and physical treatments to be suitable for the desired purposes for air remediation. After the heat treatment of fullerenes with carbon dioxide, for example, the functional groups methyl (-CH3) and hydroxyl (-OH) on the surface of fullerenes can be eliminated with the release of carbon monoxide after heat treatment with carbon dioxide (Silva et al., 1999). They can also be used to adsorb organometallics, organic compounds, 1,2-dichlorobenzene, nitrous oxide, polycyclic aromatic hydrocarbons and naphthalene (Cheng et al., 2005).

The graphene family can be specified by a hexagonal graphite monolayer, two-dimensional hybridized allotrope sp2 of carbon, and graphene oxide, graphene and reduced graphene oxide are members of this family (Bergmann and Machado, 2015).

A strategy to increase the effective surface area to improve its adsorptive capacity is the addition of metals or other compounds of pure graphene. A number of carbonyl and carboxyl groups on the leaf's edge and epoxy and hydroxyl in the leaf's basal plane also enhance their ability to adsorb pollutants into the air (Stankovich et al., 2006). The presence of these groups increases the negative charge density in graphene oxide, which contributes to the absorption of positively charged heavy metals, cationic dyes and synthetic dyes. This property can also be used for adsorbing organic compounds such as drugs, for example, antibiotics, due to its aromatic nature (Bergmann and Machado, 2015). The reduced form of graphene oxide can be used in effluents containing radionuclides, organic solvents and oils, synthetic dyes and metal ions. However, it lacks a high intensity of negative charge on its surface, which favors its activity to absorb anionic compounds, such as anionic dyes.

Currently, a NF-coated filter medium is already in use to capture polluting gases at the entrance to gasoline turbines in industrial plants. This represents a large specific surface area for adsorption. Polymer NFs are the most used. Carbon-activated NFs with groups of nitrogen-containing groups, for example, are efficient in the adsorption of formaldehyde even at low casualties. Another advantage regarding filters containing carbon-based nanomaterials is their high stability (Li et al., 2014).

CNT sensors can be used to detect traces of benzene in the air. These nanomaterials are also effective against organic compounds, NO_x and CO (Leghrib and Llobet, 2011).

3.5 FUTURE PERSPECTIVES

Environmental problems are serious threats and are present in the daily life of animal and plant species in all ecosystems worldwide. Thus, it becomes increasingly necessary to find alternatives to remedy water, soil and air pollution. However, these alternatives cannot be toxic or endanger or harm life forms in a given environment. Carbon-based nanomaterials have been shown to be effective in solving these problems. There are several options for using and modifying these tools to better suit the intended use. In this way, carbon-based nanomaterials are manufactured of preferably biodegradable materials obtained through green synthesis for the remediation of water, soil and air contamination. There is also a need to produce safe, low-cost and biocompatible nanomaterials that can be used on a large scale without posing risks. Together with the production and use of these structures, environmental awareness and sustainable development must be prioritized in human actions.

3.6 CONCLUSION

Human actions have degraded the environment to unprecedented proportions, which leads to problems in the soil, air and water. These three components are essential for life on Earth. Environmental problems have become a serious issue in every country in the world. Thus, urgent measures need to be taken to preserve the life of the planet. Carbon-based nanomaterials have proved to be a useful alternative for environmental remediation. More research must be conducted to ensure not only the safe use of these materials but also ecologically friendly and large-scale production.

REFERENCES

Achari, G.A.; Kowshik, M. Recent Developments on Nanotechnology in Agriculture: Plant Mineral Nutrition, Health, and Interactions with Soil Microflora. *J. Agric. Food Chem.* **2018**, 66(33), 8647–8661.

Ahmed, M.B.; Zhou, J.L.; Ngo, H.H.; Johir, M.A.H.; Sun, L.; Asadullah, M.; Belhaj, D. Sorption of Hydrophobic Organic Contaminants on Functionalized Biochar: Protagonist Role of Pi-Pi Electron-Donor-Acceptor Interactions and Hydrogen Bonds. *J. Hazard. Mater.* **2018**, 360, 270–278.

Ahsan, M.A.; Jabbari, V.; Imam, M.A.; Castro, E.; Kim, H.; Curry, M.L.; Valles-Rosales, D.J.; Noveron, J.C. Nanoscale Nickel Metal Organic Framework Decorated Over Graphene Oxide and Carbon Nanotubes for Water Remediation. *Sci. Total Environ.* **2020**, 698, 134214.

Alharbi, O.M.L.; Basheer, A.A.; Khattab, R.A.; Ali, I. Health and Environmental Effects of Persistent Organic Pollutants. *J. Mol. Liq.* **2018**, 263, 442–453.

Azama, M.A.; Aliasa, F.M.; Tacka, L.W.; Amalina, R.N.; Mohamad, R.S.; Taibb, F.M. Electronic Properties and Gas Adsorption Behaviour of Pristine, Silicon and Boron-Doped (8, 0) Single-Walled Carbon Nanotube: A First Principles Study. *J. Mol. Graph. Model.* **2017**, 75, 85–93.

Bai, X.; Wang, L.; Wang, Y.; Yao, W.; Zhu, Y. Enhanced Oxidation Ability of g-C3N4 Photocatalyst via C60 Modification. *Appl. Catal. B Environ.* **2014**, 152–153, 262–270.

Bergmann, C.P.; Machado, F.M. (Eds.). *Carbon Nanomaterials as Adsorbents for Environmental and Biological Applications*. Springer, Berlin, Germany, **2015**, pp. 1–122.

Biglova, Y.N.; Mustafina, A.G. Nucleophilic Cyclopropanation of [60]Fullerene by the Addition—Elimination Mechanism. *RSC Adv*. **2019**, 9, 22428–22498.

Brunet, L.; Lyon, D.Y.; Hotze, E.M.; Alvarez, P.J.J.; Wiesner, M.R. Comparative Photoactivity and Antibacterial Properties of C60 Fullerenes and Titanium Dioxide Nanoparticles. *Environ. Sci. Technol.* **2009**, 43, 4355–4360.

Chen, Z.; Mahmud, S.; Cai, L.; He, Z.; Yang, Y.; Zhang, L.; Zhao, S.; Xiong, Z. Hierarchical Poly(vinylidene Fluoride)/Active Carbon Composite Membrane with Self-Confining Functional Carbon Nanotube Layer for Intractable Wastewater Remediation. *J. Membr. Sci.* **2020**, 603, 118041.

Cheng, X.; Kan, A.T.; Tomson, M.B. Uptake and Sequestration of Naphthalene and 1,2-Dichlorobenzene by C 60. *J. Nanopart. Res.* **2005**, 7(4–5), 555–567.

Chew, J.; Zhu, L.; Nielsen, S.; Graber, E.; Mitchell, D.R.; Horvat, J.; Mohammed, M.; Liu, M.; van Zwieten, L.; Donne, S. Biochar-Based Fertilizer: Supercharging Root Membrane Potential and Biomass Yield of Rice. *Sci. Total Environ.* **2020**, 713, 136431.

Dai, Y.; Zhang, N.; Xing, C.; Cui, Q.; Sun, Q. The Adsorption, Regeneration and Engineering Applications of Biochar for Removal Organic Pollutants: A Review. *Chemosphere*. **2019**, 223, 12–27.

Djordjevic, A.; Merkulov, D.S.; Lazarevic, M.; Borisev, I.; Medic, I.; Pavlovic, V.; Miljevic, B.; Abramovic, B. Enhancement of Nano Titanium Dioxide Coatings by Fullerene and Polyhydroxy Fullerene in the Photocatalytic Degradation of the Herbicide Mesotrione. *Chemosphere*. **2018**, 196, 145–152.

Fan, H.; Jiang, L.; Lee, Y.L.; Wong, C.K.C.; Ng, E.H.Y.; Yeung, W.S.B.; Lee, K.F. Bisphenol Compounds Regulate Decidualized Stromal Cells in Modulating Trophoblastic Spheroid Outgrowth and Invasion in Vitro. *Biol. Reprod.* **2020**, 102(3), 693–704.

Fang, Z.; Gao, Y.; Bolan, N.; Shaheen, S.M.; Xu, S.; Wu, X.; Xu, X.; Hu, H.; Lin, J.; Zhang, F.; Li, J.; Rinklebe, J.; Wang, H. Conversion of Biological Solid Waste to Graphene-Containing Biochar for Water Remediation: A Critical Review. *Chem. Eng. J.* **2020**, 390, 124611.

FAO. FAO Soils Portal. **2024**. Available from: https://www.fao.org/soils-portal/en/

Gangupomu, R.H.; Sattler, M.L.; Ramirez, D. Carbon Nanotubes for Air Pollutant Control via Adsorption: A Review. *Rev. Nanosci. Nanotechnol.* **2014**, 3, 149–160.

Ghadimi, M.; Zangenehtabar, S.; Homaeigohar, S. An Overview of the Water Remediation Potential of Nanomaterials and Their Ecotoxicological Impacts. *Water*, **2020**, 12(4), 1–23.

Glaser, B. Prehistorically Modified Soils of Central Amazonia: A Model for Sustainable Agriculture in the Twenty-First Century. *Philos. Trans. R. Soc. B*. **2007**, 362(1478), 187–196.

Guo, W.; Pan, B.; Sakkiah, S.; Yavas, G.; Ge, W.; Zou, W.; Tong, W.; Hong, H. Persistent Organic Pollutants in Food: Contamination Sources, Health Effects and Detection Methods. *Int. J. Environ. Res. Public Health.* **2019**, 16, 108828.

Hammad, M.; Fortugno, P.; Hardt, S.; Kim, C.; Salamon, S.; Schmidt, T.C.; Wende, H.; Schulz, C.; Wiggers, H. Large-scale synthesis of iron oxide/graphene hybrid materials as highly efficient photo-Fenton catalyst for water remediation. *Environ. Technol. Inno.* **2020**, In Press.

Hirsch, A. Addition Reactions of Buckminsterfullerene (C60). *Synthesis*. **1995**, 8, 895–913.

Inagaki, M.; Kang, F. *Materials Science and Engineering of Carbon: Fundamentals*, 2nd ed. Elsevier, Oxford, **2014**.

Jung, C.; Park, J.; Lim, K.H.; Park, S.; Heo, J.; Her, N.; Oh, J.; Yun, S.; Yoon, Y. Adsorption of Selected Endocrine Disrupting Compounds and Pharmaceuticals on Activated Biochars. *J. Hazard. Mater.* **2013**, 263(Pt 2), 702–710.

Kabiri, S.; Degryse, F.; Tran, D.N.; da Silva, R.C.; McLaughlin, M.J.; Losic, D. Graphene Oxide: A New Carrier for Slow Release of Plant Micronutrients. *ACS Appl. Mater. Interfaces*. **2017**, 9(49), 43325–43335.

Kazemzadeh, K.; Zali, A. Pollutant-Induced DNA Damage: Mechanisms and Consequences in Cancer Development. In *Handbook of Cancer and Immunology*. Springer, Singapore, **2023**, pp. 1–23.

Keller, A.A.; Wang, H.; Zhou, D.; Lenihan, H.S.; Cherr, G.; Cardinale, B.J.; Miller, R.; Ji, Z. Stability and Aggregation of Metal Oxide Nanoparticles in Natural Aqueous Matrices. *Environ. Sci. Technol.* **2010**, 44, 1962–1967.

Khan, S.T.; Malik, A. Engineered Nanomaterials for Water Decontamination and Purification: From Lab to Products. *J. Hazard Mater.* **2019**, 363, 295–308.

Kiew, S.F.; Kiew, L.V.; Lee, H.B.; Imae, T.; Chung, L.Y. Assessing Biocompatibility of Graphene Oxide-Based Nanocarriers: A Review. *J. Control. Release.* **2016**, 226, 217–228.

Lee, J.; Mackeyev, Y.; Cho, M.; Wilson, L.J.; Kim, J.H.; Alvarez, P.J.J. C60 Aminofullerene Immobilized on Silica as a Visible-Light-Activated Photocatalyst. *Environ. Sci. Technol.* **2010**, 44, 9488–9495.

Leghrib, R.; Llobet, E. Quantitative Trace Analysis of Benzene Using an Array of Plasma-Treated Metal-Decorated Carbon Nanotubes and Fuzzy Adaptive Resonant Theory Techniques. *Anal. Chim. Acta.* **2011**, 708(1–2), 19–27.

Lei, S.; Shi, Y.; Qiu, Y.; Che, L.; Xue, C. Performance and Mechanisms of Emerging Animal-Derived Biochars for Immobilization of Heavy Metals. *Sci. Total Environ.* **2019**, 646, 1281–1289.

Li, P.; Wang, C.; Li, Z.; Zong, Y.; Zhang, Y.; Yang, X.; Li, S.; Wei, F. Hierarchical Carbon-Nanotube/Quartz-Fiber Films with Gradient Nanostructures for High Efficiency and Long Service Life Air Filters. *RSC Adv.* **2014**, 4, 54115–54121.

Lin, Y.C.; Li, J.Y.; Yen, W.T. Low Temperature ITO Thin Film Deposition on PES Substrate Using Pulse Magnetron Sputtering. *Appl. Surf. Sci.* **2008**, 254(11), 3262–3268.

Liu, H.; Liang, J.; Fu, S.; Li, L.; Cui, J.; Gao, P.; Zhao, F.; Zhou, J. N Doped Carbon Quantum Dots Modified Defect-Rich g-C3N4 for Enhanced Photocatalytic Combined Pollutions Degradation and Hydrogen Evolution. *Colloids Surf. A: Physicochem. Eng. Asp.* **2020**, 591, 124552.

Liu, R.; Hou, Y.; Jiang, S.; Nie, B. Ag(I)-Hived Fullerene Microcube as an Enhanced Catalytic Substrate for the Reduction of 4-Nitrophenol and the Photodegradation of Orange G Dye. *Langmuir.* **2020**, 36, 5236–5242.

Lu, F.; Astruc, D. Nanocatalysts and Other Nanomaterials for Water Remediation from Organic Pollutants. *Coord. Chem. Rev.* **2020**, 408.

Mezzelani, M.; Gorbi, S.; Regoli, F. Pharmaceuticals in the Aquatic Environments: Evidence of Emerged Threat and Future Challenges for Marine Organisms. *Mar. Environ. Res.* **2018**, 140, 41–60.

Mohamed, E.F. Nanotechnology: Future of Environmental Air Pollution Control. *Environ. Manag. Sustain. Dev.* **2017**, 6, 429–454.

Nasrollahzadeh, M.; Sajjadi, M.; Iravani, S.; Varma, R.S. Green-Synthesized Nanocatalysts and Nanomaterials for Water Treatment: Current Challenges and Future Perspectives. *J. Hazard. Mater.* **2021**, 401, 123401.

Novoselov, K.S.; Geim, A.K.; Morozov, S.V.; Jiang, D.; Zhang, Y.; Dubonos, S.V.; Firsov, A.A. Electric Field Effect in Atomically Thin Carbon Films. *Science.* **2004**, 306, 666–669.

Ong, Y.T.; Ahmad, A.L.; Zein, S.; Huat Tan, S. A Review on Carbon Nanotubes in an Environmental Protection and Green Engineering Perspective, Brazil. *J. Chem. Eng.* **2010**, 27(2).

Petter, F.A.; Ferreira, T.S.; Sinhorin, A.P.; Lima, L.B.; Almeida, F.A.; Pacheco, L.P.; Silva, A.F. Biochar Increases Diuron Sorption and Reduces the Potential Contamination of Subsurface Water with Diuron in a Sandy Soil. *Pedosphere.* **2019**, 29(6), 801–809.

Pulizzi, F.; Sun, W. Treating Water with Nano. *Nat. Nanotechnol.* **2018**, 13, 633.

Qi, K.; Selvaraj, R.; Al Fahdi, T.; Al-Kindy, S.; Kim, Y.; Wang, G.; Tai, C.W.; Sillanpaa, M. Enhanced Photocatalytic Activity of Anatase-TiO2 Nanoparticles by Fullerene Modification: A Theoretical and Experimental Study. *Appl. Surf. Sci.* **2016**, 387, 750–758.

Rani, U.A.; Ng, L.Y.; Ng, C.Y.; Mahmoudi, E. A Review of Carbon Quantum Dots and Their Applications in Wastewater Treatment. *Adv. Colloid Interface Sci.* **2020**, 278, 102124.

Rauti, R.; Musto, M.; Bosi, S.; Prato, M.; Ballerini, L. Carbon Review Article Properties and Behavior of Carbon Nanomaterials When Interfacing Neuronal Cells: How Far Have We Come? *Carbon.* **2019**, 143, 430–446.

Sargin, I.; Baran, T.; Arslan, G. Environmental Remediation by Chitosan-Carbon Nanotube supported Palladium Nanoparticles: Conversion of Toxic Nitroarenes into Aromatic Amines, Degradation of Dye Pollutants and Green Synthesis of Biaryls. *Sep. Purif. Technol.* **2020**, 247, 116987.

Semeniuk, M.; Yi, Z.; Poursorkhabi, V.; Tjong, J.; Jaffer, S.; Lu, Z.H.; Sain, M. Future Perspectives and Review on Organic Carbon Dots in Electronic Applications. *ACS Nano.* **2019**, 13, 6224–6255.

Seng, R.X.; Tan, L.L.; Lee, W.P.C.; Ong, W.J.; Chai, C.P. Nitrogen-Doped Carbon Quantum Dots-Decorated 2D Graphitic Carbon Nitride as a Promising Photocatalyst for Environmental Remediation: A Study on the Importance of Hybridization Approach. *J. Environ. Manag.* **2020**, 255, 109936.

Silva, S.A.M., Perez, J., Torresi, R.M., Luengo, C.A., Ticianelli, E.A. Surface and Electrochemical Investigations of a Fullerene Soot. *Electrochim. Acta.* **1999**, 44(20), 3565–3574.

Smrithi, S.P.; Kottam, N.; Arpitha, V.; Narula, A.; Anilkumar, G.N.; Subramanian, K.R.V. Tungsten Oxide Modified with Carbon Nanodots: Integrating Adsorptive and Photocatalytic Functionalities for Water Remediation. *J. Sci.: Adv. Mater. Dev.* **2020**, 5, 73–83.

Stankovich, S., Dikin, D.A., Dommett, G.H.B., Kohlhaas, K.M., Zimney, E.J., Stach, E.A., Piner, R.D., Nguyen, S.B.T., Ruoff, R.S. Graphene-Based Composite Materials. *Nature.* **2006**, 442(7100), 282.

Su, F.; Lu, C.; Cnen, W.; Bai, H.; Hwang, J.F. Capture of CO2 from Flue Gas via Multiwalled Carbon Nanotubes. *Sci. Total Environ.* **2009**, 407(8), 3017–3023.

Taniguchi, N. On the Basic Concept of 'Nano Technology'. *Proc. Intl. Conf. Prod. Eng. Tokyo, Part II.* Tokyo: Japan Society of Precision Engineering, **1974**.

Tara, N.; Siddiquia, S.I.; Bach, Q.V.; Chaudhry, S.A. Reduce Graphene Oxide-Manganese Oxide-Black Cumin Based Hybrid Composite (rGO-MnO2/BC): A Novel Material for Water Remediation. *Mater. Today Commun.* **2020**, 25, 101560.

Tiwari, S.K.; Sahoo, S.; Wang, N.; Huczko, A. Graphene Research and Their Outputs: Status and Prospect. *J. Sci. Adv. Mat. Dev.* **2020**, 5, 10–29.

Tonelli, F.C.P.; Tonelli, F.M.P. Concerns and Threats of Xenobiotics on Aquatic Ecosystems. In *Bioremediation and Biotechnology Vol 3: Persistent and Recalcitrant Toxic Substances.* Springer, New York, **2020**, pp. 15–23.

Tonelli, F.C.P.; Tonelli, F.M.P. Biocompatibility of Green Synthesized Nanomaterials. In *Synthesis of Bionanomaterials for Biomedical Applications Micro and Nano Technologies.* Elsevier, Amsterdam, **2023**, pp. 209–223.

Tonelli, F.M.P.; Lacerda, S.M.S.N.; Paiva, N.C.O.; Lemos, M.S.; Jesus, A.C.; Pacheco, F.G.; Correa Junior, J.D.; Ladeira, L.O.; Furtado, C.A.; Franca, L.R.; Resende, R.R. Efficiently and Safely in Gene Transfection in Fish Spermatonial Stem Cells Using Nanomaterials. *RSC Adv.* **2016**, 58, 1058.

Tonelli, F.M.P.; Lacerda, S.M.S.N.; Paiva, N.C.O.; Pacheco, F.G.; Scalzo Junior, S.R.A.; Macedo, F.H.P.; Cruz, J.S.; Pinto, M.C.X.; Correa Junior, J.D.; Ladeira, L.O.; França, L.R.; Guatimosim, S.C.; Resende, R.R. Functionalized Nanomaterials: Are They Effective to Perform Gene Delivery to Difficult-to-Transfect Cells with No Cytotoxicity? *Nanoscale.* **2015**, 7, 18036–18043.

Tonelli, F.M.P.; Tonelli, F.C.P.; Ferreira, D.R.C.; da Silva, K.E.; Cordeiro, H.G.; Ouchida, A.T.; Nunes, N.A.M. Biocompatibility and Functionalization of Nanomaterials. In *Intelligent Nanomaterials for Drug Delivery Applications.* Elsevier, Amsterdam, **2020**, pp. 85–103.

Trivedi, M.; Johri, P.; Singh, A.; Singh, R.; Tiwari, R.K. Latest Tools in Fight Against Cancer: Nanomedicines. In *NanoBioMedicine.* Springer, Singapore, **2020**, pp. 139–164.

Ukaogo, P.O.; Ewuzie, U.; Onwuka, C.V. Environmental Pollution: Causes, Effects, and the Remedies. In *Microorganisms for Sustainable Environment and Health.* Elsevier, Amsterdam, **2020**, pp. 419–429.

Venkatraman, G.; Giribabu, N.; Mohan, P.S.; Muttiah, B.; Govindarajan, V.K.; Alagiri, M.; Rahman, P.S.A.; Karsani, S.A. Environmental Impact and Human Health Effects of Polycyclic Aromatic Hydrocarbons and Remedial Strategies: A Detailed Review. *Chemosphere* **2024**, 351, 141227.

Vithanage, M.; Herath, I.; Almaroai, Y.A.; Rajapaksha, A.U.; Huang, L.; Sung, J.K.; Lee, S.S.; Ok, Y.S. Effects of Carbon Nanotube and Biochar on Bioavailability of Pb, Cu and Sb in Multi-Metal Contaminated Soil. *Environ. Geochem. Health.* **2017**, 39(6), 1409–1420.

Wang, Q.; Cai, J.; Biesold-McGee, G.V.; Huang, J.; Ng, Y.H.; Sun, H.; Wang, J.; Lai, Y.; Lin, Z. Silk Fibroin-Derived Nitrogen-Doped Carbon Quantum Dots Anchored on TiO2 Nanotube Arrays for Heterogeneous Photocatalytic Degradation and Water Splitting. *Nano Energy* **2020**, 78, 105313.

Wang, S.; Sun, H.; Ang, H.M.; Tadé, M.O. Adsorptive Remediation of Environmental Pollutants Using Novel Graphene-Based Nanomaterials. *Chem. Eng. J.* **2013**, 226, 336–347.

Wu, X.; Chen, L.; Zheng, C.; Yan, X.; Dai, P.; Wang, Q.; Li, W.; Chen, W. Bubble-Propelled Micromotors Based on Hierarchical MnO2 Wrapped Carbon Nanotube Aggregates for Dynamic Removal of Pollutants. *RSC Adv.* **2020**, 10, 14846–14855.

Wu, Y.; Wang, F.; Jin, X.; Zheng, X.; Wang, Y.; Wei, D.; Zhang, Q.; Feng, Y.; Xie, Z.; Chen, P.; Liu, H.; Liu, G. Highly Active Metal-Free Carbon Dots/g-C3N4 Hollow Porous Nanospheres for Solar-Light-Driven PPCPs Remediation: Mechanism Insights, Kinetics and Effects of Natural Water Matrices. *Water Res.* **2020**, 172, 115492.

Xie, H.; Li, P.; Shao, J.; Huang, H.; Chen, Y.; Jiang, Z.; Chu, P.K.; Yu, X.F. Electrostatic Self-Assembly of $Ti_3C_2T_x$ MXene and Gold Nanorods as an Efficient Surface-Enhanced Raman Scattering Platform for Reliable and High-Sensitivity Determination of Organic Pollutants. *ACS Sens.* **2019**, 4, 2303–2310.

Xiong, J.; Zhang, D.; Lin, H.; Chen, Y. Amphiprotic Cellulose Mediated Graphene Oxide Magnetic Aerogels for Water Remediation. *Chem. Eng. J.* **2020**, 400, 125890.

Xu, C.; Nasrollahzadeh, M.; Sajjadi, M.; Maham, M.; Luque, R.; Puente-Santiago, A.R. Benign-by-Design Nature-Inspired Nanosystems in Biofuels Production and Catalytic Applications. *Renew. Sustain. Energy Rev.* **2019**, 112, 195–252.

Yan, J.; Gong, J.L.; Zeng, G.M.; Song, B.; Zhang, P.; Liu, H.Y.; Huan, S.Y.; Li, X.D. Carbon Nanotube-Impeded Transport of Non-Steroidal Anti-Inflammatory Drugs in Xiangjiang Sediments. *J. Colloid Interface Sci.* **2017**, 498, 229–238.

Yeung, B.H.Y.; Wan, H.T.; Law, A.Y.S.; Wong, C.K.C. Endocrine Disrupting Chemicals: Multiple Effects on Testicular Signaling and Spermatogenesis. *Spermatogenesis* **2011**, 1(3), 231–239.

Ying, Y.; Ying, W.; Li, Q.; Meng, D.; Ren, G.; Yan, R.; Peng, X. Recent Advances of Nanomaterial-Based Membrane for Water Purification. *Appl. Mater. Today* **2017**, 7, 144–158.

Yuan, C.; Hung, C.-H.; Huang, W.-L. Enhancement with Carbon Nanotube Barrier on 1,2-Dichlorobenzene Removal from Soil by Surfactant-Assisted Electrokinetic (Saek) Process—the Effect of Processing Fluid. *Sep. Sci. Technol.* **2009**, 44(10), 2284–2303.

Zhang, X.X., Bing, Y., Dai, Z.Q., Luo, C.C. The Gas Response of Hydroxyl Modified SWCNTs and Carboxyl Modified SWCNTs to H2S and SO2. *Prz. Elektrotech.* **2012**, 88, 311–314.

Zhang, Y.; Yang, M.; Portney, N.G.; Cui, D.; Budak, G.; Ozbay, E.; Ozkan, M.; Ozkan, C.S. Zeta Potential: A Surface Electrical Characteristic to Probe the Interaction of Nanoparticles with Normal and Cancer Human Breast Epithelial Cells. *Biomed. Microdevices.* **2008**, 10, 321–328.

Zhao, L.; Yang, S.-T.; Yilihamu, A.; Wu, D. Advances in the Applications of Graphene Adsorbents: From Water Treatment to Soil Remediation. *Rev. Inorg. Chem.* **2019**, 39(1), 47–76.

Zhou, Y.; Gao, B.; Zimmerman, A.R.; Fang, J.; Sun, Y.; Cao, X. Sorption of Heavy Metals on Chitosan-Modified Biochars and Its Biological Effects. *Chem. Eng. J.* **2013**, 231, 512–518.

4 Non-Carbon-Based Nano-Materials Used for Environmental Pollution Remediation

Bhat Mohd Skinder and Masarat Nabi

4.1 INTRODUCTION

Our environment is surrounded by various harmful contaminants that need to be treated by environmentally friendly approaches or techniques (Özkara et al. 2016). Innovative approaches for the remediation of air, water, and soil pollutants are continually being developed (Manisalidis et al. 2020). In recent decades, "nanotechnology" has received a great deal of interest because of its distinctive characteristics (Guerra et al. 2018). Nanotechnology specifically means any technology on the "nanometer scale" (a nano-metre is 10^{-9} of a meter), which has numerous applications across the world (Capek 2019). Because of their higher surface/volume proportion, nano-particles offer improved reactivity and, therefore, enhanced efficiency compared to their chunkier equivalents. Furthermore, as opposed to conventional methods, nano-particles provide the ability to exploit specific surface morphology so that they can be conjugated or bonded with functional groups that might bind particular atoms of interest (contaminants) for effective remediation (Coetzee et al. 2020). In addition, deliberately modifying some characteristics of nano-particles (such as the particle size, morphology, porosity, elemental makeup, etc.) may impart enhanced promising applications that have a significant impact on the efficiency of the contaminant remediating substance. Mechanisms that are produced as a fusion of many various components (hybrids, composites, etc.), extracting unique desirable characteristics from each of its constituents, are significantly quite productive, versatile, and reliable, unlike techniques that focus on a limited nano-platform. The specificity and performance of the sample could be enhanced by fabricating the substance with particular molecules appropriate to target specific pollutants (Campbell et al. 2015). It is critical that after they have been used, the pollution remediation products should not result in some other contaminant. Thus, for this area of use, bio-degradable substances are highly significant. The incorporation of these substances can improve shareholder trust and appreciation of a specific technique in the sense that no waste production needs to be discarded, but it can also provide a cleaner and more sustainable solution for pollutant remediation in the environment. Moreover, emerging advancements based on pollutant-specific detection are particularly appealing, as poor effectiveness resulting from non-specific can be resolved. Furthermore, many naturally occurring and crafted nano-particles, including chitosan, silver nano-particles (nAg), and photocatalytic titanium dioxide (TiO_2), have shown to possess significant antibacterial effects. Thus, research efforts have been oriented to employing nano-technology principles and incorporating them with physio-chemical changes of the target particle surface to achieve crafted materials that can resolve various key difficulties in pollutant abatement (Pandey and Fulekar 2012; Guerra et al. 2018). Target-based detection, low cost, excellent thermal stability, non-toxicity, good biocompatibility, biodegradability, renewability,

DOI: 10.1201/9781003186298-4

and reusability are among the primary concerns that need to be addressed in the development of innovative sustainable remediation nano-particles (Mishra et al. 2018). Some are fundamentally unstable in ordinary operating circumstances considering the possible benefits of the earlier discussed nano-particles, so their process entails specialised methods for nano-particle development. To deter accumulation, increase uniformity, and improve consistency, extra processes are required. A further consideration that can sometimes place limits on their usage is the potential toxicity of the metal-specific nano-particles used in the treatment methods including their derivatives and recovery expenses from the clean-up area. Therefore, to design potential nano-particles that can resolve environmental issues, detailed knowledge of substance-specific technologies, manufacturing operations, and quality improvement is needed.

4.2 CHEMICAL COMPOSITION

Nano-particles are made up of a particular component or can be a combination of a variety of substances. Naturally found nano-particles are mostly aggregations of substances with different formulations, while a range of methods can effectively synthesise highly pure substances with a single constituent. Three key forms of molecular sorting in hybrid nano-particles explain how the element atoms are ordered in the nano-particle.

4.3 MIXED NANO-PARTICLES

Mixed nano-particles can be either random or ordered. Randomly mixed alloys correspond to solutions of solids, while ordered ones correspond to ordered arrangements of atoms.

4.4 CORE-SHELL NANO-PARTICLES

Core-shell nano-particles can have variable sizes, thicknesses, and shapes of the core and shell with varying surface morphologies. They may be of different forms viz., spherical, centric, eccentric, star-like, or tubular. Based on these forms, their functions change from material to material. In various disciplines of medicinal biotechnology, core-shell nano-particles have different uses, such as molecular bio-imaging, drug delivery, cancer therapy, and so on. When some functional group replaces the nano-particle surface or is covered with a thin coating with materials of altered different components, the functions of the nano-particles are improved relative to the non-functional uncoated ones. There are multiple kinds of core/shell combinations, such as metalcore; metal shells, metalcore; non-metal shells, metalcore; polymer shells, non-metal core; non-metal shells, polymer core; non-metal shells and polymer core; and polymer shells, and vice versa. Core-shell nano-particles specifically made up of metals such as platinum, palladium, and gold have gained much interest as their characteristics vary substantially from their mass (Khan et al. 2019). These nano-particles display electron containment and surface effects based on quantum-size effects and can be used as sensors, electronics, and catalysts for a variety of specialized functional purposes. Core-shells based on materials such as semiconductors and polymers have gained significant attention as a range of novel technologies with potential uses in different electronics, such as organic light-emitting diodes (OLEDs), organic photovoltaics (OPVs), and sensors and organic field-effect transistors (OFETs), because of their attributes, namely, low cost, effectiveness, and simplistic manufacturing capabilities (Watson et al. 2016). Furthermore, they exhibit applications in biomedical domains such as polymer-based drug carriers, extra-corporeal tools, prostheses, and dental products. These nano-materials are synthesised via various fabrication techniques such as hydro; solvothermal synthesis, sol-gel technique, and emulsion; and micro-emulsion polymerisation. The synthesis methods, attributes, and morphologies can be altered based on the materials of the core and shell.

4.5 LAYERED NANO-PARTICLES

Layered nano-particles are onion or dumb shell-like formations (Benelmekki 2015). They comprise a versatile and considerable unexplored range of two-dimensional structures of novel electrical characteristics and increased surface areas that are critical in features such as signalling, catalysis, and storing energy. Layered hydroxides and carbon-based layered substances such as graphene and graphite are widely known layered substances. However, some other types of layered substances, such as transition dichalcogenides, transition metal oxides, and two-dimensional substances, for example, bismuth selenide, boron nitride, and bismuth telluride, have also been fabricated (Yaya et al. 2012). Exfoliation is considered to be the basic technique to develop layered nano-particles in which a single layer/sheet from a voluminous product are removed either physically or chemically (Le et al. 2020). To achieve their maximum efficiency, layered structures should be exfoliated. For instance, exfoliated Bi2Te3 surfaces can demonstrate improved thermoelectric performance by suppressing thermal conductivity. Residual bulk conductance can be minimized by the process of the exfoliation of two-dimensional topological insulators such as Bi2Te3 and Bi2Se3, illustrating the surface effects. Furthermore, as the number of layers is decreased, we can anticipate modifications in electrical features, for example, the implicit conduction band of molybdenum disulfide (MoS_2) bulk appears to be direct in some layers/sheets. Hexagonal boron nitride is yet another structured substance with a graphite-like arrangement consisting of hexagonal ring structures divided by 3.33 Å wherein each boron atom is bound to 3 atoms of nitrogen, and vice-versa; the range between boron and nitrogen is 1.44 Å. Each boron atom associates with a nitrogen atom between the layers/sheets via electrostatic interaction and van der Waal forces. Such adaptability allows these substances to be exceedingly functional for electrical uses (Yaya et al. 2012).

4.6 TYPES OF NANO-MATERIALS

Nano-particles have been employed in different fields such as the environmental, industrial, and biomedical domains. They are categorized into three groups:

(i) Incidental nano-particles **(ii)** Naturally occurring nano-particles and **(iii)** Crafted nano-particles. Naturally occurring nanoparticles are usually present in soil and perform a significant part in various biogeochemical processes viz. clays, organic matter, and iron oxides. Incidental nano-particles reach the surroundings via air pollution, solid or liquid waste systems from nano-particle processing plants, farming activities, burning of fuels and weathering. Crafted nano-particles are developed with unique features and can be discharged into the air via various industrial activities. Zero valent metals (zero-valent iron (nZVI), quantum dots, dendrimers, and composite nano-particles are some significant illustrations of engineered nano-particles, these have been found highly applicable in areas like remediation of different components of the environment (soil, water, sediment etc.); photovoltaics; telecommunications; drug carrier; sensors; cancer detection etc (U.S. EPA 2007, 2008).

4.7 DIMENSION-BASED NANO-PARTICLES

4.7.1 One-Dimensional Nano-Particles

Small films or surface coatings are usually substances with a single dimension on the nano-metre scale. Examples include nano-tubes, nano-fibres, nano-wires, nano-rods, and nano-filaments (Bashir and Liu 2015). Over the years, thin films have been designed and employed in several areas, namely, electronics, data storage systems, chemical and biological sensors, fibre-optic systems, and magnetic and optical devices. They can be accumulated using multiple techniques and can be stably developed at the atomic stage (a monolayer) Benelmekki (2015).

4.7.2 Two-Dimensional Nano-Particles

These nano-particles possess two dimensions on the nano-metre scale viz., fibres, nano-films, nano-layers, nano-coatings, nano-wires, nano-tubes, fibrils, and dendrimers. Particles having a wide aspect proportion on the nano-metre scale are often called two-dimensional nano-particles. The features of two-dimensional nano-particles are less known, and their production capacities are also less developed (Benelmekki 2015).

4.7.3 Three-Dimensional Nano-Particles

Three-dimensional nano-particles are known to be particles that are at nano-scale in all three dimensions, for instance, quantum dots, nano-crystals, fullerenes, particles, precipitates, and colloids. Many three dimensional structures are widely recognized including natural nano-materials and combustion products, metal oxides, black carbon, titanium oxide, and zinc oxide. However, some pose major obstacles in terms of development and the interpretation of characteristics, such as dendrimers, fullerenes, and quantum dots Benelmekki (2015).

4.8 POLYMER-BASED NANO-MATERIALS

Even though the broad surface area and volume proportion of nano-materials lead to high reactivity with increased efficiency, at the same time, because of the paucity of usability, the phenomena of accumulation, non-specificity, and poor stability can restrict the usage of such nano-technologies (Guerra et al. 2018). The use of a host material, which acts as a medium or aid for other kinds of materials, is a substitute for improving the strength of nano-scale materials such as nano-particles (Zhao et al. 2011; Guerra et al. 2018). Polymers are predominantly used for the identification and elimination of residues such as organic contaminants (viz., aromatic and aliphatic hydrocarbons, volatile organic carbons, pharmaceuticals, etc.), heavy metals (viz., manganese (Mn), iron (Fe), arsenic (Ar), and other heavy metals), harmful gases (viz., carbon monoxide (CO), sulphur dioxide (SO_2), and nitrogen oxide (Nox)), and a diverse range of pathogens (viz., viruses, bacteria, and parasites). Polymer-specific hosts such as emulsifying agents, surface-functionalized ligands, and stabilizers are being used to improve the strength and durability, to resolve some disadvantages of immaculate nano-particles, and to provide other potential benefits, such as improving the compressive performance, thermal stability, and reusability of the substance. For the cleanup of poly-nuclear aromatic hydrocarbons (PAHs) from soils, amphiphilic polyurethane (APU) nano-particles have been produced, authenticating the concept that organic nano-particles can be developed with desirable characteristics (Tungittiplakorn et al. 2004; Guerra et al. 2018). The nano-particles hydrophilic surface facilitates movement in the soil, whereas affinity for hydrophobic pollutants is conferred by the hydrophobic interior of the substance. Almost eighty per cent recovery of phenanthrene has been reported from polluted aquifer sand by using amphiphilic polyurethane nano-particles (Tungittiplakorn et al. 2004; Guerra et al. 2018). Various formulations studied have revealed that as the size of the hydrophobic base increases, the tendency of amphiphilic polyurethane nano-particles to interact with phenanthrene also increases. Some polymer-based nano-particles are used for the sensing and identification of pollutants, which is summarized in Table 4.1.

4.9 METAL OXIDE-BASED NANO-PARTICLES

Inorganic nano-particles are oxide-specific nano-particles that are typically made up of non-metals and metals. They comprise semiconductor crystals that are densely packed consisting of hundreds or thousands of atoms. These are widely used in the treatment of harmful wastewater pollutants (Das et al. 2015). Nano-materials have shown different behaviour in an aqueous environment, where some nano-particles demonstrated the positive potential for the remediation of water contamination. Some of the nano-materials comprise TiO_2 (Gao et al. 2008; Anaya-Esparza et al. 2019), dendrimers (Barakat et al. 2013; Anjum et al. 2019), ZnO ((Tuzen and Soylak 2007; Anaya-Esparza et al. 2019; Anjum et al. 2019), MgO (Anaya-Esparza et al. 2019; Anjum et al. 2019), MnO (Feng et al. 2012; Anjum et al. 2019), and Fe2O3 (Xu et al. 2008; Anjum et al. 2019). Oxide-specific nano-particles are categorised with an elevated Brunauer—Emmett—Teller (BET) surface area, reduced solubility, low environmental effects, and zero secondary contaminants (Anjum et al. 2019). Some metal-specific nano-particles/materials and their environmental remediation uses are summarised in Table 4.2.

4.9.1 Iron-Specific Nano-Particles

The inherent presence of iron and its easy method of transformation make ferric oxide a cost-effective material for toxic metal adsorption. It is an environmentally friendly commodity that can be used explicitly in a polluted area with a much lower probability of secondary pollution (Guerra et

TABLE 4.1
Polymer-Specific Nano-Particles for Contaminant Sensing and Identification

Nano-Particle	Polymer medium	Technique	Specific contaminant
Iron oxide	Polypyrrole	Instantaneous gelation and polymerisation	Carbon dioxide, methane, nitrogen
Titanium dioxide	Polyaniline (PANI)	Chemical polymerisation and a sol-gel method	Trimethylamine (C_3H_9N)
Tin(IV) oxide	Polystyrene/polyaniline (PSS/PANI)	In-situ self-assembly	Carbon monoxide
Palladium	Polyaniline	Oxidative polymerisation of the solution with Pd NPs	Methanol (CH_3OH)
Gold	Chitosan	Mixed in solution	Zn^{2+}, Cu^{2+}
Tin(IV) oxide	Polyaniline (PANI)	Hydrothermal method	Ethanol (C_2H_5OH), acetone (C_3H_6O)

Source: Khin et al. 2012

TABLE 4.2
Nano-Materials for the Disinfection and Bacterial Control of Water

S. No	Nano-Materials	Removed microorganisms	Removal efficiency	Reference
1	Silver nano-materials loaded in kaolin clay	E. coli, Salmonella spp.	80% for E. coli; 9% for Salmonella spp. (concentration: 0.1 ppm)	Hassouna et al. (2017)
2	Zn_3P_2 (Zinc phosphide)-nano-wires	E. coli	Greater than a 4 log reduction	Vance et al. (2018)
3	Ag—nano-material in polysulfone membranes	E. coli	90% efficiency (silver leaching 2 μg L^{-1})	Andrade et al. (2015)
4	Ag—nano-material loaded in chitosan cryogels	B. subtilis and E. coli	3 log reduction (silver content—7.5 mg/g)	Fan et al. (2018)
5	TiO_2—iron oxide nano-composite	E. coli	99.28% removal efficiency (initial concentration of bacteria: 10 $mgmL^{-1}$)	Sharma et al. (2018)

al. 2018). Numerous iron-containing compounds, including iron sulphide, oxyhydroxide-containing iron, and alumino-silicate minerals, have been employed successfully in reducing and depositing metal ions. Among the iron-specific substances, the most effective for groundwater remediation has been observed to be elemental iron. Also with the development of nanotechnology, the use of bulk iron-specific methods for water decontamination has been substituted by iron-specific nano-particles. Various factors influence the adsorption of heavy metals on ferric oxide nano-particles viz., the incubation period, adsorbent dose, temperature, and pH. Several studies have been carried out to enhance the surface adsorption capability of ferric oxides. For instance, 3-aminopropyltrimethoxysilane has been used to alter the surface of ferric oxide nano-particles (Palimi et al. 2014). Furthermore, the alteration of such nano-adsorbents indicates a better tendency for the elimination from wastewater of toxic pollutants such as Cr^{3+}, Co^{2+}, Ni^{2+}, Cu^{2+}, Cd^{2+}, Pb^{2+}, and As^{3+}(Gupta et al. 2015).

4.9.2 Manganese Oxide-Specific Nano-Particles

Because of their large surface area (BET) and polymorphic nature, manganese oxide nano-particles exhibit higher adsorption power (Luo et al. 2010). Manganese oxide nano-particles have succeeded in eliminating a range of heavy metals such as arsenic from wastewater (Wang et al. 2011). Nano-porous and nano-tunnel manganese oxides are among the significant examples of altered manganese oxides. Adsorption on hydrous manganese oxide of different heavy metals such as divalent Pb, Cd, and Zn typically occurs due to the inner sphere-forming pathway that can be described by the principle of ion exchange (Zaman et al. 2009). Nevertheless, in two phases, the process of adsorption takes place in two stages, metal ions initially adsorb on the outer surface of hydrous manganese dioxides (HMOs), and then, intra-particle diffusion proceeds.

4.9.3 Zinc Oxide (ZnO)-Based Nano-Particles

Zinc oxide (ZnO) has permeable nano-structures for the adsorption of heavy metals with an elevated BET surface region. For instance, nano-adsorbents such as nano-assemblies, nano-plates, nano-sheet micro-spheres, and nano-rods are commonly employed for heavy metal remediation from wastewater. Compared to the conventional zinc oxides, these altered types of nano-adsorbents display an excellent elimination capacity of heavy metals (Kumar et al. 2013). Nano-plates and permeable nano-sheets have been employed to strip Cu (II) from wastewater. These altered nano-adsorbents display an elevated removal performance of divalent copper because of their unique nano-structures compared to conventional zinc oxide. In addition, nano-assemblies are being employed to extract various forms of heavy metals, including divalent metals such as nickel, copper, cobalt, lead, mercury, and cadmium (Singh et al. 2013). Nano-assemblies (micro-porous) have a strong tendency for Pb^{2+}, Hg^{2+}, and As^{3+} adsorption due to their electropositive character (Gupta et al. 2015).

4.9.4 Magnesium Oxide (MgO)-Specific Nano-Particles

Magnesium oxide (MgO) is being employed to clean up waters contaminated with various heavy metals. The MgO micro-sphere is a unique design that can enhance the tendency of adsorption for heavy metal elimination (Gupta et al. 2015). Different types of alterations have been conducted in the structure of nano-particles to maximise the adsorption ability of MgO such as 3D units, nano-cubes, nano-rods, nano-belts, and nano-wires (Zhu et al. 2001; Yin et al. 2002; Klug and Dravid 2002; Mo et al. 2005; Engates and Shipley 2011; Guerra et al. 2018).

4.10 SILVER NANO-PARTICLES

Silver can be used to produce nanosized materials (Chou et al. 2005). The effective bactericidal activity of silver nano-particles against a broad variety of species (for instance, viruses, bacteria, fungi, etc.) has been established. Thus, they are also commonly employed for water disinfection. In proteins, silver (Ag^+) ions interact with thiol groups that contribute to the inhibition of respiratory enzymes and the development of a species of reactive oxygen. These ions also prevent the replication of DNA and impact the cell membrane's structure and permeability. Ag+ ions are ultraviolet active ions that lead to an increase in bacterial and viral ultraviolet (UV) inactivation. Numerous pathways for their anti-bacterial effects have been proposed to date. The binding of nano-particles to the surface modifies the characteristics of the membrane. To degrade lipo-polysaccharides, these particles have been observed to accumulate within the membrane by developing holes, resulting in increased membrane permeability. The bacterial cell can be penetrated by silver nano-particles, causing DNA damage. Anti-microbial Ag+ ions are released through the breakdown of silver nano-particles, and bacterial strains such as *P. aeruginosa* and *E. coli* are sensitive to silver particles of a size less than 10 nm (Gogoi et al. 2006; Guerra et al. 2018). One to 10 nm silver nano-particles link with the glycoproteins of viruses, thus preventing them to bind with the host cells. Moreover, it has been observed that triangular-shaped nano-plates (nAg) are much more noxious than silver ions, rods (nAg), and spheres (nAg). Anti-bacterial effects against *Salmonella entrica*, *S. aureus E. Coli*, and *A. niger* have also been demonstrated by the integration of silver nano-particles into polymer products, including poly-methoxybenzyl and poly(L-lactic acid)-co-poly(3-caprolactone) nano-fibres (Nair et al. 2009; Jin et al. 2012).

4.11 ENVIRONMENTAL REMEDIATION METHODS: ENVIRONMENTAL REMEDIATION THROUGH CHEMICAL DEGRADATION

Chemical degradation is among the most employed techniques for environmental remediation. This comprises (i) ozone, UV radiation, hydrogen peroxide oxidation, (ii) photo-catalytic degradation, (iii) super-critical water oxidation, (iv) the Fenton technique, (v) sonochemical degradation, (vi) the electrochemical method, (vii) the electron beam process, (viii) solvated electron reduction, (ix) permeable reactive barriers of iron and other zero-valent metals, and (x) enzymatic treatment techniques. Some of these are discussed below.

Technologies relying on ozone or UV radiations are chemical oxidative processes relevant to treat water for the deterioration of specific contaminants or to reduce organic pollutants such as chemical oxygen demand; they can also be used to increase the bio-degradability of wastewater. Furthermore, they can be used for disinfection purposes. These methods commonly include photolytic and oxidative processes to eradicate foreign material present in water. Accordingly, they can interact independently or photolyse the organic material in water.

Similarly, in conjunction with UV radiation, when ozone or hydrogen peroxide is employed, the contaminants may be deteriorated by in-situ hydroxyl-based free radicals via an oxidative reaction. Phenols and certain pesticides have been the most prevalent aquatic pollutants. Some materials interact easily, while organo-chlorine derivatives show low reactivity, with hydroxyl radicals. A further characteristic of these oxidative reactions is that, because of the reaction of the contaminants with hydroxyl free radicals, it is a damaging method of elimination of water/air contamination (Khin et al. 2012).

Photo-catalytic degradation comprises a catalyst and photons. Photo-catalytic reactions of nano-particles are focused on the association of light energy and metal-specific nano-particles and are of keen importance to different contaminants because of their vast and strong photo-catalytic processes. These nano-particles typically consist of semiconductor metals that can degrade various permanent organic contaminants including dyes, solvents, detergents, pesticides, and hazardous chemical/biological compounds. In addition, they are often incredibly efficient in particular circumstances for the deterioration of halogenated and non-halogenated compounds and toxic metals. They have minimal operating requirements and even at low quantities, are quite efficient. The basic process of photo-catalysis is premised upon electron photoexcitation in the catalyst. Holes and excited electrons in the band gap are produced by photoluminescence (UV light in the context of TiO_2). The free radicals are produced due to the capturing of holes by molecules of water in an aquatic system. These radicals are potent oxidizers; they oxidize contaminants into water and gas and are deteriorated upon reaction. Because of its unique characteristics including being highly reactive in UV radiation and strength, titanium dioxide is by far the most commonly used photo-catalyst among the several nano-photo-catalysts produced so far. Likewise, because of their photo-catalytic behaviour, zinc oxides have been thoroughly investigated as they comprise a broad energy gap similar to that of titanium dioxide. Various studies have demonstrated the photo-catalytic behaviour of different synthetic photo-catalyses. The effectiveness relies on multiple parameters, for instance, the energy band, density, dosage, contaminant composition, and pH. Furthermore, cadmium sulphide nano-particles have also been employed to treat industrial wastes such as dyes. These photo-catalysts are UV active catalysts because of the presence of broad energy bands. Therefore, more improvements are required to enhance their operation in visible radiance for organic contaminant deterioration (Beltran 2003; Khin et al. 2012).

4.12 APPLICATIONS OF NANO-PARTICLES FOR ENVIRONMENTAL REMEDIATION

Over recent years, a growing range of nano-particles with environmental implications has been created. For instance, nano-particles are being employed to remediate polluted soil and groundwater at hazardous dumpsites, such as sites polluted by chlorinated solvents or chemical accidents. In different fields of science and technology, several types of nano-materials are used. Nano-particles are of great significance to environmental remediation, as compared to their size, the particle surface area is substantial. Consequently, their reactivity in chemical/biological surface-mediated reactions could increase compared to similar substances at larger volumes (Das et al. 2019). For particular purposes, they can be engineered to establish unique characteristics that rarely exist on a small or large scale in molecules of the same substance. Their high reactivity and prevalence of a greater number of reactive sites enhance the chance of their contact with the specific pollutants, thus reducing pollutant concentrations rapidly. In addition, nano-particles encompass extremely tiny areas due to their minute volume and reside suspended in groundwater if suitable coatings are applied.

Effective coatings permit particles to migrate farther than larger particles and attain broader dissemination, which enhances the elimination of pollutants. A few nano-technologies for environmental remediation are still in the testing process, while others are advancing progressively from the experimental phase to the application phase. For instance, in approaching difficult sites, such as polluted areas (chlorinated solvents), some nano-materials retain the potential for environmental applications. Currently, research is being conducted to examine particles viz., TiO_2, self-assembled monolayers on mesoporous supports (SAMMS^{TM}), dendrimers, metalloporphyrinogens, and swellable organically modified silica (SOMS). Moreover, further investigation is required to explain the environmental existence and distribution of unrestricted nano-particles, if they are permanent, if they have toxicity impacts on multiple biological processes, or whether the hypothetical advantages of nano-particles can be achieved in widespread practical products. Moreover, nano-particles are already being explored to be used in the detection and control of environmental pollutants, but nano-sensor research and development are still on-going.

4.13 CONCLUSION

Nanotechnology has proved to be an effective tool for environmental pollution remediation in a sustainable manner and has the power to impart enhanced promising applications that have a significant impact on the efficiency of the contaminant remediating substance. Extensive research is being performed for the development and advancement of such technologies for the conservation of environs from pollution.

REFERENCES

Anaya-Esparza, L. M., E. Montalvo-González, N. González-Silva, M. D. Méndez-Robles, R. Romero-Toledo, E. M. Yahia, and A. Pérez-Larios. 2019. Synthesis and characterization of TiO2-ZnO-MgO mixed oxide and their antibacterial activity. *Materials*, *12*(5):698.

Andrade, P. F., A. F. de Faria, S. R. Oliveira, Z. Arruda MA, and M. do Carmo Gonçalves. 2015. Improved antibacterial activity of nanofiltration polysulfone membranes modified with silver nanoparticles. *Water Research*, *81*:333–342.

Anjum, M., R. Miandad, M. Waqas, F. Gehany, and M. A. Barakat. 2019. Remediation of wastewater using various nanomaterials. *Arabian Journal of Chemistry*, *12*(8):4897–4919.

Barakat, M. A., R. I. Al-Hutailah, M. H. Hashim, E. Qayyum, and J. N. Kuhn. 2013. Titania supported silver-based bimetallic nanoparticles as photocatalysts. *Environmental Science and Pollution Research*, *20*(6):3751–3759.

Bashir, S., and J. L. Liu. 2015. Nanomaterials and their application. In *Advanced Nanomaterials and Their Applications in Renewable Energy*. Elsevier Inc., Amsterdam, The Netherlands, pp. 1–50.

Beltran, F. J. 2003. *Chemical Degradation Methods for Wastes and Pollutants Environmental and Industrial Applications*. Ed. M. A. Tarr. Marcel Dekker Inc., New York, pp. 18–70.

Benelmekki, M. 2015. An introduction to nanoparticles and nanotechnology. In *Designing Hybrid Nanoparticles*. Morgan and Claypool Publishers, Kentfield, CA.

Campbell, M. L., F. D. Guerra, J. Dhulekar, F. Alexis, and D. C. Whitehead. 2015. Target-specific capture of environmentally relevant gaseous aldehydes and carboxylic acids with functional nanoparticles. *Chemistry: A European Journal*, *21*:14834–14842.

Capek, I. (2019). *Nanocomposite Structures and Dispersions*. 2nd edition. Elsevier, Amsterdam, 447 p.

Chou, K. S., Y. C. Lu, and H. H. Lee. 2005. Effect of alkaline ion on the mechanism and kinetics of chemical reduction of silver. *Materials Chemistry and Physics*, *94*:429–433.

Coetzee, J. J., N. Bansal, and E. M. N. Chirwa. 2020. Chromium in environment, its toxic effect from chromite-mining and ferrochrome industries, and its possible bioremediation. *Exposure and Health*, *12*:51–62.

Das, A., M. Kamle, A. Bharti, and P. Kumar. 2019. Nanotechnology and its applications in environmental remediation: An overview. *Vegetos*, *32*(3):227–237.

Das, S., B. Sen, and N. Debnath. 2015. Recent trends in nanomaterials applications in environmental monitoring and remediation. *Environmental Science and Pollution Research*, *22*:18333–18344.

Engates, K. E., and H. J. Shipley. 2011. Adsorption of Pb, Cd, Cu, Zn and Ni to titanium dioxide nanoparticles: Effect of particle size, solid concentration and exhaustion. *Environmental Science and Pollution Research*, *18*:386–395.

Fan, M., L. Gong, Y. Huang, D. Wang, and Z. Gong. 2018. Facile preparation of silver nanoparticle decorated chitosan cryogels for point-of-use water disinfection. *Science of the Total Environment*, *613*:1317–1323.

Feng, L., M. Cao, X. Ma, Y. Zhu, and C. Hu. 2012. Superparamagnetic high-surface-area Fe3O4 nanoparticles as adsorbents for arsenic removal. *Journal of Hazardous Materials*, *217*:439–446.

Gao, C., W. Zhang, H. Li, L. Lang, and Z. Xu. 2008. Controllable fabrication of mesoporous MgO with various morphologies and their absorption performance for toxic pollutants in water. *Crystal Growth & Design*, *8*:3785–3790.

Gogoi, S. K., P. Gopinath, A. Paul, A. Ramesh, S. S. Ghosh, and A. Chattopadhyay. 2006. Green fluorescent protein-expressing escherichia coli as a model system for investigating the antimicrobial activities of silver nanoparticles. *Langmuir*, *22*:9322–9328. [CrossRef] [PubMed]

Guerra, F. D., M. F. Attia, D. C. Whitehead, and F. Alexis. 2018. Nanotechnology for environmental remediation: Materials and applications. *Molecules*, *23*(7):1760.

Gupta, V. K., I. Tyagi, H. Sadegh, R. Shahryari-Ghoshekand, A. S. H. Makhlouf, and B. Maazinejad. 2015. Nanoparticles as adsorbent; a positive approach for removal of noxious metal ions: A review. *Science, Technology and Development*, *34*:195.

Hassouna, M. E. M., M. A. ElBably, A. N. Mohammed, and M. A. G. Nasser. 2017. Assessment of carbon nanotubes and silver nanoparticles loaded clays as adsorbents for removal of bacterial contaminants from water sources. *Journal of Water and Health*, *15*:133–144.

Jin, G., M. P. Prabhakaran, B. P. Nadappuram, G. Singh, D. Kai, and S. Ramakrishna. 2012. Electrospun poly (L-lactic acid)-co-poly (ϵ-caprolactone) nanofibres containing silver nanoparticles for skin-tissue engineering. *Journal of Biomaterials Science, Polymer Edition*, *23*(18):2337–2352.

Khan, I., K. Saeed, and I. Khan. 2019. Nanoparticles: Properties, applications and toxicities. *Arabian Journal of Chemistry*, *12*(7):908–931.

Khin, M. M., A. S. Nair, V. J. Babu, R. Murugan, and S. Ramakrishna. 2012. A review on nanomaterials for environmental remediation. *Energy & Environmental Science*, *5*(8):8075–8109.

Klug, K. L., and V. P. Dravid. 2002. Observation of two-and three-dimensional magnesium oxide nanostructures formed by thermal treatment of magnesium diboride powder. *Applied Physics Letters*, *81*:1687–1689.

Kumar, K. Y., H. B. Muralidhara, Y. A. Nayaka, J. Balasubramanyam, and H. Hanumanthappa. 2013. Hierarchically assembled mesoporous ZnO nanorods for the removal of lead and cadmium by using differential pulse anodic stripping voltammetric method. *Powder Technology*, *239*:208–216.

Le, T. H., Y. Oh, H. Kim, and H. Yoon. 2020. Exfoliation of 2D materials for energy and environmental applications. *Chemistry—A European Journal*, *26*(29):6360–6401.

Luo, T., J. Cui, S. Hu, Y. Huang, and C. Jing. 2010. Arsenic removal and recovery from copper smelting wastewater using TiO2. *Environmental Science & Technology*, *44*(23):9094–9098.

Manisalidis, I., E. Stavropoulou, A. Stavropoulos, and E. Bezirtzoglou. 2020. Environmental and health impacts of air pollution: A review. *Frontiers in Public Health*, *8*:14. https://doi.org/10.3389/fpubh.2020.00014

Mishra, R. K., S. K. Ha, K. Verma, and S. K. Tiwari. 2018. Recent progress in selected bio-nanomaterials and their engineering applications: An overview. *Journal of Science: Advanced Materials and Devices*, *3*(3):263–288.

Mo, M., J. C. Yu, L. Zhang, and S. K. Li. 2005. Self-assembly of ZnO nanorods and nanosheets into hollow microhemispheres and microspheres. *Advanced Materials*, *17*(6):756–760.

Nair, A. S., N. P. Binoy, S. Ramakrishna, T. R. R. Kurup, L. W. Chan, C. H. Goh, M. R. Islam, T. Utschig, and T. Pradeep. 2009. Organic-soluble antimicrobial silver nanoparticle-polymer composites in gram scale by one-pot synthesis. *ACS Applied Materials & Interfaces*, *1*:2413–2419. [CrossRef] [PubMed]

Özkara, A., D. Akyıl, and M. Konuk. 2016. *Pesticides, Environmental Pollution, and Health, Environmental Health Risk—Hazardous Factors to Living Species, Marcelo L. Larramendy and Sonia Soloneski*. IntechOpen. https://doi.org/10.5772/63094. Available from: www.intechopen.com/books/environmental-health-risk-hazardous-factors-to-living-species/pesticides-environmental-pollution-and-health

Palimi, M. J., M. Rostami, M. Mahdavian, and B. Ramezanzadeh. 2014. Surface modification of Fe2O3 nanoparticles with 3-aminopropyltrimethoxysilane (APTMS): An attempt to investigate surface treatment on surface chemistry and mechanical properties of polyurethane/Fe2O3 nanocomposites. *Applied Surface Science*, *320*:60–72.

Pandey, B., and M. H. Fulekar. 2012. Nanotechnology: Remediation technologies to clean up the environmental pollutants. *Research Journal of Chemical Sciences*, *2*:90–96.

Sharma, B., P. K. Boruah, A. Yadav, and M. R. Das. 2018. TiO2—Fe2O3 nanocomposite heterojunction for superior charge separation and the photocatalytic inactivation of pathogenic bacteria in water under direct sunlight irradiation. *Journal of Environmental Chemical Engineering*, *6*:134–145.

Singh, S., K. C. Barick, and D. Bahadur. 2013. Fe3O4 embedded ZnO nanocomposites for the removal of toxic metal ions, organic dyes and bacterial pathogens. *Journal of Materials Chemistry A*, *1*:3325–3333.

Tungittiplakorn, W., L. W. Lion, C. Cohen, and J. Y. Kim. 2004. Engineered polymeric nanoparticles for soil remediation. *Environmental Science & Technology*, *38*:1605–1610. [CrossRef] [PubMed]

Tuzen, M., and M. Soylak. 2007. Multiwalled carbon nanotubes for speciation of chromium in environmental samples. *Journal of Hazardous Materials*, *147*:219–225.

U.S. Environmental Protection Agency. 2008. Nanotechnology for Site Remediation Fact Sheet. *Solid Waste and Emergency Response*. EPA 542-F-08-009.

U.S. Environmental Protection Agency. Science Policy Council. 2007. Nanotechnology White Paper. U.S. *Environmental Agency*. Available from: www.epa.gov/ncer/nano/publications/whitepaper12022Z005

Vance, C. C., S. Vaddiraju, and R. Karthikeyan. 2018. Water disinfection using zinc phosphide nanowires under visible light conditions. *Journal of Environmental Chemical Engineering*, *6*:568–573.

Wang, H. Q., G. F. Yang, Q. Y. Li, X. X. Zhong, F. P. Wang, Z. S. Li, and Y. H. Li. 2011. Porous nano-MnO2 Large scale synthesis via a facile quick-redox procedure and application in a supercapacitor. *New Journal of Chemistry*, *35*:469–475.

Watson, B. W., L. Meng, C. Fetrow, and Y. Qin. 2016. Core/shell conjugated polymer/quantum dot composite nanofibers through orthogonal non-covalent interactions. *Polymers*, *8*(12):408.

Xu, D., X. Tan, C. Chen, and X. Wang. 2008. Removal of Pb (II) from aqueous solution by oxidized multi-walled carbon nanotubes. *Journal of Hazardous Materials*, *154*:407–416.

Yaya, A., B. Agyei-Tuffour, D. Dodoo-Arhin, E. Nyankson, E. Annan, D. S. Konadu, and C. P. Ewels. 2012. Layered nanomaterials-a review. *Global Journal of Engineering Design and Technology*, *2*:32–41.

Yin, Y., G. Zhang, and Y. Xia. 2002. Synthesis and characterization of MgO nanowires through a vapor-phase precursor method. *Advanced Functional Materials*, *12*:293–298.

Zaman, M. I., S. Mustafa, S. Khan, and B. Xing. 2009. Effect of phosphate complexation on Cd2+ sorption by manganese dioxide (b-MnO2). *Journal of Colloid and Interface Science*, *330*:9–19.

Zhao, X., L. Lv, B. Pan, W. Zhang, S. Zhang, and Q. Zhang. 2011. Polymer-supported nanocomposites for environmental application: A review. *Chemical Engineering Journal*, *170*:381–394.

Zhu, Y. Q., W. K. Hsu, W. Z. Zhou, M. Terrones, H. W. Kroto, and D. R. M. Walton. 2001. Selective Co-catalysed growth of novel MgO fishbone fractal nanostructures. *Chemical Physics Letters*, *347*:337–343.

5 Green-Synthesis of Nanomaterials for Environmental Remediation

Swapnali Jadhav, Himanshu Yadav, Mahipal Singh Sankhla, Swaroop S Sonone, Kapil Parihar, and Prashant Singh

5.1 INTRODUCTION

Pollutants introduced into the environment cause adverse effects to both the environment and human beings. Pollutants may be harmful trash, poisonous gases, the smoke that has been released into the environment due to rapid industrialization, and man-made and some natural activities (Khin et al., 2012). Hazardous waste, heavy metals, chemical compounds (fertilizers, herbicides, and pesticides), oil spilling, noxious gases, manufacturing wastes, manure, and organic composites are a few instances of pollutants (Vaseashta et al., 2007; Khan and Ghoshal, 2000). Released pollutants are highly unstable and greatly complex, and they have low responsiveness. Hence, the dilapidation of these contaminants can stimulate the soil, water, and air (Khin et al., 2012; Mohan et al., 2021). Nanotechnology has increasingly been a focus because of the physical characteristics of nanomaterials, which can help to reduce contaminants from the environment and help to maintain environmental sustainability. Nanotechnology is the conversion of material with a biological, physical, and chemical procedure through methods with or without another component to manufacture materials that have specific functionality, improved features, and distinct properties that can be applied in several areas (Goswami et al., 2012; Taghizadeh et al., 2013). Nanotech has implemented new nanomaterials (NMs) for the redressal of heavy metals, organic-inorganic compounds, and microorganisms from contaminated water (Kumar et al., 2017; Awual et al., 2015). Nanoparticles (NPs) that have dimensions from 1 to 100 nm show specific physical and chemical features compared to macro-scale components (Li et al., 2001). These NPs have attracted numerous researchers because of their morphologic, physical, and chemical characteristics including their miniature measurements, distribution, and unique configuration. They also have catalytic, magnetic, electronic, optical, and mechanical properties, which provide distinct ways for their development (Bardos et al., 2015). NMs have high reactivity and a specific surface that help to render them unique with better functionality, which makes them good adsorbents, catalysts, and sensors. The large surface area-to-mass ratio can considerably improve the adhesion potential of the components. These different characteristics can be employed for the degeneration of pollutants from the environment (Sánchez et al., 2011).

Nanoscience is the study of components and structures on a miniature scale with the broader and safer utilization of green chemistry. Remediation with NMs includes the overall degradation of pollutants that have been released in the environment from various sources (Bardos et al., 2015). Nanotechnology can help to prevent, mitigate, and minimize the harm affecting individual well-being and the environment (Hood, 2004). Green chemistry is a growing research area that focuses on sustainable chemical strategies in the modern period of sustainable development goals (SDGs) (Matlin et al., 2015). Green chemistry principles can be employed to minimize the usage and production of harmful substances with the production of clean, safe, and enhanced NMs. The

 DOI: 10.1201/9781003186298-5

principles of green chemistry are described in Figure 5.1 (Anastas and Warner, 1998a, 1998b). The utilization of the values of green chemistry in nanotechnology will help the production of greener and more sustainable NMs with improved efficiency. Green technology uses renewable substances to reduce waste and helps to maintain environmental sustainability (Anastas and Eghbali, 2010). Green technology is unquestionably the most significant area of nanotechnology, employing the 12 green chemistry principles for numerous uses, and it produces sustainability and protects and advance occupational security and well-being throughout production and process utilization (Matos et al., 2010; Anderson et al., 2010). Green synthesis is a process of synthesizing NPs by maintaining the sustainability of the environment and minimizing its hazards (Fleischer and Grunwald, 2008). Green synthesis principles that are practised to synthesize NPs has been found to be highly assuring. Green chemistry is directing sustainable chemical strategies towards accomplishing the SDGs. It is an ecologically sound technique of generating substances that are less damaging due to the use of biodegradable reactants. Such a green strategy for the production of NMs is discovered to decrease the significant dangers regarding pollutants. Green technology is a merging of nanotechnology with green chemistry that showcases the purposes of producing eco-friendly NMs and determining their applicability to reducing environmental contaminants (Matlin et al., 2015). Integrated NPs create a limited harmful consequence on the soil, water, air, and individual wellness compared with those formed by chemical methods. (Anthony et al., 2013). The approach has environmental advantages, such as nanocatalysts for remediation, thermoelectric substances, low-weight nanocomposite substances, and small gadgets to decrease the utilization of material (Liang and Guo, 2009). This chapter focuses on the greener routes of production of various metallic, carbon-based, and polymeric NPs with various natural sources and their use in the depletion of environmental pollutants.

5.2 APPROACHES OF NP SYNTHESIS

Conventional ways are still utilized, but it has been demonstrated that green approaches in the production of NPs have the added benefits of fewer odds of disappointment, a low price, and an affluence of description (Zhang et al., 2016). To obtain NMs of the required dimensions, form, and capabilities, there are two main methods to manufacture them, namely, the 1) top-down and 2) bottom-up approaches. In the top-down approach, the appropriate majority substance is fragmented

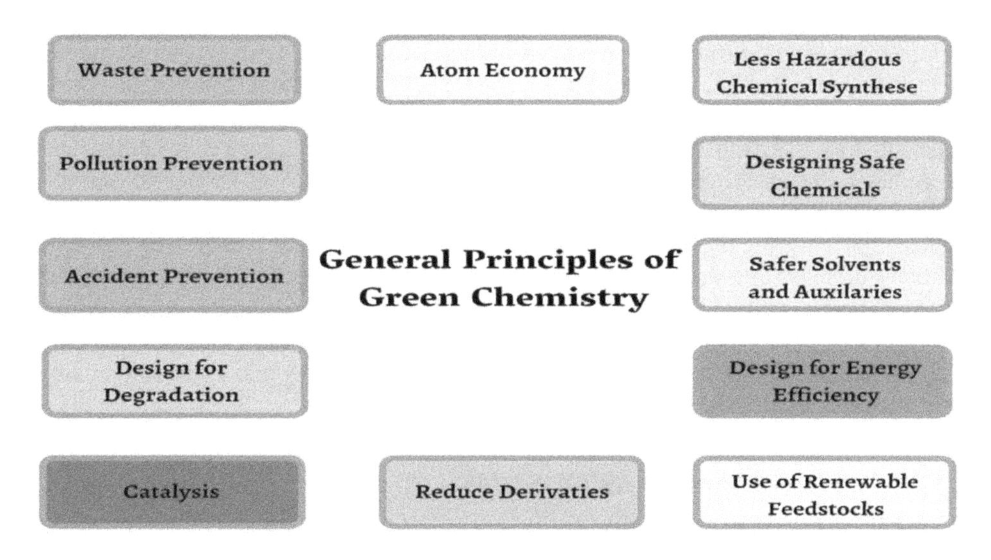

FIGURE 5.1 Twelve general principles of green chemistry.

into tiny elements by dimension reduction comprising lithographic methods, sputtering, etching, and grinding in a ball mill. In contrast, in the bottom-up method, NPs are produced by utilizing biochemical and organic approaches through a flat pack of atoms to new nuclei, which rise in nanosized elements. The bottom-up approach involves biochemical decrease, electrochemical approaches, and sono-decomposition. Bottom-up methods comprise rising NPs from simple units, specifically, reaction predecessors. They regulate the dimension and form of NPs that utilize target predecessors' attention, reaction conditions (temperature and pH), and functionalization of external models (Virkutyte and Varma, 2013; Mathur et al., 2017). The creation of NPs through traditional chemical and physical methods creates a lethal secondary product that causes ecological threats. Traditional methods can be utilized to produce NPs in greater amounts with limited dimensions and figures in a smaller period; however, such approaches are complex, expensive, ineffective, and out-of-date. Currently, there has been increasing consideration in making ecologically approachable NPs that do not create deadly leftover materials during the procedure (Parashar et al., 2009; Li et al., 2011). Old-style approaches are still utilized, but researchers have demonstrated that green approaches are highly effective concerning the production of NPs with the benefits of lesser probabilities of disaster, an economical price, and ease of description (Abdelghany et al., 2018). Chemical and physical methods of producing NPs have modeled numerous pressures on the atmosphere because of their lethal metabolites. The production of NPs with green approaches can be easily accomplished, and they are also economical. In light of their excellent characteristics, green-produced NPs are presently favoured over conventionally created NPs (Hussain et al., 2016).

Overcoming the limitation of this new age of green producing approaches is receiving great focus in the recent R&D on resource technology and science. Fundamentally, the green production of NMs generated by control, clean up, regulation, and remedial procedures will no doubt help their ecological sociability (Singh et al., 2018). The biogenesis method involves the maximum number of green chemistry principles by utilizing biologically approachable, harmless, and safe substances and making it profitable, with little upkeep, and an effortless increase in the large-scale generation of NPs, in contrast to the biochemical methods that employ greater radiation and highly concentrated reductants and stabilizing agents that damage the atmosphere and social well-being (Narayanan and Sakthivel, 2011; Kharissova et al., 2013). A few fundamental principles of green production can thus be described by numerous aspects such as the inhibition of waste, a decrease of byproducts/contamination, the usage of non-toxic mixtures/aides, and inexhaustible raw material. The green environmental methods in chemistry and biochemicals are gradually becoming well known and necessary because of global complications related to ecological worries (Ahmed et al., 2016). Green production procedures grounded in biological predecessors rely on varied reaction agents including temperature, solvents, pressure, and the pH environment (basic, acidic, or neutral) (Doble et al., 2010). A biotic-dependent generation of NPs employs a bottom-up approach in which arrangements transpire with the assistance of stabilizing and reducing agents. Three significant stages in the production of NPs from a chemistry viewpoint consist of a safer solvent phase, a non-lethal reducing agent, and ecologically gentle stabilizing factors. Another important agent is the choice of the covering materials utilized to passivate the surface of the NP. The capping agent demonstrates an important impact on the geomorphologies, subjected implementations, and size ranges (Mohanpuria et al., 2008; Raveendran et al., 2003; Annu and Ahmed, 2018). Newer studies have showcased the production, application, and characterization of NMs. Accordingly, eco-friendly resolutions are acquiring acknowledgement in the modern world (Machado et al., 2013; Huang et al., 2014; Luo et al., 2016).

5.3 GREEN SYNTHESIS OF VARIOUS NMS

5.3.1 Metal and Metal Oxide NMs

For environmental remediation, the most widely studied inorganic NMs are especially metal NMs and metal oxides NMs, which include silver, iron, gold, iron oxide, and titanium oxide. Metal NPs

(MNPs) have shown more immeasurable catalytic properties than bulk compounds. The individual characteristics of MNPs depend on their appearance, size, crystalline structure, and composition (Selvan et al., 2003). Metals and metal oxide NMs have a higher functionality and lower cost for the elimination of pollutants; hence, inorganic NMs have interested researchers (Saravanan et al., 2011) in numerous areas such as biomedicine, sensors, catalysis, fuel cells, and magnetic data storage (Wu et al., 2016).

5.3.1.1 Biological Components

Recently, significant research has been carried out on biological systems involving bacteria, fungi, yeasts, and plants, and this has directed the transformation of inorganic metal ions into metal NPs (Devatha et al., 2018). For NM synthesis with the green approach, compounds from the biological system can act as a reducing and capping agent. Enzymes, polysaccharides, vitamins, and amino acids from natural entities help degrade metallic ions (Sahayaraj and Rajesh, 2001). For the green production of metal and metal oxide NPs, plants are considered because plants have active phytochemicals, particularly in the leaves, such as ketones, aldehydes, flavones, amides, terpenoids, carboxylic acids, phenols, and ascorbic acids. These phytochemicals act as reducing agents that help transform the metal salts into metal NPs (Doble et al., 2010). Plant biodiversity has attracted high attention because of its biodegradability and availability. Metal oxide NPs have also been integrated with a green approach with plant parts and microorganism-based components (Annu and Ahmed, 2018). Bacteria can decrease metallic ions and are possible mortals in manufacturing NPs (Iravani, 2014). Similar to the bacteria and microorganism-mediated method, fungi-mediated synthesis is also highly effective in generating monodispersed inorganic NPs. With this method, several intracellular enzymes act as more active biological agents to generate NMs (Chen et al., 2009; Mohanpuria et al., 2008).

5.3.1.2 Metal and Metal Oxide NMs for Environmental Remediation

Metals have been used for thousands of years as antimicrobial agents. Although the antimicrobial properties of metal have been used for various purposes in the past, their application in the medicinal field decreased after the development of antibiotics in 1920. Egyptians initially reported the application of copper as an adstringent. Indians, Egyptians, Persian kings, Phoenicians, Greeks, and Romans also utilized copper and silver for food preservation and water disinfection. Silver has also been utilized as suture and infection remedies (Gold et al., 2018).

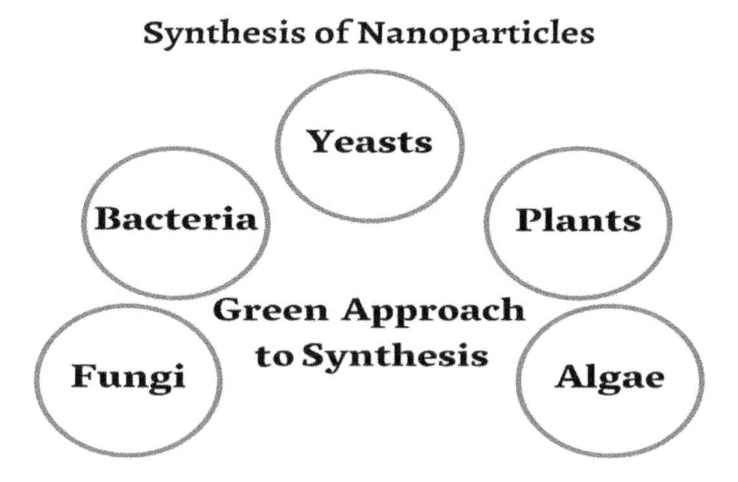

FIGURE 5.2 Biological sources for the green synthesis of NPs.

Metal oxides are a class of chemical composites with a varied structure and allied characteristics. These properties are a crystal structure, thermodynamic endurance, morphology, connected electronic arrangement, and electric conductivity (Pal, 2020). There has been a remarkable increase in the improvement and manufacture of NMs in recent years with a wide extent of uses, from antimicrobials and drug transporters to next-generation computer chips and superior substances, to name but a few. Determined yearly global generation quantities for normal NMs, such as zinc oxide (ZnO), copper (Cu), Ag, titanium dioxide ($TiO2$), and iron (Fe), were recorded to be 34,000, 200, 450, 88,000, and 42,000 tons, respectively, as of 2013 (Zhang et al., 2019). NMs as antibacterial factors are not only reliant on material chemistry but also strongly connected to their physical characteristics, such as their shape, size, solubility, agglomeration, and surface charge. For example, the morphology of NPs has been established to perform a role in the antiseptic activity of NPs (Morones et al., 2005).

Green production of Fe NPs is the procedure in which plant excerpts, microorganisms such as fungi, yeast, algae, and bacteria, and an enzyme play an important part in the production of a natural decomposable substance. Amongst such methods, production with plant extract is beneficial as it reduces the danger of additional pollution by preserving the cell assembly and lessening the response time (Ajitha et al., 2015). Fe is easily available, and its magnetic nature has attracted the production of NPs that have a great opportunity in ecological usage. Fe NPs show great dimensional constancy and harmlessness and lean towards producing oxides. Separately from having a greater surface range and thermal and electrical conduction, Fe NPs also possess magnetic characteristics and are called magnetic NPs (Arabi et al., 2016). Spherical Fe NPs were produced by a facile one-step green procedure utilizing Eucalyptus extract for the treatment of eutrophic wastewater (Wang et al., 2014), and it was described that 30% of complete phosphorus, 72% of complete nitrogen, and 85% of chemical oxygen demand (COD) were eliminated by the generated NPs. Thus, NPs might play a significant role in the remediation of wastewater. Fe NPs—trimetallic, bimetallic, metallic, and oxides—have been extensively described as hopeful mediators for ecological remedy (Sharma et al., 2018). There are 16 identified Fe oxides (Wu et al., 2015). They hold antimicrobic characteristics and have been utilized to decompose colourants (by an absorption procedure) (Sharma et al., 2017), treat wastewater (Fazlzadeh et al., 2018), and eliminate metal pollutants and other substances (Mondal and Purkait, 2018) through ecological methods. The rise of Fe NPs as a remedial process is because of their magnetic vulnerability, harmlessness, and dual oxidation-reduction ability upon reaction with water. Furthermore, they have a greater surface area and reaction rate. Fe NPs are widely reported for eliminating various heavy metals such as Ni2+, Co2+ (Hooshyar et al., 2013; Ebrahim et al., 2016), Cu2+ (Poguberovic et al., 2016), and Cd2+ (Ebrahim et al., 2016), along with the remediation of chlorinated organic solvents (Guo et al., 2017; Han et al., 2016). Fe oxide NPs

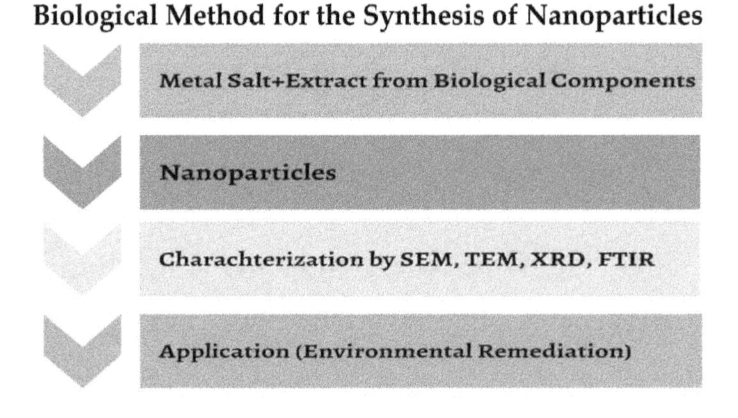

FIGURE 5.3 Bio-based method for the production of NPs.

have been broadly researched for wastewater remediation, air purification, self-purgation of surfaces, and as photocatalysts in wastewater remedy utilization because of their characterized cheap price and non-toxic, photocatalytic, semiconducting, gas sensing, energy converting, and electronic properties (Adesina, 2004; Li et al., 2008). Sankar et al. (2014) also described the probability of carboxyl, nitro, and amide groups from an Azadirachta indica leaf excerpt to produce TiO2 NPs of an average size of 124 nm with a spherical form. Moreover, they studied NPs as an active photocatalyst for the treatment of contamination. TiO2 NPs are adaptable; they can be used as reductive and oxidative catalysts for inorganic and organic contaminants (Hernández-Alonso et al., 2009; Nakata and Fujishima, 2012). Rajakumara et al. (2012) discussed the production of TiO2 NPs that have dimensions of 2–74 nm and an oval or spherical shape through Aspergillus flavus as a capping and reducing agent. The resultant NPs are an antimicrobic natural counter to Klebsiella pneumoniae, B. subtilis, Streptomyces aureus, and E. coli. Rice straw as a lignocellulosic waste material has been used for the production of TiO2 NPs by a sol-gel process (Ramimoghadam et al., 2014).

TiO2 NPs have been widely used in various fields including photo-catalysis (Fujishima et al., 2000), sensors (Bai and Zhou, 2014), and antimicrobial agents (Kubacka et al., 2014). Semiconductor NMs such as TiO2 and ZnO NPs that have a broad gap of 3.2 eV are generally studied because of their higher photocatalyst activity. They are utilized not only for the dilapidation of organic pollutants in air and water (Armon et al., 2004; Keller et al., 2005) but also for the application of antimicrobic factors (Pathakoti et al., 2014). Ag NPs are commonly recognized for their noteworthy antiviral, antibacterial, and antifungal actions and are thus used as wastewater decontaminators (Chou et al., 2005; Gupta et al., 1998; Bosetti et al., 2002). The usage of the yeast *R. mucilaginosa* to, through bioremediation, remove Cu-consisting wastewater is common through the production of Cu NPs (Salvadori et al., 2014). The Cu oxide (CuO) nanostructure was implied for the absorption of Pb (II) and organic dyes from aqueous solution (Farghali et al., 2013; Mustafa et al., 2013). CuO NPs have the possibility to adsorb both heavy metal ions and organic dyes. Moreover, the usage of gold and silver metallic NPs has been observed in ecological purification (Sundarajan et al., 2013).

5.3.1.3 Characterization of NPs

After the biosynthesis of NMs, there is a need to describe NMs with characterization techniques. This includes ultraviolet (UV)—Vis absorption spectroscopy, dynamic light scattering (DLS), Fourier transform infrared spectroscopy (FTIR), X-ray diffraction (XRD), Fourier energy dispersive X-ray analysis (EDAX), scanning electron microscopy (SEM), and transmission electron microscopy (TEM). Characterization deals with measurements of the physical and chemical characteristics of NPs. Conformational features about the appearance, size, uniformity, dispersion state, and surface properties are determined with different techniques.

TABLE 5.1
Techniques Used for Characterizing Nanomaterials for Environmental Remediation

Techniques	Applications	References
UV—Vis absorption spectroscopy	Determines the size and shape of NPs in aqueous suspension	(Rajesh et al., 2009)
Scanning electron microscopy and transmission electron microscopy	Determines the surface morphology and dimension of NPs	(Schaffer et al., 2009)
X-ray diffraction	Identifies the size and phase of metallic NPs	(Sun et al., 2000)
Fourier transmission infrared spectroscopy	Determines the functional groups, and its metabolite is present on the NPs	(Sankar et al., 2014)
Dynamic light scattering energy dispersive X-ray analysis	Determines the size distribution of NPs in a liquid and their elemental components	(Jiang et al., 2009; Strasser et al., 2010)

5.3.2 Carbon-Based NMs

Carbon-based NMs are better than conventional materials as they have a better surface area. Moreover, they are good adsorbents because of their individual electronic and structural characteristics (Lianchao et al., 2006). The composition of carbon-based NMs and their various hybridization states is because of the physical, chemical, and electronic capabilities that are unique to carbon-based NMs (Ren et al., 2011). As nanotechnology is evolving, various types of carbon- and metal-based NMs have evolved. Some examples include NMs that have applications in water purification, for example, multiwalled carbon nanotubes (MWNTs) and graphene oxides (Perreault et al., 2015). Most of the properties of these materials are dependent on their size. Carbon-based NMs are preferred to decontaminate the environment because

(1) They can inculcate various pollutants,
(2) They have an immediate reaction rate,
(3) They have an enhanced surface area of absorption, and
(4) They have an affinity for absorbing aromatic substances (Bosso and Enzweiler, 2002).

As these carbon-based NMs and their variants have a better surface area, flexible properties, ability to decontaminate foreign microbes, and efficient physio-chemical properties and because they are good decontaminators of pollutants, they have extensive applications in water purification (Smith and Rodrigues, 2015). Examples include C60, C540, nanotubes, and graphene (Mauter and Elimelech, 2008). Small nanoscale carbon NMs have high adsorption capabilities as they have micropores on their internal topology, which is accessible to humic acid, whereas larger carbon black-based NMs fail at this capability (Hochella, 2002). These carbon-based NMs also have photocatalytic techniques, and when these materials are radiated with UV light, the quanta greater or equal to the bandgap create band holes and electrons from the conductive bands. Hydroxyl radicles are formed because of these holes that participate in the oxidation of chlorinated compounds through these reactions; superoxide intermediates are formed that contribute to the decontamination of heavy metals. Graphene has been reported in different studies to be used for photocatalytic activity (Yang et al., 2013). TiO2 NPs in graphene have a better photocatalytic effect due to their enhanced conductive properties (Zhang et al., 2010).

5.3.3 NMs Based on Silica

Silica has been very popular in nanosciences due to its dynamic applications such as catalysis and adsorption. Silica materials have great importance in environmental decontamination because of their large surface area, flexible pore sizes, and better pore volumes, which makes them an effective material against environmental decontamination (Tsai et al., 2016).

TABLE 5.2
Silica Nanomaterials and Applications in the Environmental Remediation of Contaminants

Material	Application	References
Thiol-functionalized mesoporous silica	Wastewater—Heavy metals	(Arencibia et al., 2010)
Silicates of aluminum that are modified amines	Gaseous—CO_2, aldehydes, ketones	(Nomura and Jones, 2013)
Amino-functionalized mesoporous silica	Heavy metals in contaminated water	(Wang et al., 2015)
Amine-modified xerogels	Gaseous—CO_2 and H_2S	(Huang et al., 2003)
Silica mesopores of carboxylic acids	Wastewater—Cationic dyes and heavy metals	(Bruzzoniti et al., 2007)

5.3.4 NMs Based on Polymers

NMs have an improved surface area, and because of this, they have enhanced performance, improved reactivity, and a nonspecific nature, but with these properties, the overall functionality of the NM reduces significantly, as it becomes unstable. To overcome this limitation and improve the stability ratio of the NM, host materials are preferred as they act as a matrix for other materials (Zhao et al., 2011). Examples include amphiphilic polyurethane-based NMs that are used to remove polynuclear aromatic hydrocarbons (PAHs) from earthly materials such as soils (Tungittiplakorn et al., 2004).

5.4 APPLICATION OF GREEN NANOTECHNOLOGY IN ENVIRONMENTAL REMEDIATION

In the 21st century, the major focus of researchers has been on nanotechnology to improve the environment (green nanotechnology), and considerable studies and development are focused on this evolving area. These NMs are preferred more than metal-based NMs. In addition, green NMs have found their utility, specifically in the pharmaceutical sector. These NMs have also been extensively used in testing (in-vitro), clinical utilities, and even in medicines (Gunalan et al., 2012). NMs synthesized for decontamination have been proven to possess antibacterial effects (Mahdavi et al., 2013). They can destroy the bacteria by causing damage to its cell covering, which produces free radicles and inflicts oxidative damage on DNA (Chaloupka et al., 2010). These materials have therefore been essential for environmental decontamination (Njagi et al., 2011). Furthermore, green NMs are used as eco-friendly fertilizers that enhance crop production and have replaced pesticides in pest control and prevention (Kottegoda et al., 2011). They also have a broad scale of applicability in the treatment and decontamination of water from toxic biological and physical agents (Dhandapani et al., 2012).

5.5 CONCLUSION

Today, environmental contamination is one of the prime problems of the world, and there is a great necessity to maintain its sustainability. To overcome this problem, nanoscience plays a crucial role in environmental bioremediation. Progress in nanotechnology and green chemistry offers a simple, reproducible, and cost-effective method for synthesizing various NPs. Unique characteristics of NPs and the eco-friendly approach of their synthesis make them efficient and effective over the NPs that are synthesized with a chemical approach or commercially existing NPs. Considering the research, NPs, such as metal and metal oxide NPs, carbon-based, silica-based, and polymeric NPs synthesized from biological agents, plants, or microbial agents, play a significant role in environmental remediation. Metallic, carbonaceous, and silica-based NMs synthesized with principles of green chemistry have gained attention for their ability to eliminate contaminants from the soil, water, and air because of their excellent characteristics and applications, some of which are discussed in this chapter. This might help researchers to develop an advanced method for synthesising NPs with a

TABLE 5.3
Polymeric Material and Its Application in Environmental Remediation

Material	Application	References
Nanocomposites based on polymers	Water—Metal ions, dyes, and microorganisms	(Khare et al., 2016)
Amphiphilic polyurethane NPs	Soil—Polynuclear aromatic hydrocarbons	(Tungittiplakorn et al., 2004)
Polyamine-modified cellulose	Volatile organic compounds (VOCs)	(Guerra et al., 2018)
PAMAM dendrimers	Heavy metals in contaminated water	(Diallo et al., 2005)
Amine-modified PDLLA-PEG	Gaseous—VOCs	(Campbell et al., 2015)

greener technique, and bio-based synthesis offers a cost-effective, sustainable way to degrade contaminants in the environment.

REFERENCES

Abdelghany, T.M., Al-Rajhi, A.M.H., & Al Abboud, M.A. Recent advances in green synthesis of silver nanoparticles and their applications: About future directions. A review. *Bionanoscience*, 2018, *8*, 5–16.

Adesina, A.A. Industrial exploitation of photocatalysis: Progress, perspectives and prospects. *Catalysis Surveys from Asia*, 2004, *8*, 265–273.

Ahmed, S., Saifullah, Ahmad, M., Swami, B.L., & Ikram, S. Green synthesis of silver nanoparticles using Azadirachta indica aqueous leaf extract. *Journal of Radiation Research and Applied Sciences*, 2016, *9*(1), 1–7.

Ajitha, B., Reddy, Y.A.K., & Reddy, P.S. Green synthesis and characterization of silver nanoparticles using Lantana camara leaf extract. *Materials Science and Engineering: C*, 2015, *49*, 373–381.

Anastas, P.T., & Eghbali, N. Green chemistry: Principles and practice. *Chemical Society Reviews*, 2010, *39*(1), 301–312. https://doi.org/10.1039/B918763B

Anastas, P.T., & Warner, J.C. *Green: Chemistry. Frontiers*. Oxford University Press, New York, 1998a.

Anastas, P.T., & Warner, J.C. Principles of green chemistry. In: *Green Chemistry: Theory and Practice*. Oxford University Press, New York, 1998b, p. 29.

Anderson, A.A., Brossard, D., & Scheufele, D.A. The changing information environment for nanotechnology: Online audiences and content. *Journal of Nanoparticle Research*, 2010, *12*(4), 1083–1094. https://doi.org/10.1007/s11051-010-9860-2

Annu, A.A., & Ahmed, S. Green synthesis of metal, metal oxide nanoparticles, and their various applications. In: *Handbook of Ecomaterials*. Springer International Publishing, Cham, 2018, pp. 1–45.

Anthony, K.J.P., Muralidharan, M., Muniyandi, J., & Sangiliyandi, G. Multiple strategic approaches for green synthesis and application of silver and gold nanoparticles. *Green Biosynthesis of Nanoparticles: Mechanisms and Applications*, 2013, 13–30.

Arabi, S., Javar, H. A., & Khoobi, M. Preparation and characterization of modified polyethyleneimine magnetic nanoparticles for cancer drug delivery. *Journal of Nanomaterials*, 2016, 1–6.

Arencibia, A., Aguado, J., & Arsuaga, J.M. Regeneration of thiol-functionalized mesostructured silica adsorbents of mercury. *Applied Surface Science*, 2010, *256*(17), 5453–5457.

Armon, R., Weltch-Cohen, G., & Bettane, P.G. Disinfection of Bacillus spp. Spores in drinking water by TiO2 photocatalysis as a model for Bacillus anthracis. *Water Science and Technology*, 2004, *4*, 7–14.

Awual, M.R., Eldesoky, G.E., & Yaita, T. Schiff based ligand containing nano-composite adsorbent for optical copper (II) ions removal from aqueous solutions. *Chemical Engineering Journal*, 2015, *279*, 639–647. https://doi.org/10.1016/j.cej.2015.05.049

Bai, J., & Zhou, B. Titanium dioxide nanomaterials for sensor applications. *Chemical Reviews*, 2014, *114*, 10131–10176.

Bardos, P., Bone, B., Černík, M., Elliott, D.W., Jones, S., & Merly, C. Nanoremediation and international environmental restoration markets. *Remediation Journal*, 2015, *25*(2), 83–94. https://doi.org/10.1002/rem.21426

Bosetti, M., Masse, A., Tobin, E., & Cannas, M. Silver coated materials for external fixation devices: In vitro biocompatibility and genotoxicity. *Biomaterials*, 2002, *23*, 887–892.

Bosso, S.T., & Enzweiler, J. Evaluation of heavy metal removal from aqueous solution onto scolecite. *Water Research*, 2002, *36*(19), 4795–4800.

Bruzzoniti, M.C., Prelle, A., Sarzanini, C., Onida, B., Fiorilli, S., & Garrone, E. Retention of heavy metal ions on SBA-15 mesoporous silica functionalised with carboxylic groups. *Journal of Separation Science*, 2007, *30*(15), 2414–2420.

Campbell, M.L., Guerra, F.D., Dhulekar, J., Alexis, F., & Whitehead, D.C. Target-specific capture of environmentally relevant gaseous aldehydes and carboxylic acids with functional nanoparticles. *Chemistry—A European Journal*, 2015, *21*(42), 14834–14842.

Chaloupka, K., Malam, Y., & Seifalian, A.M. Nanosilver as a new generation of nanoproduct in biomedical applications. *Trends in Biotechnology*, 2010, *28*(11), 580–588.

Chen, Y.L., Tuan, H.Y., Tien, C.W., Lo, W.H., Liang, H.C., & Hu, Y.C. Augmented biosynthesis of cadmium sulfide nanoparticles by genetically engineered Escherichia coli. *Biotechnology Progress*, 2009. https://doi.org/10.1002/btpr.199

Chou, K.-S., Lu, Y.-C., & Lee, H.-H. Effect of alkaline ion on the mechanism and kinetics of chemical reduction of silver. *Materials Chemistry and Physics*, 2005, *94*, 429–433.

Devatha, C.P., & Thalla, A.K. Green Synthesis of Nanomaterials. In: *Synthesis of Inorganic Nanomaterials: Advances and Key Technologies*. Woodhead Publishing, Cambridge, 2018, pp. 169–184.

Dhandapani, P., Maruthamuthu, S., & Rajagopal, G. Bio-mediated synthesis of TiO2 nanoparticles and its photocatalytic effect on aquatic biofilm. *Journal of Photochemistry and Photobiology*, 2012, *110*, 43–49.

Diallo, M.S., Christie, S., Swaminathan, P., Johnson, J.H., & Goddard, W.A. Dendrimer enhanced ultrafiltration. 1. Recovery of Cu (II) from aqueous solutions using PAMAM dendrimers with ethylene diamine core and terminal NH2 groups. *Environmental Science & Technology*, 2005, *39*(5), 1366–1377.

Doble, M., Rollins, K., & Kumar, A. *Green Chemistry and Engineering*. Academic Press, New York, 2010.

Ebrahim, S.E., Sulaymon, A.H., & Saad Alhares, H. Competitive removal of Cu2+, Cd2+, Zn2+, and Ni2+ ions onto iron oxide nanoparticles from wastewater. *Desalination and Water Treatment*, 2016, *57*, 20915–20929.

Farghali, A.A., Bahgat, M., Allah, A.E., & Khedr, M.H. Adsorption of Pb (II) ions from aqueous solutions using copper oxide nanostructures. *Beni-Suef University Journal of Basic and Applied Sciences*, 2013, *2*(2), 61–71.

Fazlzadeh, M., Ansarizadeh, M., & Leili, M. Data of furfural adsorption on nano zero valent iron (NZVI) synthesized from Nettle extract. *Data in Brief*, 2018, *16*, 341–345. https://doi.org/10.1016/j.dib.2017.11.035

Fleischer, T., & Grunwald, A. Making nanotechnology developments sustainable. A role for technology assessment? *Journal of Cleaner Production*, 2008, *16*(8–9), 889–898.

Fujishima, A., Rao, T.N., & Tryk, D.A. Titanium dioxide photocatalysis. *Journal of Photochemistry and Photobiology C: Photochemistry Reviews*, 2000, *1*, 1–21.

Gold, K., Slay, B., Knackstedt, M., & Gaharwar, A.K. Antimicrobial activity of metal and metal-oxide based nanoparticles. *Advanced Therapeutics*, 2018, *1*(3), 1700033.

Goswami, A., Raul, P.K., & Purkait, M.K. Arsenic adsorption using copper (II) oxide nanoparticles. *Chemical Engineering Research and Design*, 2012, *90*(9), 1387–1396.

Guerra, F.D., Campbell, M.L., Attia, M.F., Whitehead, D.C., & Alexis, F. Capture of aldehyde VOCs using a series of amine-functionalized cellulose nanocrystals. *ChemistrySelect*, 2018, *3*(20), 5495–5501.

Gunalan, S., Sivaraj, R., & Rajendran, V. Green synthesized ZnO nanoparticles against bacterial and fungal pathogens. *Progress in Natural Science: Materials International*, 2012, *22*(6), 693–700.

Guo, M., Weng, X., Wang, T., & Chen, Z. Biosynthesized iron-based nanoparticles used as a heterogeneous catalyst for the removal of 2,4-dichlorophenol. *Separation and Purification Technology*, 2017, *175*, 222–228.

Gupta, A., & Silver, S. Molecular genetics: Silver as a biocide: Will resistance become a problem? *Nature Biotechnology*, 1998, *16*, 888.

Han, Y., & Yan, W. Reductive dechlorination of trichloroethene by zero-valent iron nanoparticles: Reactivity enhancement through Sulfidation treatment. *Environmental Science & Technology*, 2016, *50*, 12992–13001.

Hernández-Alonso, M.D., Fresno, F., Suárez, S., & Coronado, J.M. Development of alternative photocatalysts to TiO2: Challenges and opportunities. *Energy & Environmental Science*, 2009, *2*(12), 1231–1257.

Hochella Jr, M.F. There's plenty of room at the bottom: Nanoscience in geochemistry. *Geochimica et Cosmochimica Acta*, 2002, *66*(5), 735–743.

Hood, E. Nanotechnology: Looking as we leap. *Environ Health Perspect*, 2004, *112*(13), A740. https://doi.org/10.1289/ehp.112-a740

Hooshyar, Z., Rezanejade Bardajee, G., & Ghayeb, Y. Sonication enhanced removal of nickel and cobalt ions from polluted water using an iron based sorbent. *Journal of Chemistry*, 2013, 786954.

Huang, H.Y., Yang, R.T., Chinn, D., & Munson, C.L. Amine-grafted MCM-48 and silica xerogel as superior sorbents for acidic gas removal from natural gas. *Industrial & Engineering Chemistry Research*, 2003, *42*(12), 2427–2433.

Huang, L., Weng, X., Chen, Z., Megharaj, M., & Naidu, R. Green synthesis of iron nanoparticles by various tea extracts: Comparative study of the reactivity. *Spectrochimica Acta Part A: Molecular and Biomolecular Spectroscopy*, 2014, *130*, 295–301.

Hussain, I., Singh, N.B., Singh, A., et al. Green synthesis of nanoparticles and its potential application. *Biotechnology Letters*, 2016, *38*, 545–560.

Iravani, S. Bacteria in nanoparticle synthesis: Current status and future prospects. *International Scholarly Research Notices*, 2014. https://doi.org/10.1155/2014/35931622

Jiang, J., Oberdörster, G., & Biswas, P. Characterization of size, surface charge, and agglomeration state of nanoparticle dispersions for toxicological studies. *Journal of Nanoparticle Research*, 2009, *11*, 77–89.

Keller, V., Keller, N., Ledoux, M.J., & Lett, M.C. Biological agent inactivation in a flowing air stream by photocatalysis. *Chemical Communications*, 2005, *23*, 2918–2920.

Khan, F.I., & Ghoshal, A.K. Removal of volatile organic compounds from polluted air. *Journal of Loss Prevention in the Process Industries*, 2000, *13*, 527–545.

Khare, P., Yadav, A., Ramkumar, J., & Verma, N. Microchannel-embedded metal—carbon—polymer nanocomposite as a novel support for chitosan for efficient removal of hexavalent chromium from water under dynamic conditions. *Chemical Engineering Journal*, 2016, *293*, 44–54.

Kharissova, O.V., Dias, H.V.R., Kharisov, B., Perez, B.O., & Perez, V.M.J. The greener synthesis of nanoparticles. *Trends in Biotechnology*, 2013, *31*, 240.

Khin, M.M., Nair, A.S., Babu, V.J., Murugan, R., & Ramakrishna, S. A review on nanomaterials for environmental remediation. *Energy & Environmental Science*, 2012, *5*(8), 8075–8109.

Kottegoda, N., Munaweera, I., Madusanka, N., & Karunaratne, V. A green slow-release fertilizer composition based on urea-modified hydroxyapatite nanoparticles encapsulated wood. *Current Science*, 2011, 73–78.

Kubacka, A., Diez, M.S., Rojo, D., Rafael Bargiela, R., Ciordia, S., Zapico, I., Albar, J.P., Barbas, C., Martins dos Santos, V.A.P., Fernández-García, M., & Ferrer, M. Understanding the antimicrobial mechanism of TiO2-based nanocomposite films in a pathogenic bacterium. *Natural Science Reports*, 2014. https://doi.org/10.1038/srep04134

Kumar, A., Kumar, A., & Sharma, G. Sustainable nano-hybrids of magnetic biochar supported g-C3N4/FeVO4for solar powered degradation of noxious pollutants-synergism of adsorption, photocatalysis & photo-ozonation. *Journal of Cleaner Production*, 2017, *165*, 431–451. https://doi.org/10.1016/j.jclepro.2017.07.117

Li, L.S., Hu, J., Yang, W., & Alivisatos, A.P. Band gap variation of size-and shape-controlled colloidal CdSe quantum rods. *Nano Letters*, 2001, *1*(7), 349–351.

Li, Q., Mahendra, S., Lyon, D.Y., Brunet, L., Liga, M.V., Li, D., & Alvarez, P.J.J. Antimicrobial nanomaterials for water disinfection and microbial control: Potential applications and implications. *Water Research*, 2008, *42*, 4591–4602.

Li, X., Xu, H., Chen, Z.S., & Chen, G. Biosynthesis of nanoparticles by microorganisms and their applications. *Journal of Nanomaterials*, 2011, *2011*, 270974.

Lianchao, L., Baoguo, W., Huimin, T., Tianlu, C., & Jiping, X. A novel nanofiltration membrane prepared with PAMAM and TMC by in situ interfacial polymerization on PEK-C ultrafiltration membrane. *Journal of Membrane Science*, 2006, *269*(1–2), 84–93.

Liang, M., & Guo, L.H. Application of nanomaterials in environmental analysis and monitoring. *Journal of Nanoscience and Nanotechnology*, 2009, *9*(4), 2283–2289. https://doi.org/10.1166/jnn.2009.SE22

Luo, F., Yang, D., Chen, Z., Megharaj, M., & Naidu, R. One-step green synthesis of bimetallic Fe/Pd nanoparticles used to degrade Orange II. *Journal of Hazardous Materials*, 2016, *303*, 145–153.

Machado, S., Pinto, S.L., Grosso, J.P., Nouws, H.P.A., Albergaria, J.T., & Delerue-Matos, C. Green production of zero-valent iron nanoparticles using tree leaf extracts. *Science of the Total Environment*, 2013, *445*, 1–8.

Mahdavi, M., Namvar, F., Ahmad, M.B., & Mohamad, R. Green biosynthesis and characterization of magnetic iron oxide (Fe3O4) nanoparticles using seaweed (Sargassum muticum) aqueous extract. *Molecules*, 2013, *18*(5), 5954–5964.

Mathur, P., Jha, S., Ramteke, S., et al. Pharmaceutical aspects of silver nanoparticles. *Artif Cells Nanomedicine Biotechnology*, 2017, *46*, 1–12.

Matlin, S.A., Mehta, G., Hopf, H., & Krief, A. The role of chemistry in inventing a sustainable future. *Nature Chemistry*, 2015, *7*(12), 941–943.

Matos, J., García, A., & Poon, P.S. Environmental green chemistry applications of nanoporous carbons. *Journal of Materials Science*, 2010, *45*(18), 4934–4944. https://doi.org/10.1007/s10853-009-4184-2

Mauter, M.S., & Elimelech, M. Environmental applications of carbon-based nanomaterials. *Environmental Science & Technology*, 2008, *42*(16), 5843–5859.

Mohan, H., Rajput, S.S., Jadhav, E.B., Sankhla, M.S., Sonone, S.S., Jadhav, S., & Kumar, R. Ecotoxicity, occurrence, and removal of pharmaceuticals and illicit drugs from aquatic systems. *Biointerface Research in Applied Chemistry*, 2021, *11*(5), 12530–12546. https://doi.org/10.33263/BRIAC115.1253012546

Mohanpuria, P., Rana, N.K., & Yadav, S.K. Biosynthesis of nanoparticles: Technological concepts and future applications. *Journal of Nanoparticle Research*, 2008, *10*(3), 507–517.

Mondal, P., & Purkait, M.K. Green synthesized iron nanoparticles supported on pH responsive polymeric membrane for nitrobenzene reduction and fluoride rejection study: Optimization approach. *Journal of Cleaner Production*, 2018, *170*, 1111–1123. https://doi.org/10.1016/j.jclepro.2017.09.222

Morones, J.R., Elechiguerra, J.L., Camacho, A., Holt, K., Kouri, J.B., Ramírez, J.T., & Yacaman, M.J. The bactericidal effect of silver nanoparticles. *Nanotechnology*, 2005, *16*(10), 2346.

Mustafa, G., Tahir, H., Sultan, M., & Akhtar, N. Synthesis and characterization of cupric oxide (CuO) nanoparticles and their application for the removal of dyes. *African Journal of Biotechnology*, 2013, *12*(47), 6650–6660.

Nakata, K., & Fujishima, A. TiO2 photocatalysis: Design and applications. *Journal of Photochemistry and Photobiology C: Photochemistry Review*, 2012, *13*(3), 169–189.

Narayanan, K.B., & Sakthivel, N. Green synthesis of biogenic metal nanoparticles by terrestrial and aquatic phototrophic and heterotrophic eukaryotes and biocompatible agents. *Advances in Colloid and Interface Science*, 2011, *169*, 59.

Njagi, E.C., Huang, H., Stafford, L., Genuino, H., Galindo, H.M., Collins, J.B., . . . & Suib, S.L. Biosynthesis of iron and silver nanoparticles at room temperature using aqueous sorghum bran extracts. *Langmuir*, 2011, *27*(1), 264–271.

Nomura, A., & Jones, C.W. Amine-functionalized porous silicas as adsorbents for aldehyde abatement. *ACS Applied Materials & Interfaces*, 2013, *5*(12), 5569–5577.

Pal, D. Synthesis of metal oxide nanoparticles—A general overview. *Indian Journal of Chemistry—Section A*, 2020, *59A*, 1513–1528. http://nopr.niscair.res.in/handle/123456789/55450

Parashar, V., Parashar, R., Sharma, B., & Pandey, A.C. Parthenium leaf extract mediated synthesis of silver nanoparticles: A novel approach towards weed utilization. *Digest Journal of Nanomaterials & Biostructures (DJNB)*, 2009, *4*(1).

Pathakoti, K., Huang, M.-J., Watts, J.D., He, X., & Hwang, H.-M. Using experimental data of Escherichia coli to develop a QSAR model for predicting the photo-induced cytotoxicity of metal oxide nanoparticles. *Journal of Photochemistry and Photobiology B: Biology*, 2014, *130*(Suppl. C), 234–240.

Perreault, F., De Faria, A.F., Nejati, S., & Elimelech, M. Antimicrobial properties of graphene oxide nanosheets: Why size matters. *ACS Nano*, 2015, *9*(7), 7226–7236.

Poguberović, S.S., Krčmar, D.M., Maletić, S.P., Kónya, Z., Pilipović, D.D.T., Kerkez, D.V., & Rončević, S.D. Removal of as (III) and Cr (VI) from aqueous solutions using "green" zero-valent iron nano particles produced by oak, mulberry and cherry leaf extracts. *Ecological Engineering*, 2016, *90*, 42–49.

Rajakumara, G., Rahumana, A.A., Roopan, S.M., Khannac, V.G., Elangoa, G., & Kamaraja, C. Fungus-mediated biosynthesis and characterization of TiO2 nanoparticles and their activity against pathogenic bacteria. *Spectrochimica Acta Part A*, 2012, *9*, 123–129.

Rajesh, W.R., Jaya, R.L., Niranjan, S.K., Vijay, D.M., & Sahebrao, B.K. Phytosynthesis of silver nanoparticle using Gliricidia sepium. *Current Nanoscience*, 2009, *5*, 117–122.

Ramimoghadam, D., Bagheri, S., Bee, S., & Hamid, A. Biotemplated synthesis of anatase titanium dioxide nanoparticles via lignocellulosic waste material. *BioMed Research International*, 2014, *2014*, 205636.

Raveendran, P., Fu, J., & Wallen, S.L. Completely "green" synthesis and stabilization of metal nanoparticles. *Journal of the American Chemical Society*, 2003, *125*(46), 13940–13941.

Ren, X., Chen, C., Nagatsu, M., & Wang, X. Carbon nanotubes as adsorbents in environmental pollution management: A review. *Chemical Engineering Journal*, 2011, *170*(2–3), 395–410.

Sahayaraj, K., & Rajesh, S. Bionanoparticles: Synthesis and antimicrobial applications. *Science Against Microbial Pathogens: Communicating Current Research Technology Advanced*, 2001, *23*, 228–244.

Salvadori, M.R., Ando, R.A., do Nascimento, C.A.O., & Corrêa, B. Intracellular biosynthesis and removal of copper nanoparticles by dead biomass of yeast isolated from the wastewater of a mine in the Brazilian Amazonia. *PLoS ONE*, 2014, *9*(1), e87968.

Sánchez, A., Recillas, S., Font, X., Casals, E., González, E., & Puntes, V. Ecotoxicity of and remediation with, engineered inorganic nanoparticles in the environment. *TrAC Trends in Analytical Chemistry*, 2011, *30*(3), 507–516.

Sankar, R., Rizwana, K., Shivashangari, K.S., Ravikumar, V. Ultra-rapid photocatalytic activity of Azadirachta indica engineered colloidal titanium dioxide nanoparticles. *Applied Nanoscience*, 2014, *5*, 731–736.

Saravanan, R., Shankar, H., Prakash, T., Narayanan, V., & Stephen, A. ZnO/CdO composite nanorods for photocatalytic degradation of methylene blue under visible light. *Materials Chemistry and Physics*, 2011, *125*(1–2), 277–280.

Schaffer, B., Hohenester, U., Trugler, A., & Hofer, F. High resolution surface plasmon imaging of gold nanoparticles by energy-filtered transmission electron microscopy. *Physical Review B*, 2009, *79*, 0414011–0414014.

Selvan, R.K., Augustin, C.O., Berchmans, L.B., & Sarawathi, R. Combustion synthesis of CuFe2O4. *Materials Research Bulletin*, 2003, *38*, 41–54.

Sharma, G., Kumar, A., Naushad, M., & Kumar, A. Photoremediation of toxic dye from aqueous environment using monometallic and bimetallic quantum dots-based nanocomposites. *Journal of Cleaner Production*, 2018, *172*, 2919–2930. https://doi.org/10.1016/j.jclepro.2017.11.122

Singh, J., Dutta, T., Kim, K.H., Rawat, M., Samddar, P., & Kumar, P. 'Green' synthesis of metals and their oxide nanoparticles: Applications for environmental remediation. *Journal of Nanobiotechnology*, 2018, *16*(1), 1–24.

Smith, S.C., & Rodrigues, D.F. Carbon-based nanomaterials for removal of chemical and biological contaminants from water: A review of mechanisms and applications. *Carbon*, 2015, *91*, 122–143.

Strasser, P., Koh, S., Anniyev, T., Greeley, J., More, K., & Yu, C. Lattice-strain control of the activity in de alloyed core—shell fuel cell catalysts. *Nature Chemistry*, 2010, 2, 454–460.

Sun, S., Murray, C., Weller, D., Folks, L., & Moser, A. Monodisperse Fe Pt nanoparticles and ferromagnetic Fe Pt nanocrystal superlattices. *Science*, 2000, *287*, 1989–1992.

Sundarajan, S., Sameem, S.M., Sankaranarayanan, S., & Ramaraj, S. Synthesis, characterization and application of zero-valent silver nano adsorbents. *Organization (WHO)*, 2013, *2*(12).

Taghizadeh, F., Ghaedi, M., Kamali, K., Sharifpour, E., Sahraie, R., & Purkait, M.K. Comparison of nickel and/or zinc selenide nanoparticle loaded on activated carbon as efficient adsorbents for kinetic and equilibrium study of removal of Arsenazo (III) dye. *Powder Technology*, 2013, *245*, 217–226.

Tsai, C.H., Chang, W.C., Saikia, D., Wu, C.E., & Kao, H.M. Functionalization of cubic mesoporous silica SBA-16 with carboxylic acid via one-pot synthesis route for effective removal of cationic dyes. *Journal of Hazardous Materials*, 2016, *309*, 236–248.

Tungittiplakorn, W., Lion, L.W., Cohen, C., & Kim, J.Y. Engineered polymeric nanoparticles for soil remediation. *Environmental Science & Technology*, 2004, *38*(5), 1605–1610.

Vaseashta, A., Vaclavikova, M., Vaseashta, S., Gallios, G., Roy, P., & Pummakarnchana, O. Nanostructures in environmental pollution detection, monitoring, and remediation. *Science and Technology of Advanced Materials*, 2007, *8*, 47–59.

Virkutyte, J., & Varma, R.S. Green synthesis of nanomaterials: Environmental aspects. In: *Sustainable Nanotechnology and the Environment: Advances and Achievements*. American Chemical Society, Washington, DC, 2013, pp. 11–39.

Wang, S., Wang, K., Dai, C., Shi, H., & Li, J. Adsorption of Pb2+ on amino-functionalized core—shell magnetic mesoporous SBA-15 silica composite. *Chemical Engineering Journal*, 2015, *262*, 897–903.

Wang, T., Lin, J., Chen, Z., Megharaj, M., & Naidu, R. Green synthesis of Fe nanoparticles using Eucalyptus leaf extracts for treatment of eutrophic wastewater. *Science of the Total Environment*, 2014, *466–467*, 210–213.

Wu, W., Wu, Z., Yu, T., & Jiang, C. Recent progress on magnetic iron oxide nanoparticles: Synthesis, surface functional strategies and biomedical applications. *Science and Technology of Advanced Materials*, 2015, *16*.

Wu, Z., Yang, S., & Wu, W. Shape control of inorganic nanoparticles from solution. *Nanoscale*, 2016, *8*(3), 1237–1259.

Yang, M.Q., Zhang, N., & Xu, Y.J. Synthesis of fullerene—, carbon nanotube—, and graphene—TiO2 nanocomposite photocatalysts for selective oxidation: A comparative study. *ACS Applied Materials & Interfaces*, 2013, *5*(3), 1156–1164.

Zhang, P., Misra, S., Guo, Z., Rehkämper, M., & Valsami-Jones, E. Stable isotope labeling of metal/metal oxide nanomaterials for environmental and biological tracing. *Nature Protocols*, 2019, *14*(10), 2878–2899.

Zhang, X.F., Liu, Z.G., Shen, W., & Gurunathan, S. Silver nanoparticles: Synthesis, characterization, properties, applications, and therapeutic approaches. *International Journal of Molecular Sciences*, 2016, *17*(9), 1534.

Zhang, Y., Tang, Z.R., Fu, X., & Xu, Y.J. TiO2– graphene nanocomposites for gas-phase photocatalytic degradation of volatile aromatic pollutant: Is TiO2– graphene truly different from other TiO2– carbon composite materials? *ACS Nano*, 2010, *4*(12), 7303–7314.

Zhao, X., Lv, L., Pan, B., Zhang, W., Zhang, S., & Zhang, Q. Polymer-supported nanocomposites for environmental application: A review. *Chemical Engineering Journal*, 2011, *170*(2–3), 381–394.

6 Comparison between Traditional and Nanoremediation Technology with a Special Reference to Soil and Heavy Metal Contamination

J. Immanuel Suresh and A. Judith

6.1 INTRODUCTION

Soil is found to be differentiated in the horizon of minerals and organic constituents that usually differs in physical properties, chemical properties, composition and biological characteristics from the parent material. It is composed of many elements such as mineral particles, humus, soil atmosphere, soil water and soil microorganisms. Therefore, soil plays a major role in maintaining a balanced environment. The life of many plants, animals, humans and microbes are dependent on soil. Soil performs essential functions that include regulating water, sustaining plant and animal life, filtering and buffering potential pollutants, cycling nutrients and providing physical stability and support. It contains both dynamic and inherent qualities. Thus, healthy soil provides access to clean air and water, healthy food crops, diverse wildlife and beautiful landscapes. Soil as the "universal sink" suffers heavily from the release and accumulation of toxic contaminants. Moreover, soil quality is being contaminated in various ways that includes the discharge of industrial waste, pesticides, heavy metals, organic waste and agricultural chemicals into soil. Soil contamination can be regarded as the accumulation of toxic compounds, radioactive elements and salts that result in adverse effects on plants and animals (Okrent, 1999).

Developing countries with industrialized economies are seriously affected by soil contamination due to heavy metals. Heavy metals accumulated in the soil are taken up by plants. Since plants serve as a major source of food for many living beings, including humans, heavy metals tend to get transported through the food chain, causing a major threat to human and animal health (Mohamed et al., 2010). A study undertaken on soil contamination by heavy metals in India revealed that the soil is significantly contaminated and shows higher levels of toxic elements. Some common heavy metals found in soil are cadmium, cobalt, chromium and copper. Heavy metals are also found in soil that is not contaminated by industrial effluents due to the weathering from their parent materials. They originate from mining, the combustion of fuels, industrial waste and agricultural chemicals. These heavy metals can lead to reduced soil fertility, ecological imbalance, increased salinity and health disorders.

DOI: 10.1201/9781003186298-6

6.2 HEAVY METAL CONTAMINATION

In industrial areas, not only industrial pollutants, including carbon monoxide, chlorofluorocarbons and various heavy metals such as arsenic, lead and cadmium, but also hydrocarbons are released into the air. Nitrogen and sulfur-dioxide contaminants in the air result in acid rain, which ultimately pollutes the water and soil. Industrial effluents and sewage content released into water bodies also tend to settle in river beds, resulting in water and soil pollution. The range of toxicity or the toxic level differs based upon the various contaminants. Sometimes, even a minute quantity of a contaminant can cause a major threat (for example, arsenic exhibits toxicity when present at 10 ppm in soil). Heavy metal contamination occurs through various sources. Batteries are a major source of the generation of lead, nickel and cadmium contamination. Fertilizers, pesticides, paint, timber treatment, dyes and fumigants release heavy metal contaminants including cadmium, arsenic, chromium, copper, manganese, zinc, mercury, nickel and molybdenum (Malik and Biswasa, 2012).

Usually, contaminants are present in mixtures; therefore, it becomes ideal to develop various techniques that favor the reduction, identification, quantification and treatment of hazardous environmental contaminants present in mixtures (Filipponi and Sutherland, 2010). Even though traditional methods are widely in use, due to increasing challenges, the field of nanotechnology is being exploited for environment remediation. The distinguishing properties of nanoparticles can serve as a promising approach in achieving a safer environment through nanoremediation.

6.3 CONTAMINATED SITE ASSESSMENT

It is very important to assess a contaminated site before beginning the remediation process. Through assessment, the concentration and spatial distribution of harmful pollutants can be determined. An environmental and human health risk assessment is conducted by identifying the source, type and physical form of the contaminants, the extent of contamination and the potential receptors of contamination. The extent of site remediation must be determined. The various factors that influence the risk assessment are toxicity, reactivity, corrosivity and ignitability (U.S. EPA, 1989; Asante-Duah, 1996). These will be useful in choosing the method of remediation. The various data required for planning contaminated site remediation are as follows: site history and land use pattern, geologic and hydrologic properties (pH, ionic strength, dissolved oxygen, oxidation-reduction potential (ORP) and concentrations of nitrate, nitrite and sulfate), geotechnical information, the source and extent of contamination and hazard assessment and zoning.

6.4 TREATMENT OF HEAVY METAL-CONTAMINATED SOIL

Environmental remediation is the process of removing toxic pollutants from the soil, air and water to protect the environment and the living organisms that rely on it. The process of remediation is chosen after assessing the contaminated site and determining the human health and ecological risks. Based upon the intensity of contamination and other assessment results, the method of remediation is determined. The various approaches used for remediation can be broadly classified into physical remediation, chemical remediation, bioremediation and nanoremediation.

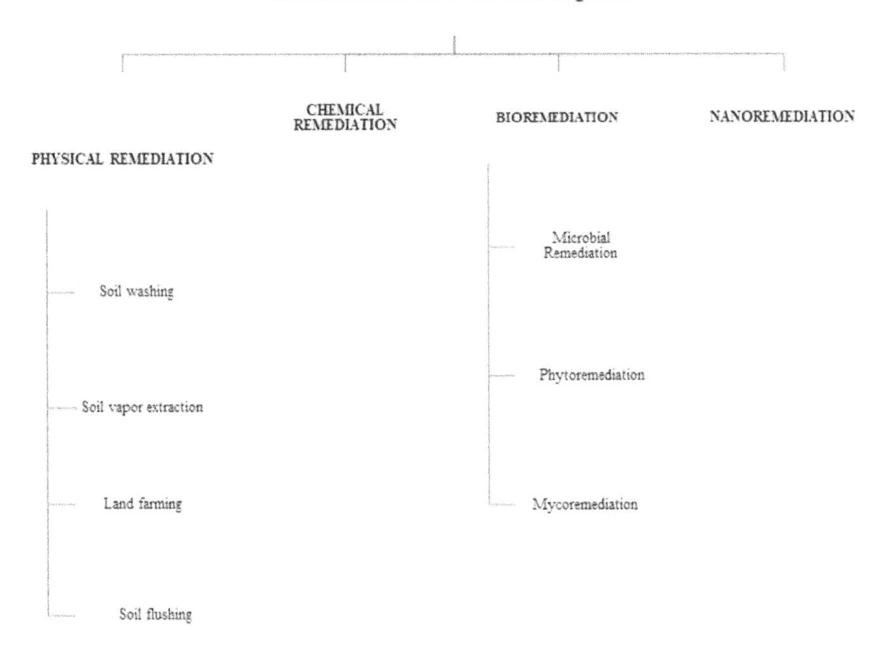

6.4.1 Traditional Methods of Soil Remediation

6.4.1.1 Soil Isolation and Containment

This remediation method includes barriers and hydraulic containment. In this method, barriers are installed, and the contaminated soil is capped (Figure 6.1) and contained within the barriers. The barrier walls otherwise regarded as slurry walls are made up of impermeable materials such as grout, cement, clay, bentonite and steel (Mulligan et al., 2001). These materials can be used for horizontal or vertical containment. This method is useful for heavy metal-contaminated soil. Capping the contaminated soil within impermeable barriers can prevent the contaminated groundwater from spreading to nearby water bodies. A vertical wall prevents the mixing of contaminated soil with uncontaminated water, whereas a horizontal wall restricts the downward movement of metal contaminants. Hydraulic containment is often preferred for groundwater contamination. It is not a direct remediation process but can widely be used to restrict the movement of hazardous heavy metals towards groundwater (ICS, 2005; Jankaite and Vasarevičius, 2005).

6.4.1.2 Solidification and Stabilization

The mobility of metal contaminants is reduced with the help of binders and additives in the solidification and stabilization method. It is suitable for highly toxic pollutants. The immobilized toxic contaminants do not have any spatial or temporal effect. Commonly used binders include cement, fly ask and zeolites; for soil contaminated with organic pollutants, polyethylene epoxy and resins (organic binders) are used (ICS, 2005). This method can be performed in a single step or in two steps. In the single-step process, to fix and render the polluted soil insoluble, it is mixed with a special binder. Meanwhile, in the two-step process, the polluted soil is made insoluble and non-reactive in the first step, and later, in the second step, it is solidified. In the event of solidification, the contaminated soil is primarily mixed with physical binding agents (cement, bitumen, thermoplastic binders, etc.), resulting in a crystalline glassy or polymeric mass. During stabilization, chemical reactions are initiated between the stabilizing agent and metal contaminant (Figure 6.2). This further promotes the binding of contaminants to substrate thereby making it less mobile. Stabilization yields a chemically stable compound that can be removed later from the contaminated environment (Jankaite and Vasarevičius, 2005).

6.4.1.3 Vitrification

This method is a solidification and stabilization method that requires thermal energy. In this method, graphite electrodes are inserted with appropriate space in between and are energized with high resistance electrical heating to heat the contaminated soil at a high temperature, thereby melting it (Dada et al., 2015). The metals are immobilized by the glassy solid formed after cooling. These immobilized metals can then be taken out of the region under remediation when desired. It is highly applicable for polluted areas that have a shallow depth and for treating a large volume (Mulligan et al., 2001).

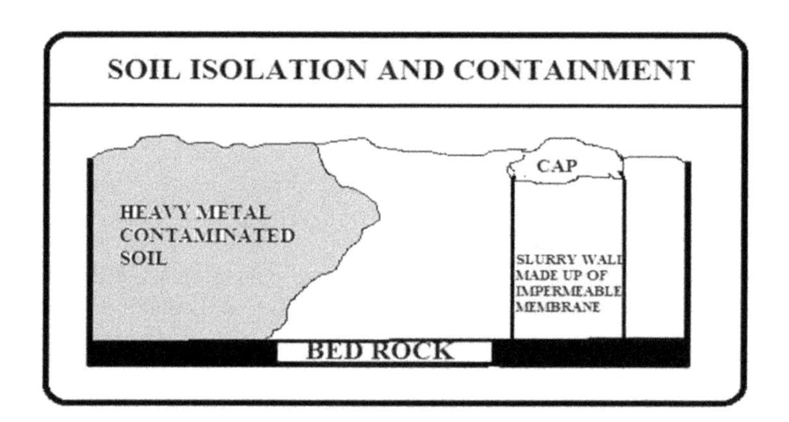

FIGURE 6.1 Soil isolation and containment.

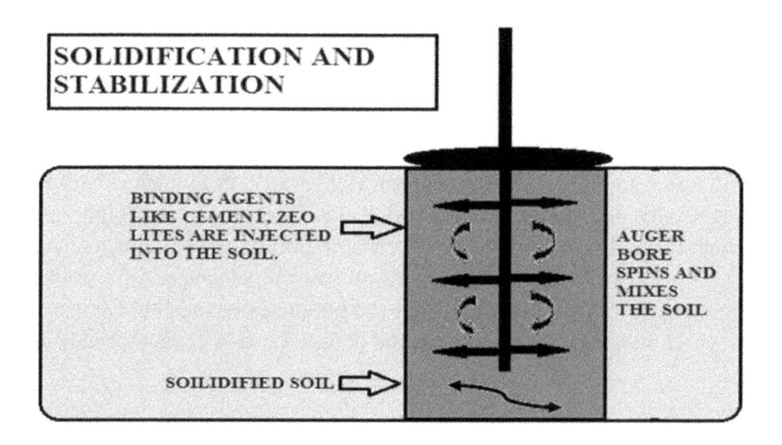

FIGURE 6.2 Solidification and stabilization.

6.4.1.4 Soil Flushing

This method involves flushing the soil with water or any other suitable washing solution. Organic and oil contaminants such as phenol are eliminated by adding acids, bases and surfactants to water that are used as extraction or flushing solutions to recover metals (ICS, 2005; Jankaite and Vasarevičius, 2005). Surface flooding, sprinklers, surface trenches, horizontal drains or vertical drains are used to directly inject or infiltrate the soil with the flushing solution. The contaminants react with the flushing solution and become mobilized by wither solubilization and by the formation of an emulsion or chemical reaction between them. This process further dissolves the contaminants in the solution used for flushing and leaches them into groundwater. The contaminant containing water and the flushing solution are collected with the help of wells or trenches that are placed strategically for further treatment and reuse (Dada et al., 2015).

6.4.1.5 Electrokinetics/Electroreclamation

This method utilizes electrical principles to decontaminate particular sites. This method is suitable for granular and clay types of soil, especially where the groundwater flow rate is low, to remove uranium, mercury and other heavy metals. In this technology, two metal/ceramic electrodes are inserted into the soil. The two electrodes act as a cathode and anode and are positioned on each side (array) of the contaminated soil; meanwhile, its pH is maintained using buffer solutions. Electronic conduction, along with the charge transfer between electrodes and solids in the soil, occurs upon the establishment of an electric field (a low intensity direct current) across the electrodes. When the electric current is applied, it results in electro-migration (movement of charged chemicals), electroosmosis (movement of fluid), electrophoresis (movement of charged particles) and electrolysis (chemical reactions due to an electric field) (Rodsand and Acar, 2000; Mulligan et al., 2001). This leads to the movement of contaminants from one electrode to another based upon their charges.

Occasionally, the process of movement is facilitated by the use of surfactants and complexing agents. The contaminants are deposited at the electrodes. They are then recovered by electroplating, precipitation, adsorption, complexing with ion exchange resins or the pumping of water or any other fluid near the electrodes by using ion exchange resistance.

Advantages

- Remediation takes place between the electrodes. Hence, the process is target specific.
- Metal contaminants are moved as they carry a charge. Induced flow moves the non-charged contaminants.
- It can be used as both in-situ and ex-situ methods.
- It can be tailored to site-specific conditions.

Disadvantages

- It is a time-consuming process.
- The current flow may get diverted.
- The electrolysis reaction may change the soil pH.
- A change in pH may have an adverse effect upon the microbial wealth of the soil.

6.4.1.6 Soil Vapor Extraction

This technique is also known as soil venting or vacuum extraction. It is used for remediating unsaturated soils contaminated with diversified contaminants by installing either vertical or horizontal wells in the contaminated zone (Zhan and Park, 2002). The evaporation is facilitated by air blowers. The vacuums through the wells evaporate the volatile substances of the contaminated mass. They are further withdrawn using an extraction well. The vapors are then treated and released into the atmosphere.

6.4.1.7 Soil Washing

This method involves the use of water in combination with solvents and mechanical processes to scrub the soil. Depending upon the nature of the contaminant, the solvents are selected (Chu and Chan, 2003). Pollutants such as hydrocarbon are usually found to be bound with clay and fine soil; therefore, in the soil washing process, clay (fine soil) is separated from sand and gravel (coarse soil). This process reduces the volume of polluted soil (Urum et al., 2003; Riser-Roberts, 1998). The polluted fine soil is isolated and subjected to other treatment methods.

Advantages

- It is cost effective.
- When performed under ideal conditions, it can reduce about 90% of the originally contaminated soil (Sharma and Reddy, 2004).
- It is performed on-site. Therefore, the treated soil can be reused as backfill at the site.
- It is performed in a closed system where the pH and other factors involved in remediation can be easily monitored.

Disadvantages

- It requires a large area to set up the system.
- It is predominantly effective with coarse soil, and at a clay site, this technique will not work well.
- It is ineffective for soil containing more than 30 to 50% silt.

6.4.1.8 Land Farming

Land farming, an above-ground remediation technology, reduces the concentration of pollutants such as petroleum present in soil. Land farming is performed in association with bioremediation. Over the ground surface of the treatment site, the excavated contaminated soil is spread out as a thin layer. Later, the contaminated soil is subjected to bioremediation by stimulating aerobic microbial activity (Hejazi et al., 2003).

Advantages

- It requires low capital input.
- It has a simple design and is easy to implement.
- It is energy efficient, and even a large volume of soil can be treated.

Disadvantages

- Only biodegradable elements can be efficiently eliminated.
- A large area is needed for treatment.

6.4.1.9 Bioremediation

The organic pollutants in soil are biologically treated in the process of bioremediation. It can be in situ (treatment of contaminated soil at the site) or ex situ (excavation of contaminated soil and treatment at some other place). This method involves the use of natural or genetically engineered microorganisms (Vidali, 2001) and plants. Bioremediation is a natural process and usually occurs on its own as nature's repair mechanism. It tends to metabolize organic chemical compounds and degrade them via reactions such as microbial degradation, hydrolysis and redox reactions (Dada et al., 2015). Bioremediation can be classified into microbial bioremediation and phytoremediation (Gupta et al., 2003).

Microbial remediation involves the use of a highly diverse microbial community (Giller et al., 1998) to degrade the organic contaminants into less toxic substances (Garbisu and Alkorta, 2003).

Solvents, polyaromatic hydrocarbons and pesticides are some organic pollutants (Adeniji, 2004). Depending on the contaminant present at the site, specific microorganisms can be used to enhance the remediation process. Microorganisms used in the process of bioremediation are known as bioremediators. Various microorganisms including bacteria, yeast, fungi and actinomycetes facilitate bioremediation. Since the contaminants present in the soil are generally present in the mixture, a combination of microorganisms can be used for efficient treatment as it shows synergistic interactions. Bacteria, for example, *Pseudomonas*, *Modococci* and *Streptomyces*, can be used for the removal of hydrocarbon contaminants from soil (Thapa et al., 2012).

There is limited evidence on the bioremediation of heavy metal-contaminated soil using microbes (Sinha et al., 2010). However, some metal-tolerant bacteria carry out various mechanisms to survive, resist and cope with heavy metal-contaminated soil. The mechanisms used by several microbes are as follows: a few microorganisms tend to bind the metals to proteins or extracellular polymers; some even compartmentalize the metals within the cells; they form insoluble metal sulphides; they reduce heavy metals by synthesizing complex enzymes in a process known as enzyme reduction; and a few others produce metal chelators (Giller et al., 1998; Sinha et al., 2010) to resist and cope with heavy metals. *Pseudomonas* are known to reduce chromium toxicity, and *Staphylococcus*, *Bacillus*, *Pseudomonas* and *Citrobacteia* are some of the commonly used bacteria for soil remediation. Bioremediation using microorganisms is performed through two mechanisms: bioaugmentation and biostimulation (Adeniji, 2004; Garbisu and Alkorta, 2003). While using microorganisms for remediation, the complexation and competition effects on the site have to be assessed. A study using the metal-resistant bacterium *Cupriavidus metallidurans* revealed that the accumulation of cadmium was affected by competition with calcium, magnesium and zinc. The accumulation of lead was influenced by complexation with humic acids. Therefore, the assessment of chemical site specificity is very important (Hajdu and Slaveykova, 2012; Dhal et al., 2013).

Mycoremediation involves fungi mycelia and mushrooms for remediating contaminants, including heavy metals. The mycelium performs decomposition by secreting several enzymes and acids that degrade organic pollutants. Mushrooms possess an effective mechanism that enables them to readily take up metals from the ecosystem (Turkekul et al., 2004). Mushrooms can also bioaccumulate heavy metals (Asiriuwa et al., 2013). A determination of the right fungal species is necessary for significant remediation effects.

Phytoremediation is a method that involves the use of plants to remove pollutants from the ecosystem. It eliminates the need to excavate the polluted material for treatment (U.S. EPA, 1989). Some plant species have the potential to eliminate a wide range of organic, inorganic and heavy metal contaminants by degrading, metabolizing or immobilizing them (Vidali, 2001; Garbisu and Alkorta, 2003; Rahimi et al., 2012). Phytoextraction (phytoaccumulation) (Rahimi et al., 2012; Njoku et al., 2009; Sinha et al., 2010; Gupta et al., 2003), phytovolatilization (Jankaite and Vasarevičius, 2005), phytostabilization, rhizofiltration and phytostimulization are the various techniques employed by plants in phytoremediation.

Phytoextraction is the hyperaccumulation of heavy metals from the soil in the harvestable parts of plants. Plants such as *Brassica juncea*, *Thalspi caerulescens* and *Helianthus annus* are involved in the phytoextraction of inorganic metals including cobalt, chromium, nickel, lead, zinc, gold, mercury, molybdenum, silver and cadmium. The plants that act as hyperaccumulators exhibit slow growth and development. Hence, the bioproductivity is less. After the process has been carried out, the plant mass should be disposed of properly.

Rhizofiltration involves the accumulation of heavy metals in the rhizosphere through sorption, concentration and precipitation. This technology is observed in plants such as *Brassica juncea*, *Helianthes annus*, tobacco, rye, spinach and corn against cadmium, copper, nickel, zinc and chromium.

Phytovolatilization is performed by the leaves of plants including *Arabidopsis thaliana*, poplars, alfalfa and *Brassica juncea* through transpiration for the removal of selenium, mercury and arsenic.

The plants transform the pollutants to a less toxic form. This includes the risk of transfer of heavy metals through the food chain.

Phytostabilization involves the complexation, sorption and precipitation of inorganic metals such as arsenic, cadmium, copper, chromium, lead, zinc and hassium by *Brassica juncea*, hybrid poplars and grasses. The extra steps involved in conventional methods, for example, the removal of soil, are eliminated. Therefore, it is cost effective. However, the plants usually require modified soil specific for their growth.

One environmentally friendly and cost-effective method to clean up toxic metal-contaminated soil and water is by using metal accumulating plants (Raskin et al., 1997; Ali et al., 2013). A study revealed that plants such as *Brassica juncea*, *Zea mays* and *Ambrosia artemisiifilia* accumulate metals including lead, copper and nickel by concentrating it in their harvestable dried shoot biomass. Almost 400 species of plants have been identified to be capable of removing toxic contaminants from soil and water.

Advantages:

- It is a natural process.
- The microbes increase in number at a high concentration of contaminant and decline when the contaminant quantity decreases.
- The residues of the treatment are usually harmless products.
- The complete destruction of the target pollutants is possible.

Disadvantages:

- It is limited to compounds that are biodegradable.
- The biological processes are highly specific. All essential factors such as suitable environmental conditions, metabolically capable microbes and an appropriate level of nutrients are required for effective results.
- It is difficult to extrapolate from a pilot-scale to full-scale studies.
- The process is more time consuming.

6.4.1.10 Chemical Remediation

The addition of chemicals to heavy metal-contaminated soil can transform the heavy metals into a less toxic form. As a result, the heavy metals remain in the soil but in a less harmful form. This method is becoming popular due to its high rate of success, but it requires special equipment and operators. It is a periodical treatment. Large-scale heavy metal remediation is currently not possible with this method. Various methods include remediation using actinide chelators, chemical immobilization and critical fluid extraction (Czupyrna et al., 1989; Gopalan et al., 1993).

Cyanide can be transformed into cyanate upon treatment with oxidants that have high electron affinity. Therefore, cyanide contamination in soil can be reduced by the addition of oxidants (Young et al., 1995).

Chromium (Cr) has applications in various industries, for example, leather tanning and wood preserving. Cr(VI) is a very toxic contaminant found in soil and water. It causes multiple serious health issues. On the contrary, Cr(III) is non-toxic. By reducing Cr(VI) to Cr(III), the load of toxic contaminant in soil can be significantly reduced. This can be achieved by the inexpensive chemical precipitation method. It is a two-step process: first, Cr(VI) is reduced under acidic conditions, and then, the trivalent chromium is precipitated as hydroxyl species.

6.4.2 Nanoremediation for Heavy Metal Contamination

Nanotechnology, an emerging branch of science, has a wide range of applications in pharmaceuticals, the medical field, agriculture, etc. (Hasnain et al., 2013). Nanostructured materials are studied to be used as adsorbents and catalysts for the transformation and detoxification of toxic environmental pollutants (Agarwal and Joshi, 2010). Nanoremediation is a process involving nano-sized materials that enable both chemical reduction and catalysis to eliminate pollutants. The use of nanoremediation methods prevents the pumping of groundwater and transport of soil for treatment and disposal (Otto et al., 2008; Karn et al., 2009). The valuable properties of nanoparticles, including their minute size, alterable surface coatings and various other physical and chemical properties, enable them to pass through less space and facilitate remediation. They can also retain groundwater for a longer period of time, thereby providing a long-term solution for soil remediation (Zhang, 2003). Even though the nanomaterials currently explored for remediation do not travel far from the site of injection, their size still gives hope for development in this aspect (Tratnyek and Johnson, 2006).

Nanomaterials such as nanoscale zeolites, metal oxides, carbon nanotubes and fibers, enzymes, various noble metals and titanium dioxide are used for nanoremediation. Iron nanoparticles have been shown to be effective against common environmental contaminants. These efficient nanoparticles can be further catalyzed and modified for enhanced function in remediation. Titanium dioxide (TiO_2) is also a leading candidate for nanoremediation and wastewater treatment, although it is reported to have not yet been expanded to full-scale commercialization. TiO_2 and zinc oxide (ZnO) are capable of semiconducting, photolcatalysis and energy conversion. These materials are low cost, less toxic and easily available. They have grabbed the attention of researchers with their fascinating properties. Their photocatalytic property has been widely studied for its application in water (Mansoori et al., 2008). Magnetic nanoparticles enable the separation of contaminants using the magnetic field, and this is considered a quality step for nanoremediation. Polymer brush-grafted magnetic nanoparticles are efficient in removing mercury (II) ions from water (Farrukh et al., 2013). Biologically synthesized magnetite (Fe_3O_4) nanoparticles have been proven to reduce toxic Cr(VI) into a stable, nontoxic form, such as a Cr31 spinel layer when studied using X-ray absorption and X-ray magnetic circular dichroism (Telling et al., 2009).

Nanoscale zero-valent iron (nZVI), the most commonly used nanomaterial (Theron et al., 2008), ranges in size from 10 to 100 nm in diameter. The distribution of nZVI and its mobility can be further enhanced upon creating a catalytic synergy by adding noble metals (which promotes iron oxidation or electron transfer) such as palladium (which catalyzes dechlorination and hydrogenation) and silver (U.S. EPA, 2008). These metals are generally less reactive, but they make the reaction more efficient (Tratnyek and Johnson, 2006).

Nanoparticles attach to soil and aquifer materials via various aspects such as through the surface chemistry, groundwater chemistry and hydrodynamic conditions (U.S. EPA, 2007, 2008). nZVI and carbon nanotubes can be used to clean up and reuse contaminated groundwater (Rajan, 2011). Hexavalent chromium occupies the top priority position in the US Environmental Protection Agency (EPA)'s defined list of toxic and environmental pollutants. It is likely a carcinogen and a mutagen. Such a toxic compound can be eliminated by using nZVI. A study proved that 1.5g of nZVI can eliminate 98% of hexavalent chromium in a limited time (Singh et al., 2011).

TABLE 6.1
Example of Nanomaterials for Heavy Metal Ion Remediation

Carbonaceous nanomaterials	CeO_2 carbon nanotubes (CNTs)
	CNTs functionalized with polymers
	CNTs functionalized with iron
	Multi-walled CNTs
	Biopolymers
	Zero-valent iron nanoparticles (nZVI)
TiO_2 photocatalyst	Nanocrystalline TiO_2

6.4.2.1 Advantages of Nanoremediation

Nanotechnology in environmental remediation makes the process cost effective and reduces the effort involved in cleaning up contaminated sites. Its ability to be used as an on-site or in-situ treatment is the major advantage of nanoremediation. Through this method, tedious processes, such as removing and transporting soil (involving excessive time and effort), can be avoided (Karn et al., 2009). Nanotechnology can destruct organic contaminants and transform or precipitate inorganic contaminants. nZVI is also known to treat radionuclides including radium and uranium. Their particle size enables closer contact with the contaminant, thereby increasing the rate of contaminant destruction. A study revealed that degradation was much higher with the use of microscale iron particles in combination with bimetallic nanoparticles than the use of microscale nanoparticles alone. nZVI is found to be operative across a wide range of pHs and temperatures.

6.4.2.2 Disadvantages of Nanoremediation

Risk factors cannot be completely eliminated when new compounds are introduced into the environment. Even though nanoparticles used for remediation are designed to reduce the contaminated toxicity, they may generate certain byproducts that have toxic effects and high mobility or they may stay at the site of injection for a longer duration. Recent research indicates that the nanoparticles used for remediation can enter the food chain and bioaccumulate (Karn et al., 2009).

6.5 TRADITIONAL REMEDIATION METHODS VERSUS NANOREMEDIATION

Traditional physical and chemical remediation methods are time consuming and expensive. Bioremediation specifically needs organisms that will successfully reduce pollutant levels. It is time-consuming that appropriate conditions empirically needed for the organism's growth have

TABLE 6.2
Nanoremediation versus Traditional Remediation Methods

Nanoremediation	Traditional remediation techniques
It reduces the effort involved in cleaning up the contaminated site.	Traditional remediation methods involve tedious and complex steps.
It facilitates in-situ treatment. Therefore, excessive effort and time consuming steps such as removing and transporting soil can be avoided.	In a traditional remediation method such as soil washing, the soil from the contaminated site has to be removed and transported for ex-situ treatment.
The developing field of nanotechnology enables the smart design of nanoparticles. Accordingly, nanoparticles can be designed in such a way that they can treat an extended range of contaminants.	Every traditional remediation method focuses only upon a specific range of contaminants.
The changes in environmental factors will not pose a problem because nanoparticles can be designed to survive such changes.	In the case of bioremediation, the survival of microorganisms is highly favored by diverse environmental factors.
It also helps in improving the extent and speed of contaminant destruction.	The removal, transportation and treatment of soil, the time taken for the growth and multiplication of microorganisms and the setting up of the instruments delay the remediation process.
The particle size enables closer contact with the contaminant, thereby increasing the rate of contaminant destruction.	Traditional treatments including vitrification, electrokinetic remediation, solid isolation and containment aim to treat only a specific site at a time, leaving the adjacent area untreated.
Properly designed nanoparticles will not alter the nature of the soil.	The chemical agents used in traditional methods may change the nature of the soil and contribute toxicity. Microbes used in the remediation and electrolysis processes tend to alter the pH of the soil.

to be strictly maintained. Non-biodegradable contaminants such as polymers and plastic will still remain in the soil. In contrast, nanotechnology helps design nanoscale particles that can not only reach inaccessible areas such as crevices and aquifers but also degrade the contaminant. This will further eliminate costly traditional operations. Moreover, nanotechnology enables us to design nanoparticles specific to certain contaminants, therefore increasing the affinity, selectivity and sensitivity of the technique. Therefore, nanoremediation is more specific and cost effective (Filipponi and Sutherland, 2010).

6.6 CONCLUSION

Air, water and soil are contaminated with different types of pollutants or contaminants generated by human activity. Heavy metal ions are one of the major pollutants that are hazardous to living beings. The traditional methods available for the removal of pollutants are generally labor intensive, consume more time and are expensive. Apart from these techniques, the remediation of toxic compounds can be achieved by emerging nanotechnology. Various nanoparticles have been found to be very effective for the elimination of contaminants from the environment. This approach is comparatively cheaper and more efficient. Therefore, nanotechnology serves as a better alternative for remediating our environment (Ingle et al., 2014).

REFERENCES

Adeniji A. *Bioremediation of Arsenic, Chromium, Lead, and Mercury*. A document submitted to US Environmental protection Agency Office of Solid Waste and Emergency Response Technology Innovation Office. Vol. C. Washington, DC, 2004, p. 43.

Agarwal A, Joshi H. Application of nanotechnology in the remediation of contaminated groundwater: A short review. *Recent Res Sci Technol*. 2010;2:51–57.

Ali H, Khan E, Sajad MA. Phytoremediation of heavy metals: Concepts and applications. *Chemosphere*. 2013;91:869–881.

Asante-Duah DK. *Management of Contaminated Site Problems*. New York: Lewis Publication; 1996.

Asiriuwa OD, Ikhuoria JU, Ilori EG. Myco-remediation of heavy metals from contaminated soil. *Bull Env Pharmacol Life Sci*. 2013;2(5).

Chu W, Chan KH. The mechanism of the surfactant-aided soil washing system for hydrophobic and partial hydrophobic organics. *Sci Total Environ*. 2003;307(1–3):83–92.

Czupyrna G, Levy RD, MacLean AI, Gold H. *In-Situ Immobilization of Heavy Metal Contaminated Soils*. Park Ridge, NJ: Noyes Data Corporation; 1989.

Dada EO, Njoku KI, Osuntoki AA and Akinola MO. A review of current techniques of in-situ Physico-chemical and biological remediation of heavy metals polluted soil. *Ethiop J Environ Stud Manag*. 2015;8(5):606.

Dhal B, Thatoi HN, Das NN, Pandey BD. Chemical and microbial remediation of hexavalent chromium from contaminated soil and mining/metallurgical solid waste: A review. *J Hazard Mater*. 2013;250–251:272–291.

Farrukh A, Akram A, Ghaffar A, Hanif S, Hamid A, Duran H, et al. Design of polymer-brush-grafted magnetic nanoparticles for highly efficient water remediation. *ACS Appl Mater Interfaces*. 2013;5(9):3784–3793.

Filipponi L, Sutherland D. Environment: Application of nanotechnologies. *Nanoyou Teachers Training Kit Nanotechnol*. 2010;1–26.

Garbisu C, Alkorta I. Basic concepts on heavy metal soil bioremediation. *Eur J Mineral Process Environ Prot*. 2003;3(1):58–66.

Giller KE, Witter E, Mcgrath SP. Toxicity of heavy metals to microorganisms and microbial processes in agricultural soils: A review. *Soil Biol Biochem*. 1998;30(10–11):1389–1414.

Gopalan A, Zincircioglu O, Smith P. Minimization and remediation of DOE nuclear waste problems using high selectivity actinide chelators. *Radioactive Waste Manage Nucl Fuel Cycle*. 1993;17:161–175.

Gupta AK, Yunus M, Pandey PK. Bioremediation: Ecotechnolgoy for the present century. *Enviro New, Bull Int Soc Environ Botan*. 2003;9(2).

Hajdu R, Slaveykova VI. Cd and Pb removal from contaminated environment by metal resistant bacterium Cupriavidus metallidurans CH34: Importance of the complexation and competition effects. *Environ Chem*. 2012;9(4):389.

Hasnain S, Ali SS, Uddin Z, Zafar R. Application of nanotechnology in health and environmental research: A review. *Res J Environ Earth Sci.* 2013;5(3):160–166.

Hejazi R, Husain T, Khan FI. Land farming operation in arid region-human health risk assessment. *J Hazard Mater.* 2003;99:287–302.

Ingle AP, Seabra AB, Duran N, Rai M. Nanoremediation. In: *Microbial Biodegradation and Bioremediation.* Amsterdam: Elsevier; 2014, pp. 233–250.

International Centre for Science and High Technology ICS. Trieste, Italy: ICS; 2005.

Jankaite A, Vasarevičius S. Remediation technologies for soils contaminated with heavy metals. *J Environ Eng Landsc Manage.* 2005;13(2):109–113.

Karn B, Kuiken T, Otto M. Nanotechnology and in situ remediation: A review of the benefits and potential risks. *Environ Health Perspect.* 2009;117(12):1813–1831.

Malik N, Biswasa AK. Role of higher plants in remediation of metal contaminated sites. *Sci Rev Chem Commun.* 2012;2(2):141–146.

Mansoori GA, Rohani T, Bastami A, Ahmadpour Z, Eshaghi. Environmental application of nanotechnology. *Annu Rev Nano Res.* 2008;2(2):1–73.

Mohamed I, Ahamadou B, Li M, Gong C, Cai P, Liang W, et al. Fractionation of copper and cadmium and their binding with soil organic matter in a contaminated soil amended with organic materials. *J Soils Sediments.* 2010;10(6):973–982.

Mulligan CN, Yong RN, Gibbs BF. Remediation technologies for metal-contaminated soils and groundwater: An evaluation. *Eng Geol.* 2001;60(1–4):193–207.

Njoku KL, Akinola MO, Oboh BO. Phytoremediation of crude oil contaminated soil: The effect of growth of Glycine max on the physico-chemistry and crude oil contents of soil. *Nat Sci.* 2009;12:22–30.

Okrent D. On intergenerational equity and its clash with intragenerational equity and on the need for policies to guide the regulation of disposal of wastes and other activities posing very long-term risks. *Risk Anal.* 1999;19(5):877–901.

Otto M, Floyd M, Bajpai S. Nanotechnology for site remediation: Nanotechnology for Site Remediation. *Remediation (N Y).* 2008;19(1):99–108.

Rahimi M, Farhardi R, Yusef PH. Phytoremediation of arsenic. *Int J Agron Plant Prod.* 2012;3(7):230–233.

Rajan CSR. Nanotechnology in groundwater remediation. *Int J Environ Sci Dev.* 2011;182–187.

Raskin II, Smith RD, Salt DE. Phytoremediation of metals: Using plants to remove pollutants from the environment. *Curr Opin Biotechnol.* 1997;8(2):221–226.

Riser-Roberts E. *Remediation of Petroleum Contaminated Soils.* Boca Raton: Lewis Publishers; 1998.

Rodsand T, Acar YB. Electrokinetic extraction of lead from spiked Norwegian marine clay. *Geoenvironmental.* 2000;2:1518–1534.

Sharma HD, Reddy KR. *Soil Remediation Technologies. Geoenvironmental Engineering: Site Remediation, Waste Containment, and Emerging Waste Management Technologies.* Hoboken, NJ: Wiley; 2004.

Singh R, Misra V, Singh RP. *Remediation of Cr(VI) Contaminated Soil by Zerovalent Iron Nanoparticles (nZVI) Entrapped in Calcium Alginate Beads.* Singapore: IACSIT Press; 2011.

Sinha RK, Chauhan K, Valani D, Chandran V, Soni BK, Patel V. Earthworms: Charles Darwin's 'unheralded soldiers of mankind': Protective & productive for man & environment. *J Environ Prot (Irvine Calif).* 2010;1(3):251–260.

Telling ND, Coker VS, Cutting RS, van der Laan G, Pearce CI, Pattrick RAD, et al. Remediation of Cr(VI) by biogenic magnetic nanoparticles: An x-ray magnetic circular dichroism study. *Appl Phys Lett.* 2009;95(16):163701.

Thapa B, Kc AK, Ghimire A. A review on bioremediation of petroleum hydrocarbon contaminants in soil. *Kathmandu Univ J Sci Eng Technol.* 2012;8(1):164–170.

Theron J, Walker JA, Cloete TE. Nanotechnology and water treatment: Applications and emerging opportunities. *Crit Rev Microbiol.* 2008;34(1):43–69.

Tratnyek PG, Johnson RL. Nanotechnologies for environmental cleanup. *Nano Today.* 2006;1(2):44–48.

Turkekul I, Elmastas M, Tüzen M. Determination of iron, copper, manganese, zinc, lead, and cadmium in mushroom samples from Tokat, Turkey. *Food Chem.* 2004;84(3):389–392.

Urum K, Pekdemir T, Gopur M. Optimum conditions for washing of crude oil contaminated soil with biosurfactant solutions. *Proc Safety Environ Prot Trans Inst Chem Eng Part B.* 2003;81:203–209.

U.S. EPA. *Risk Assessment Guidance for Superfund.* Cincinnati, OH: U. S. Environmental Protection Agency; 1989, pp. 540/1–89/002.

U.S. EPA. *Nanotechnology White Paper*. Washington, DC: U. S. Environmental Protection Agency; 2007.

U.S. EPA. *Nanotechnology for Site Remediation: Fact Sheet*. Washington, DC: U. S. Environmental Protection Agency; 2008.

Vidali M. Bioremediation: An overview. *Pure Appl Chem*. 2001;73(7):1163–1172.

Young CA, Jordan TS, Erickson LE, Tillison DL, Grant SC, McDonald J. Cyanide remediation: Current and past technologies. *Proceedings of the 10th Annual Conference on Hazardous Waste Research*; 1995. Available from: https://engg.k-state.edu/HSRC/95Proceed/young.pdf

Zhan H, Park E. Vapor flow to horizontal wells in unsaturated zones. *Soil Sci Soc Am J*. 2002;66(3):710–721.

Zhang W-X. Nanoscale iron particles for environmental remediation: An overview. *J Nanopart Res*. 2003;5: 323–332.

7 Joint Use of Nanomaterials and Plants for the Remediation of Inorganic Pollutants

Aqsa Nadeem Shah, Naila Safdar, Gul-e-Saba Chaudhry, and Azra Yasmin

7.1 INORGANIC POLLUTANTS

Waste from industries contains a variety of pollutants that are both inorganic and organic in nature. Many inorganic contaminants that can cause harm to the environment are produced and used in different ways and can cause serious health hazards. A wide range of inorganic pollutants is produced from industrial sources, for example, the pulp and paper industry (Tavangar et al., 2020). The pharmaceutical industry is the main source of inorganic pollutants (Singh & Chandra, 2019). These inorganic pollutants mainly include heavy metals and are released worldwide. According to one estimate, 50,000 tons/year of chromium are discharged from refuse incineration and the burning of coal and wood, and 60,000 tons/year of nickel is released by the combustion of coal, whereas the major portion of it remains in ash. In addition, 7,000 tons/year of cadmium are emitted from the combustion of coal and sewage sludge incineration (Ali et al., 2019). Some well-known inorganic pollutants include the following:

- Cadmium: It appears in rechargeable batteries, coating, pigments, plating, plastic stabilizers and tobacco smoke.
- Chromium: Many manufacturers use it for leather, brick, wood, pigments and dyes.
- Lead: It is utilized in paints, batteries, gasoline and explosives.
- Mercury: It is used in thermometers, incandescent lights, barometers, fossil fuel emissions and batteries.
- Arsenic: This element is employed as an alloy of lead, for example, in car batteries and ammunition (Saxena & Bharagava, 2017)

7.2 CRITICAL THRESHOLD VALUE OF INORGANIC POLLUTANTS

Allowable exposure concentrations are also known as the exposure limit, toxicity reference value or critical threshold. These threshold values are determined from toxicity experiments. A few heavy metals are known as essential heavy metals, and they are very important from a biological point of view. In contrast, if their dosage and contact time is increased, then they become dangerous to life. Non-essential heavy metals that include cadmium, lead and mercury and metalloids that include arsenic are toxic even at lower concentrations. Our body needs very small quantities of essential elements. However, if their concentration is increased above a specified limit, then they becomes toxic. As reported by Saxena et al. (2019), the highest threshold value in water for copper (Cu) is 2 mg/L, and the limit value for cadmium's threshold in water is 0.003 mg/L. In addition, in soil, the highest threshold value is reported for chromium (Cr) at 1–1,000 mg/kg, and the lowest is reported for mercury (Hg) at 0.001–3 mg/kg. Moreover, these heavy metals can act in the body like a metabolic

DOI: 10.1201/9781003186298-7

poison. Heavy metals' reaction with the sulfhydryl enzyme system results in their inhibition as these enzymes are involved in cellular energy production; therefore, the metabolic system becomes disturbed (Ali et al., 2019).

7.3 INORGANIC POLLUTANTS IN AIR

Inorganic pollutants in gaseous form become part of the atmosphere because of human activities. The pollutants that are added to the atmosphere in large quantities include carbon monoxide (CO), sulfur dioxide (SO_2), nitrogen dioxide (NO_2) and nitric oxide (NO). However, the quantity of carbon dioxide (CO_2) in the atmosphere is more than these pollutants. Significant quantities of these pollutants are added yearly to the atmosphere due to human activities. The worldwide emission of carbon, sulfur and nitrogen oxides are almost several hundred million tons each year (Manahan, 2017).

Many inorganic pollutants become part of the air, and their presence requires air treatment. These pollutants commonly include carbon, nitrogen and sulfur oxides. Volatile organic compound (VOC) removal from the air is easier than the removal of inorganic pollutants, and a considerable data exist about their removal. Compounds comprising nitrogen and oxygen are known as nitrogen oxides (NOs). Among them, the most toxic are nitric oxide (NO) and nitrogen oxide (NO_2). Another toxic oxide of nitrogen is nitrogen pentaoxide (N_2O_5). The main sources of NOs are anthropogenic sources that include aircrafts, vehicles, power plants and volcanic eruptions, along with some other natural causes. NOs that are present in a sizable amount in large cities can be the contributing factor in monitoring the air quality. Along with their negative effects on human health, they also contribute to acid rain and global warming and damaging the tropospheric ozone (Ângelo et al., 2013). Sulfur oxide is also known as a common atmospheric pollutant (Boyjoo et al., 2017). It is responsible for causing not only damage to buildings and vegetation but also acid rain (Burns et al., 2016). When present in the atmosphere as airborne elements, sulfur oxides can be the source of a reduction in visibility. When inhaled for a longer duration, sulfur oxides can cause many problems in the respiratory tract, including asthma and the constriction of bronchi (Krishnan et al., 2013).

7.4 INORGANIC POLLUTANTS IN WATER

Water that is free from toxins and pathogens, that is, clean water, is important for human health. Clean water is also important in a variety of industries such as electronics, food and pharmaceuticals. However, rapid industrialization has resulted in the production of hazardous waste, which is released directly in water. This includes arsenides, heavy metal ions, fluorides and inorganic toxic substances (Cao & Li, 2014).

The pulp and paper industry is the major contributor in releasing multifaceted pollutants in nature, namely, organic and inorganic pollutants. They are also released at the time of pulping and bleaching in the paper manufacturing procedure. The inorganic pollutants that are released during this process include Cu, nickel (Ni), zinc (Zn), magnesium and ferrous, which has been shown to be a cause of neurotoxicity in catfish. It accumulates in muscles, gills, ovaries and the liver. Other basic contaminants include hexadecanoic acids, octacosane, benzoic acid, terpenes, phenols and decalones. Of these, a few chemicals are known as endocrine-disrupting chemicals (EDCs), and they cause harm to the reproductive systems of aquatic life (Singh & Chandra, 2019).

The waste water released from the paint industry also contains many pollutants that have an organic and inorganic nature. Like organic and inorganic pigments, emulsion paint consists of emulsifying agents and non-cellulosic and cellulosic thickeners, extenders and latexes. The chemicals used in paint manufacturing are responsible for the presence of hazardous pollutants, which are mostly heavy metals in released water. Heavy metals are persistent, and their uptake by aquatic life results in different health problems that include gene mutations, deformities, kidney problems and cancer (Aniyikaiye et al., 2019).

7.5 INORGANIC POLLUTANTS IN SOIL

In soil, the major toxicants found consist of arsenic (As), Cu, Zn, Hg, cadmium (Cd) and Cr. These pollutants become part of the soil through poor management practices in agriculture, for example, industrial waste not being discharged properly, leakage of landfills, the unlawful discarding of household wastes, wet and dry depositions and industrial emissions, and through natural occurrences such as volcanic eruptions (Saha et al., 2017).

Paddy soils have a complex nature and are a source of the nutrients required for plant growth. The nutrient level of paddy soils is dependent upon factors including the application of fertilizer, tillage and movement of nutrients in the soil. Paddy soils become polluted with either industrial waste water or sewage waste water, which contain many pollutants, namely, heavy metals, fertilizers, pesticides and the leakage of petrochemicals. Inorganic pollutants that include heavy metals, for example, lead (Pb), Zn, Cd, As, Hg and Cu, cause serious health hazards to humans, plants and animals and also affects soil fertility (Akram et al., 2018). According to Li et al. (2018), urban soil, also known as anthropic soil, has been disturbed due to human activities such as importing, exporting and mixing, which causes contamination with many organic and inorganic pollutants. Urban soil contamination is because of emissions in urban areas through transport, the combustion of coal and industrial waste. Urban soil includes the soil in urban green areas such as gardens and parks. Therefore, the contamination of soil can directly affect human health when people come into contact with it. Urban soil contains non-biodegradable heavy metals. Moreover, they are difficult to remove.

7.6 HEALTH HAZARDS FROM INORGANIC POLLUTANTS

The global update 2005, air quality guidelines of the World Health Organization (WHO) provided an assessment of the health effects of air pollution. It also provided a threshold pollution level that is harmful to health. In 2016, 91% of the world's population lived where the WHO guidelines for air quality were not satisfying. In 2016, city and rural air pollution triggered 4.2 million premature deaths all over the world. Moreover, in low- and middle-income countries, 91% of premature deaths occurred due to "Ambient (outdoor) air pollution" (2018).

Toxic heavy metals such as Cr, Cd and Pb are found in higher concentrations in cancer and diabetic patients than people without these diseases in the City of Lahore in Pakistan (Ali et al., 2019). Diseases including cholera, diarrhea, hepatitis A and dysentery are mostly the result of unsanitary and unclean potable water. According to one estimate, every year, more than 842,000 deaths occur globally from diarrhea. In addition, as pollution also affects groundwater, this impacts 70 million people worldwide (Hasan et al., 2019). Inorganic pollutants that mainly include heavy metals result in various health hazards that are briefly described in Table 7.1.

7.7 EMERGING POLLUTANTS

Other than inorganic pollutants, some pollutants have been found in nature and can enter the environment. These pollutants are called emerging pollutants (EPos), and they have been part of our environment for a long time but have been ignored. These pollutants can affect human health, ecology and wildlife. EPos can be released from using inorganic chemicals or from dumping used chemicals. These pollutants are not even part of national and international routine monitoring programs, and their adverse effects and consequences are not well known. EPos can be organic or inorganic in nature. Inorganic EPos include selenium, thallium, Cu, silver, perfluoroalkyl compounds and chlorinated paraffin. These pollutants arrive through urban, industrial, transport and agricultural sources and are released by runoff, leaching and erosion (Geissen et al., 2015). EPos can also

TABLE 7.1
List of Inorganic Pollutants along with their Health Hazards

Inorganic pollutants	Health hazards	References
Cadmium	Abdomen pain, Burning sensation, Vomiting, Salivation, Nausea, Muscle cramps	Singh and Chandra (2019)
Chromium	Skin irritation, Nasal irritation, Perforation in eardrum, Ulceration, Lung carcinoma	RoyChowdhury et al. (2018)
Arsenic	Disturbance in the cardiovascular system and CNS, Hemolysis, Bone marrow depression, Melanosis, Polyneuropathy, Encephalopathy	Saxena and Bharagava (2017)
Lead	Headache, Poor attention span Irritability, Loss of memory Psychosis, Adverse effects to kidneys, liver, endocrine system, reproductive system	Das et al. (2017)
Mercury	Mental retardation, Blindness Loss of hearing, Neurological deficits, Developmental defects, Abnormal muscle tone	Manahan (2017)
Copper	Cancer in nose, larynx and lungs, Failure of respiratory tract, Asthma, Birth defects, Heart disease, Renal and Skin diseases	Madhav et al. (2020)
Nickel	Wilson's disease, Risk of lung cancer in humans	

be termed as materials that result from many daily uses whose nature can be biological or chemical (Pal et al., 2014). Different EPos are as follows.

7.7.1 Pharmaceuticals

The use of veterinary and human medicines, which includes antibiotics, analgesics and antidepressants, is extensive and continuously enters aquatic environments; a majority of these compounds are not metabolized and are therefore toxic to living beings. In healthcare facilities, pharmaceuticals are consumed at different concentrations. In the case of Germany, the total consumption of pharmaceuticals for hospitals including the psychiatric, nursing home and general hospital was estimated, and it ranges from 32 for psychiatric hospitals and 1,263 kg per year for general hospitals, whose regular annual usage for single pharmaceuticals ranged from 0.1 to 1,000 g/bed. Upon consumption, the excretion of these drugs takes place through urine at about 55–80%, and a lower rate is excreted through feces at 4–30%. These drugs are either released as metabolites, in conjunction with inactive substances, or as non-metabolized substances (Oliveira et al., 2017).

7.7.2 Personal Care Products

Substances of common use such as shampoo, toothpaste, perfumes, sunscreens and soaps become a part of our environment and are detected in groundwater and waste water because they are not completely degraded. These products release several pollutants that are termed EPos (Rodriguez-Narvaez et al., 2017).

Archer *et al.* (2017) conducted a study at a South African waste water treatment work (WWTW) and verified the existence of pharmaceuticals and personal care products (PPCPs). About 90 emerging contaminants were detected in this study. The treatment site was the Province of Gauteng in South Africa. Water samples were collected and passed through treatment steps. Out of 90 EPos, 19 classes of PPCPs were detected. Most of them were reduced in effluent, but some were retained and even showed higher concentrations.

7.7.3 Sunscreen Products

The use of sunscreen products has increased over recent years because of the popular trend of coastal sight-seeing all over the world. This has caused sunscreens to be a source of harmful chemicals to marine life. Sunblock/sunscreen products comprise both organic and inorganic ingredients/pollutants, which have not been addressed thus far. Among them are titanium oxide and Zn oxide that protect the human skin from direct ultraviolet (UV) radiation by scattering UV rays. In addition, these sunscreens also include certain products such as preservatives, different kinds of coloring agents and viscosity controllers and scents. These additives are fatal to phytoplankton in marine ecosystems; therefore, their removal is quite important, which can be performed either by biodegradation processes or absorption sedimentation. Sunscreen products have accumulated easily in the food chain because of their stability against biodegradation.

7.8 STRATEGIES TO REMEDIATE INORGANIC POLLUTANTS

7.8.1 Bioremediation

Bioremediation is the process in which biological causes, namely, microbes (algae, plants, fungi and bacteria), are used to remove toxins from the environment. In a polluted environment, one of the most efficient tools includes bioremediation to manage the organic and inorganic environmental contaminants. The harmful wastes generated from chemical operations are treated with the help of physical or chemical methods by industries to meet the standards of cleanliness as set by Environmental Protection Act, 1986 (Singh, 2014). Moreover, these physical and chemical methods are not zero waste and pose various other environmental problems.

7.8.2 Types of Bioremediation

Bioremediation approaches to treat pollutants include microbial bioremediation in which bacteria, fungi and algae are used to treat both inorganic and organic contaminants (Bharagava et al., 2020). Another technique involves phytoremediation, which controls pollutants in an economical, environmentally friendly and energy-efficient way. With this technology, plant parts and microbes interact together and mitigate the contaminants in the soil and air. The rhizosphere of plants is involved in the remediation of pollutants because pollutants accumulate in the soil when leaves fall off and precipitation occurs (Lee et al., 2020).

7.8.3 Phytoremediators in Pollutant Remediation

Plants can accept heavy metals from the soil in their leaves and shoots, without showing any toxicity, but only 0.2% of all plant species have this ability. The amount of metals that they can accumulate is 100–1,000 times greater than plants that are from a non-hyperaccumulator group. They also take up to 0.1% of metals including Cu, Pb, Cd, Cr, Ni and cobalt, whereas 1% of Zn and manganese are taken up by these plants. Plants such as *Brassica campestris L.*, *Brassica carinata*, *Brassica juncea L.* and *Brassica nigra L.* can collect a noticeable quantity of Pb in their dry biomass. *Cynodon dactylon* is a plant species that can accumulate metals such as Zn, Pd and Cd in quantities of 14 mg/kg, 658 mg/kg and 828 mg/kg, respectively (RoyChowdhury et al., 2018).

Arundo donax is an invasive weed, but its role in phytoremediation has also been identified. This plant is also named for a giant reed, and it has a prominent role in the phytoremediation of selenium, which is a toxic element. This plant is used for the phytoremediation of soils rich in selenium in China and the USA (Fiorentino et al., 2017). *A. Donax* is tolerant against As in higher concentrations. It not only bioaccumulates but also volatilizes As (7.2–22%) in concentrations from 300 to 1,000 ug/L. Due to this property, it is used to re-establish As-contaminated soil (Prabakaran et al.,

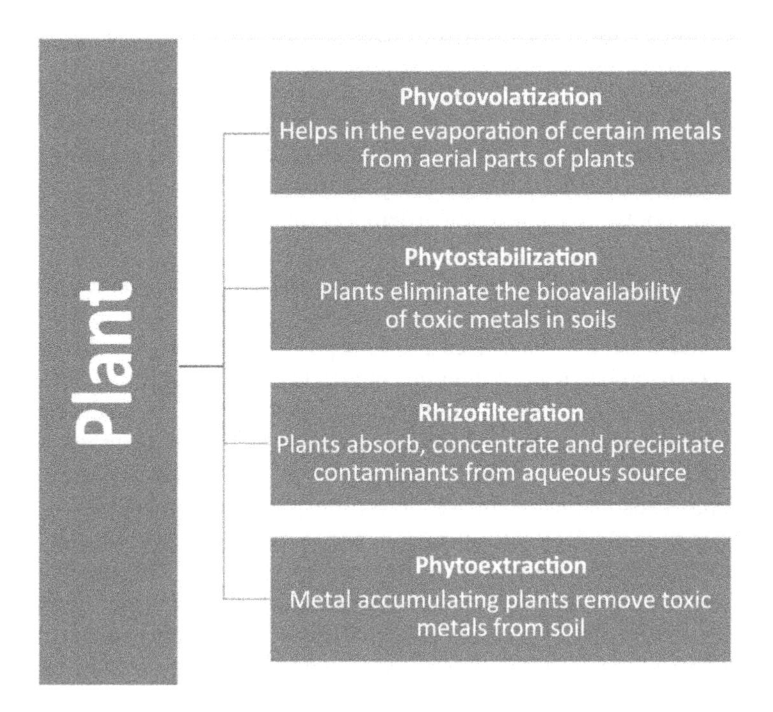

FIGURE 7.1 Modes of phytoremediation.

FIGURE 7.2 Various approaches in bioremediation.

2019). Awa and Hadibarata (2020) described the well-known mechanisms used in the phytoremediation of heavy metal-contaminated soil and water, which are shown in Figure 7.1. Common modes of phytoremediation include phytovolatalization, phytoextraction, phytostabilization and rhizofilteration. Some other common techniques involved in bioremediation strategies are shown in Figure 7.2.

7.9 NANO TREATMENTS OF INORGANIC POLLUTANTS

One of the greatest challenges of the present century is environmental pollution. Different technologies have been invented to mitigate pollutants in water, soil and air. Among these, nano treatments are newly emerging technology adapted for the treatment of waste and pollutants in the environment. Nano treatments, also referred to as nano-therapy, involve the use of nanoparticles with a nanoscale (1–100 nm) for the remediation of different pollutants regarding their differentiated environments of soil, water or air.

Nanoparticles that use a chemical method have been synthesized and employed to mitigate. Lim et al. (2018) reported that heavy metals and dyes can be removed from water by incorporating the usage of graphene and its oxides. Nano-sized graphene possesses a large surface area. It is possible to prepare it through both top-down and bottom-up approaches. Examples are the well-known method of chemical vapor deposition (CVD) and chemical exfoliation. Because of their porous structure, nanoparticles adsorb contaminants present in water. Huang et al. (2018) described nanoscale zerovalent iron (nZVI) nanoparticles as significant material for the removal of azo dyes, chlorinated and nitro aromatic compounds and heavy metals. The principle of this technology is based on a thin iron oxide core that through electronic interaction and surface complexation can quickly stop the uptake of pollutants. Nanofiber technology as reported by Lim (2017) has emerged because of its high specific surface area and high surface area-to-volume ratio. Nanofibers can be synthesized by different means through solution blow spinning, centrifugal jet spinning and electro hydrodynamic direct writing. Pollutants in water can be removed with particles that have a distinct pore size. The efficient removal of methylene blue from water has been conducted with hybrid MnO_2-coated cellulose nanofibers. These nanofibers showed efficient adsorption and oxidation in the decolonization of methylene blue. Cu and Ni from aqueous solutions can be successfully removed by using magnetic hydroxyapatite nano rods as indicated by Thanh et al. (2018). Nanomaterials of silica are utilized in eliminating cationic dyes, heavy metals and hydrogen sulfide from water, as reported by Guerra et al. (2018). Other various types of these compounds used in the mitigation of inorganic pollutants are listed in Table 7.2.

TABLE 7.2
Types of Nanomaterials Involved in the Remediation of Inorganic Pollutants

Nanoparticles	Pollutant	Function	Reference
Silver Nanoparticles	Pb	Excellent in removing Pb and other waste water and groundwater metals	Guerra et al. (2018)
Titanium Oxides	Pb, Hg, Cd and Cr	Extensively used to treat waste, purify air, clean surfaces and as water photocatalysts because of their cost effectiveness and non-toxic properties	Guerra et al. (2018)
Zinc Oxide Nanoparticles	Nitrogen oxide (NO_2) and sulfur oxides (SO_2)	Effectively used for water disinfection and the removal of foul-smelling pollutants in waste water	Dimapilis et al. (2018)
Single-Walled Carbon Nanotubes	N_2, CH_4, CO and CO_2	Excellent for the removal of pollutants from water	Rahman et al. (2017)
Sodium Titanate Nanobelt Membrane (Na-TNB)	Oil elimination and radioactive Cs+ ions and Sr^{2+}	Utilized in eliminating radioactive pollutants and oil spills from water and is up to 23 times more effective than other remediators because of good mechanical and thermal stability and excellent recyclability	Wen et al. (2016)
Magnetic Multi-Wall Carbon Nanotubes	Cr(VI)	Because of their structural and chemical stability, these nanotubes show potential in the adsorption of heavy metal ions	Huang et al. (2018)

7.10 LIMITATIONS OF CHEMICALLY SYNTHESIZED NANO TREATMENTS

Various nanomaterials of a chemically synthesized nature are employed in the treatment of inorganic pollutants, but because of their limitations, their usage is not common. Some of the limitations of chemically synthesized nanomaterials include

- High cost and low yield
- High temperature requirement
- High technical facilities are needed
- Potentially hazardous
- Capping ligands and residual toxins add to environmental toxicity
- They threaten the environment by generating new toxins and pollutants

(Khalil et al., 2016).

7.11 GREEN NANOMATERIALS

The term "green" denotes the utilization of plant or plant parts by following green chemistry and green engineering. The main purpose of this technology is to contribute towards environmental sustainability by producing nanoproducts for the benefit of mankind and the environment (Verma et al., 2019). Well-known advantages of green nanomaterials include

- Reduction of greenhouse gases and harmful remains
- Amplified energy competence
- Reduced intake of non-renewable raw provisions
- Lessen the risk for discharging toxic residues into the environment
- Ensure sustainable production and application.

Various types of plant-based nanoparticles are employed in the treatment of inorganic pollutants. The remediation efficiency of various types of synthesized nanomaterials vary with the class and type of green material used. Different classes of green nanomaterials employed in the remediation of pollutants are discussed in Table 7.3.

7.12 MODE OF THE MECHANISM FOR NANO TREATMENTS

Nano treatments are one of the latest tools for the remediation of inorganic pollutants. Due to their small surface area, nanomaterials act as potent pollutant remediators. Common mechanisms employed for nano-remediation include

1. Adsorption (Figure 7.3) and
2. Photocatalysis.

7.12.1 ADSORPTION

Adsorption can be defined as the procedure of adsorption of heavy metals on the surface of a solid substance. When the heavy metal amount that has been adsorbed by water reaches to constant, an equilibrium is achieved. The adsorption rate is determined through different kinetic models. If the adsorption rate is fast, then this means that a shorter amount of time is required to attain the equilibrium in adsorption. The rate of adsorption is usually affected through different factors such as the pore diffusion velocity of the adsorbates, time, pollutants, concentration of the adsorbent, temperature and pH.

TABLE 7.3
Class of Green Nanomaterials with Their Properties and Applications

Groups	Plant from which synthesized	Properties	Applications	References
Carbon-Based Nanoparticles	*Solanum lycopersicum*, *Brassica napus* and *Amaranthus dubius*	They consist of fullerenes, carbon dots and nano dots. Due to their variety of shapes and properties related to their electronic and thermal conductivity, they present many applications.	Removal of akageneite and hematite from soil	Yusof et al. (2019), Zaytseva, and Neumann (2016)
Ceramic Nanoparticles	*Nyctanthes arbor-tristis*	They are nonmetallic solids of inorganic nature and are prepared by heating first and then cooling. They are present in various form such as dense, amorphous, hollow and polycrystalline.	Photodegradation of dyes	Khan et al. (2019)
Metal Nanoparticles	*Azadirachta indica*, *Avena sativa* and *Syzygium aromaticum*	These include alkali and noble metals, namely, Au, Ag and Cu. They possess unique optoelectrical properties.	Water purification	Khan et al. (2019), Das et al. (2017)
Polymeric Nanoparticles	*Xanthium strumarium* and *Taraxacum laevigatum*	They are a particle of nature from 1–1,000 nm. They are modified by entrapping active compounds in them or adsorbed on their polymeric core.	Environmental remediation	Zielińska et al. (2020), Salem and Fouda (2020)
Lipid-Based Nanoparticles	*Citrus paradisi*, *Zingiber officinale*, *Brassica oleracea* var. italica, and *Citrus limon*	These have a spherical shape and a diameter from 10–1,000 nm. They are extensively used due to their high degree of biocompatibility.	Pharmacological and pollution remediation	García-Pinel et al. (2019), Yang et al. (2018)

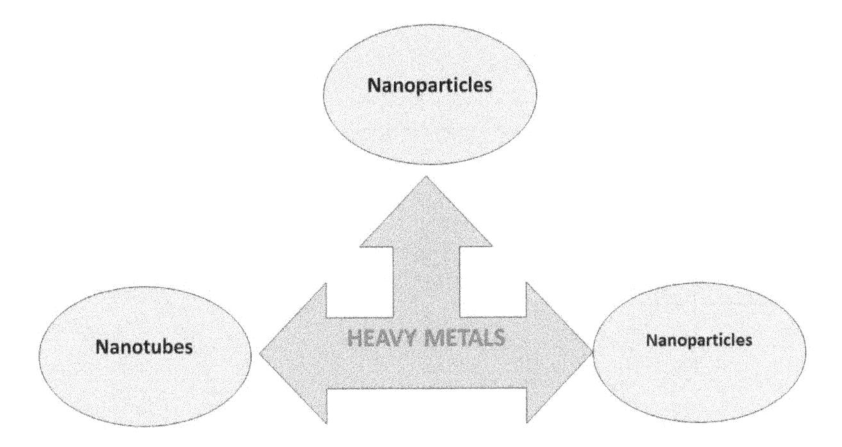

FIGURE 7.3 Nanomaterials can promote the adsorption of heavy metals.

Many nanomaterials including carbon composites and nanotubes, graphene, polymeric sorbents, metal oxides and nano metals are being used to remove inorganic pollutants in different mediums, namely, the air, soil and water. They possess an efficient adsorption ability. Major reasons for their usage are the easy removal of adsorbed contaminants from their surface and their recycling ability (Wang et al., 2020).

7.12.2 Photocatalysis

Photocatalysis is one of the evolving technologies for treating pollutants by using different types of semiconductors as a photocatalyst. Because of the electronic structure of the semiconductors, they contribute to photocatalysis. Semiconductors comprise a

- Valance Band (VB) and a
- Conduction band (CB).

The band gap is known as the difference in energy between these two bands. Its energy is > 3.5 volts. The electrons and holes in the VB before exposure to light are not in an excited state. When the semiconductor surface is exposed to the light of a specific type of wavelength, a transfer of electrons takes place from the VB to the CB. Because of this process, holes are produced in the VB that lead to the production of an electron-hole pair. Holes and electrons both reach the surface and cause oxidation and a reduction of adsorbed reactants, respectively. Additionally, from the electron-hole pair, a strong redox system comes into existence. This results in a generation of a hydroxyl radical by the photo produced holes via OH^- and H_2O oxidation that are being absorbed on the surface of the semiconductor. Meanwhile, a reduction of O_2 molecules, which come from absorbed air, takes place by the action of the electrons in the conduction band. This eventually results in the synthesis of peroxyl radicals. These produced radicals have the potential for oxidation and the degradation of organic/inorganic substances. These reactions are the basis for any photocatalytic mechanism as shown in Figure 7.4.

Semicondutors are used because they are cost efficient, stable and harmless to humans and the environment. Applications of photocatalysis include the purification of waste water, photocatalytic self-cleaning, photovoltaics and photocatalytic water splitting. Commonly used semiconductors for photocatalysis are based on titanium and Zn, which are the ideal metals for nanoparticle synthesis when the goal is pollutant treatment (Jain & Vaya, 2017).

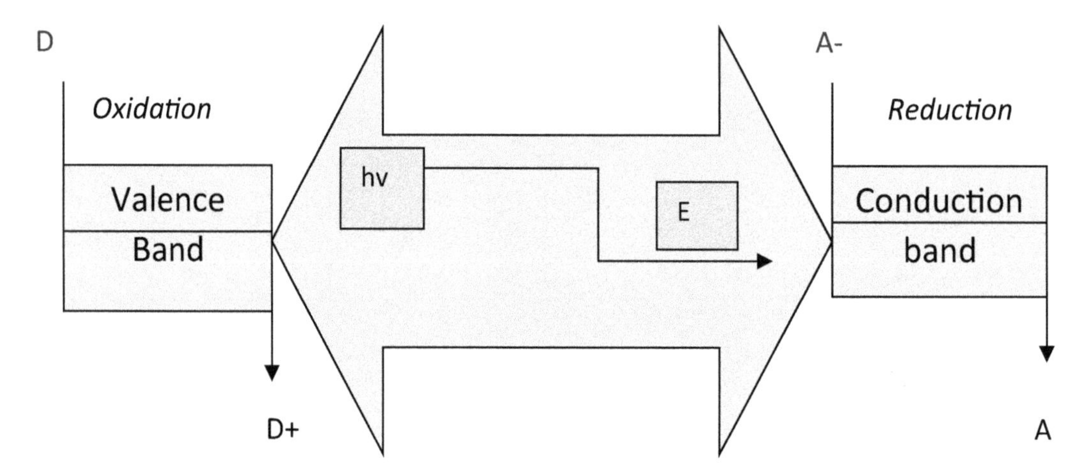

FIGURE 7.4 Mechanism of photocatalytic activity.

7.13 SUCCESS STORIES OF GREEN NANOMATERIALS IN THE TREATMENT OF INORGANIC POLLUTANTS

Nanoparticle formation by numerous biological systems has been described, but presently, the focus is on the biosynthesis of plant nanoparticles, which is considered among the most suitable methods. Listed below are some of the recent success stories of green nanomaterials used in the treatment of inorganic pollutants.

1. Shittu and Ihebunna (2017) reported the use of nanoparticles of Ag for the purification of waste water. They were synthesized from leaf extract of *Piliostigma thonningii*. These nanoparticles effectively removed heavy metals from water with a removal efficiency of 96.9% and subchronic effect of purified water in *in-vivo* rats.
2. Abril et al. (2018) described the exclusion of heavy metals (Cu, Zn) in soil and water with a removal efficiency of greater than 99% for water and 95% for soil by using the extract of mortino fruit (*Vaccinium floribundum* Kunth).
3. Cd (II) is a toxic heavy metal that affects the health of all humans, animals and plants. Silver nanoparticles synthesized from the leaf extract of the Ficus tree (*Ficus Benjamina*), as reported by Al-Qahtani (2017), have been utilized to eliminate Cd (II) in water. The removal of Cd increased with the increase in the time of the biosorbent.
4. Samrot et al. (2019) described *Araucaria heterophylla*, *Azadirachta indica* and *Prosopis chilensis* as plants whose gums were used for the synthesis of Ag nanoparticles. These nanoparticles were less than 50 nm in size. Along with anticancer and antimicrobial activity, they were effectively used for heavy metals removal.
5. Goutam et al. (2018) discussed the production of titanium oxide nanoparticles synthesized from biodiesel from a plant known as *Jatropha curcas* L. and examined its activity to treat tannery waste water (TWW). Chemical oxygen demand was improved and Cr was removed from the TWW when the nanoparticles underwent secondary treatment. Because of the treatment in the self-designed reactor, a success rate of 82.26% COD and 76.48% Cr removal was achieved. Therefore, this treatment is being used popularly as green technology for the *in-situ* remediation of TWW.
6. Yasmin et al. (2020) reported the production of Ag nanoparticles from the leaf extract of the plant *Diospyros lotus*. The best Ag nanoparticles were synthesized at a pH of 8.6 with

1.5 mM conc. of $AgNO_3$ with a 10 ml concentration of leaf extract. The photocatalytic activity of these nanoparticles was observed at 72.91% for the decolorization of industrial waste water at 54 h.

7. Magudieshwaran et al. (2019) reported that cerium oxide (CeO_2) nanoparticles synthesized from *Jatropha curcas* L. have the potential of the photocatalytic degradation of indoor gaseous pollutants. These nanoparticles were found to be of a reduced particle size (3–5nm) compared to the chemically synthesized nanoparticles of CeO_2 (18–25 nm). These green synthesized nanoparticles showed a more effective degradation of acetaldehyde through photocatalysis.

7.14 CONCLUSION

Many inorganic pollutants are released from different industries and become part of our environment. Inorganic pollutants include heavy metals, which cause serious health hazards to living beings. Inorganic pollutants are found in the air, water and soil. Some pollutants are not well known and are found in the environment as EPos. These may be inorganic or organic in nature. These pollutants can be degraded through bioremediation. Nano treatment is the emerging technology for pollutant remediation, but due to its certain limitations, a more promising technology known as green technology is evolving. This new emerging technology is the future for the remediation of pollutants from the environment. Recent studies and developments emphasize using the plant-based synthesis of nanoparticles as a cost-effective and sustainable approach.

REFERENCES

Abril, M., Ruiz, H., & Cumbal, L. H. (2018). Biosynthesis of multicomponent nanoparticles with extract of mortiño (Vaccinium floribundum Kunth) berry: Application on heavy metals removal from water and immobilization in soils. *Journal of Nanotechnology, 2018.*

Akram, R., Turan, V., Hammad, H. M., Ahmad, S., Hussain, S., Hasnain, A., & Nasim, W. (2018). Fate of organic and inorganic pollutants in paddy soils. In *Environmental Pollution of Paddy Soils* (pp. 197–214). Springer, Cham.

Ali, H., Khan, E., & Ilahi, I. (2019). Environmental chemistry and ecotoxicology of hazardous heavy metals: Environmental persistence, toxicity, and bioaccumulation. *Journal of Chemistry, 2019.*

Al-Qahtani, K. M. (2017). Cadmium removal from aqueous solution by green synthesis zero valent silver nanoparticles with Benjamina leaves extract. *The Egyptian Journal of Aquatic Research, 43*(4), 269–274.

Ambient (Outdoor) Air Pollution. (2018, May 2). Retrieved from www.who.int/news-room/fact-sheets/detail/ambient-(outdoor)-air-quality-and-health#

Ângelo, J., Andrade, L., Madeira, L. M., & Mendes, A. (2013). An overview of photocatalysis phenomena applied to NOx abatement. *Journal of Environmental Management, 129*, 522–539.

Aniyikaiye, T. E., Oluseyi, T., Odiyo, J. O., & Edokpayi, J. N. (2019). Physico-chemical analysis of wastewater discharge from selected paint industries in Lagos, Nigeria. *International Journal of Environmental Research and Public Health, 16*(7), 1235.

Archer, E., Petrie, B., Kasprzyk-Hordern, B., & Wolfaardt, G. M. (2017). The fate of pharmaceuticals and personal care products (PPCPs), endocrine disrupting contaminants (EDCs), metabolites and illicit drugs in a WWTW and environmental waters. *Chemosphere, 174*, 437–446.

Awa, S. H., & Hadibarata, T. (2020). Removal of heavy metals in contaminated soil by phytoremediation mechanism: A review. *Water, Air, & Soil Pollution, 231*(2), 1–15.

Bharagava, R. N., Saxena, G., & Mulla, S. I. (2020). Introduction to industrial wastes containing organic and inorganic pollutants and bioremediation approaches for environmental management. In *Bioremediation of Industrial Waste for Environmental Safety* (pp. 1–18). Springer, Singapore.

Boyjoo, Y., Sun, H., Liu, J., Pareek, V. K., & Wang, S. (2017). A review on photocatalysis for air treatment: From catalyst development to reactor design. *Chemical Engineering Journal, 310*, 537–559.

Burns, D. A., Aherne, J., Gay, D. A., & Lehmann, C. (2016). Acid rain and its environmental effects: Recent scientific advances. *Atmospheric Environment*, *146*, 1–4.

Cao, Y., & Li, X. (2014). Adsorption of graphene for the removal of inorganic pollutants in water purification: A review. *Adsorption*, *20*(5–6), 713–727.

Das, R. K., Pachapur, V. L., Lonappan, L., Naghdi, M., Pulicharla, R., Maiti, S., & Brar, S. K. (2017). Biological synthesis of metallic nanoparticles: Plants, animals and microbial aspects. *Nanotechnology for Environmental Engineering*, *2*(1), 1–21.

Dimapilis, E. A. S., Hsu, C. S., Mendoza, R. M. O., & Lu, M. C. (2018). Zinc oxide nanoparticles for water disinfection. *Sustainable Environment Research*, *28*(2), 47–56.

Fiorentino, N., Ventorino, V., Rocco, C., Cenvinzo, V., Agrelli, D., Gioia, L., & Fagnano, M. (2017). Giant reed growth and effects on soil biological fertility in assisted phytoremediation of an industrial polluted soil. *Science of the Total Environment*, *575*, 1375–1383.

García-Pinel, B., Porras-Alcalá, C., Ortega-Rodríguez, A., Sarabia, F., Prados, J., Melguizo, C., & López-Romero, J. M. (2019). Lipid-based nanoparticles: Application and recent advances in cancer treatment. *Nanomaterials*, *9*(4), 638.

Geissen, V., Mol, H., Klumpp, E., Umlauf, G., Nadal, M., van der Ploeg, M., & Ritsema, C. J. (2015). Emerging pollutants in the environment: A challenge for water resource management. *International Soil and Water Conservation Research*, *3*(1), 57–65.

Goutam, S. P., Saxena, G., Singh, V., Yadav, A. K., Bharagava, R. N., & Thapa, K. B. (2018). Green synthesis of TiO2 nanoparticles using leaf extract of Jatropha curcas L. for photocatalytic degradation of tannery wastewater. *Chemical Engineering Journal*, *336*, 386–396.

Guerra, F. D., Attia, M. F., Whitehead, D. C., & Alexis, F. (2018). Nanotechnology for environmental remediation: Materials and applications. *Molecules*, *23*(7), 1760.

Hasan, M., Islam, M. A., Aziz Hasan, M., Alam, M. J., & Peas, M. H. (2019). Groundwater vulnerability assessment in Savar upazila of Dhaka district, Bangladesh—A GIS-based drastic modeling. *Groundwater for Sustainable Development*, *9*, 100220.

Huang, X. Y., Ling, L., & Zhang, W. X. (2018). Nanoencapsulation of hexavalent chromium with nanoscale zero-valent iron: High resolution chemical mapping of the passivation layer. *Journal of Environmental Sciences*, *67*, 4–13.

Jain, A., & Vaya, D. (2017). Photocatalytic activity of TiO2 nanomaterials. *Journal of the Chilean Chemical Society*, *62*(4), 3683–3690.

Khalil, I., Julkapli, N. M., Yehye, W. A., Basirun, W. J., & Bhargava, S. K. (2016). Graphene—gold nanoparticles hybrid—synthesis, functionalization, and application in a electrochemical and surface-enhanced raman scattering biosensor. *Materials*, *9*(6), 406.

Khan, I., Saeed, K., & Khan, I. (2019). Nanoparticles: Properties, applications and toxicities. *Arabian Journal of Chemistry*, *12*(7), 908–931.

Krishnan, P., Zhang, M. H., Cheng, Y., Riang, D. T., & Liya, E. Y. (2013). Photocatalytic degradation of SO2 using TiO2-containing silicate as a building coating material. *Construction and Building Materials*, *43*, 197–202.

Lee, B. X. Y., Hadibarata, T., & Yuniarto, A. (2020). Phytoremediation mechanisms in air pollution control: A review. *Water, Air, & Soil Pollution*, *231*(8), 1–13.

Li, G., Sun, G. X., Ren, Y., Luo, X. S., & Zhu, Y. G. (2018). Urban soil and human health: A review. *European Journal of Soil Science*, *69*(1), 196–215.

Lim, C. T. (2017). Nanofiber technology: Current status and emerging developments. *Progress in Polymer Science*, *70*, 1–17.

Lim, J. Y., Mubarak, N. M., Abdullah, E. C., Nizamuddin, S., & Khalid, M. (2018). Recent trends in the synthesis of graphene and graphene oxide-based nanomaterials for removal of heavy metals—A review. *Journal of Industrial and Engineering Chemistry*, *66*, 29–44.

Madhav, S., Ahamad, A., Singh, A. K., Kushawaha, J., Chauhan, J. S., Sharma, S., & Singh, P. (2020). Water pollutants: Sources and impact on the environment and human health. In *Sensors in Water Pollutants Monitoring: Role of Material* (pp. 43–62). Springer, Singapore.

Magudieshwaran, R., Ishii, J., Raja, K. C. N., Terashima, C., Venkatachalam, R., Fujishima, A., & Pitchaimuthu, S. (2019). Green and chemical synthesized CeO2 nanoparticles for photocatalytic indoor air pollutant degradation. *Materials Letters*, *239*, 40–44.

Manahan, S. (2017). Gaseous inorganic air pollutants. In *Environmental Chemistry* (pp. 267–290). CRC Press, New York.

Oliveira, T. S., Al Aukidy, M., & Verlicchi, P. (2017). Occurrence of common pollutants and pharmaceuticals in hospital effluents. In *Hospital Wastewaters* (pp. 17–32). Springer, Cham.

Pal, A., He, Y., Jekel, M., Reinhard, M., & Gin, K. Y. H. (2014). Emerging contaminants of public health significance as water quality indicator compounds in the urban water cycle. *Environment International*, *71*, 46–62.

Prabakaran, K., Li, J., Anandkumar, A., Leng, Z., Zou, C. B., & Du, D. (2019). Managing environmental contamination through phytoremediation by invasive plants: A review. *Ecological Engineering*, *138*, 28–37.

Rahman, M. M., Sime, S. A., Hossain, M. A., Shammi, M., Uddin, M. K., Sikder, M. T., & Kurasaki, M. (2017). Removal of pollutants from water by using single-walled carbon nanotubes (SWCNTs) and multi-walled carbon nanotubes (MWCNTs). *Arabian Journal for Science and Engineering*, *42*(1), 261–269.

Rodriguez-Narvaez, O. M., Peralta-Hernandez, J. M., Goonetilleke, A., & Bandala, E. R. (2017). Treatment technologies for emerging contaminants in water: A review. *Chemical Engineering Journal*, *323*, 361–380.

RoyChowdhury, A., Datta, R., & Sarkar, D. (2018). Heavy metal pollution and remediation. In *Green Chemistry* (pp. 359–373). Elsevier, Amsterdam.

Saha, J. K., Selladurai, R., Coumar, M. V., Dotaniya, M. L., Kundu, S., & Patra, A. K. (2017). Major inorganic pollutants affecting soil and crop quality. In *Soil Pollution-An Emerging Threat to Agriculture* (pp. 75–104). Springer, Singapore.

Salem, S. S., & Fouda, A. (2020). Green synthesis of metallic nanoparticles and their prospective biotechnological applications: An overview. *Biological Trace Element Research*, 1–27.

Samrot, A. V., Angalene, J. L. A., Roshini, S. M., Raji, P., Stefi, S. M., Preethi, R., & Madankumar, A. (2019). Bioactivity and heavy metal removal using plant gum mediated green synthesized silver nanoparticles. *Journal of Cluster Science*, *30*(6), 1599–1610.

Saxena, G., & Bharagava, R. N. (2017). Organic and inorganic pollutants in industrial wastes: Ecotoxicological effects, health hazards, and bioremediation approaches. In *Environmental Pollutants and their Bioremediation Approaches* (pp. 23–56). CRC Press, New York.

Saxena, G., Purchase, D., Mulla, S. I., Saratale, G. D., & Bharagava, R. N. (2019). Phytoremediation of heavy metal-contaminated sites: Eco-environmental concerns, field studies, sustainability issues, and future prospects. *Reviews of Environmental Contamination and Toxicology*, *249*, 71–131.

Shittu, K. O., & Ihebunna, O. (2017). Purification of simulated waste water using green synthesized silver nanoparticles of Piliostigma thonningii aqueous leave extract. *Advances in Natural Sciences: Nanoscience and Nanotechnology*, *8*(4), 045003.

Singh, A. K., & Chandra, R. (2019). Pollutants released from the pulp paper industry: Aquatic toxicity and their health hazards. *Aquatic Toxicology*, *211*, 202–216.

Singh, R. (2014). Microorganism as a tool of bioremediation technology for cleaning environment: A review. *Proceedings of the International Academy of Ecology and Environmental Sciences*, *4*(1), 1.

Tavangar, T., Karimi, M., Rezakazemi, M., Reddy, K. R., & Aminabhavi, T. M. (2020). Textile waste, dyes/inorganic salts separation of cerium oxide-loaded loose nanofiltration polyethersulfone membranes. *Chemical Engineering Journal*, *385*, 123787.

Thanh, D. N., Novák, P., Vejpravova, J., Vu, H. N., Lederer, J., & Munshi, T. (2018). Removal of copper and nickel from water using nanocomposite of magnetic hydroxyapatite nanorods. *Journal of Magnetism and Magnetic Materials*, *456*, 451–460.

Verma, A., Gautam, S. P., Bansal, K. K., Prabhakar, N., & Rosenholm, J. M. (2019). Green nanotechnology: Advancement in phytoformulation research. *Medicines*, *6*(1), 39.

Wang, L., Shi, C., Pan, L., Zhang, X., & Zou, J. J. (2020). Rational design, synthesis, adsorption principles and applications of metal oxide adsorbents: A review. *Nanoscale*, *12*(8), 4790–4815.

Wen, T., Zhao, Z., Shen, C., Li, J., Tan, X., Zeb, A., & Xu, A. W. (2016). Multifunctional flexible free-standing titanate nanobelt membranes as efficient sorbents for the removal of radioactive 90 Sr 2+ and 137 Cs+ ions and oils. *Scientific Reports*, *6*(1), 1–10.

Yang, C., Zhang, M., & Merlin, D. (2018). Advances in plant-derived edible nanoparticle-based lipid nano-drug delivery systems as therapeutic nanomedicines. *Journal of Materials Chemistry B*, *6*(9), 1312–1321.

Yasmin, S., Nouren, S., Bhatti, H. N., Iqbal, D. N., Iftikhar, S., Majeed, J., & Rizvi, H. (2020). Green synthesis, characterization and photocatalytic applications of silver nanoparticles using Diospyros lotus. *Green Processing and Synthesis*, *9*(1), 87–96.

Yusof, N. A., Abd Rahman, S. F., & Muhammad, A. (2019). Carbon nanotubes and graphene for sensor technology. In *Synthesis, Technology and Applications of Carbon Nanomaterials* (pp. 205–222). Elsevier, Amsterdam.

Zaytseva, O., & Neumann, G. (2016). Carbon nanomaterials: Production, impact on plant development, agricultural and environmental applications. *Chemical and Biological Technologies in Agriculture*, *3*(1), 1–26.

Zielińska, A., Carreiró, F., Oliveira, A. M., Neves, A., Pires, B., Venkatesh, D. N., & Souto, E. B. (2020). Polymeric nanoparticles: Production, characterization, toxicology and ecotoxicology. *Molecules*, *25*(16), 3731.

8 Integration of Nano-Phytoremediation and Omics Technology for Sustainable Environmental Cleanup

Ashutosh Triphati and Tanveer Bilal Pirzadah

8.1 INTRODUCTION

Environmental pollution has become a major threat worldwide, and it is gaining rapid momentum due to various anthropogenic activities such as poor agricultural practices, geogenic and technogenic activities and natural calamities (Earth 88). In addition, population explosion during recent decades has become a serious issue, which forces us to rethink policy to meet our necessities, and from this point, severe challenges relating to the environment started. Unplanned colonization, a lack of proper waste disposal and unsystematic solid waste management and salvage have also kept the environment at high risk. Deforestation has occurred mainly for the purpose of industrialization, agriculture and urbanization and has proved to be a major challenge and origin of many environmental problems through its degradation, as seen by the enhanced rate of soil erosion, sedimentation in rivers, silting, rate of floods and droughts and green house effects and a decrease in the top fertile soil; this has also produced a chain reaction. Developments in the field of agriculture have also degraded the environment mainly through the excessive application of chemical fertilizers, pesticides, insecticides, herbicides and even banned chemicals for better yield to meet requirements. All of these negative practices have led to a disruption in plant-microbe dynamics. Rapid growth in industries has caused them to exploit natural resources, which creates several hazardous environmental problems and an ecological imbalance not only at the local level but also at the global level. Although these industries produce useful products, there are several unwanted byproducts that are lethal to the environment and health, and they thus pollute and degrade environment. Therefore, the natural composition of the soil, water and air are becoming imbalanced by a continuous increase in the level of polluting agents such as toxic gases, volatile organic compounds, organic substances, heavy metals, industrial dyes, solid wastes, polycyclic aromatic hydrocarbons (PAHs) and polychlorinated biphenyls (PCBs). Moreover, most of them degrade slowly and possess a longer residence time (Zhou et al. 2008; Jördening and Winter 2006; Roig et al. 2012; Cieślik et al. 2015; Liu et al. 2005; Joshi et al. 2014). Accordingly, due to these seen and unseen anthropogenic activities, the environment is now at a critical point. Pollution has affected not only food chains but also the overall food web, and as result, all elements of the environment are severely affected. Therefore, it is time to rethink and reframe environmental protection policies and to focus much more on the use of safer and more advanced tools and techniques to clean the environment by removing pollutants (Reddy 2017). However, soil has its own filtration systems, namely, physical filtration by a sieving action, chemical filtration by adsorption and precipitation, transformation and biological filtration by decomposing organic materials, but they work only up to a certain limit. Several techniques are used to remediate pollutants from the soil, water and air by using physical, chemical and biological

DOI: 10.1201/9781003186298-8

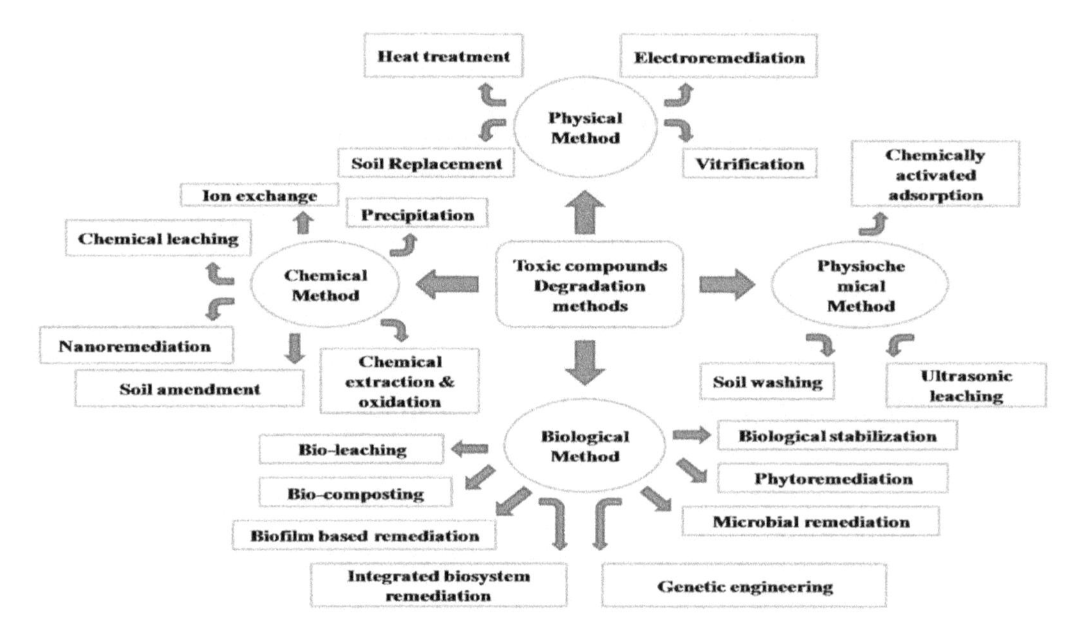

FIGURE 8.1 Various methods used for the remediation of co-contaminated soil.

systems (Figure 8.1). For example, phytoremediation is a natural, cost effective and eco-friendly technique where plants can be employed to clean up the environment by using their various mechanisms (Huang et al. 2004). In addition, nanoparticle-assisted phytoremediation is a very promising and interdisciplinary approach to detoxify toxic chemicals (Sozer and Kokini 2009). Moreover, CRISPER/Cas9—a genome editing tool—could revolutionize phytoremediation technology by designing engineered plants with immense phytoremediation potential (Mukherjee et al. 2017). In this chapter, we discuss the conventional and advanced technologies for the phytoremediation of environmental pollutants.

8.2 VARIOUS METHODS EMPLOYED IN SOIL REMEDIATION

8.2.1 Physical Method

Physical methods involve techniques including heat treatment, electroremediation, vitrification, soil flushing and replacement, air sparging and soil vapor extraction. Although these methods are helpful to decontaminate the soil from various pollutants, they have some limitations such as they require further dispensation and are costly techniques; thus, they are not suitable for a commercial scale (Shi et al. 2013).

8.2.2 Chemical Method

This method involves precipitation, ion exchange, chemical extraction and oxidation, soil amendment, chemical leaching and stabilization and nanoremediation. This is an effective method to decrease the hazardous properties of pollutants (Leštan et al. 2008). Most of the techniques in this method require high-cost chemicals, thus limiting their large-scale adaptation.

8.2.3 Physiochemical Method

This method implies both physical and chemical methods, which makes it more effective where soil washing, ultrasonic leaching and chemically activated adsorption techniques are used (Dermont et al. 2008).

8.2.4 Biological Method

The remediation of environmental contaminants by using microorganisms and plants is described as a biological method, and it has several advantages over conventional methods. It is a cost-effective, eco-friendly technique and never creates any resulting pollution. Moreover, it helps to maintain the microflora of the soil that play a pivotal role in the plant-microbe dynamics for enhancing agriculture production (Abbas et al. 2014). It includes biological stabilization, bioleaching, biocomposting, phytoremediation, biofilm-based remediation, microbial remediation, integrated biosystem remediation and genetic engineering. One potential biological method involves phytoremediation that employs the use of plants to remediate soil health. It is a cost-effective and eco-friendly remediation technique where plants metabolize compounds in their tissues and degrade them within their root rhizosphere. By nature's virtue, plants can accumulate, metabolize and degrade pollutants from the soil, water and air. Recent advances in physiological and molecular mechanisms of phytoremediation have improved its process for cleaning contaminants. Plants have various biophysical and biochemical mechanisms by which they can remediate both heavy metals and organic pollutants. Faster-growing plants have proved to be very effective for phytoremediation (Chiapusio et al. 2007; Fang et al. 2001; Wang et al. 2008). Plants with extensive root systems and a high transpiration rate such as poplar (Populus spp.) and willow (Salixspp.) have been shown to be very effective for phytoremediation (Jansson and Douglas 2007). Plants have multifarious actions against contaminants, for example, immobilization, storage, volatilization and the transformation into different levels or in combination, which mainly depend on the type of contamination, plant species and environmental conditions. Plants use various mechanisms to remediate contamination viz., phytoextraction, phytodegradation, phytostabilization, phytovolatilization and rhizodegradation (Subhashini and Swamy 2013; Rafati et al. 2011; Padmavathiamma and Li 2007; Barcelo and Poschenrieder 2003; Mukhopadhyay and Maiti 2010).

8.2.5 Phytoextraction

Phytoextraction is a phytoremediation technique where plants extract contaminants from soil and water into their roots and transfer them to their shoots or leaves. Such plants are tolerant to a higher concentration of contaminants; they have abundant biomass, grow faster and possess a profound root system. These plants are able to accumulate biomass hastily with ample storage capacity (Blaylock and Huang 2000; McGrath 1998; Akanang and Adamu 2017).

8.2.6 Phytodegradation/Phytotransformation

In phytodegradation, plants use their metabolic pathways to degrade or break down organic pollutants through their enzymes or enzyme cofactors into a simpler form (Susarla et al. 2002). Plant enzymes or enzyme cofactors phytodegrade or phytotransform pollutants such as those in the Cannas plant that detoxify different soil xenobiotics. Some enzymes from plants and soil are reported that have a role in the elimination of contaminants viz., nitrilase, laccase, peroxidase, nitroreductase and dehalogenase (Dec and Bollang 1994; Strand et al. 1995; Kvesitadze et al. 2006).

8.2.7 Phytostabilization

Phytostabilization is a phytoremediation approach where plants decrease the migration of various contaminants in soil through, for example, erosion, leaching and run off. In this way, this method restricts not only the bioavailability of contaminants in the environment but also their mixing into food chains. The roots of *Nerium oleander* and *Fetucarubra* L. can phytostabilize heavy metals (Elloumi et al. 2017; Cunningham et al. 1996).

8.2.8 Phytovolatilization

Phytovolatilization is a method where plants take pollutants from contaminated soil and convert them into a volatile form and release them in the environment. The remediation of selenium (Se) and mercury (Hg) has been shown in some studies by using this method (Bañuelos et al. 2000; Henry 2000). Volatile contaminants resolve in plants and disseminate into the environment through a gas exchange from the stomata of leaves. Higher plants may biologically demethylate methylated phenol and tetrabromobisphenol back to their close relative compounds (Fu et al. 2018; Hou et al. 2018; Sun et al. 2016; Zhu et al. 2016b). Hydrophilic compounds such as phenol and acetone may be moderately phytovolatilized compared with hydrophilic compounds viz., lower chlorinated benzene and chlorinated ethenes (Grove and Stein 2005; Polprasert et al. 1996; MacLeod 1999; Bankston et al. 2002; Ma and Burken 2003).

8.2.9 Rhizodegradation/Filtration

In this process, specific vascular plants treat contaminants by increasing a microorganism's activity in its rhizosphere. Plant growth-promoting bacteria show an increased bioavailability of heavy metals to the plant (Chen et al. 2017). Microorganisms and plants exhibit a symbiotic relationship, which makes the rhizosphere a very active microbial zone that has a dense population (Anderson et al. 1993; Anderson et al. 1994; Jordahl et al. 1997; Schwab et al. 1995; Siciliano and Germida 1998). This is because plants have very deep and wide root systems that provide them the potential to degrade pollutants in the rhizosphere, which is triggered by enhancing microbial activity (Gerhardt et al. 2009). Meanwhile, phytoremediation is a very helpful, eco-friendly and cost-effective technique, but it is somewhat slower in action than other methods of bioremediation as it requires a specific plant type, microbes, suitable weather and an appropriate growing season, which take time. Industries are discharging approximately ten million lethal chemical compounds every year (Avio et al. 2017; Alimi et al. 2018; Sousa et al. 2018; Thompson and Darwish 2019). Therefore, to enhance the phytoremediation process, a multi-interdisciplinary approach is needed that involves nanotechnology and other omics techniques to develop engineered plants with high hyperaccumulating and remediation potential for environmental cleanup. The integration of nanotechnology with phytoremediation has proved to be a potential technique for the remediation of contaminated sites.

8.3 ROLE OF NANOTECHNOLOGY IN PHYTOREMEDIATION

Nanotechnology is a very fast-growing and promising field of research that has a broad spectrum and applications. It has been found that nanotechnological approaches are more efficient than other traditional methods of remediation as nanoparticles have a higher surface area-to-volume ratio; hence, a larger quantity of the material can get in touch with nearby materials. A smaller size, superior physiochemical properties and the ability to form different shapes and sizes make them excellent tools for environmental cleanup (Baruah et al. 2019). Because they have a surface plasmon resonance property, nanoparticles can be used to detect toxic compounds. The green syntheses of nanoparticles from plants and microorganisms have shifted the focus towards eco-friendly and cost-effective technology for the remediation of contaminants. Iron nanoparticles synthesized through a green route show their efficiency in remediation due to their redox potential while reacting with water (Bolade et al. 2020). Similarly, zinc oxide (ZnO), gold chloride ($AuCl_3$), silver chloride ($AgCl_2$) and titanium dioxide (TiO_2) nanoparticles synthesized through a green route by using plants, fruit peels and vegetable extracts act as stabilizers and reducing agents to manage the growth of crystals (Elumalai and Velmurugan 2015; Mittal et al. 2013; Davar et al. 2015; Huo et al. 2017; Rosi and Kalyanasundaram 2018). Some of the potential nanoparticles that find great application in the phytoremediation process are depicted in Table 8.1. Figure 8.2 shows the application of nanotechnology in phytoremediation.

Phytoremediation efficiency can be enhanced by nanomaterials, and they can be used to remediate soil and water infected with organic, inorganic and heavy metal contamination. Several studies have found that nanoparticles have decontaminated organic pollutants including PCBs, PAHs, pharmaceuticals and personal care products (PPCPs) and solvents in soil (Crane and Scott 2012). Cd is one of the major pollutants that is widely released by industries, and it has been found that a number of nanoparticles improve its phytoextraction from soil. TiO_2 nanoparticles showed a good response on Cd uptake in soybean plants (Singh and Lee 2016). Nanomaterials generally increase the bioavailability of materials by transporting them from the environment to a cell while they are taken up by plants (Su et al. 2013). Nanomaterials must be phytoavailable and adsorb the contaminants properly for the improved phytoavailability of contaminants to plants. Various researchers have found that a number of nanomaterials probably promote plant growth viz., Zn, Ag, nanoparticles, graphene, carbon nanotubes and quantum dots. Graphene and quantum dots act as nanofertilizers and pesticides (Chakravarty et al. 2015). The plant reproductive system has been activated by carbon nanotubes to thus help in fruit production (Khodakovskaya et al. 2013). Some functional nanoparticles enhance plant metabolism and create superior nitrogen uptake, thus supplementing chlorophyll activity and biomass production (Das et al. 2018). The utilization of Zn and P has been increased by ZnO nanoparticles without toxicity (Venkatachalam et al. 2017). Moreover, these

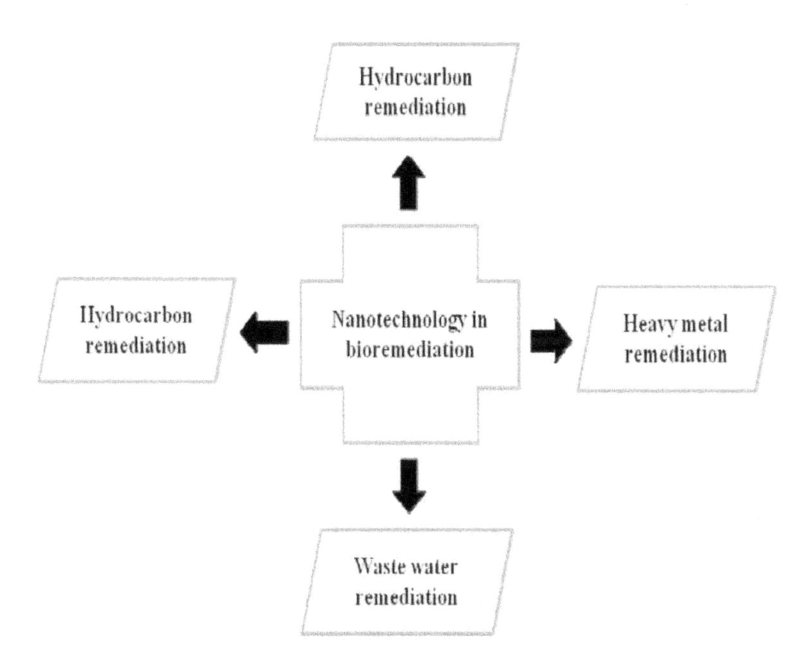

FIGURE 8.2 Application of nanotechnology in bioremediation.

TABLE 8.1
Removal of Heavy Metals by Different Types of Nanomaterials

Nanomaterial used	Target pollutant	Remarks	Reference
nZVI	Cr(VI)	Reduction of chromium from Cr(VI) to Cr(III)	Singh et al. (2012)
nZVI	Pb, Zn	Effective immobilization and reduced concentrations in leachates	Gil-Dıaz et al. (2014)
nZVI	As, Cr, Pb, Cd, Zn	Reduction in metal availability	Gil-Dıaz et al. (2016)
CMC-nZVI	Cr(VI)	Reductive immobilization of Cr(VI)	Madhavi et al. (2014)
CMC-stabilized nZVI	Cr(VI)	Reduced Cr(VI) to Cr(III) in both water and soil	Xu and Zhao (2007)
Biochar-supported nZVI	Cr(VI)	Effective immobilization of Cr and improvement in soil pH, fertility and organic matter content	Su et al. (2016)
Bimetallic nZVI/Cu	Cr(VI)	Increase in the Cr(VI) removal rate with increasing temperature	Zhu et al. (2016a)
CMC-stabilized FeS nanoparticles (NPs)	Cr(VI)	Effective immobilization of Cr(VI) via adsorption, reduction and coprecipitation	Wang et al. (2019)
Fe(II) phosphate NPs	Pb(II)	Reduction in the leachability and bioaccessibility of soil-bound Pb(II) and Cu(II)	Liu and Zhao (2007)
Ca(II) phosphate NPs	Pb(II)	In-situ immobilization of Pb(II) and the possible remediation of other hazardous heavy metals such as Cu, Cd, Zn and U	Liu and Zhao (2013)
CMC-stabilized Fe-Mn binary oxide NPs	As (III)	In-situ immobilization of arsenite	An and Zhao (2012)
Nano-Fe/Ca/CaO	As, Cd, Pb	Total immobilization of soil heavy metals As, Cd and Pb	Mallampati et al. (2013)
Na-zeolitic nanotuff	Cd	Immobilization and increased Cd sorption	Ghrair et al. (2010)
Water treatment residual NPs (nWTR)	Hg, Cr	Increased metal sorption for both Hg and Cr and a significant decrease in their soil release	Moharem et al. (2019)
Starch-stabilized nZVI, FeS, Fe_3O_4	As	Decreased As bioaccessibility and leachability with an increasing Fe/As molar ratio	Zhang et al. (2010)
Al_2O_3, SiO_2, TiO_2 NPs	Zn, Cd, Ni	Reduction in metal mobility	Peikam and Jalali (2018)

nanoparticles also promote the growth and development of plants by improving the photosynthetic rate, osmolyte production and antioxidative defense system, in addition to helping chelate these toxic pollutants within the root rhizosphere (Pirzadah et al. 2020). Nanoparticles were found to be effective in endorsing the Pb phytoextraction competency by ryegrass (Liang et al. 2017). Improved phytoextraction has been shown in *Lsatisca ppadocica* by using salicylic acid nanoparticles (Souri et al. 2017). In another study, it was reported that by providing nanoscale zero-valent iron (nZVI) particles, sunflower rhizosphere stabilized arsenic (Vıtkova et al. 2018). In addition, nanomaterials enhanced the phytoremediation of soil contaminated with trichloroethylene, endosulfan and trinitrotoluene (Jiamjitrpanich et al. 2012; Pillai and Kottekottil 2016; Ma and Wang 2010). However, certain factors affect the nano-phytoremediation process and are depicted in Figure 8.3. In nano-phytoremediation, some points must be considered: a) type of nanoparticle, b) toxicity of the nanoparticle, c) nanosystem should increase plant growth, biomass and plant height, d) nanosystem should increase plant enzymes, e) nanoparticles should enhance plant growth hormones,

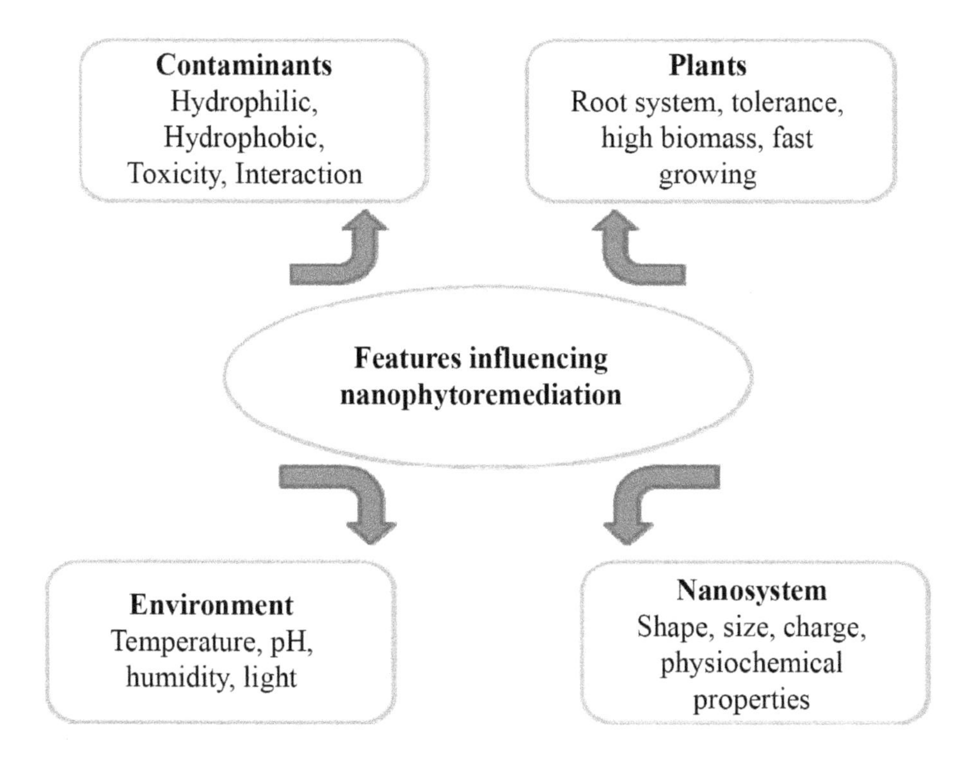

FIGURE 8.3 Features affecting nano-phytoremediation.

f) nanoparticles must be capable of binding contaminants and increase their bioavailability to plants, g) nanoparticles should speed the phytoremediation process, f) plants should have a profound root system, g) plants should be tolerant of a higher concentration of contaminants, h) plants must be able to accumulate contaminants and i) plants should be fast-growing and have a high biomass rate. Many studies are being conducted in this field to improve the process of nano-phytoremediation by improving the features of nanoparticles, plants and microbes. Still, we need to focus on the interactions among nanosystems, plants and the associated microorganisms to understand the mechanism of action, expression patterns and metabolic process for better results.

8.4 OMICS APPROACHES IN PHYTOREMEDIATION

Recent advances in the field of molecular biology have facilitated new approaches to find their applications for a better understanding of the mechanisms involved in phytoremediation processes such as genomics, metagenomics, proteomics and metabolomics. The emergence of next-generation sequencing (NGS) technology and bioinformatics tools have paved the way for environmental scientists to explore more and solve the problems associated with the phytoremediation of environmental pollutants (Maphosa et al. 2010). Sequencing techniques and gene amplification have proved to be very helpful for assessing microbial population dynamics (Gołebiewski and Tretyn 2020). The gene sequencing analysis of 16S rRNA is a proficient and frequently used technology that is easily available through different bioinformatics tools for outlining a complicated microbial population (Han et al. 2020). Handelsman and colleagues first used metagenomics to understand the chemistry of undiscovered soil microorganisms (Handelsman et al. 1998). In this method, independent culture sequencing is performed based on DNA analysis, which is separated from environmental samples (Daniel 2005). Phylogenetic analysis of sequences gives trustworthy information concerning both the diversity and function of the microbes. Several software tools are available to access metagenomics sequence data, which makes it easy to investigate metagenomic libraries (Dai et al. 2018). Transcriptome analysis is conducted by using microarray and sequencing technologies, where microarray helps to evaluate the expression of genes while next generation RNA sequencing helps to set the amount of RNA (Maroli et al. 2018). Omics technologies have revived the study of the catabolism of PAH and offer combined insight regarding the biochemical processes for the dilapidation of PAH (Amrani et al. 2015). Possible transporters for phenols and upregulated degradation pathways have been found by using comparative transcriptomics to obtain a better accumulation and phenol tolerance through lipid accumulating *Rhodococcus opacus* PD630 (Yoneda et al. 2016). *Pseudomonas aeruginosa* strains that have differentially expressed genes have been discovered by using transcriptome analysis for crude oil degradation (Das et al. 2020). Metabolites produced by the cells in reaction to changing environmental circumstances are analyzed by metabolomic studies that provide information regarding regulatory incidents in a cell; by using these metabolites as bio-indicators, biological effects can be screened for pollutants for better insight into the environment (Krumsiek et al. 2015). It is well known that the microbial community is closely associated with the rhizosphere of plants and has a symbiotic association that promotes plant growth and development; thus, these microorganisms may enhance the process of phytoremediation by improving phytoextraction, rhizodegradation and the bioavailability of contaminants.

8.5 CRISPR/CAS9 TECHNOLOGY—A GENOME EDITING TOOL TO DESIGN ENGINEERED PLANTS WITH EFFICIENT PHYTOREMEDIATION POTENTIAL

CRISPR/Cas9 is a multipurpose gene editing tool that is used for amplifying chosen traits in plants through targeting extremely precise sequences of DNA (Wolt et al. 2016). It is a very promising tool of functional genomics (Perez-Pinera et al. 2013). The gene sequencing of plants can be changed by altering DNA using this technology to improve heavy metals detoxification and to increase the bioavailability of metals for plants via plant microbial interactions (Bortesi and Fischer 2015). It is reported that genes MTA1, MT1 and MT2 encoding metallothioneins when overexpressed in tobacco and Arabidopsis showed their increased ability to accumulate and stabilize Zn, Cu and Cd (Abhilash et al. 2012; Xia et al. 2012). MT2b gene expression in *H. incana* enhanced its ability for Pb tolerance and inflation. The genes APS and SMT, which are responsible for Se tolerance, have been transferred to *B. juncea*. Several organic contaminants such as PAHs, PCBs, and explosives including RDX and TNT can be detoxified by using this gene editing tool. Several studies revealed that plants have genes to detoxify organic xenobiotics (Banerjee and Roychoudhury 2019; Jaiswal et al. 2019). In plants, enzyme systems are accountable for the removal of contaminants that may be focused for their enrichment through CRISPR technology (Pandey and Singh 2019). Gene editing has been conducted in rice and Arabidopsis, which expressed tolerance against naphthalene and phenanthrene and were accountable for naphthalene dioxygenase production (Peng et al. 2014). Expression of the bph gene has been changed in alfalfa plants to enhance tolerance against PCBs and 2,4-dichlorophenol (Wang et al. 2015). CRISPR/Cas9 technology is an emerging tool to design engineered plants that could be used to remediate contaminated sites.

8.6 CONCLUSION AND FUTURE RECOMMENDATIONS

Nano-phytoremediation is an emerging field for environmental cleanup, but it is currently in the infancy stage. Most of the current research is restricted to lab-scale and pot culture studies because of a difference in a number of parameters such as pH, temperature, soil texture, nutrients and other edaphic factors that play a vital role in the phytoremediation process. Plant-mediated synthesis of nanoparticles is a promising approach that could assist and enhance the phytoremediation potential of plants growing at contaminated sites. Although nanotechnology has the potential to revolutionize the phytoremediation process, at the same time, the application of nanoparticles could lead to their deposition in the soil, and they could enter the food chain, which may be detrimental to human health and agricultural production. Therefore, using nanoparticles for remediation purposes should be seriously debated by the scientific community. A comprehensive life cycle assessment of these engineered particles needs to be performed to ensure the safety of these materials. Thus, in employing nanotechnology in environmental cleanup on a commercial scale, scientists should focus on the pros and cons of this technology to prevent any potential hazardous environmental effects that they might cause.

REFERENCES

Abbas, S.H., Ismail, I.M., Mostafa, T.M. and Sulaymon, A.H. 2014. Biosorption of heavy metals: A review. *J Chem Sci Tech* 3(4):74–102.

Abhilash, P.C., Powell, J.R., Singh, H.B. and Singh, B.K. 2012. Plant microbe interactions: Novel applications for exploitation in multipurpose remediation technologies. *Trends Biotechnol* 30(8):416–420.

Akanang, H. and Adamu, H. 2017. The potential of cowpea (*Vigna Unguiculata*) as bioremediation tool of heavy metals contaminated soil. *J Chem Soc Niger* 40(1).

Alimi, O.S., Farner Budarz, J., Hernandez, L.M. and Tufenkji, N. 2018. Microplastics and nanoplastics in aquatic environments: Aggregation, deposition, and enhanced contaminant transport. *Environ Sci Technol* 52:1704–1724.

Amrani, E.L., Dumas, A. S., Wick, L.Y., Yergeau, E. and Berthomé, R. 2015. "Omics" insights into PAH degradation toward improved green remediation biotechnologies. *Environ Sci Technol* 49:11281–11291.

An, B. and Zhao, D. 2012. Immobilization of as (III) in soil and groundwater using a new class of polysaccharide stabilized Fe—Mn oxide nanoparticles. *J Hazard Mater* 211:332–341.

Anderson, T.A., Guthrie, E.A. and Walton, B.T. 1993. Bioremediation in the rhizosphere. *Environ Sci Technol* 27:2630–2636.

Anderson, T.A., Kruger, E.L. and Coats, J.R. 1994. Enhanced degradation of a mixture of three herbicides in the rhizosphere of a herbicide-tolerant plant. *Chemosphere* 28:1551–1557.

Avio, C.G., Gorbi, S. and Regoli, F. 2017. Plastics and microplastics in the oceans: From emerging pollutants to emerged threat. *Mar Environ Res* 128:2–11.

Banerjee, A. and Roychoudhury, A. 2019. Genetic engineering in plants for enhancing arsenic tolerance. In *Transgenic Plant Technology for Remediation of Toxic Metals and Metalloids*. Academic Press, 463–475.

Bankston, J.L., Sola, D.L., Komor, A.T. and Dwyer, D.F. 2002. Degradation of trichloroethylene in wetland microcosms containing broad-leaved cattail and eastern cottonwood. *Water Res* 36:1539–1546.

Bañuelos, G., Zambrzuski, S. and Mackey, B. 2000. Phytoextraction of selenium from soils irrigated with selenium-laden effluent. *Plant Soil* 224(2):251–258.

Barcelo, J. and Poschenrieder, C. 2003. Phytoremediation: Principles and perspectives. *Contrib Sci* 2:333–344.

Baruah, A., Chaudhary, V., Malik, R. and Tomer, V.K. 2019. Nanotechnology based solutions for wastewater treatment. *Nanotechnol. Water Wastewater Treat* 2019:337–368.

Blaylock, M.J. and Huang, J.W. 2000. Phytoremediation of toxic metals: Using plants to clean up the environment. In *Phytoextraction of Metals*. I. Raskin and B. D. Ensley (eds.). John Wiley and Sons, Inc., 53–70.

Bolade, O.P., Williams, A.B. and Benson, N.U. 2020. Green synthesis of iron-based nanomaterials for environmental remediation: A review. *Environ Nanotechnol Monit Manag* 13:100279.

Bortesi, L. and Fischer, R. 2015. The CRISPR/Cas9 system for plant genome editing and beyond. *Biotechnol Adv* 33(1):41–52.

Chakravarty, D., Erande, M.B. and Late, D.J. 2015. Graphene quantum dots as enhanced plant growth regulators: Effects on coriander and garlic plants. *J Sci Food Agric* 95:2772–2778.

Chen, X., Liu, X., Zhang, X., Cao, L. and Hu, X. 2017. Phytoremediation effect of Scirpustriqueter inoculated plant-growth-promoting bacteria (PGPB) on different fractions of pyrene and Ni in co-contaminated soils. *J Hazard Mater* 325:319–326.

Chiapusio, G., Pujol, S., Toussaint, M.L., Badot, P.M. and Binet, P. 2007. Phenanthrene toxicity and dissipation in rhizosphere of grassland plants (*Loliumperenne* L. and *Trifolium pretense* L.) in three spiked soils. *Plant Soil* 294:103–112.

Cieślik, B.M., Namieśnik, J. and Konieczka, P. 2015. Review of sewage sludge management: Standards, regulations and analytical methods. *J Clean Prod* 90:1–15.

Crane, R.A. and Scott, T.B. 2012. Nanoscale zero-valent iron: Future prospects for an emerging water treatment technology. *J Hazard Mater* 211:112–125.

Cunningham, S.D., Anderson, T.A., Schwab, P. and Hsu, F.C. 1996. Phytoremediation of soils contaminated with organic pollutants. *Adv Agron* 56:55–114.

Dai, Z., Zhang, S., Yang, Q., Zhang, W., Qian, X., Dong, W., et al. 2018. Genetic tool development and systemic regulation in biosynthetic technology. *Biotechnol Biofuels* 11:152.

Daniel, R. 2005. The metagenomics of soil. *Nat Rev Microbiol* 3:470–478.

Das, D., Mawlong, G.T., Sarki, Y.N., Singh, A.K., Chikkaputtaiah, C. and Boruah, H.P.D. 2020. Transcriptome analysis of crude oil degrading *Pseudomonas aeruginosa* strains for identification of potential genes involved in crude oil degradation. *Gene* 755:144909.

Das, P., Barua, S., Sarkar, S., Karak, N., Bhattacharyya, P., Raza, N., et al. 2018. Plant extract—mediated green silver nanoparticles: Efficacy as soil conditioner and plant growth promoter. *J Hazard Mater* 346:62–72.

Davar, F., Majedi, A. and Mirzaei, A. 2015. Green synthesis of ZnO nanoparticles and its application in the degradation of some dyes. *J Am Ceram Soc* 98(6):1739–1746.

Dec, J. and Bollag, J.M. 1994. Use of plant material for the decontamination of water polluted with phenols. *Biotechnol Bioeng* 44:1132–1139.

Dermont, G., Bergeron, M., Mercier, G. and Richer-Lafèche, M. 2008. Soil washing for metal removal: A review of physical/chemical technologies and field applications. *J Hazard Mater* 152(1):1–31.

Elloumi, N., Belhaj, D., Mseddi, S., Zouari, M., Abdallah, F.B., Woodward, S. and Kallel, M. 2017. Response of Nerium oleander to phosphogypsum amendment and its potential use for phytoremediation. *Ecol Eng* 99:164–171.

Elumalai, K. and Velmurugan, S. 2015. Green synthesis, characterization and antimicrobial activities of zinc oxide nanoparticles from the leaf extract of Azadirachta indica (L.). *Appl Surf Sci* 345:329–336.

Fang, C., Radosevich, M. and Fuhrman, J.J. 2001. Atrazine and phenanthrene degradation in grassrhizosphere soil. *Soil Biol Biochem* 33:671–678.

Fu, Q.G., Liao, C.Y., Du, X.Y. and Schlenk, D.J. 2018. GanBack conversion from product to parent: Methyl triclosan to triclosan in plants. *Environ Sci Technol Lett* 5:181–185.

Gerhardt, K.E., Huang, X.D., Glick, B.R. and Greenberg, B.M. 2009. Phytoremediation and rhizoremediation of organic soil contaminants: Potential and challenges. *Plant Sci* 176:20–30.

Ghrair, A.M., Ingwersen, J. and Streck, T. 2010. Immobilization of heavy metals in soils amended by nanoparticulate zeolitic tuff: Sorption-desorption of cadmium. *J Plant Nutr Soil Sci* 173(6):852–860.

Gil-Dıaz, M., Ortiz, L.T., Costa, G., Alonso, J., Rodríguez-Membibre, M.L., Sanchez-Fortun, S., et al. 2014. Immobilization and leaching of Pb and Zn in an acidic soil treated with zerovalent iron nanoparticles (nZVI): Physicochemical and toxicological analysis of leachates. *Water Air Soil Poll* 225(6):1990.

Gil-Dıaz, M., Pinilla, P., Alonso, J. and Lobo, M.C. 2016. Viability of a nanoremediation process in single or multi-metal (loid) contaminated soils. *J Hazard Mater* 321:812–819.

Gołebiewski, M. and Tretyn, A. 2020. Generating amplicon reads for microbial community assessment with next generation sequencing. *J Appl Microbiol* 128:330–354.

Grove, J.K. and Stein, O.R. 2005. Polar organic solvent removal in microcosm constructed wetlands. *Water Res* 39:4040–4050.

Han, D., Gao, P., Li, R., Tan, P., Xie, J., Zhang, R., et al. 2020. Multicenter assessment of microbial community profiling using 16S rRNA gene sequencing and shotgun metagenomic sequencing. *J Adv Res* 26:111–121.

Handelsman, J., Rondon, M.R., Brady, S.F., Clardy, J. and Goodman, R.M. 1998. Molecular biological access to the chemistry of unknown soil microbes: A new frontier for natural products. *Chem Biol* 5:R245–R249.

Henry, J.R. 2000. Overview of the Phytoremediation of lead and mercury. In *Overview of the Phytoremediation of Lead and Mercury*. EPA.

Hou, X.W., Yu, M., Liu, A.F., Li, Y.L., Ruan, T., Liu, J.Y., Schnoor, J.L. and Jiang, G.B. 2018. Biotransformation of tetrabromobisphenol A dimethyl ether back to tetrabromobisphenol A in whole pumpkin plants. *Environ Pollut* 241:331–338.

Huang, X.D., El-Alawi, Y., Penrose, D.M., Glick, B.R. and Greenberg, B.M. 2004. A multi-process phytoremediation system for removal of polycyclic aromatic hydrocarbons from contaminated soils. *Environ Pollut* 130(3):465–476.

Huo, Y., Singh, P., Kim, Y.J., Veronika, S.J.K. and Josua, M. 2017. Biological synthesis of gold and silver chloride nanoparticles by *Glycyrrhizauralensis* and in vitro applications. *Artif Cells Nanomed Biotechnol* 4:1–13.

Jaiswal, S., Singh, D.K. and Shukla, P. 2019. Gene editing and systems biology tools for pesticide bioremediation: A review. *Front Microbiol* 10:87.

Jansson, S. and Douglas, C.J. 2007. Populus: A model system for plant biology. *Annu Rev Plant Biol* 58:435–458.

Jiamjitrpanich, W., Parkpian, P., Polprasert, C. and Kosanlavit, R. 2012. Enhanced phytoremediation efficiency of TNT-contaminated soil by nanoscale zero valent iron. *2nd International Conference on Environment and Industrial Innovation*. IACSIT Press, Singapore, Vol. 35, 82–86.

Jordahl, J.L., Foster, L., Schnoor, J.L. and Alvarez, P.J.J. 1997. Effect of hybrid poplar trees on microbial populations important to hazardous waste bioremediation. *Environ Toxicol Chem* 16:1318–1321.

Jördening, H.J. and Winter, J. 2006. *Environmental Biotechnology: Concepts and Applications*. John Wiley & Sons, p. 488.

Joshi, M.N., Dhebar, S.V., Bhargava, P., Pandit, A.S., Patel, R.P., Saxena, A.K., et al. 2014. Metagenomic approach for understanding microbial population from petroleum muck. *Genome Announc* 2:e00533–14.

Khodakovskaya, M.V., Kim, B.S., Kim, J.N., Alimohammadi, M., Dervishi, E., Mustafa, T. and Cernigla, C.E. 2013. Carbon nanotubes as plant growth regulators: Effects on tomato growth, reproductive system, and soil microbial community. *Small* 9(1):115–123.

Krumsiek, J., Mittelstrass, K., Do, K.T., Stückler, F., Ried, J., Adamski, J., et al. 2015. Gender-specific pathway differences in the human serum metabolome. *Metabolomics* 11:1815–1833.

Kvesitadze, G., Khatisashvili, G., Sadunishvili, T. and Ramsden, J.J. 2006. *Biochemical Mechanisms of Detoxification in Higher Plants: Basis of Phytoremediation*. Springer Science & Business Media. ISBN 978-3-540-28997-5

Leštan, D., Luo, C. and Li, X. 2008. The use of chelating agents in the remediation of metal-contaminated soils: A review. *Environ Pollut* 153(1):3–13.

Liang, J., Yang, Z., Tang, L., Zeng, G., Yu, M., Li, X., Luo, Y. et al. 2017. Changes in heavy metal mobility and availability from contaminated wetland soil remediated with combined biochar-compost. *Chemosphere* 181:281–288.

Liu, H., Probst, A. and Liao, B. 2005. Metal contamination of soils and crops affected by the Chenzhou lead/zinc mine spill (Hunan, China). *Sci Total Environ* 339(1–3):153–166.

Liu, R. and Zhao, D. 2007. *In situ* immobilization of Cu (II) in soils using a new class of iron phosphate nanoparticles. *Chemosphere* 68(10):1867–1876.

Liu, R. and Zhao, D. 2013. Synthesis and characterization of a new class of stabilized apatite nanoparticles and applying the particles to *in situ* Pb immobilization in a fire-range soil. *Chemosphere* 91(5):594–601.

Ma, X. and Burken, J.G. 2003. TCE diffusion to the atmosphere in phytoremediation applications. *Environ Sci Technol* 37:2534–2539.

Ma, X. and Wang, C. 2010. Fullerene nanoparticles affect the fate and uptake of trichloroethylene in phytoremediation systems. *Environ Eng Sci* 27:989–992.

MacLeod, C.J.A. 1999. The fate of chlorinated organic pollutants in a reed-bed system. In *Phytoremediation and Innovative Strategies for Specialized Remedial Applications: The Fifth International In Situ and On-Site Bioremediation Symposium*. A. Leeson and B.C. Alleman (eds.). Batelle Press, 19–22.

Madhavi, V., Prasad, T.N.V.K.V., Reddy, B.R., Reddy, A.V.B. and Gajulapalle, M. 2014. Conjunctive effect of CMC—zero-valent iron nanoparticles and FYM in the remediation of chromium-contaminated soils. *Appl Nanosci* 4(4):477–484.

Mallampati, S.R., Mitoma, Y., Okuda, T., Sakita, S. and Kakeda, M. 2013. Total immobilization of soil heavy metals with nano-Fe/Ca/CaO dispersion mixtures. *Environ Chem Lett* 11(2):119–125.

Maphosa, F., de Vos, W.M. and Smidt, H. 2010. Exploiting the ecogenomics toolbox for environmental diagnostics of organohalide-respiring bacteria. *Trends Biotechnol* 28:308–316.

Maroli, A.S., Gaines, T.A., Foley, M.E., Duke, S.O., Doğramacı, M., Anderson, J.V., et al. 2018. Omics in weed science: A perspective from genomics, transcriptomics, and metabolomics approaches. *Weed Sci* 66:681–695.

McGrath, S.P. 1998. Phytoextraction for soil remediation. In *Plants that Hyperaccumulate Heavy Metals*. R.R. Brooks (ed.). CAB International, 109–128.

Mittal, A.K., Chisti, Y. and Banerjee, U.C. 2013. Synthesis of metallic nanoparticles using plant extracts. *Biotechnol Adv* 31(2):346–356.

Moharem, M., Elkhatib, E. and Mesalem, M. 2019. Remediation of chromium and mercury polluted calcareous soils using nanoparticles: Sorption—desorption kinetics, speciation and fractionation. *Environ Res* 170:366–373.

Mukherjee, A., Chettri, B., Langpoklakpam, J.S., Basak, P., Prasad, A., Mukherjee, A.K., et al. 2017. Bioinformatic approaches including predictive metagenomic profiling reveal characteristics of bacterial response to petroleum hydrocarbon contamination in diverse environments. *Sci Rep* 7:1–22.

Mukhopadhyay, S. and Maiti, S.K. 2010. Phytoremediation of metal enriched mine waste: A review. *Glo J Environ Res* 4:135–150.

Padmavathiamma, P.K. and Li, L.Y. 2007. Phytoremediation technology: Hyper-accumulation metals in plants. *Water Air Soil Pollut* 184(1–4):105–126.

Pandey, V.C. and Singh, V. 2019. Exploring the potential and opportunities of current tools for removal of hazardous materials from environments. In *Phytomanagement of Polluted Sites*. Elsevier, 501–516.

Peikam, E.N. and Jalali, M. 2018. Application of three nanoparticles (Al_2O_3, SiO_2 and TiO_2) for metal contaminated soil remediation (measuring and modeling). *Int J Environ Sci Technol*:1–14.

Peng, R.H., Fu, X.Y., Zhao, W., Tian, Y.S., Zhu, B., Han, H.J., et al. 2014. Phytoremediation of phenanthrene by transgenic plants transformed with a naphthalene dioxygenase system from Pseudomonas. *Environ Sci Technol* 48(21):12824–12832.

Perez-Pinera, P., Kocak, D.D., Vockley, C.M., Adler, A.F., Kabadi, A.M., Polstein, L.R., et al. 2013. RNA-guided gene activation by CRISPR-Cas9 based transcription factors. *Nat Methods* 10(10):973.

Pillai, H.P. and Kottekottil, J. 2016. Nano-phytotechnological remediation of endosulfan using zero valent iron nanoparticles. *J Environ Prot* 7:734–744.

Pirzadah, T.B., Pirzadah, B., Jan, A., Dar, F.A., Hakeem, K.R., Rashid, S., Salam, S.T., Dar, P.A. and Fazili, M.A. 2020. Development of nano-formulations via green synthesis approach. In *Nanobiotechnology in Agriculture: An Approach Towards Sustainability*. T.B. Pirzadah and K.R. Hakeem (eds.). Springer Nature Switzerland, 171–183. https://doi.org/10.1007/978-3-030-39978-8

Polprasert, C., Dan, N.P. and Thayalakumaran, N. 1996. Application of constructed wetlands to treat some toxic wastewaters under tropical conditions. *Water Sci Technol* 34(11):165–171.

Rafati, M., Khorasani, N., Moattar, F., Shirvany, A., Moraghebi, F. and Hosseinzadeh, S. 2011. Phytoremediation potential of *Populusalba* and *Morus alba* for cadmium, chromium and nickel absorption from polluted soil. *Int J Environ Res* 5:961–970.

Reddy, R. A. 2017. Environmental pollution causes and consequences: A study. *NAIRJSSH* 3(8): ISSN: 2454–9827.

Roig, N., Sierra, J., Martí, E., Nadal, M., Schuhmacher, M., et al. 2012. Longterm amendment of Spanish soils with sewage sludge: Effects on soil functioning. *Agric Ecosyst Environ* 158:41–48.

Rosi, H. and Kalyanasundaram, S. 2018. Synthesis, characterization, structural and optical properties of titanium-dioxide nanoparticles using Glycosmisco chinchinensis leaf extract and its photocatalytic evaluation and antimicrobial properties. *WNOFNS* 17:1–15.

Schwab, A.P., Banks, M.K. and Arunachalam, M. 1995. Biodegradation of polycyclic aromatic hydrocarbons in rhizosphere soil. In *Bioremediation of Recalcitrant Organics*. R.E. Hinchee, D.B. Anderson and R.E. Hoeppel (eds.). Battelle Press.

Shi, W., Liu, C., Ding, D., Lei, Z., Yang, Y., Feng, C. and Zhang, Z. 2013. Immobilization of heavy metals in sewage sludge by using subcritical water technology. *Bioresour Technol* 137:18–24.

Siciliano, S.D. and Germida, J.J. 1998. Bacterial inoculants of forage grasses enhance degradation of 2-chlorobenzoic acid in soil. *Environ Toxicol Chem* 16:1098–1104.

Singh, J. and Lee, B.K. 2016. Influence of nano-TiO2 particles on the bioaccumulation of Cd in soybean plants (Glycine max): A possible mechanism for the removal of Cd from the contaminated soil. *J Environ Manage* 170:88–96.

Singh, R., Misra, V. and Singh, R.P. 2012. Removal of Cr (VI) by nanoscale zero-valent iron (nZVI) from soil contaminated with tannery wastes. *Bull Environ Contam Toxicol* 88(2):210–214.

Souri, Z., Karimi, N., Sarmadi, M. and Rostami, E. 2017. Salicylic acid nanoparticles (SANPs) improve growth and phytoremediation efficiency of Isatiscappadocica Desv., under as stress. *IET Nanobiotechnol* 11:650–655.

Sousa, J.C.G., Ribeiro, A.R., Barbosa, M.O., Pereira, M.F.R. and Silva, A.M.T. 2018. A review on environmental monitoring of water organic pollutants identified by EU guidelines. *J Hazard Mater* 344:146–162.

Sozer, N. and Kokini, J.L. 2009. Nanotechnology and its applications in the food sector. *Trends Biotechnol* 27(2):82–89.

Strand, S.E., Newman, L., Ruszaj, M., Wilmoth, J., Shurtleff, B., Brandt, M., Choe, N., Ekuan, G., Duffy, J., Massman, J.W., Heilman, P.E. and Gordon, M.P. 1995. Removal of trichloroethylene from aquifers using trees. In *Innovative Technologies for Site Remediation and Hazardous Waste Management, Proceedings of the National Conference of the Environmental Engineering*. R.D. Vidic and F.G. Pohland (eds.). Division of the American Society of Civil Engineers.

Su, H., Fang, Z., Tsang, P.E., Zheng, L., Cheng, W., Fang, J., et al. 2016. Remediation of hexavalent chromium contaminated soil by biochar-supported zero-valent iron nanoparticles. *J Hazard Mater* 318:533–540.

Su, Y., Yan, X., Pu, Y., Xiao, F., Wang, D. and Yang, M. 2013. Risks of single-walled carbon nanotubes acting as contaminants-carriers: Potential release of phenanthrene in Japanese medaka (Oryziaslatipes). *Environ Sci Technol* 47:4704–4710.

Subhashini, V. and Swamy, A.V.V.S. 2013. Phytoremediation of Pb and Ni contaminated soils using *Catharanthusroseus* (L.). *Univer J Environ Res Technol* 3:465–472.

Sun, J.T., Pan, L.L., Su, Z.Z., Zhan, Y. and Zhu, L.Z. 2016. Interconversion between methoxylated and hydroxylated polychlorinated biphenyls in rice plants: An important but overlooked metabolic pathway. *Environ Sci Technol* 50:3668–3675.

Susarla, S., Medina, V.F. and McCutcheon, S.C. 2002. Phytoremediation: An ecological solution to organic chemical contamination. *Ecol Eng* 18:647–658.

Thompson, L.A. and Darwish, W.S. 2019. Environmental chemical contaminants in food: Review of a global problem. *J Toxicol* 234:5283.

Venkatachalam, P., Priyanka, N., Manikandan, K., Ganeshbabu, I., Indiraarulselvi, P., Geetha, N., et al. 2017. Enhanced plant growth promoting role of phycomolecules coated zinc oxide nanoparticles with P supplementation in cotton (Gossypiumhirsutum L.). *Plant Physiol Biochem* 110:118–127.

Vıtkova, M., Puschenreiter, M. and Komarek, M. 2018. Effect of nano zero-valent iron application on As, Cd, Pb, and Zn availability in the rhizosphere of metal (loid) contaminated soils. *Chemosphere* 200:217–226.

Wang, J., Zhang, Z., Su, Y., He, W., He, F. and Song, H. 2008. Phytoremediation of petroleum polluted soil. *Pet Sci* 5:167–171.

Wang, T., Liu, Y., Wang, J., Wang, X., Liu, B. and Wang, Y. 2019. *In-situ* remediation of hexavalent chromium contaminated groundwater and saturated soil using stabilized iron sulfide nanoparticles. *J Environ Manage* 231:679–686.

Wang, Y., Ren, H., Pan, H., Liu, J. and Zhang, L. 2015. Enhanced tolerance and remediation to mixed contaminates of PCBs and 2,4-DCP by transgenic alfalfa plants expressing the 2,3-dihydroxybiphenyl-1,2-dioxygenase. *J Hazard Mater* 286:269–275.

Wolt, J.D., Wang, K. and Yang, B. 2016. The regulatory status of genome-edited crops. *Plant Biotechnol J* 14(2):510–518.

Xia, Y., Qi, Y., Yuan, Y., Wang, G., Cui, J., Chen, Y., et al. 2012. Over expression of Elsholtzia Chinensis metallothionein 1 (EhMT1) in tobacco plants enhances copper tolerance and accumulation in root cytoplasm and decreases hydrogen peroxide production. *J Hazard Mater* 233:65–71.

Xu, Y. and Zhao, D. 2007. Reductive immobilization of chromate in water and soil using stabilized iron nanoparticles. *Water Res* 41(10):2101–2108.

Yoneda, A., Henson, W.R., Goldner, N.K., Park, K.J., Forsberg, K.J., Kim, S.J., et al. 2016. Comparative transcriptomics elucidates adaptive phenol tolerance and utilization in lipid-accumulating *Rhodococcusopacus* PD630. *Nucleic Acids Res* 44:2240–2254.

Zhang, M., Wang, Y., Zhao, D. and Pan, G. 2010. Immobilization of arsenic in soils by stabilized nanoscale zerovalent iron, iron sulfide (FeS), and magnetite (Fe_3O_4) particles. *Chin Sci Bull* 55(4–5):365–372.

Zhou, Q., Zhang, J., Fu, J., Shi, J. and Jiang, G. 2008. Biomonitoring: An appealing tool for assessment of metal pollution in the aquatic ecosystem. *Anal ChimActa* 606(2):135–150.

Zhu, F., Li, L., Ma, S. and Shang, Z. 2016a. Effect factors, kinetics and thermodynamics of remediation in the chromium contaminated soils by nanoscale zero valent Fe/Cu bimetallic particles. *J Chem Eng* 302:663–669.

Zhu, H.K., Sun, H.W., Zhang, Y.W., Xu, J.X. and Zhou, Q.X. 2016b. Uptake pathway, translocation, and isomerization of hexabromocyclododecanediastereoisomers by wheat in closed chambers. *Environ Sci Technol* 50:2652–2659.

9 Water Nano-Phytoremediation

Mahipal Singh Sankhla, Ekta B Jadhav, Kapil Parihar, Gaurav Kumar Singh, Rohit Kumar Verma, Swaroop S Sonone, Rajeev Kumar, Ankita, and Ashutosh Tripathi

9.1 INTRODUCTION

Water is the most plentiful and essential compound for all known forms of life, including humankind. Thus, it is vital to supply water to small and large cities and towns. Since water is directly connected with human welfare and its extent, water quality is the essential concern (Mishra et al., 2006). Globally increasing industrialization and urbanization causes damage to and contaminates natural water ecosystems with their wastes. This may include different kinds of toxic pollutants such as domestic waste, industrial chemical waste, heavy metals, pesticides, fertilizers, and many more, which create potentially serious implications on human and animal health (Sankhla et al., 2016, 2018a, 2018b, 2019; Gulia et al., 2020). Waste contaminates water bodies and water channels or directly seeps into the soil itself, and from there, it enters the food chain and adversely affects human life directly or indirectly (Gupta et al., 2015). Nanotechnology is termed as the study, design, fabrication, and manipulation of resources at the nanometric scale, the transition scale among atoms and molecules to micro and bulk resources. At the nanoscale, materials exhibit different properties than the bulk state (Nanjwade et al., 2019; Devasahayam, 2019). Nanomaterial applications are not incomplete to cleansing; sensing rudiments such as probes have increased sensitivity and selectivity by applying nanomaterial probes in water monitoring devices. Although these applications are a part of nanotechnology, the green technology approach for sustainable development is also important. Nanotechnologies create immense environmental benefits in terms of water management and treatment through convalescing, filtering, decontamination, desalination, conservation, recycling, sewerage systems, and developing sensitive analytics or monitoring systems (Bhati and Rai, 2018). Phytoremediation is an effective technique to eliminate dangerous heavy metals from the polluted atmosphere. The term phytoremediation consists of the Greek prefix "phyto" which means plants that are attached to, and the Latin root medium means to correct or remove an evil (Tangahu et al., 2011). Because of the presence of many components in polluted water, toxic metals are sources of many diseases and effects in humans and plants (Rasheed et al., 2018). Phytoremediation comprises the approaches accepted by plants to neutralize soil, sludge, residue, and wastewater. Some of the recognized approaches of phytoremediation are phytoextraction; the elimination of heavy metal/toxic substances from soil/water; the phytotransformation/breakdown of organic contaminants; the phytovolatilization of contaminants taken up by tissue and then volatilized into the environment; rhizofiltration, a technique of clarifying water concluded in a mass of roots; phytostimulation, which inspires microbial degradation in roots or plant parts; phytostabilization in which released composites upsurge microbial action in the rhizosphere; and phytoscreening where plants act like biosensors. Phytoremediation approaches are economical, can be functional in situ, and are solar-driven (Bhati and Rai, 2018).

9.2 WATER CONTAMINATION AND HEALTH IMPACTS

The occurrence of unwanted materials in water produces water contamination and water quality changes and creates an atmosphere very harmful to human health (Alrumman et al., 2016; Briggs,

 DOI: 10.1201/9781003186298-9

2003). Water is a vital natural resource for intake and other developmental purposes in life (Bibi et al., 2016). All over the world, safe drinking water is essential for human health. As a universal solvent, water is a major source of infection. According to the World Health Organization (WHO), 80% of syndromes are waterborne. The drinking water in various countries does not meet WHO standards (Khan et al., 2013; Sankhla et al., 2018b). Phytoremediation has used green florae and their related microorganisms, soil amendments, and agronomic methods to eliminate, absorb, or reduce offensive ecological contaminants. This occurs through the usage of living green florae and waste of the fruit/vegetal for the in-situ elimination or degradation of pollutants in soils, slurry residues, surface water, and groundwater. Phytoremediation allows cost-saving and solar energy-driven attack methods. It is a lucrative method for remediation that involves plants metabolizing particles and reducing the damage to the environment. It is the usual aptitude of plants to accrue, degrade, or distillate pollutants from the soil, water, and air. Contaminants and poisonous metals are the main aims of phytoremediation. In recent studies, physiological and molecular mechanisms of phytoremediation have been designed to improve the phytoremediation process for environmental cleanup. It is also supportive of managing the utilization of food/vegetable/fruit waste (Srivastav et al., 2019). At the same time, an increase in urbanization has been observed, which affects the quality and availability of freshwater; meanwhile, the need for water for agriculture purposes, household consumption, and industrial use is also increasing. The result of this overuse is the depletion and pollution of surface water and groundwater. In particular, waste is dumped into lakes and rivers, including untreated or partially treated municipal sewage, industrial poisons, and harmful chemicals that leach into surface and groundwater during these anthropological activities. Polluted water, water shortages, and unsanitary living might cause illnesses such as cholera, hepatitis A, dysentery, dengue, and malaria. These contaminants worsen the value of water even with remarkably low absorption and may instigate dangerous effects on human health, animals, plants, and aquatic organisms. Moreover, a significant amount of water is wasted because of inefficient irrigation systems, poor watershed management, and inappropriate agricultural subsidies (Hinrichsen and Tacio, 2020). Meanwhile, an important food for many inhabitants of the globe is fish. Global per capita fish intake has increased to above 20 kg/year−1 (FAO, 2016). Most people who live near river banks are dependent on fish as a source of protein. In India, the annual per capita fish consumption is 5–6 kg for the general population and 8–9 kg for the fish-eating population, which is about 50% of global consumption (Salim, 2016). The current inland fish manufacture donates 6.57 million tonnes. The Ganga River provides extensive fish manufacture to its populations. The L. rohita, C. catla, and C. mrigala fish species originate in this watercourse. All of these fish are the main sources of protein for the people who live there. The health risks arising from the toxicity of metals mostly comprise kidney and emaciated injuries, neurological syndromes, endocrine distraction, cardiovascular dysfunction, and special carcinogenic effects (Renieri et al., 2019). Dietetic contact with numerous heavy metals has been recognized as a risk to human health through the intake of polluted food. Many heavy metals create an impasse with the sulfur existing in enzymes, thus disrupting their purpose (Renieri et al., 2019; Ali and Khan, 2018). Currently, the major and complex difficulty that directly adversely affects human health is air pollution. The growth of phytotechnologies for removing contaminants from the atmosphere to advance air excellence is obligatory for a quality life. Current studies are focused on mechanisms by which florae may constrain the air-polluted contaminants from the atmosphere. Because of anthropogenic actions, the attentiveness of rudiments and their composites is endlessly cumulative in soils. The physical examination and annihilation of all soil pollutants are expensive and time-consuming. Consequently, it is significant to distinguish the soil-polluted contact to avert the endorsement of pollutants from the soil to a plant (Henry et al., 2013). Previous examinations examined the metal accretion in florae in the examination of hyperaccumulators; after that, the notion of consuming plants for the remediation method was advanced. Many researchers investigated other plant contaminant interactions, rhizosphere degradation, and the uptake of metals. The progression of research led to a full-scale phytoremediation process at various contaminated sites (Burken, 2001). Although plants may expel poisonous waste, there is a chance of nanomaterials entering the food chain through these plants. The main challenge facing the nano-phytoremediation process is how to dispose of the accumulated heavy metals in plants, as they are considered toxic waste (Verma et al., 2021). Currently, most of the attention has been given to ecological issues, and a specific anxiety about safe food has developed that concludes with the appearance of various dangerous disorders. The maltreatment caused by heavy metal contamination to agricultural activities is very illustrative. This concludes the food chain after which the contamination accumulates in the body, generating a dangerous risk to human vigor. The soil area contaminated by heavy metals in China is about 20 million hm^2, accounting for 1/5 of the total refined terrestrial parts. Heavy metal soil contamination causes crop loss of over 10 million tons annually, and the loss of heavy metal-contaminated grains is about 12 million tons annually, with a yearly economic loss of over 20 billion (Wei and Chen, 2001).

9.3 REMEDIATION OF CONTAMINATED WATER

About 71% of the Earth's surface is covered by water, and about 95.6% of the Earth's water is held by the oceans only. Water is a vital resource for human life. There are two kinds of water, namely, groundwater and natural water. Water is polluted due to human activities and industrialization, which further leads to causing disturbances in the ecosystem. The waste discharged from various industries contains both toxic and non-toxic metallic ions and is then dumped into water bodies leading to water pollution and many health effects for both humans and plants. Many harmful components present in polluted water are the source of many diseases in both plants and humans. The concentrations of toxic metals can be determined by using techniques such as atomic absorption spectroscopy (AAS) and some other chromatographic techniques. Among all methods, only AAS and inductively coupled plasma mass spectrometry (ICP-MS) provide effective outcomes. There are also many other biological, chemical, and physical methods for cleaning polluted water. After comparing all methods, phytoremediation is found to be more effective. This method is time-consuming, but it is the only permanent solution for the removal of contaminants compared to other methods (Jeevanantham et al., 2019). Excess use of pesticides is one of the greatest causes of water pollution, and its remains are a threat to the water supply and human health. Therefore, there is a need for a suitable approach for the remediation of such pesticides from drinking water (Romeh, 2020).

9.3.1 Nanoremediation

Materials having a size of 100 nm or less than 100 nm are known as nanomaterials. These materials can be categorized into organic and inorganic nanoparticles. Organic nanoparticles are derived from dead organic matter such as potato and orange peelings, whereas inorganic nanoparticles comprise magnetic nanoparticles, noble metal nanoparticles, and semiconductor nanoparticles. Due to their specific size and brilliant properties, these nanomaterials have applications in manufacturing, medicine, energy, and electronics. Nanomaterials having magnetic properties are called magnetic nanomaterials and are widely used in remediation methods especially for the removing the contaminants from aqueous solutions due to their magnetic properties, less toxicity, high chemical stability, easy synthesis, and excellent recycling capability. Due to their unique structural and morphological features, carbon nanomaterials are considered excellent adsorbents for the removal of both organic and inorganic pollutants from wastewater. These widely used nanomaterials are discharged into water resources and precipitate in river sediments. Nanomaterials' main role is in shaping agriculture with the controlled release of nutrients, fertilizers, and pesticides to promote the growth of plants and prevent diseases.

The use of nanomaterials as alterations to fortify the method of phytoremediation is a new finding that is holding environmental significance. The following characteristics should be present for the application of nanoparticles:

i. The nanoparticles should not be toxic.
ii. Sorbents should have the capability of getting recycled infinitely.
iii. A low interaction between sorbents and contaminants should occur for the easy removal of contaminants later.
iv. The nanoparticles should have high sorption capacity and selectivity (Jeevanantham et al., 2019; Gong et al., 2019; Srivastav et al., 2019).

Nanomaterial-assisted phytoremediation comprises plants, pollutants, and nanomaterials. Nanomaterials can make phytoremediation better by directly acting through the redox and reactions or adsorption of both plants and pollutants. A number of studies have found the application of nanomaterials to be very beneficial in plant systems for plant protection, nutrition, and the recognition of plant diseases and contaminants. It has been found that approximately 40% of the contributors prefer carbon-based nanomaterials such as carbon nanotubes, liposomes, and organic polymers, as well as titanium dioxide, silica, and alumina nanomaterials (Srivastav et al., 2019). The application of nanoparticles is clean, non-toxic, eco-friendly, quick, and cost-effective and has the features of dynamic morphology, the required dimensions, and the desired nature for the treatment of wastewater. This green technology is a new approach, and from a technical point of view, it is economically sustainable and reliable, but the techniques are still developing in the fields of fabrication mechanism, regeneration, and reusability (Bhati and Rai, 2018). Chlorfenapyr poses a considerable risk to the ecosystem and to the reproductive ability of birds. The implementations of green nanotechnology and phytoremediation by Plantago major have a significant role in soil remediation and water polluted with chlorfenapyr. By using Plantago major and nanoparticles supported by charcoal, chlorfenapyr was removed rapidly. This approach is environment friendly, time-efficient, and less expensive and comprises bio-organic compound coatings that can be used for multiple purposes. Studies have found that nanoparticles such as gold (Ag) and iron (Fe) can remediate contaminants such as heavy metals, pesticides, nitro-aromatic compounds, and nitrates (Ali et al., 2020).

9.3.2 Phytoremediation

There are many physical, biological, and chemical approaches for the removal of pollutants in polluted water. Comparing all of these methods, the phytoremediation approach has been found to be very significant and appropriate for pollutant removal by using plants. In this remediation method, plants are allowed to grow in the contaminated area, and during their growth, the plants will absorb the heavy metals from the soil and water. In this remediation approach, the mechanism that takes place changes the contaminated metals into a non-hazardous form, ultimately making the environment toxic-free. Two types of contaminants are present in the environment, specifically, organic and inorganic pollutants, and they are detoxified in several ways. Microbes are added to promote the growth of plants to let them survive for a longer time and to increase the tolerance of plants in the contaminated area; this is even possible in saltwater (Jeevanantham et al., 2019). Phytoremediation is an eco-friendly and low-cost in-situ remediation method for the removal of heavy metals from contaminated areas to treat contaminated sediments such as cadmium-polluted sediments (Gong et al., 2019). Over recent years, the increased deposition of hazardous metals and pollutants in soil has been found due to increased industrialization and anthropogenic activities. Many approaches and techniques have been established to decrease soil pollution. Among these, phytoremediation is a popular bioremediation technique that is inexpensive and simple, with an esthetic appearance, broad compliance, less damage to the soil configuration, and a high acceptance. This technique has been successfully implemented worldwide and has a broad global market. Many other techniques including agronomic management, treatment with chemical additives, the addition of rhizospheric microorganisms, and genetic engineering are also introduced to improve phytoremediation effectiveness. Currently, nanotechnology is bringing more novel ideas to the phytoremediation of polluted soil. This approach uses plants to remove the contaminants present in the environment such as air, water, soil, and sediments. Heavy metal pollution is a serious issue and a significant challenge to human health and food protection across the world. Studies have found that this approach of phytoremediation could be further improved for the remediation of soil polluted with heavy metals such as cadmium, lead, nickel, and zinc by applying nanomaterials (Song et al., 2019).

9.3.2.1 Mechanism of Phytoremediation

A number of mechanisms are involved in tropical plants for their metal or mineral uptake in the remediation of pollutants to convert them into non-hazardous substances, and they vary accordingly, for example, organic and inorganic, as shown in Figure 9.1. The mechanism followed by plants for remediation comprises phytoextraction, phytotransformation, phytodegradation, phytostabilization, phytostimulation, phytovolatilization, and rhizodegradation, as described in Figures 9.2 and 9.3.

9.3.2.1.1 Phytoextraction

The capability of plants to absorb contaminants into their roots and translocate the contaminants above to the shoot system is known as phytoextraction. Briefly, this process is the removal of contaminants from either groundwater or surface water and soil by using live plants. The plants selected for this process should have the following characteristics for the phytoextraction mechanism:

- Tolerance to high concentration metals,
- High metal holding potential in the shoot tissues,
- Great biomass,
- Rapid growth capability, and
- Abundant root system.

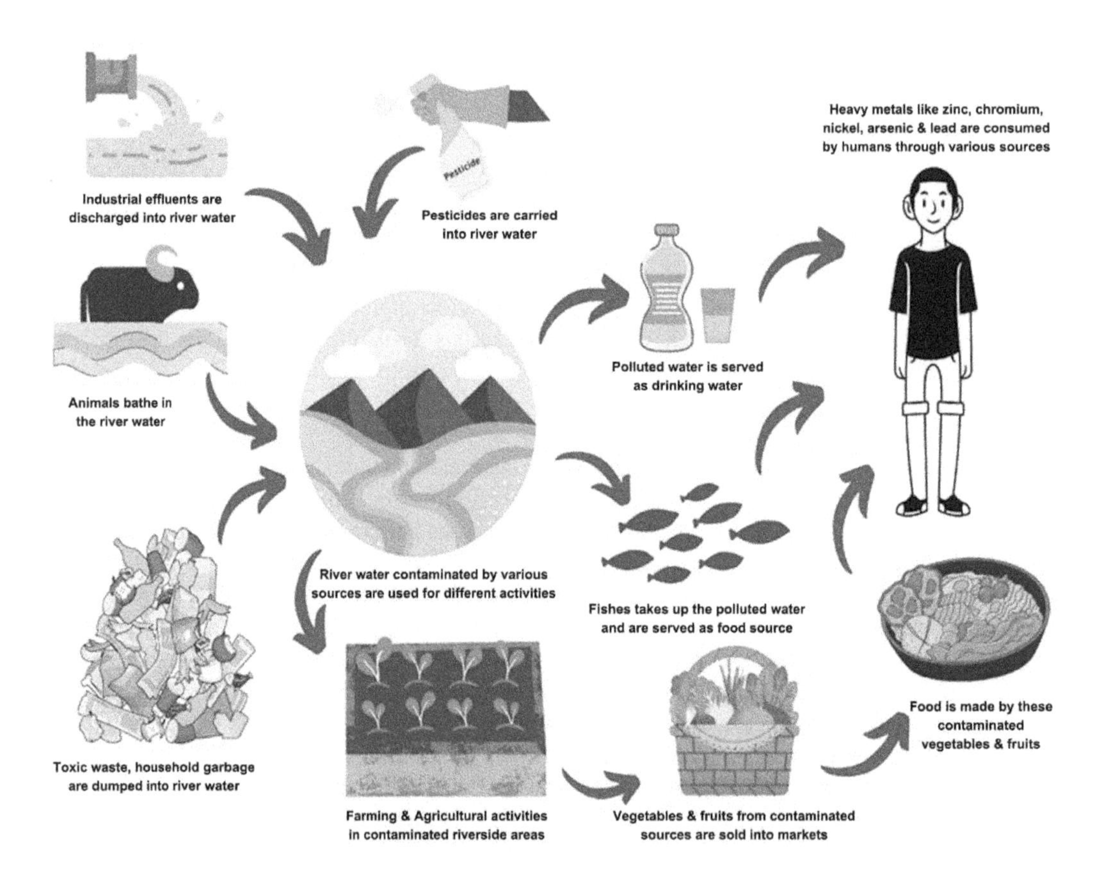

FIGURE 9.1 Sources and transport of pollutants into aquatic systems.

Phytoremediation mechanism
Upgraded take-up by transporters
Bio activation of metals
Detoxification by dispersion
Vacuole sequestration

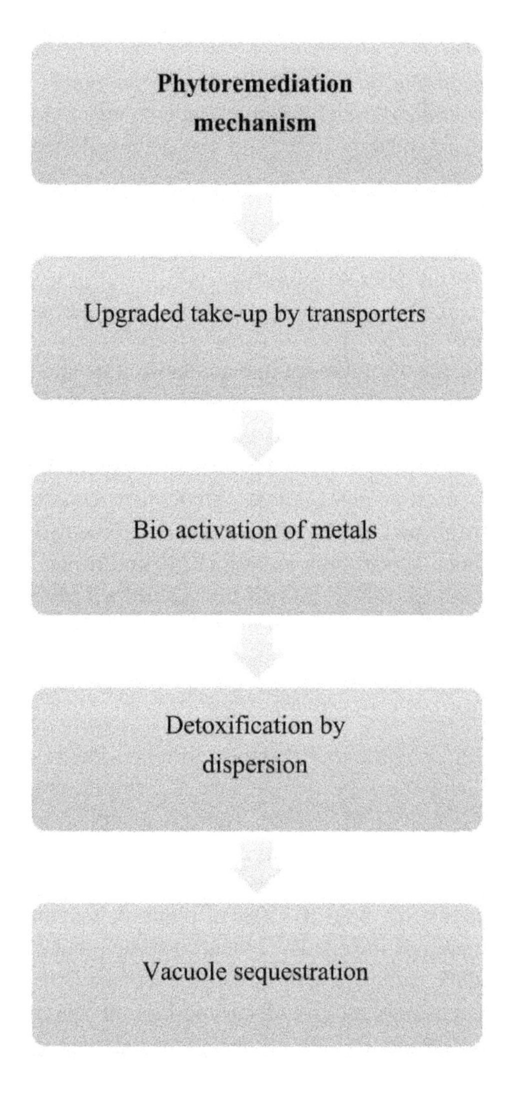

FIGURE 9.2 The mechanism of phytoremediation.

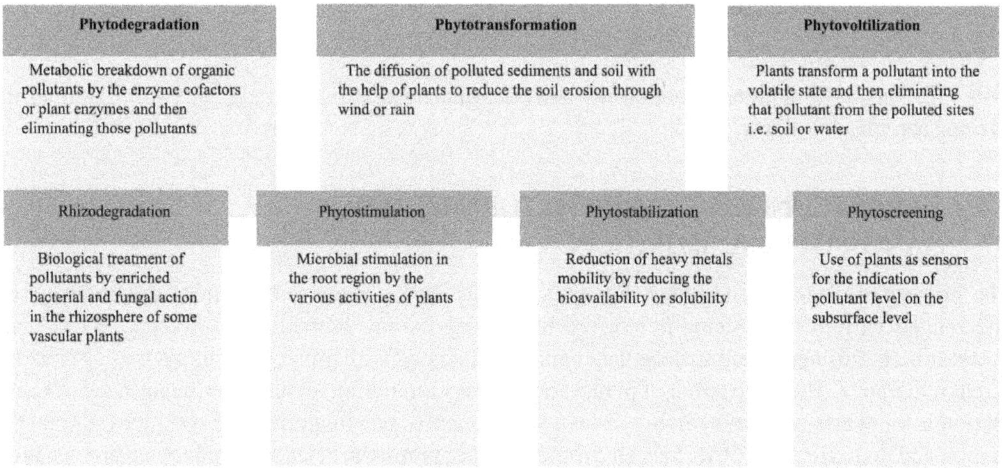

Phytodegradation
Metabolic breakdown of organic pollutants by the enzyme cofactors or plant enzymes and then eliminating those pollutants
Phytotransformation
The diffusion of polluted sediments and soil with the help of plants to reduce the soil erosion through wind or rain
Phytovoltilization
Plants transform a pollutant into the volatile state and then eliminating that pollutant from the polluted sites i.e. soil or water
Rhizodegradation
Biological treatment of pollutants by enriched bacterial and fungal action in the rhizosphere of some vascular plants
Phytostimulation
Microbial stimulation in the root region by the various activities of plants
Phytostabilization
Reduction of heavy metals mobility by reducing the bioavailability or solubility
Phytoscreening
Use of plants as sensors for the indication of pollutant level on the subsurface level

FIGURE 9.3 List of the various mechanisms in the phytoremediation method.

9.3.2.1.2 Phytodegradation

This method includes the metabolic breakdown of organic pollutants by the enzyme cofactors or plant enzymes and then eliminating these pollutants. The enzymes found are dehalogenase, nitroreductase, peroxidase, laccase, and nitrilase in the soil and plant sediments.

9.3.2.1.3 Phytostabilization

Phytostabilization is a mechanism of phytoremediation comprising the decline of heavy metals' mobility by reducing the bioavailability or solubility.

9.3.2.1.4 Phytovolatilization

In this mechanism, plants transform a pollutant into a volatile state and then eliminate this pollutant from the polluted sites of soil or water. It is a natural phenomenon in plants to evaporate any pollutant or contaminant uptake through their root system, which is developed as a natural air stripping pump system. Then, these volatile pollutants diffuse from the plant into the atmosphere through the open stomata. The processes such as plant uptake, phytovolatilization, and metabolic conversion may be applicable to a few plants and organic chemicals. For some hydrophilic compounds such as acetone and phenol, the mechanism of phytovolatilization and direct volatilization are moderate. To the contrary, this mechanism of volatilization is considered a significant elimination method for volatile hydrophobic compounds, for example, benzene, toluene, ethylbenzene, and xylene (BTEX) compounds, lower chlorinated benzene, and ethenes.

9.3.2.1.5 Rhizodegradation

This is a biological treatment of pollutants through enriched bacterial and fungal action in the rhizosphere of some vascular plants. The rhizosphere is a region present at the root of legumes that has increased microbial activity and density. There is a symbiotic association between plants and microorganisms that increases the microbial activity in the root zone. Plant roots are usually dispersed miles away per acre, and this signifies the capability of contaminant reduction in the rhizosphere. There is a deposition of about 10–50% of plant photosynthate, that is, sugars, organic acids, and larger organic compounds in soil. The essential component of phytoremediation, namely, rhizoremediation, which occurs naturally, can be further enhanced by the addition of particular contaminant-degrading microbes or plant growth-promoting microorganisms.

9.3.2.1.6 Phytoscreening

This mechanism of phytoremediation involves the use of plants as sensors to indicate pollutant levels on the subsurface level. This mechanism is simple, non-invasive, and less expensive. (Srivastav et al., 2019; Bhati and Rai, 2018).

9.3.2.1.7 Phytostimulation

This mechanism of phytoremediation involves the microbial stimulation in the root region by the various activities of plants.

9.4 NANOTECHNOLOGY AND PHYTOREMEDIATION: NANO-PHYTOREMEDIATION

The process of cleaning and removing contaminants, or decontamination, from the environment and related various ecosystems is referred to as remediation. Remediation protects the soil, air, groundwater, sediments, and surface water and is associated with protecting human beings from the threat of damage. The application of plants and their accumulating features for the purpose of remediation is known as phytoremediation. Meanwhile, reactive nanomaterials are used for the transformation and detoxification of various environmental contaminants. Nanotechnology allows for more selective, cost-effective, and sensitive remediation tools (Karn et al., 2009). Nano-phytoremediation

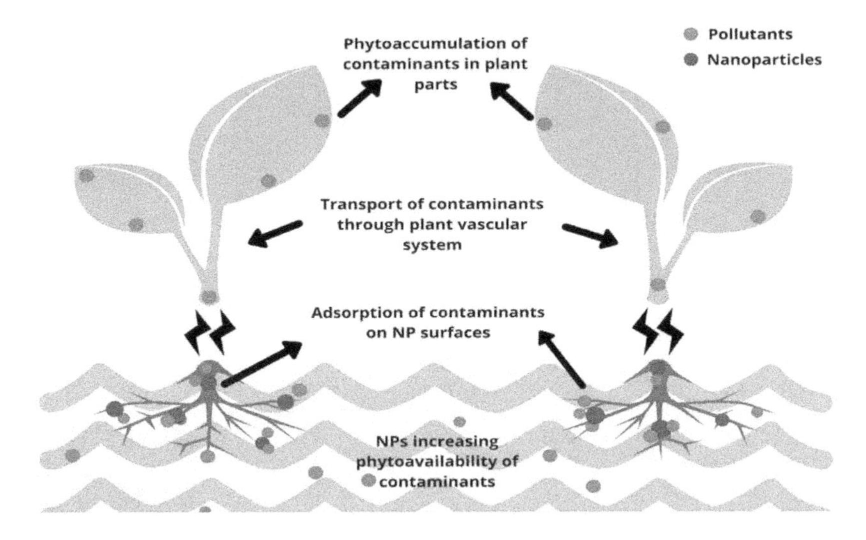

FIGURE 9.4 Nano-phytoremediation: The phytoremediation of contaminants using nanomaterials.

is employed to remediate environmental contaminants by applying both nanotechnology and phytotechnology together. That is, nano-phytoremediation is a technique of detoxifying and decontaminating environmental contaminants by using both nanotechnology and phytotechnology. Sometimes, the application of nano-phytoremediation is more effective than applying the individual techniques of nano-remediation or phytoremediation (Ingle et al., 2014; Srivastav et al., 2019).

9.4.1 Selection of Plant Species for Nano-Phytoremediation

The feasible plant species for the nano-phytoremediation of contaminated water depends on several external factors such as the type of water and the quality of contaminants present in it. The plants utilized for nano-phytoremediation of contaminated water should have the following features (Ahmadpour, 2012; Jasrotia et al., 2017; Srivastav et al., 2019):

- A fast and specific growth rate in contaminated water,
- High root-shoot elongation, plant height, and productivity of biomass;
- A widespread and branched root system with a large root surface area and a large specific surface area of the other portion that is in contact with water,
- High translocation potential,
- High tolerance or accumulation potential for a variety of contaminants,
- Easy to harvest depending on the use of the sink organ,
- High hyperaccumulation potential for contaminants mainly in the aboveground parts,
- Susceptible and sensitive to genetic modification, and
- Non-consumable by humans and animals.

9.4.2 Ideal Nanoparticle Features for Nano-Phytoremediation

Several nanomaterials have been designed and employed for the purpose of environmental remediation and clean-up. The significant features of nanomaterials such as a large surface area, and consequently, a higher reactivity than their bulk form, and the potential to penetrate contaminated sites make them an ideal component of the remediation process (Davis et al., 2017). Some ideal features of nanomaterials (Figure 9.3) are especially employed for nano-phytoremediation (Jeevanantham et al., 2019; El-Ramady et al., 2020; Kumar et al., 2020).

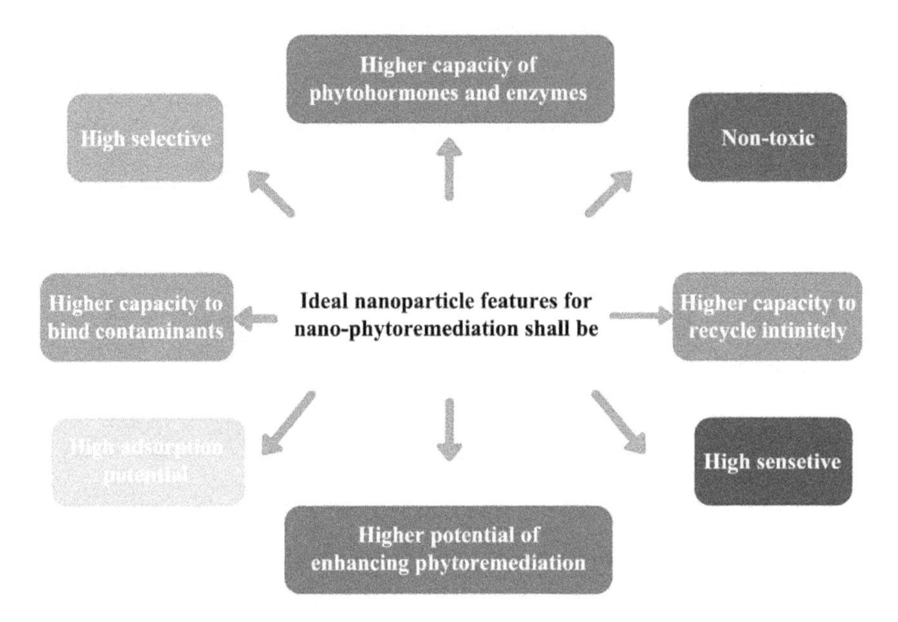

FIGURE 9.5 Ideal features of nanoparticles designed for nano-phytoremediation.

9.4.3 Factors Influencing the Nano-Phytoremediation of Contaminants

The process of nano-phytoremediation depends on several factors including the nature of the nanoparticle, the plant species and its physiology, and the interaction between the nanomaterial and contaminants. Moreover, the size, morphology, type of nanomaterial, and chemical composition have a special influence on nano-phytoremediation. The presence of other organisms, such as bacteria and fungi, influences the plant uptake of nanoparticles (Feng et al., 2013; Wang et al., 2016). Recent advancements and new perspectives of nano-phytoremediation nanomaterial surface functionalization and coating are determining factors for the penetration of nanomaterials into plant tissue (Judy et al., 2012; Raliya et al., 2016; Mohammadi et al., 2019). The factors affecting nano-phytoremediation are as follows (Srivastav et al., 2019):

- Physical and chemical properties of the contaminants, for example, their reactivity, toxicity, water-solubility, flexibility, type of bonds, vapor pressure, mobility, molecular weight, and hydrophilic and hydrophobic interactions,
- Environmental factors such as pH, oxygen level, light, temperature, micro-organisms, and toxic components,
- Plant characteristics such as the plant species, its root system, a high growth rate, large productivity, enzymes, the potential for tolerance, and the accumulation of contaminants, and
- The structure, composition, size, shape, concentration, surface activity, and physical and chemical properties of the nanomaterials.

9.5 ADVANTAGES OF NANO-PHYTOREMEDIATION

Nanotechnology has the potential for environmental monitoring and remediation without harming nature by using eco-friendly and green technology approaches (Srivastav et al., 2019). Along with the purification and removal of contaminants from various ecosystems, nanotechnology also aids in the probe-like sensing of water contaminants with high sensitivity and selectivity. Therefore, nanomaterial probes have been successfully employed as water monitoring devices. The use of

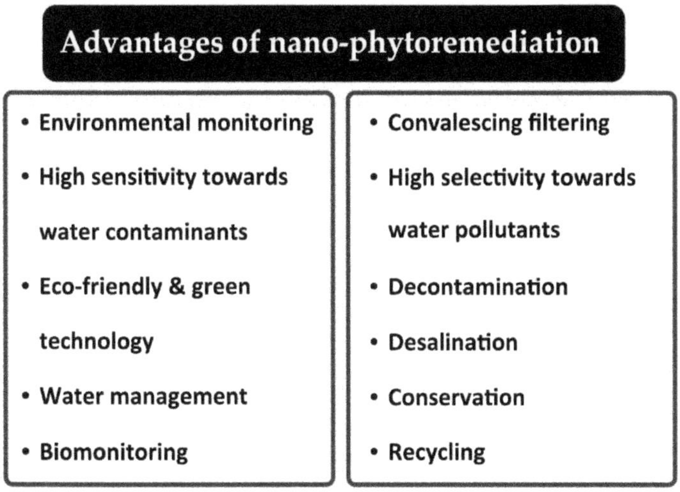

Advantages of nano-phytoremediation

- Environmental monitoring
- High sensitivity towards water contaminants
- Eco-friendly & green technology
- Water management
- Biomonitoring

- Convalescing filtering
- High selectivity towards water pollutants
- Decontamination
- Desalination
- Conservation
- Recycling

FIGURE 9.6 Advantages of nano-phytoremediation.

nanotechnologies, along with phytoremediation, has provided immense environmental benefits, particularly the decontamination of water systems. This technology of nano-phytoremediation is significantly involved in water biomonitoring, management, and treatment by convalescing, filtering, decontamination, desalination, conservation, recycling, and sewerage systems and by developing sensitive analytics or monitoring systems. Nano-phytoremediation includes the potential adopted by plants and nanomaterials to decontaminate soil, sludge, sediment, and wastewater (Bhati and Rai, 2018). Phytoremediation has the potential for environmental monitoring, detoxification, and decontamination with effective efficiency, eco-friendliness, and cost-effectiveness. However, this technique is entirely influenced by the plant species, contaminant bioavailability, water quality, and weather conditions. Therefore, phytoremediation is a time-consuming and seasonal technique that is limited by the plant species and bioavailability of contaminants in water (Song et al., 2019). These limitations can be fulfilled with the help of nanomaterials. Nanoparticles increase the phytoremediation efficiency of plants through nanoparticles' versatile and nano nature such as their dynamic morphology, desired size, superparamagnetic behavior, and high saturation magnetization. These characteristics of nanoparticles make them effective for the remediation of various water ecosystems including groundwater and wastewater treatment (WWT). Nanomaterials provide green techniques for water environment biomonitoring and decontamination with their non-toxic, cost-effective, eco-friendly, and less time-consuming approach (Bhati and Rai, 2018).

9.6 ROLE OF NANOPARTICLES IN PHYTOREMEDIATION

9.6.1 Plant Growth

Phytoremediation is a very effective, eco-friendly, cost-effective, and popular technique of environmental remediation. This technique is also user-friendly, less damaging to analytical systems, and universally accepted, with an esthetic appearance and wide adaptability (Macek et al., 2000; Glick, 2003; Gong et al., 2018). Phytoremediation has been designed and greatly utilized for the remediation of sites and ecosystems polluted by radionuclides, heavy metals, crude oil, chlorinated solvents, pesticides, and explosives (Sharma and Pandey, 2014; Dubchak and Bondar, 2019). Phytoremediation technologies used for the remediation of water contamination include rhizofiltration, phytodegradation, phytostabilization, photodegradation, phytoextraction, and phytovolatilization (Sakakibara et al., 2010). This method uses the metabolic system of plants for the uptake of pollutants from their surroundings and accumulates them in the plant biomass or organs. Moreover, in other cases, the branched roots of plants remove the contaminants in water by sorbing them (Hanks et al., 2015). Aquatic plants are the primary stage of the aquatic food web and contribute to a large portion of primary productivity in their ecosystem. Several aquatic plants have the potential to rapidly produce large quantities of biomass. Furthermore, many of them have been reported to grow rapidly in the presence of contaminants in water. Several aquatic plants have a hyperaccumulator capacity of metal uptake and accumulation; therefore, they have the ability to remediate water and sediment contaminated with metals (Kaewtubtim et al., 2017b).

9.6.2 Phytoavailability of Contaminants

The physicochemical factors such as metal speciation, soil pH and texture, organic matter and clay content, cation exchange capacity, and soil's redox status determine the bioavailability of metals to plants (Planquart et al., 1999; Notten et al., 2005). Phytoremediation is a green and environmentally friendly technology for environmental remediation. Several types of plants act as significant hyperaccumulators and play a major role in contaminant uptake from groundwater, surface water, and sediment. Phytoremediation has been employed for the remediation of water contaminated with organic compounds, heavy metals such as lead, mercury, and arsenic, and toxic chemicals such as trichloroethylene, ethylbenzene, benzene, toluene, 2,4,6-trinitrotoluene (TNT), xylene, and radionuclides. Although, phytotechnology is cost-effective and environmentally friendly compared with other technologies, it is a long-term treatment technology and takes comparatively more time for remediation (Paquin et al., 2004).

9.6.3 Nano-Phytoremediation of Contaminated Water

The nano-phytoremediation technique can be successfully employed for the remediation of wastewater, sludge, sediment, groundwater, and surface water. The desired size, superparamagnetic behavior, high saturation magnetization value, and dynamic morphology of phytogenic magnetic nanoparticles (PMNPs) play a significant role in water treatment and WWT. This eco-friendly technology is more suitable than other methods and is economically sustainable. The optimization of fabrication protocol solution parameters such as the solvent type, pH, extract volume, physical parameters including temperature, and strength of the precursor is required for the development of such PMNPs. The stability of PMNPs after pollutant removal and the yielding biomass separation can be enhanced by improving the morphology of the particles. The long-term health impact and sustainability of using such techniques are still under study (Ali et al., 2013, 2017).

9.6.3.1 Remediation of Groundwater and Freshwater Ecosystems

Polluted aquatic environments are a growing concern regarding the health of associated aquatic organisms, plants, microorganisms, and consequently, the whole food chain, which are ultimately posing a risk for ecosystems and human health (Batayneh, 2012). Carbon nanotubes have been significantly employed for the remediation of contaminated water in recent years due to their unique adsorption properties and their affinity towards the target analyte (Savage and Diallo, 2005). Efforts for decreasing the release of pollutants and minimizing their effects as mediated by living organisms such as plants are approached with the phytoremediation technique (Pilon-Smits, 2005; Conesa et al., 2012). This is the best solution for pollution control and the most effective technology that uses biological systems for the treatment of contaminants (Dua et al., 2002; Cardenas et al., 2008). Bioremediation is eco-friendly and utilizes natural biological processes to remove toxic pollutants. This technique employs microorganisms and their enzymes to remediate the environment altered by pollutants and restore it to its original condition (Kensa, 2011; Chakraborty et al., 2012).

9.6.3.1.1 Removal of Inorganic Contaminants

The remediation efficiency and decontamination potential of phytoremediation can be boosted with the gene manipulation of plants by enhancing their anti-oxidative enzyme systems (Mani and Kumar, 2014; Gomes et al., 2016). The modified or transferred genes in wild plants can improve the homeostasis mechanisms towards metal stress (Antosiewicz et al., 2014). Some plants can uptake and accumulate various inorganic environmental contaminants, which depends on their phytoremediation potential (Islam et al., 2015; Chen et al., 2019). Phytoremediation processes are directly associated with photosynthetic mechanisms and sequestering pollutants into the above-ground organs of the same plant (Pietrini et al., 2009; Kofroňová et al., 2018; Lima et al., 2019; Tang et al., 2019). Due to various pollutant doses and accumulation, the photosynthesis process in hyperaccumulators becomes affected by impacting the phytoremediation mechanisms such as phytoextraction, phytodegradation, phytostabilization, and rhizofiltration (Pietrini et al., 2009).

9.6.3.1.2 Removal of Organic Contaminants

Hazardous organic contaminants have been identified as significant causative agents of environmental degradation. Organic xenobiotics can harm environmental quality and affect the food chain by impacting the health of its biotic components (Tang et al., 2019; Hussain et al., 2009; Kang, 2014). Persistent organic pollutants (POPs) have a high affinity for fatty acids, and consequently, they resist photochemical degradation (Rai, 2018a; Hussain et al., 2009). Such POPs with high toxicity also possess the tendency not only for bioaccumulation but also for biomagnification across different trophic levels (Lima et al., 2019; Matteucci et al., 2015; Hussain et al., 2009). Chlorinated organics are particularly persistent in human environments, and polychlorinated biphenyl (PCB) is also utilized as a raw material for insecticides in the agriculture sector (Husain et al., 2018; Hussain et al., 2009). Organophosphates such as paraoxon and parathion are a class of pesticides that are neurotoxic in nature and have the capability to degrade acetylcholinesterase; consequently, they pose serious health risks by perturbing systematic nerve transmission (Fernández et al., 2017). Moreover, pesticides, petroleum, explosives, polycyclic aromatic hydrocarbons, and their derivatives are common organic pollutants in both the soil and water, and it is crucial to decontaminate such affected sites (Martin et al., 2016).

9.6.3.2 Wastewater Treatment (WWT)

Phytoremediation is an organic method that eliminates pollutants from a water resource by utilizing plant life. This procedure applies the plant's metabolic structure to remove pollutants and nutrients from their neighboring areas and stock these in its biomass. Perfect plants for phytoremediation need great root structures, as the root extent determines the closeness of the plant parts that can interrelate with the polluted water resources. To guarantee the active elimination of metals and pollutants from an aquatic source, the florae that work in phytoremediation must be copious metal and nutrient accumulators that will attain real pollutant exclusion. These florae require rapid duplication to ensure that the scheme for water remediation is self-sufficient, therefore reducing the treatment expenditure (Kang, 2014). These plants are presently employed in numerous parts of the world for heavy metal water purification drives and have shown hopeful outcomes in the elimination of silver and other metal ions from polluted water resources (Harris and Bali, 2008; Odjegba and Fasidi, 2004). However, few studies are available that assesses the efficiency of these plants at eliminating metal nanoparticles from a water source compared to their ionic counterparts.

9.7 CHALLENGES OF NANO-PHYTOREMEDIATION

Nanomaterial-aided phytoremediation studies are very rare. Only small-scale experiments have been conducted to date, so there is a requirement to use extra-accurate revisions in upcoming studies, and better strategies at the practical stage are highly required. Long-standing tests are needed to understand the real effects of nanomaterials in phytoremediation procedures and in the alteration of the soil productiveness status. Nanomaterials can cumulate this and may decrease their movement, so polymer or other coverings are needed to improve their bioavailability or flexibility. An assessment of the influences and security of nanomaterial implementation in polluted soil is needed. Maintainable nano-phytoremediation relies primarily on climatological circumstances; thus, ecologically unchanging nanomaterials should be recognized (Gupta et al., 2012). Nano-phytoremediation is a suitable technique for places that have average amounts of contamination due to unmaintainable plant development in an extremely polluted land. An improved understanding of pollutant acceptance by plants from the soil will also help in stimulating agro-mining, which could be utilized to remove pollutants even from harvestable plant biomass. Nano-phytoremediation includes both nanotechnology and phytoremediation collectively to clean up the atmosphere. Nanomaterials are extensively used currently in the arenas of make-up, dyes, drugs, and fabrics. Nanotechnology increases phytoremediation's efficacy, and nanoparticles can also be utilized to remedy soils and water that are polluted with metals and inorganic and organic contaminants. The current research has showcased that carbon-based pollutants, including chlorpyrifos, moline, and atrazine, can be tarnished with nano-sized zerovalent irons. Nanoparticles in enzyme-dependent bioremediation can also be used in a mixture with phytoremediation (Srivastav et al., 2019; Yadav et al., 2017b). Numerous nanomaterials have been technologically advanced for ecological utilization. Nanomaterials are beneficial for remediation procedures due to their large surface area; hence, such resources are more sensitive than their bulk form, and they can effortlessly enter polluted locations (Singh and Walker, 2006). The usage of nanomaterials for ecological remediation is rapidly developing. Iron nanomaterials are the most extensively utilized for remedy purposes. Some academics have found that nanomaterials disinfect organic contaminants (PPCPs, PAHs, PCBs, and carbon-based solvents) in land (Davis et al., 2017). According to Cooley et al. (2012) and Jiamjitrpanich et al. (2013), a mixture of nanotechnology and phytotechnology for TNT remediation from land is more operative than land without nanomaterials. Nanomaterials not only unswervingly catalyze the reduction of excess and deadly supplies but also help recover the efficacy of microorganisms in the degradation of waste and toxic materials. Bioremediation uses living organisms to break up or eliminate poisons and damaging materials from farming water and soil. Thus, with bioremediation, heavy metals can be separated from water and soil ecologically and professionally by microbes (Jiamjitrpanich et al., 2013). Nano-structured materials, such as polyacrylic acid, zeolite, chitosan, hydroxyapatite, and clay minerals, are utilized to advance enrichers to be used for soil and/or foliar utilization. The greater surface area of hydroxyapatite and its strong connections with urea lead to the slow release of nitrogen (N) from urea (Dixit et al., 2015). Urea modified with hydroxyapatite nanoparticles release N for up to 60 days of plant growth compared to other conventional fertilizers (urea and ammonium nitrate), which release N till 30 days (Dixit et al., 2015). Kottegoda et al. (2011) described an increase in the chlorophyll content of soybeans by using superparamagnetic iron oxide. A 10% increase in chlorophyll content was measured in cowpea (Vigna unguiculata (L. Walp.)) by applying iron (Fe) nanoparticles on foliage as a 0.5 g L−1 solution compared with the same solution of common forms of Fe (Ghafariyan et al., 2013). Phytoremediation is a skill that uses plants and the related rhizosphere microbes to eliminate, alter, or cover toxic substances situated in soils, residues, surface water, groundwater, and even air (Karami et al., 2014). Numerous species of florae are utilized for the action or elimination of a diversity of pollutants, for example, radionuclides, chlorinated solvents, explosives, oil, nutrients, metals, and pesticides. The phyto-toxicity of the contaminants can disturb the phytoremediation procedure as florae that tolerate the occurrence of extreme quantities of toxic mixtures are more vulnerable to illnesses and stress circumstances (Susarla et al., 2002; Gulia et al., 2020). A particular type of phytoremediation grounded on swamp plants rising in a soil or grit substrate, usually known as constructed wetland systems (CWs), has been observed as a promising method for the elimination of drugs from wastewater (Datta et al., 2013; Matamoros and Bayona, 2006). The elimination of ecological pollutants (including heavy metals) from polluted sites using nanoparticles/nanomaterial shaped by fungi, plant, and bacteria with the assistance of nanotechnology is called nano-bioremediation (NBR). NBR is a novel upcoming method for the elimination of contaminants through ecological cleaning. The present methodologies for the remediation of polluted locations are physical and chemical remediation, incineration, and bioremediation, which have their own benefits and drawbacks. With new progressions, bioremediation offers an ecologically approachable and sparingly practicable choice to eliminate pollutants from the atmosphere (Singh and Walker, 2006).

9.8 CONCLUSION

Nano-phytoremediation is a very effective technique and can be successfully employed for the remediation of various types of contaminants from aquatic ecosystems. Although several studies have been conducted over the utilization of different nanomaterials, typically nanoparticles and nanotubes coupled with the technique of phytoremediation, limited studies have been conducted concerning the nano-phytoremediation of water and sediments. As with a number of processes in phytoremediation, nano-phytoremediation can also involve processes such as nono-phytostabilization, nano-phytofiltration, nano-phytovolatilization, nano-phytoextraction, and nano-phytovolatilization. As aquatic plants are the base of the aquatic food chain, there is a need to analyze the toxicity of components produced during nano-phytoremediation.

REFERENCES

Ahmadpour, P. 2012. Phytoremediation of heavy metals: A green technology. *African Journal of Biotechnology*, 11(76), 14036–14043.

Ali, A., Ahmed, R., & Saber, I. 2020. Green nano-phytoremediation and solubility improving agents for the remediation of chlorfenapyr contaminated soil and water. *Journal of Environmental Management*, 260, 110104.

Ali, H., & Khan, E. 2018. Bioaccumulation of non-essential hazardous heavy metals and metalloids in freshwater fish. Risk to human health. *Environmental Chemistry Letters*, 16(3), 903–917.

Ali, H., Khan, E., & Sajad, M.A. 2013. Phytoremediation of heavy metals—concepts and applications. *Chemosphere*, 91, 869–881.

Ali, I., Peng, C., Naz, I., Khan, Z.M., Sultan, M., Islam, T., & Abbasi, I.A. 2017. Phytogenic magnetic nanoparticles for wastewater treatment: A review. *RSC Advances*, 7(64), 40158–40178.

Alrumman, S.A., El-kott, A.F., & Keshk, S.M.A.S. 2016. Water pollution: Source and treatment. *American Journal of Environmental Engineering*, 6(3), 88–98.

Antosiewicz, D.M., Barabasz, A., & Siemianowski, O. 2014. Phenotypic and molecular consequences of overexpression of metal-homeostasis genes. *Frontiers in Plant Science*, 5, 1–7.

Batayneh, A.T. 2012. Toxic (aluminum, beryllium, boron, chromium and zinc) in groundwater: Health risk assessment. *International Journal of Environmental Science and Technology*, 9, 153–162.

Bhati, M., & Rai, R. 2018. Nano-phytoremediation application for water contamination. In *Phytoremediation* (pp. 441–452). Springer, Cham.

Bibi, S., Khan, R.L., Nazir, R., Khan, P., Rehman, H.U., Shakir, S.K., . . . & Jan, R. 2016. Heavy metals analysis in drinking water of Lakki Marwat District, KPK, Pakistan. *World Applied Sciences Journal*, 34, 15–19.

Briggs, D. 2003. Environmental pollution and the global burden of disease. *British Medical Bulletin*, 68(1), 1–24.

Burken, J.G. 2001. Advancement of phytoremediation. *Practice Periodical of Hazardous, Toxic, and Radioactive Waste Management*, 5(3), 120.

Cardenas, E., Wu, W.M., Leigh, M.B., Carley, J., Carroll, S., Gentry, T., Luo, J., Watson, D., Gu, B., Ginder-Vogel, M., et al. 2008. Microbial communities in contaminated sediments, associated with bioremediation of uranium to submicromolar levels. *Applied and Environmental Microbiology*, 74, 3718–3729.

Chakraborty, R., Wu, C.H., & Hazen, T.C. 2012. Systems biology approach to bioremediation. *Current Opinion in Biotechnology*, 23, 1–8.

Chen, S., Zhang, M., Feng, Y., Sahito, Z.A., Tian, S., & Yang, X. 2019. Nicotianamine Synthase Gene 1 from the hyperaccumulator Sedum alfredii Hance is associated with Cd/Zn tolerance and accumulation in plants. *Plant and Soil*, 443(1–2), 413–427.

Conesa, H.M., Evangelou, M.W.H., Robinson, B.H., & Schulin, R. 2012. A critical view of current state of phytotechnologies to remediate soils: Still a promising tool? *The Scientific World Journal*. http://doi.org/10.1100/2012/173829

Cooley, A., Rexroad Crane, R.A., & Scott, T.B. 2012. Nanoscale zero-valent iron: Future prospects for an emerging water treatment technology. *Journal of Hazardous Materials*, 211, 112–125.

Datta, R., Das, P., Smith, S., Punamiya, P., Ramanathan, D.M., Reddy, R., & Sarkar, D. 2013. Phytoremediation potential of vetiver grass [Chrysopogon zizanioides (L.)] for tetracycline. *International Journal of Phytoremediation*, 15(4), 343–351.

Davis, A.S., Prakash, P., & Thamaraiselvi, K. 2017. Nanobioremediation technologies for sustainable environment. In Prashanthi, M. (ed.) *Bioremediation and Sustainable Technologies for Cleaner Environment. Environmental Science* (pp. 13–33). Springer International Publishing AG, Cham.

Devasahayam, S. 2019. Nanotechnology and nanomedicine in market: A global perspective on regulatory issues. In *Characterization and Biology of Nanomaterials for Drug Delivery* (pp. 477–522). Elsevier, Amsterdam.

Dixit, R., Wasiullah, Malaviya, D., Pandiyan, K., Singh, U.B., Sahu, A., et al. 2015. Bioremediation of heavy metals from soil and aquatic environment: An overview of principles and criteria of fundamental processes. *Sustainability*, 7, 2189–2212.

Dua, M., Sethunathan, N., & Johri, A.K. 2002. Biotechnology bioremediation success and limitations. *Applied Microbiology and Biotechnology*, 59(2–3), 143–152.

Dubchak, S., & Bondar, O. 2019. Bioremediation and phytoremediation: Best approach for rehabilitation of soils for future use. In Gupta, D.K., & Voronina, A. (eds.) *Remediation Measures for Radioactively Contaminated Areas* (pp. 201–221). Springer International Publishing, Cham.

El-Ramady, H., El-Henawy, A., Amer, M., Omara, A E.-D., Elsakhawy, T., Salama, A.-M., Ezzat, A., Elsherif, A., Elmahrouk, M., & Shalaby, T. 2020. Agro-pollutants and their nano-remediation from soil and water: A mini-review. *Environment, Biodiversity and Soil Security*, 4.

FAO, Food and Agriculture Organisation of United Nations. 2016. www.fao.org/news/story/en/item/421871/icode.

Feng, Y., Cui, X., He, S., Dong, G., Chen, M., Wang, J., & Lin, X. 2013. The role of metal nanoparticles in influencing arbuscular mycorrhizal fungi effects on plant growth. *Environmental Science and Technology*, 47(16), 9496–9504.

Fernández-Cruz, T., Martínez-Carballo, E., & Simal-Gándara, J. 2017. Perspective on pre- and post-natal agro-food exposure to persistent organic pollutants and their effects on quality of life. *Environment International*, 100, 79–101.

Ghafariyan, M.H., Malakouti, M.J., Dadpour, M.R., Stroeve, P., & Mahmoudi, M. 2013. Effects of magnetite nanoparticles on soybean chlorophyll. *Environmental Science & Technology*, 47, 10645–10652.

Glick, B.R. 2003. Phytoremediation: Synergistic use of plants and bacteria to clean up the environment. *Biotechnology Advances*, 21, 383–393.

Gomes, M., Hauser-Davis, R.A., de Souza, A.N., & Vitória, A.P. 2016. Metal phytoremediation: General strategies, genetically modified plants and applications in metal nanoparticle contamination. *Ecotoxicology and Environmental Safety*, 134, 133–147.

Gong, X., Huang, D., Liu, Y., Zeng, G., Wang, R., Wei, J., . . . & Zhang, C. 2018. Pyrolysis and reutilization of plant residues after phytoremediation of heavy metals contaminated sediments: For heavy metals stabilization and dye adsorption. *Bioresource Technology*, 253, 64–71.

Gong, X., Huang, D., Liu, Y., Zeng, G., Wang, R., Xu, P., & Chen, S. 2019. Roles of multiwall carbon nanotubes in phytoremediation: Cadmium uptake and oxidative burst in Boehmeria nivea (L.) Gaudich. *Environmental Science: Nano*, 6(3), 851–862.

Gulia, S., Rohilla, R., Sankhla, M.S., Kumar, R., & Sonone, S. 2020. Impact of pesticide toxicity in aquatic environment. *Biointerface Research in Applied Chemistry*, 11, 10131–10140.

Gulia, S., Sankhla, M.S., Kumar, R., & Sonone, S.S. 2020. Phytomicrobiome studies for combating the abiotic stress. *Biointerface Research in Applied Chemistry*, 11, 10493–10509.

Gupta, N., Yadav, K.K., & Kumar, V. 2015. A review on current status of municipal solid waste management in India. *Journal of Environmental Sciences*, 37, 206–217.

Gupta, P., Roy, S., et al. 2012. Treatment of water using water hyacinth, water lettuce and vetiver grass—a review. *Resources Environment*, 2(5), 202–215.

Hanks, N.A., Caruso, J.A., & Zhang, P. 2015. Assessing *Pistia stratiotes* for phytoremediation of silver nanoparticles and Ag(I) contaminated waters. *Journal of Environmental Management*, 164, 41–45.

Harris, A., & Bali, R. 2008. On the formation and extent of uptake of silver nanoparticles by live plants. *Journal of Nanoparticle Research*, 10(4), 691–695.

Henry, H.F., Burken, J.G., Maier, R.M., Newman, L.A., Rock, S., Schnoor, J.L., & Suk, W.A. 2013. Phytotechnologies: Preventing exposures, improving public health. *International Journal of Phytoremediation*, 15, 889–899.

Hinrichsen, D., & Tacio, H. 2020. *The Coming Freshwater Crisis Is Already Here*. www.wilsoncenter.org/sites/default/files/media/documents/publication/popwawa2.pdf

Husain, I., Aleti, G., Naidu, R., Puschenreiter, M., Mahmood, Q., Rahman, M.M., Wang, F., Shaheen, S., Syed, J.H., & Reichenauer, T.G. 2018. Microbe and plant assisted-remediation of organic xenobiotics and its enhancement by genetically modified organisms and recombinant technology: A review. *Science of the Total Environment*, 628–629, 1582–1599.

Hussain, S., Siddique, T., Arshad, M., & Saleem, M. 2009. Bioremediation and phytoremediation of pesticides: Recent advances. *Critical Reviews in Environmental Science and Technology*, 39(10), 843–907.

Ingle, A.P., Seabra, A.B., Duran, N., & Rai, M. 2014. Nanoremediation: A new and emerging technology for the removal of toxic contaminant from environment. In *Microbial Biodegradation and Bioremediation* (pp. 233–250). Elsevier, Amsterdam.

Islam, E., Khan, M.T., & Irem, S. 2015. Biochemical mechanisms of signaling: Perspectives in plants under arsenic stress. *Ecotoxicology and Environmental Safety*, 114, 126–133.

Jasrotia, S., Kansal, A., & Mehra, A. 2017. Performance of aquatic plant species for phytoremediation of arsenic-contaminated water. *Applied Water Science*, 7(2), 889–896.

Jeevanantham, S., Saravanan, A., Hemavathy, R.V., Kumar, P.S., Yaashikaa, P.R., & Yuvaraj, D. 2019. Removal of toxic pollutants from water environment by phytoremediation: A survey on application and future prospects. *Environmental Technology and Innovation, 13*, 264–276.

Jiamjitrpanich, W., Parkpian, P., Polprasert, C., & Kosanlavit, R. 2013. Trinitrotoluene and its metabolites in shoots and roots of Panicum maximum in nano-phytoremediation. *International Journal of Environmental Science and Development*, 4(1), 7.

Judy, J.D., Unrine, J.M., Rao, W., Wirick, S., & Bertsch, P.M. 2012. Bioavailability of gold nanomaterials to plants: Importance of particle size and surface coating. *Environmental Science and Technology*, 46(15), 8467–8474.

Kaewtubtim, P., Meeinkuirt, W., Seepoma, S., & Pichtel, J. 2017b. Radionuclide (226 Ra, 232 Th, 40 K) accumulation among plant species in mangrove ecosystems of Pattani Bay, Thailand. *Marine Pollution Bulletin*, 115(1–2), 391–400.

Kang, J.W. 2014. Removing environmental organic pollutants with bioremediation and phytoremediation. *Biotechnology Letters*, 36, 1129–1139. http://doi.org/10.1007/s10529-014-1466-9.

Karami, M., Bahabadi, M.A., Delfani, S., & Ghozatloo, A. 2014. A new application of carbon nanotubes nanofluid as working fluid of low-temperature direct absorption solar collector. *Solar Energy Materials and Solar Cells*, 121, 114–118.

Karn, B., Kuiken, T., & Otto, M. 2009. Nanotechnology and in situ remediation: A review of the benefits and potential risks. *Environmental Health Perspectives*, 117(12), 1823–1831.

Kensa, M.V. 2011. Bioremediation: An overview. *Journal of Industrial Pollution Control*, 27(2), 161–168.

Khan, N., Hussain, S.T., Saboor, A., et al. 2013. Physiochemical investigation of the drinking water sources from Mardan, Khyber Pakhtunkhwa, Pakistan. *International Journal of Physical Sciences*, 8(33), 1661–1671.

Kofroňová, M., Mašková, P., & Lipavská, H. 2018. Two facets of world arsenic problem solution: Crop poisoning restriction and enforcement of phytoremediation. *Planta*, 248, 19–35.

Kottegoda, N., Munaweera, I., Madusanka, N., & Karunaratne, V. 2011. A green slow-release fertilizer composition based on urea-modified hydroxyapatite nanoparticles encapsulated wood. *Current Science*, 101, 73–78.

Kumar, S., Kumari, N., Karmakar, S., Ankit, Singh, R., Behera, M., Rani, A., & Kumar, N. 2020. Advances in plant—microbe-based remediation approaches for environmental cleanup. In Bharagava, R.N. (ed.) *Emerging Eco-friendly Green Technologies for Wastewater Treatment, Microorganisms for Sustainability 18* (pp. 103–128). Springer Nature Singapore Pte Ltd., Singapore.

Lima, L.R., Silva, H.F., Brignoni, A.S., Silva, F.G., Camargos, L.S., & Souza, L.A. 2019. Characterization of biomass sorghum for copper phytoremediation: Photosynthetic response and possibility as a bioenergy feedstock from contaminated land. *Physiology and Molecular Biology of Plants*, 25(2), 433–441.

Macek, T., Mackova, M., & Kas, J. 2000. Exploitation of plants for the removal of organics in environmental remediation. *Biotechnology Advances*, 18(1), 23–34.

Mani, D., & Kumar, C. 2014. Biotechnological advances in bioremediation of heavy metals contaminated ecosystems: An overview with special reference to phytoremediation. *International Journal of Environmental Science and Technology*, 11, 843–872.

Martin, B.C., George, S.J., Price, C.A., Shahsavari, E., Ball, A.S., Tibbett, M., & Ryan, H. 2016. Citrate and malonate increase microbial activity and alter microbial community composition in uncontaminated and diesel-contaminated soil microcosms. *Soil*, 2(3), 487.

Matamoros, V., & Bayona, J.M. 2006. Elimination of pharmaceuticals and personal care products in subsurface flow constructed wetlands. *Environmental Science & Technology*, 40(18), 5811–5816.

Matteucci, F., Ercole, C., & Del Gallo, M. 2015. A study of chlorinated solvent contamination of the aquifers of an industrial area in central Italy: A possibility of bioremediation. *Frontiers in Microbiology*, 6, 924.

Mishra, S., Reddy-Noone, K., Jain, A., & Verma, K.K. 2006. Monitoring organophosphorus pesticides and their degradation products formed by Fenton's reagent using solid-phase extraction—gas chromatography—mass spectrometry. *International Journal of Environment and Pollution*, 27(1–3), 49–63.

Mohammadi, P., Hesari, M., Samadian, H., Hajialyani, M., Bayrami, Z., Farzaei, M.H., & Abdollahi, M. 2019. Recent advancements and new perspectives of phytonanotechnology. *Comprehensive Analytical Chemistry*, 84, 1–22.

Nanjwade, B.K., Sarkar, A.B., & Srichana, T. 2019. Design and characterization of nanoparticulate drug delivery. In *Characterization and Biology of Nanomaterials for Drug Delivery* (pp. 337–350). Elsevier, Amsterdam.

Notten, M., Oosthoek, A., Rozema, J., & Aerts, R. 2005. Heavy metal concentrations in a soileplantesnail food chain along a terrestrial soil pollution gradient. *Environmental Pollution*, 138, 178–190.

Odjegba, V.J., & Fasidi, I.O. 2004. Accumulation of trace elements by Pistia stratiotes: Implications for phytoremediation. *Ecotoxicology*, 13(7), 637–646.

Paquin, D.G., Campbell, S., & Li, Q.X. 2004. Phytoremediation in subtropical Hawaii—a review of over 100 plant species. *Remediation Journal: The Journal of Environmental Cleanup Costs, Technologies & Techniques*, 14(2), 127–139.

Pietrini, F., Zacchini, M., Iori, V., Pietrosanti, L., Bianconi, D., & Massacci, A. 2009. Screening of poplar clones for cadmium phytoremediation using photosynthesis, biomass and cadmium content analyses. *International Journal of Phytomedicine*, 12(1), 105–120.

Pilon-Smits, E. 2005. Phytoremediation. *Annual Review of Plant Biology*, 56, 15–39.

Planquart, P., Bonin, G., Prone, A., & Massiani, C. 1999. Distribution, movement and plant availability of trace metals in soils amended with sewage sludge composts: Application to low metal loadings. *Science of the Total Environment*, 241, 161–179.

Rai, P.K. 2018a. Heavy metal phyto-technologies from Ramsar wetland plants: Green approach. *Chemistry and Ecology*, 34(8), 786–796.

Raliya, R., Franke, C., Chavalmane, S., Nair, R., Reed, N., & Biswas, P. 2016. Quantitative understanding of nanoparticle uptake in watermelon plants. *Frontiers in Plant Science*, 7(August).

Rasheed, T., Chuanlong, L., Bilal, M., Chunyang, Y., & Iqbal, H.M.N. 2018. Potentially toxic elements and environmentally-related pollutants recognition using colorimetric and ratiometric fluorescent probes. *Science of the Total Environment*, 640(641), 174–193.

Renieri, E.A., Safenkova, I.V., Alegakis, A.K., Slutskaya, E.S., Kokaraki, V., Kentouri, M., . . . & Tsatsakis, A.M. 2019. Cadmium, lead and mercury in muscle tissue of gilthead seabream and seabass: Risk evaluation for consumers. *Food and Chemical Toxicology*, 124, 439–449.

Romeh, A.A. 2020. Synergistic effect of Ficus-zero valent iron supported on adsorbents and Plantago major for chlorpyrifos phytoremediation from water Plantago major for chlorpyrifos phytoremediation from water. *International Journal of Phytoremediation*, 1–11.

Sakakibara, M., Watanabe, A., Inoue, M., Sano, S., & Kaise, T. 2010. Phytoextraction and phytovolatilization of arsenic from As-contaminated soils by Pteris vittata. *Proceedings of the Annual International Conference on Soils, Sediments, Water and Energy*, 12, Article 26.

Salim, S. 2016. *Fish Consumption Pattern in India and Exports-overview*, 8th ed. (pp. 25–28). Food and Beverage News Foodex. https://eprints.cmfri.org.in/10991/1/Food%20and%20Beverage%20News_Shyam%20Salim%202016.pdf

Sankhla, M.S., Kumar, R., & Agrawal, P. 2018a. Arsenic in water contamination & toxic effect on human health: Current scenario of India. *Journal of Forensic Sciences & Criminal Investigation*, 10, 1–5.

Sankhla, M.S., Kumar, R., & Prasad, L. 2019. Zinc impurity in drinking water and its toxic effect on human health. *Indian Internet Journal of Forensic Medicine & Toxicology*, 17, 84–87.

Sankhla, M.S., Kumari, M., Nandan, M., Kumar, R., & Agrawal, P. 2016. Heavy metals contamination in water and their hazardous effect on human health-a review. *International Journal of Current Microbiology and Applied Sciences*, 5, 759–766.

Sankhla, M.S., Kumari, M., Sharma, K., Kushwah, R.S., & Kumar, R. 2018b. Water contamination through pesticide & their toxic effect on human health. *International Journal for Research in Applied Science & Engineering Technology (IJRASET)*, 6, 967–970.

Savage, N., & Diallo, M.S. 2005. Nanomaterials and water purification: Opportunities and challenges. *Journal of Nanoparticle Research*, 7(4), 331–342.

Sharma, P., & Pandey, S. 2014. Status of phytoremediation in world scenario. *International Journal of Environmental Bioremediation & Biodegradation*, 2, 178–191.

Singh, B.K., & Walker, A. 2006. Microbial degradation of organophosphorus compounds. *FEMS Microbiology Reviews*, 30, 428–471.

Song, B., Xu, P., Chen, M., Tang, W., Zeng, G., Gong, J., . . . & Ye, S. 2019. Using nanomaterials to facilitate the phytoremediation of contaminated soil. *Critical Reviews in Environmental Science and Technology*, 49(9), 791–824.

Srivastav, A., Yadav, K.K., Yadav, S., Gupta, N., Singh, J.K., Katiyar, R., & Kumar, V. 2019. Nano-phytoremediation of pollutants from contaminated soil environment: Current scenario and future prospects. *Phytoremediation: Management of Environmental Contaminants*, 6, 383–401.

Susarla, S., Medina, V.F., & McCutcheon, S.C. 2002. Phytoremediation: An ecological solution to organic chemical contamination. *Ecological Engineering*, 18(5), 647–658.

Tang, L., Hamid, Y., Gurajala, H.K., He, Z., & Yang, X. 2019. Effects of CO2 application and endophytic bacterial inoculation on morphological properties, photosynthetic characteristics and cadmium uptake of two ecotypes of Sedum alfredii Hance. *Environmental Science and Pollution Research*, 26, 1809–1820.

Tangahu, B.V., Abdullah, S.R.S., Basri, H., Idris, M., Anuar, N., & Mukhlisin, M. 2011. A Review on Heavy Metals (As, Pb, and Hg) uptake by plants through phytoremediation. *International Journal of Chemical Engineering*, 2011, 939161–939192.

Verma, A., Roy, A., & Bharadvaja, N. 2021. Remediation of heavy metals using nanophytoremediation. In Shah, M.P. (ed.) *Advanced Oxidation Processes for Effluent Treatment Plants* (pp. 273–296). https://doi.org/10.1016/B978-0-12-821011-6.00013-X.

Wang, F., Liu, X., Shi, Z., Tong, R., Adams, C.A., & Shi, X. 2016. Arbuscular mycorrhizae alleviate negative effects of zinc oxide nanoparticle and zinc accumulation in maize plants—A soil microcosm experiment. *Chemosphere*, 147, 88–97.

Wei, C.Y., & Chen, T.B. 2001. Hyperaccumulators and phytoremediation of heavy metal contaminated soil: A review of studies in China and abroad. *ACTA Ecologica Sinica*, 21(7), 1196–1204.

Yadav, K.K., Singh, J.K., Gupta, N., & Kumar, V. 2017b. A review of nanobioremediation technologies for environmental cleanup: A novel biological approach. *Journal of Materials and Environmental Science*, 8(2), 740–757.

10 Nanotools-Coupled Phytoremediation

Auspicious Technology for the Detoxification of Contaminated Pedospheric Matrices

Khuram Shahzad Ahmad, Shaan Bibi Jaffri, and Asma Jabeen

10.1 INTRODUCTION

Human beings have been striving for the development of products and lifestyles that assures maximum facilitation (Jaffri and Ahmad 2020b). In such efforts aimed at the attainment of excellence in lifestyle and the fulfillment of the demands of human survival, many synthetic chemicals have been added to ecological settings, in addition to the inherently present chemicals, for example, heavy metals, in the environment. With the passage of time, as we are advancing towards a modernized and urbanized form of globalization, the contamination of different environmental compartments has been augmenting in a consequential manner (Jaffri and Ahmad 2018d; Jaffri et al. 2019b). In this regard, all aspects of the environment are at stake, inclusive of the hydrospheric, atmospheric and lithospheric regions. However, since the lithosphere is an immediately available base for the reception of different types of pollutants, its exposure and contamination is therefore comparatively higher than that of other compartments. Another considerable factor here is the transboundary nature of different pollutants and contaminants. This is true in terms not only of the geographical transferal of the pollutants but also of the transport of pollutants from one ecological compartment to another (Naeem et al. 2020). For instance, agrochemicals can be transferred to the soils where they are employed on crops, but there is a chance of the residues and daughter products of these agrochemicals to be transferred to the atmosphere via volatilization or leaching to groundwater reservoirs to cause a disturbance in the overall ecological functioning (Iram et al. 2020b, 2020a).

The contamination of the pedospheric region is not limited to only agrochemicals, and there are a variety of other anthropogenic sources, for example, waste produced from industries. In addition, numerous other human activities are responsible for the release of contaminants in soils, including nutrients, solvents, organic and inorganic substances and heavy metals (Iram et al. 2018; Shazia et al. 2018). The pedospheric zone of the environment is representative of the most important functionalities that it carries out in terms of supporting the life of the floral and faunal species residing inside and over it, production of food for various species, storage of carbon and acting as a carbon sink, purification of water, and conservation of biological diversity (Shaheen et al. 2016; Zahra et al. 2017). Such significant functionalities of soils are expressive of their essential nature for humans and other forms of biological diversity. Therefore, safer and healthier soils constitute an integral part of sustainable development for human society (Jaffri et al. 2019a). However, to our dismay, soils on a global scale are under a constant serious threat due to the enhanced extent of urbanization and the industrial and agricultural orientations being dependent on the utilization of synthetic chemicals for higher output (Tahir et al. 2020a, 2020b). Recent decades have especially

 DOI: 10.1201/9781003186298-10

seen an exhausting trend in the release of such contaminants in soils through different activities. These chemicals are marked not only by their noxiousness towards human and other floral/faunal forms of biodiversity but also by their equal damage to environmental integrity (Amjad et al. 2019; Ali et al. 2019a).

The problem of the degradation and contamination of soils has adverse impacts on the agricultural productivity and security of food. Degradation of soils is further worsened by the occurrence of other factors, such as the reduced fertility of soils, overgrazing and erosion. The global significance of soils and their essentiality in overall global functioning makes their remediation an urgent task (Iram et al. 2019; Ishtiaq et al. 2020). A myriad of physical, chemical and biological approaches have been adopted for the remediation of soils and restoration of their ecological integrity. However, all kinds of methods have been known for their limitations and adoption at the field level. Such limitations identify the efforts to be exerted to improve these methods for the effective remediation of soils (Iftikhar et al. 2019). Biological methods are preferred over chemical and physical ones due to their eco-friendliness, facile nature and economic viability (Okaiyeto et al. 2021). In contrast to the chemical methods where synthetic substances further deteriorate environmental quality and the physical methods characterized by the need for complicated executional requirements, biological methods are actually based on the utilization of nature's tools for the sustainable remediation of soils.

Nanoscience and nanotechnology are known for nanoscale materials possessing unique characteristics, which have revolutionized human life in recent decades. A great deal of nanomaterial-based products have been developed using different types, morphologies and size ranges of nanomaterials aimed at human welfare (Shaheen et al. 2021; Ahmad et al. 2020). For instance, nanomaterials have now become part of our lives in the form of drug delivery agents, pesticides' nano-encapsulation, medical diagnostic materials and pharmaceuticals. The unique characteristics associated with nanomaterials in contrast to their bulk counterparts make them further attractive for initialization in various applications ranging from biomedical employments and photovoltaics to agriculture and environmental remediation.

In terms of environmental remediation, these nanoparticles have been extensively used for the detoxification of persistent organic pollutants, including dyes (Cittrarasu et al. 2021; Uribe-López et al. 2021). The recent era has also seen the utilization of these nanomaterials in improving agricultural productivity by the development of these nanomaterials into different products, for example, nano-pesticides (El Gohary et al. 2021), nano-fertilizers (Ahmadian et al. 2021) and nano-sensors (Verma et al. 2021). Nanomaterials can be altered through modifications in the synthesis route and manipulation at different scales, including the atomic, molecular, macromolecular and micromolecular scales; therefore, they are often referred to as engineered nanoparticles (ENPs) or just nanoparticles (NPs) (Tahir et al. 2020a, 2020b; Afsheen et al. 2020). In a more advanced manner, environmentalists and other members of the scientific community in the present era have been inculcating novel materials that have dimensions less than 100 nm and are referred to as nanomaterial inside biological methods for effective remediation. In this collaborative approach, nanomaterials in close association with hyper-accumulator plants detoxify soil matrices in an eco-friendly manner; thus, this concept of "nano-phytoremediation" is gaining further interest among the scientific community (Romeh and Saber 2020). Phytoremediation is an effective phyto-technology and an extension of bioremediation that is highly appreciated due to its cost effectiveness, facile operation, wide-scale adaptability and aesthetic appeal, and it has also been publicly accepted on several occasions (Dal Ferro et al. 2021). Therefore, the research on phytoremediation has been ongoing, resulting in further advancements aimed at the detoxification of soil matrices from various toxic and persistent pollutants, such as chlorinated solvents, heavy metals, explosive materials, agrochemicals, radionuclides and crude oil (Moradi et al. 2021; Yan et al. 2021). The success stories based on the utilization of the phytoremediation technique through field trials have further encouraged the scientific community to adopt this method on a global scale, and numerous such cases have been reported (Amjad et al. 2021).

The utilization of ENMs and NPs in association with phytoremediation is seen as an effective solution for the detoxification of the environment. However, some critical analyses also consider this advancement in terms of impacting the overall health of soils in different indigenous ecosystems (Shrivastava et al. 2019). There has been an increment in the concerns over the health implications of these ENMs and NPs if utilized for pedospheric detoxification, inclusive of the impacts on the sub-surficial environment and ecological settings of soils (O'Connor and Hou 2018). However, the greater efficiencies obtained for such nano-phytoremediation approaches make them worthy of utilization after proper optimization for the concerns raised. Nano-phytoremediation can be successfully engineered for optimum output without impacting plant germination, roots/shoots development, microbial species growth, maturation and metabolic activities and non-target species, and the contaminants do not obtain access to the food chain (Chen et al. 2018). Therefore, during field application processes, such considerations need to be taken into account to derive the maximum benefit from nano-phytoremediation. The potential of nano-phytoremediation for the pedospheric matrix cannot be overlooked and thus needs to be studied comprehensively by considering a variety of NPs and plants (Liu et al. 2018). Although some research has been conducted using nano-phytoremediation for the detoxification of the pedospheric matrix, to date, no comprehensive studies have been published that analyze the advancements made in this field. Therefore, the current chapter considers this knowledge gap and provides a comprehensive review and summarization of the potential of nano-phytoremediation in the detoxification of the pedospheric zone. The chapter specifically aims at the potential of phytoremediation for environmental cleanup, the interplay of different mechanisms in phytoremediation and nano-phytoremediation and the challenges and future prospects associated with the nano-phytoremediation carried out with different types of NPs and plants.

10.2 METHODS

10.2.1 Search Strategy

This chapter is representative of the exploration of nanotechnology in the detoxification of the pedospheric region in collaboration with phytoremediation. Data regarding the subject matter have been obtained by a meticulous assortment of the limited materials from specific articles. The data were obtained from various electronic platforms and databases. Such databases are inclusive of Science Direct, Google Scholar, Web of Science, Cochrane, Embase, ProQuest and PubMed. Different types of articles were selected that encompass full-length research articles, reviews, perspectives, chapters of different books from renowned publishers and reports produced by different institutes. The materials studied for the current chapter were produced by various publishers, for example, Taylor & Francis, Springer, Jstore, Elsevier, Springer Nature, the American Chemical Society, etc. Generally, publications produced from 2000–2021 were chosen for comprehensive studies of the subject matter, but those between 2015 and 2021 were specifically emphasized. The searching approach was varied by the insertion of different terms into electronic search engines, such as "Soil pollution", "Soil remediation", "Nanotechnology based remediation", "Phytoremediation", "Hyper-accumulators", "Nano-phytoremediation" and "NPs in soil remediation".

10.3 SUSTAINABLE PHYTOREMEDIATION AND ECOLOGICAL DETOXIFICATION

In the contemporary world, contaminated pedospheric and hydrospheric regions are being remediated with a variety of physicochemical methods that are aimed at the extraction of pollutants. More specifically, groundwater contamination is often remediated through precipitation–filtration (Kasaishi et al. 2021), ion exchange (Gao et al. 2021), flocculation (Naruka et al. 2021), reverse–osmosis (Lee et al. 2021), coagulation (Campinas et al. 2021), electro-dialysis (Ramasamy et al.

2021), carbon adsorption (Park et al. 2021), chelate-flushing (Omran 2021) and oxidation–reduction. In the case of soils, the extraction of contaminants is often performed through soil flushing, solidification, electro-kinetic treatment, stabilization and vitrification (Zhang et al. 2020; Vidal et al. 2021). However, such methods have been known for their negative aspects of higher costs and their impacts on agricultural soils, which alter the soil structure and make it unsuitable for future agricultural activities (Ye et al. 2017). Such reservations have given rise to the advanced obstructive protocols and demanded fabrication of sustainable methods that could detoxify the environment without causing permanent changes in the environmental integrity of different areas.

Environmental quality maintenance achieved via phytoremediation signifies a sustainable and cleaner approach. Ecological detoxification using phytoremediation is in an emerging phase and needs further studies in an optimized manner for practical adoption in real life (Ahila et al. 2020). However, the concept of sustainability and sustainable development needs to be closely considered if we want future and present generations to be benefitted from sustainability. In this regard, sustainable phytoremediation can be easily accomplished through the utilization of the plants that have the potential of natural colonization in terms of the unprompted development of vegetation. The pollutants and contaminants present in pedospheric matrices are well known for their persistence on a global scale (Brereton et al. 2020). In most cases, they are also resistant to any kind of degradation or destruction. In fact, they undergo the phenomenon of bioaccumulation via the food chain, which leads to negative impacts on the health of different species including humans and other forms of biodiversity.

Research on phytoremediation with different types of plant species has expressed a greater candidacy of these plants in terms of environmental friendliness, economic viability and facileness of use. Especially naturally colonizing plants can further add to the sustainability aspect in addition to these features since they are mostly perennial, economically viable and not easily palatable. Furthermore, such plants are also naturally equipped with the potential of growing even if ecological succession takes place (Chiudioni et al. 2021). Many studies also support the fact that any plant species under consideration for phytoremediation needs to be beneficial from the point of view of society's welfare and goods' productivity. Phytoremediation is also considered an auspicious phytotechnology due to its aesthetic advantages and long-term beneficial implications. It represents an in-situ detoxification system dependent upon solar irradiance and the plant species (Yadav et al. 2018). In-situ detoxification via phytoremediation is a comparatively simpler technique marked by meagre labor intensiveness (Jaskulak et al. 2019).

10.4 CATEGORIZATION OF PHYTOREMEDIATION

Phytoremediation can be used to clean the environment either in in-situ or ex-situ mode. In most cases, in-situ phytoremediation is preferred over ex-situ phytoremediation. In the in-situ mode, environmental remediation is performed on site, and no excavation is conducted for the transferal of the contaminated samples to other sites or a relevant laboratory. Therefore, this method is often preferable since it remediates the soil without causing further disturbance and labor exertion. More specifically, this method is often associated with the transfiguration or extermination of pollutants and contaminants; it is also based on the segregation or isolation of the pollutants or contaminants from larger clumps of soil. Furthermore, in-situ phytoremediation is often known for the immobilization of contaminants to alleviate their bioavailability (Nejad et al. 2018). In contrast, ex-situ phytoremediation is associated with the displacement of polluted soils for the purpose of treatment away from the locality of contamination; upon treatment, the treated soils are returned to the site of excavation for the purpose of reversion of the resorted site. In this type of remediation, the efficiency of the removal of contaminants is based on several factors such as the excavation, transferal and recreation of the resorted site. Furthermore, the contaminated soil is transferred to a laboratory and several procedures are adopted for detoxification, for example, immobilization, stabilization, incineration and solidification. Phytoremediation has been classified into different categories depending upon

TABLE 10.1
Different Classes of Phytoremediation Technology Using Different Kinds of Plant Species for the Decontamination of Pedospheric Matrices

Phytoremediation type	Plant species	Common name	Image	Reference
Phyto-stimulation	*Morus alba L.*	Mulberry stakes		Borges et al. (2016)
	Zea mays L.	Maize		Hajabbasi et al. (2009)
	Musa paradisiaca	Banana		Kumar Savani et al. (2021)
	Sagittaria platyphylla	Arrowhead		Kwong et al. (2018)
	Millettia pinnata	Panigrahi		Kumar et al. (2017)

(Continued)

TABLE 10.1 *(Continued)*
Different Classes of Phytoremediation Technology Using Different Kinds of Plant Species for the Decontamination of Pedospheric Matrices

Phytoremediation type	Plant species	Common name	Image	Reference
Phyto-extraction or phyto-accumulation	*Linum usitatissimum*	Flaxseed		Amna et al. (2015)
	Salicornia europaea	Glasswort	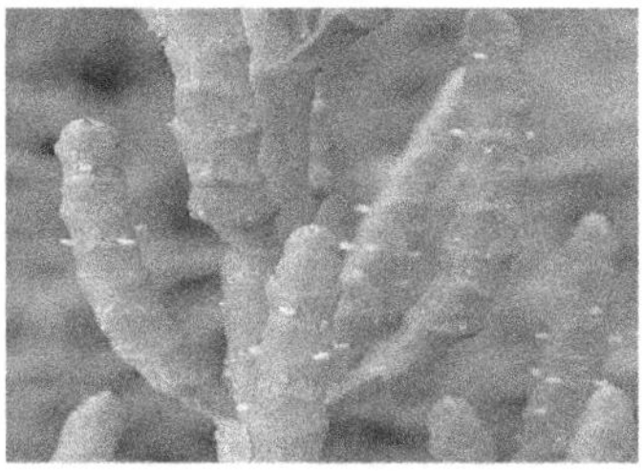	Khanlarian et al. (2020)
	Sorghum bicolor	Sorghum		Bacaha et al. (2015)
	Cynodon nlemfuensis	African Bermuda-grass		Madyiwa et al. (2003)
	Brassica juncea arawali	Indian mustard		Kaur et al. (2013)

(Continued)

TABLE 10.1 *(Continued)*
Different Classes of Phytoremediation Technology Using Different Kinds of Plant Species for the Decontamination of Pedospheric Matrices

Phytoremediation type	Plant species	Common name	Image	Reference
Phyto-degradation	*Phragmites australis*	Common reed		He et al. (2017)
	Azolla Filiculoides	Mosquito fern		Zazouli et al. (2014)
	Erythrina crista-galli	Coral tree		de Farias et al. (2009)
	Blumea malcolmii Hook.	Malcom's blumea		Kagalkar et al. (2011)

(Continued)

TABLE 10.1 *(Continued)*
Different Classes of Phytoremediation Technology Using Different Kinds of Plant Species for the Decontamination of Pedospheric Matrices

Phytoremediation type	Plant species	Common name	Image	Reference
Phyto-degradation	*Eichhornia crassipes*	Water hyacinth		Muthunarayanan et al. (2011)
Phyto-rhizofiltration	*Carex pendula*	Hanging sedge		Yadav et al. (2011)
	Helianthus annuus and *Phaseolus vulgaris*	Sunflower and kidney bean	 	Lee and Yang (2010)
	Plectranthus amboinicus	Indian borage		Ignatius et al. (2014)

(Continued)

TABLE 10.1 *(Continued)*
Different Classes of Phytoremediation Technology Using Different Kinds of Plant Species for the Decontamination of Pedospheric Matrices

Phytoremediation type	Plant species	Common name	Image	Reference
Phyto-rhizofiltration	*Pistia stratiotes*	Water lettuce		Galal et al. (2018)
	Brassica juncea and *Chenopodium amaranticolor*	Brown mustard and Goosefoot		Eapen et al. (2003)

the type of procedure, choice of plants and method of pollutant uptake as shown in Table 10.1. These categories are broadly classified as follows.

- *Phyto-stimulation*: Phyto-stimulation is a type of phytoremediation in which various kinds of enzymes derived from microbial species, including bacteria, fungi, etc., are stimulated inside the rhizosphere zone, which leads to the bioremediation of the pollutants and contaminants. Such enzymes are actually released from the root exudates (Kumar et al. 2020).
- *Phyto-extraction or phyto-accumulation*: In this type of phytoremediation, pollutant and contaminant species are eliminated by using different hyper-accumulator species that possess an inherent capacity of accumulating these pollutants inside them when grown in a contaminated pedospheric region. Such plants are actually known for storing these contaminants inside their shoots. The shoots of these phyto-extracted plants can then be used upon harvesting (Roshanfar et al. 2020).
- *Phyto-volatilization*: Different hyper-accumulator plants take up chemicals from the environment and deposit them in their leaves. Chemicals or contaminants that are often being phyto-volatilized possess a volatile nature and are produced by different anthropogenic activates (Osama et al. 2021).
- *Phyto-degradation*: During the phyto-degradation process, the obnoxious chemicals and pollutants are broken down to less-toxic daughter products by the action of some specialized enzymes. This technique is often used in the case of persistent organic pollutants (POPs), namely, agrochemicals, trichloroethylene, etc. Another interesting feature associated with phyto-degradation is that its occurrence can either be internal or external

depending upon the type of plant species since some plants are also specialized in secreting external enzymes (Mohanty et al. 2021).

- *Phyto-rhizofiltration*: Phyto-rhizofiltration is another advanced form of phytoremediation used for the detoxification of pedospheric matrices and protection of groundwater reservoirs from receiving different pollutants and contaminants produced due to anthropogenic activities. Phyto-rhizofiltration is accomplished with the help of plants that bear an extensive system of roots, which can easily carry out the filtration of the contaminants through different mechanisms, such as adsorption and absorption (Castro-Castellon et al. 2021).
- *Phyto-transformation*: In phyto-transformation, complex and large-sized pollutants and contaminants are converted to smaller products through degradation accomplished by hyper-accumulator plants that have an inherent potential of doing this. Reaction products released in such reactions are then taken up by the plants and incorporated inside plant tissues.
- *Phyto-stabilization*: Phyto-stabilization stops all kinds of movement of organic and inorganic pollutants and contaminants by stabilizing them inside pedospheric matrices. Prevention of the movement of such contaminants in a timely manner is also known for stopping the entrance of these pollutants into food chains and disrupting consumer health.

10.5 NANO-ENABLED PHYTOREMEDIATION

Nano-remediation is a new technology that is in an emerging stage and is associated with the implementation of nanoscale materials having size ranges less than 100 nm; it is aimed at the effective detoxification of the pedospheric and hydrospheric zones. A great deal of nanomaterials are being synthesized and employed for the environmental remediation of different pollutants, for example, dyes, pesticides, pharmaceuticals and solvents (Jaffri and Ahmad 2018a; Jaffri and Ahmad 2018b). Table 10.2 expresses an account of nano-enabled phytoremediation. Zero-valent iron is known for the effective and rapid removal of dissolved heavy metals through immobilization. Previous studies support the remediation of heavy metals via adsorptive mechanisms over the surficial region of iron NPs (update). Another site of adsorption for heavy metals and other contaminants is created by the process of corrosion, in which the products act as an attachment site for pollutants. The recent era has seen an immense augmentation in the application of NPs in different sectors, such as the cosmetics, agriculture and biomedicine industries (Iqbal et al. 2019; Khan et al. 2019b). A variety of NPs can be fabricated through different physicochemical and biological methods, and they possess different size ranges, geometries and functionalities depending upon the type of application in which they are intended for utilization. NPs have advantageous features compared to other conventional materials in terms of higher surficial activity, greater surficial reaction sites, remarkable catalytic performance and exceptional opto-magnetic characteristics (Ijaz et al. 2020c, 2020a, 2020b).

The utilization of NPs in agriculture has extensively emerged, giving rise to the field of agri-nanotechnology. Previous studies have discussed the implications associated with the use of such NPs in agriculture. For instance, in a study by Hao et al. (2020), carbon-based nanomaterials were used, and the results were indicative of the sensitivity of rice fungal endophytes towards a meagre dose of 10 mg/L. In the case of exposing peanut to the minimum dose of 50 mg/kg of silver (Ag) NPs, negative impacts were exerted on the biomass and quality of peanuts (Rui et al. 2017). The impacts of such NPs have been studied on plants and have also been investigated for other fau3nal life forms. For example, nickel oxide (NiO) NPs in lower doses of 5 and 50 mg/kg were found to not affect the existence, reproduction and development rate of the test animals, namely, adult earthworms; however, when the dose was augmented to 200 and 500 mg/kg, comparatively negative impacts were exerted on the physiological and biochemical makeup of the earthworms Adeel et al. 2019. Nevertheless, other studies also support the effective results in the case of agriculture and the environment obtained as a result of NP utilization. Applications of NPs are associated with a considerable improvement in the germination of seeds, photosynthetic rate, resistance towards oxidative

TABLE 10.2
Utilization of Different Types of Nanomaterials for Targeting Various Kinds of Pollutants by Coupling Them with Hyper-Accumulator Plants in Nanotools-Coupled Phytoremediation (NCP)

Nanomaterial employed	Target pollutant	Plant specie	Common name	References
Nano-silicon	Cd	*Oryza sativa*	Rice	Chen et al. (2018)
Nano-silicon	Cd	*Oryza sativa*	Rice	Wang et al. (2015)
ZnO NPs	Cd	*Triticum aestivum*	Wheat	Rizwan et al. (2019b)
Nano-silicon	As	*Oryza sativa*	Rice	Liu et al. (2014)
TiO_2 and Si NPs	Cd	*Oryza sativa*	Rice	Rizwan et al. (2019c)
Mercapto-functionalized nano-silica (MPTS/nano-silica)	Cd	*Triticum aestivum*	Wheat	Wang et al. (2020)
Nano-silica	Cd	*Triticum aestivum* and *Zea mays*	Wheat and corn	Wang et al. (2020a)
ZnO NPs	Cd	*Triticum aestivum*	Wheat	Bashir et al. (2020)
TiO_2 and SiO_2 NPs	Cd	*Triticum aestivum*	Wheat	Dai et al. (2019)
Nanoscale zero-valent iron (nZVI)	Pentachlorophenol	*Oryza sativa*	Rice	Liu et al. (2021)
Fe NPs	Cd	*Oryza sativa*	Rice	Rizwan et al. (2019a)
ZnO NPs	Cd	*Triticum aestivum*	Wheat	Khan et al. (2019a)
ZnO and Fe NPs	Cd	*Triticum aestivum*	Wheat	Rizwan et al. (2019d)
Fe_2O_3 NPs	NaCl	*Trachyspermum ammi L.*	Ajowan	Abdoli et al. (2020)
ZnO NPs	Cd	*Triticum aestivum*	Wheat	Hussain et al. (2018)
Fe NPs	Cd	*Triticum aestivum*	Wheat	Adrees et al. (2020)
Fe NPs	Cd	*Triticum aestivum*	Wheat	Hussain et al. (2019a)
nZVI	Cd	*Oryza sativa*	Rice	Guha et al. (2020)
CuO	As	*Oryza sativa japonica*	Rice	Liu et al. (2018)
Fe_3O_4	Pb, Zn, Cd and Cu	*Triticum aestivum*	Wheat	Konate et al. (2017)
Citrate-coated Fe_3O_4 NPs	Cr	*Triticum aestivum*	Wheat	López-Luna et al. (2016)
ZnO NPs	Cd	*Oryza sativa*	Rice	Ali et al. (2019b)
Se and Si NPs	Cd and Pb	*Oryza sativa*	Rice	Hussain et al. (2020)
Fe_3O_4 NPs	Cd and Na	*Oryza sativa*	Rice	Sebastian et al. (2019)
Si NPs	Cd	*Triticum aestivum*	Wheat	Hussain et al. (2019b)
ZnO NPs	Cd	*Oryza sativa*	Rice	Li et al. (2020a)
Fe and hydrogel NPs	Cd	*Oryza sativa*	Rice	Ahmed et al. (2021)
ZnO NPs	Cd and Pb	*Leucaena leucocephala*	White leadtree	Venkatachalam et al. (2017)
ZnO NPs	NaCl	*Lupinus termis*	Lupine	Latef et al. (2017)
CeO_2 NPs	Cd	*Glycine max*	Soyabean	Rossi et al. (2018)

stress, maturation and development of the rhizome and overall yield and quality of crops (Kah et al. 2019; Li et al. 2016). Applications of NPs are far ranging. For instance, they are employed on crops for fertility augmentation in the form of nanofertilizers (Li et al. 2020) and for exterminating pests in the form of nanopesticides (Adeel et al. 2020). Such nanofertilizers and nanopesticides are marked by the easier absorption of plants and comparatively slower emission into the target ecological setting than that of conventionally employed fertilizers. Contrariwise, other NPs, including CeO_2, TiO_2 and Mn_3O_4, are known for their enhancement of the potential of antioxidant enzymes. Such enzymes then subsequently alleviate the aggregation of the reactive oxygen species (ROS) in different types of plants, giving rise to a considerable reduction in the phyto-stress and improvement in the overall crop yield and quality (Usman et al. 2020). In a recent study by Zhou et al. (2021), a comprehensive study was conducted that reviewed the literature addressing nano-enabled agriculture and discussing the mechanisms of the uptake and transport of different pollutants by plants. The analysis was supportive of the remarkable performance of NPs on the development of plants under the stress of different pollutants and heavy metals. Furthermore, the employment of NPs in collaboration with phytoremediation needs systematic research to achieve sustainable nano-enabled agriculture (Zhou et al. 2021).

Phytoremediation performed with the assistance of various types of nanomaterials is actually based on three major components, specifically, plant species specialized in the hyper-accumulation of the pollutants and heavy metals, target pollutants and contaminants and nanomaterials. The employment of NPs of different metals and other substances has a dual functionality of the direct action on the pollutants and the plant species through the development of the interaction. In addition, these NPs serve to develop an indirect association with the plants and pollutants, thus causing an indirect impact over the final detoxification efficiency (Jaffri and Ahmad 2018c; Ahmad and Jaffri 2018b). In this type of nano-phytoremediation, environmental decontamination is conducted by using the rhizospheric region of the plant that contains beneficial microbial species, inclusive of bacteria, that are specialized in plant growth-promoting microorganisms (PGPM), nitrogen-fixation and fungal species, namely, arbuscular mycorrhizal fungi (AMF). Nano-enabled phytoremediation is performed to meet the challenges of the alleviated agri-productivity caused by some biotic and abiotic stresses. In this case, biotic stresses are inclusive of the phyto-pathogens and pests that cause different diseases, and abiotic stresses might be exerted due to the pedospheric matrix contamination caused by heavy metals and other types of pollutants (Shang et al. 2019). Nano-enabled phytoremediation is advantageous over other detoxification methods for the remediation of toxic contaminants and heavy metals. Some of the famous nanomaterials in this regard are nanocarbon black, nanohydroxyapatite modified with biochar, Ag, zinc (Zn), Ni, tin oxide (SnO_2), titanium dioxide (TiO_2) and nano zero-valent iron (nZVI) (Yadav et al. 2017; Mousavi et al. 2018; Zhang et al. 2018; Romeh and Saber 2020). Variable results are obtained for each type of nanomaterial when coupled with phytoremediation, depending upon the plant type, nanomaterial size, morphology, etc. In fact, the size tuning of these NPs can be performed since this factor has a real influence on the immobilization of the contaminants and heavy metals followed by the restoration of the affected soils (Shrivastava et al. 2019).

Among the different types of nanomaterials studied for phytoremediation coupling for soil cleanup, the role of TiO_2 NPs has been quite influential (Waani et al. 2021). The treatment of different plants with TiO_2 NPs is expressive of the significant improvement in the performance of floral species in terms of their growth, maturation, physiological aspects and agricultural yield. The improvement in this regard has been achieved under both normal and stressful circumstances and is used for the growth of the target plants. In a study conducted on spinach plants being treated with TiO_2 NPs (Gordillo-Delgado et al. 2020), considerable improvement was observed in the dry weight of spinach, biosynthesis extent of the chlorophyll, overall rate of photosynthesis and activities of different enzymes. Such improvement subsequently led to the alleviation in the deposition of the ROS and other free radicals, augmented the extent of the enzymatic activities, for example, malionaldehyde and antioxidative actions, and regulated the chloroplast morphology and stability of the membranous structure in the target spinach plants (Gordillo-Delgado et al. 2020). Another interesting feature associated with the utilization of TiO_2 NP-based phytoremediation is that it is known not only for alleviating the toxic impacts of contaminants and heavy metals and other disturbances in plants, such as stress inflicted due to UV-B, but also restricting the toxicity related to heavy metals and accelerating the rate of photosynthesis and growth/maturation in plants. More specifically, TiO_2 NPs confine the toxicity caused by Cd and augment the growth of plants in addition to enhancing the content of the water and chlorophyll in soyabean (El-Gazzar et al. 2020).

NPs and other forms of nanomaterials are significant in the transportation of heavy metals and other contaminants inside floral samples; particularly, the transferal of these pollutants inside the root cell cytoplasm of the plants is highly impacted by nanomaterials. Plants coupled with NPs often exhibit varied responses in terms of the up taking of the heavy metals and the interplay of the rhizosphere's chemico-biological reactions taking place in a complex manner. However, advanced research is needed to gain a complete and mechanistic understanding of the modified heavy metals and contaminants inside plant tissues coupled with nanomaterials. Many concurrent procedures are taking place in ecological compartments, especially in the pedospheric matrices at different profiles, which affects the association created between NPs and the co-existence of the ecological contaminants inside a nano-enabled phytoremediation system. Ma and Wang (2018) emphasized the investigation of relevant concepts of nano-phytoremediation since it is a highly understudied domain despite possessing remarkable potential. For instance, the concepts addressing different factors such as root exudates emitting phyto-chelating substances in the rhizospheric region and other types of enzymes and hormones need to be studied to comprehend the types of interactions developed with NPs and the consequent remediation of the environment. Nevertheless, the obstacles in this regard cannot be overlooked in terms of the adoption and implementation of nano-phytoremediation in pedospheric detoxification in a sustainable manner. One major objection and concern raised by conventional environmentalists is the inherent toxicity associated with the utilization of NPs synthesized via the green route. However, this problem can be easily solved through the adoption of the facile, green and sustainable method of NP synthesis based on the utilization of reducing cum stabilizing agents from nature, specifically, plant phyto-chemicals. Through the utilization of biogenic NPs, including, nZVI (Martins et al. 2017), Ag (Bafana et al. 2018) and Au (Pati et al. 2014), ecologically friendly and economical nanotools can be used to detoxify the pedospheric matrices by coupling them with phytoremediation technology (Ahmad and Jaffri 2018a; Jaffri and Ahmad 2020a; Jaffri et al. 2020).

10.6 CHALLENGES AND RECOMMENDATIONS

The utilization of nanotools for coupling with phytoremediation is an emerging concept that has been gaining attention over time and as the domains of nanotechnology and bioremedial technology are revolutionizing. Despite its auspiciousness in its different aspects of economic benefits, such as its facileness, simplicity and eco-friendly quality, nano-phytoremediation has faced a great deal of criticism and challenges in terms of its practical adoption at the field level. In this regard, the aspect that hinders the practical adoption of nano-enabled phytoremediation is the noxious toxicity associated with the use of NPs and other nanomaterials. Since the pedospheric matrices represent an ecological niche of several species that is inclusive of different floral, faunal and microbial communities and because NPs are mostly known for their toxic impacts, their use is always therefore dealt with from a critical point of view. However, because of the practical results associated with nano-phytoremediation in terms of environmental decontamination, the scientific community is forced to explore the risks associated with NPs, work on improving them and then use them in nano-phytoremediation. Therefore, the approach that needs to be adopted in such a scenario must be dually addressed in terms of the investigation of the ecological risk associated with the utilization of NPs. Moreover, the process of phytoremediation and NP coupling needs to be regularized so that the maximum advantage is accomplished with minimum or no risks at all.

Currently, the concept of nano-enabled phytoremediation is still at the inception and growth stage, and it is quite fortunate that several studies have expressed positive results; however, we need long-term experience in this domain through rigorous studies for further comprehension. Additional investigations in this regard will enable us to easily evaluate the causes and conditions that lead to the failure or minimum performance of nanotools coupled with phytoremediation. In this regard, the research areas that need immediate attention include the response mechanism of different plants towards different NPs, the variety of pollutants and contaminants, the types of pedospheric samples and the climatic factors involved in the nano-phytoremediation conglomeration system. Furthermore, the results obtained for nanotools-coupled phytoremediation can also be enhanced by inculcating different approaches, for example, agronomic managerial strategies, chemical additives incorporation, modification via microbial species through inoculation in the rhizosphere and genetic engineering.

10.7 CONCLUSION AND FUTURE PERSPECTIVES

Environmental remediation has become an integral part of the Sustainable Development Goals (SDGs) in the current era of an increased and unregulated pattern of urbanization and industrialization. The decontamination of pedospheric and hydrospheric matrices has occurred with a variety of physicochemical techniques. More specifically, groundwater contamination is often remediated through precipitation–filtration, ion exchange, flocculation, reverse–osmosis, coagulation, electro-dialysis, carbon adsorption, chelate-flushing and oxidation–reduction. Meanwhile, in the case of soils, the extraction of contaminants is often performed through soil flushing, solidification, electro-kinetic treatment, stabilization and vitrification. However, such methods have been known for their negative aspects of higher costs and their impacts on agricultural soils, as they alter the soil structure and make it unsuitable for future agricultural activities. Nanotools coupled with the green phyto-technology of phytoremediation have been seen as an attractive and advantageous methodology marked by eco-friendliness, cheaper costs and simpler operation at the field level. Hyper-accumulator plants, known for their potential of taking up the contaminants and heavy metals, are especially emphasized in this type of cleanup process. Based on an extensive survey and comprehension of the literature analyzed in this regard, the current chapter points towards the need for comprehending the basic mechanism behind the transferal of contaminants inside hyper-accumulator plants facilitated by the utilization of nanotools so that these procedures can be maximized and greater benefits can be accomplished from nano-enabled phytoremediation. There has been increasing interest in this cross-disciplinary domain due to its auspicious and future candidacy for sustainable environmental detoxification.

CONFLICTS OF INTEREST

None

FUNDING

None

REFERENCES

Abdoli, S., Ghassemi-Golezani, K., and Alizadeh-Salteh, S. 2020. Responses of ajowan (*Trachyspermum ammi* L.) to exogenous salicylic acid and iron oxide nanoparticles under salt stress. *Envir. Sci. Poll. Res.* 27:36939–36953.

Adeel, M., Farooq, T., White, J., Hao, Y., He, Z., and Rui, Y. 2020. Carbon-based nanomaterials suppress Tobacco Mosaic Virus (TMV) infection and induce resistance in *Nicotiana benthamian*. *J. Hazard. Mater.*:1–10.

Adeel, M., Ma, C., Ullah, S., Rizwan, M., Hao, Y., Chen, C., Jilani, G., Shakoor, N., Li, M., and Wang, L. 2019. Exposure to nickel oxide nanoparticles insinuates physiological, ultrastructural and oxidative damage: A life cycle study on *Eisenia fetida*. *Environ. Pollut.* 1:254.

Adrees, M., Khan, Z. S., Ali, S., Hafeez, M., Khalid, S., Ur Rehman, M. Z., and Rizwan, M. 2020. Simultaneous mitigation of cadmium and drought stress in wheat by soil application of iron nanoparticles. *Chemosphere* 238:124681.

Afsheen, S., Naseer, H., Iqbal, T., Abrar, M., Bashir, A., and Ijaz, M. 2020. Synthesis and characterization of metal sulphide nanoparticles to investigate the effect of nanoparticles on germination of soybean and wheat seeds. *Mat. Chem. Phys.* 252:123216.

Ahila, K. G., Ravindran, B., Muthunarayanan, V., Nguyen, D. D., Nguyen, X. C., Chang, S. W., and Thamaraiselvi, C. 2020. Phytoremediation potential of freshwater macrophytes for treating dye-containing wastewater. *Sustain.* 13:329.

Ahmad, K. S., and Jaffri, S. B. 2018a. Phytosynthetic Ag doped ZnO nanoparticles: Semiconducting green remediators: Photocatalytic and antimicrobial potential of green nanoparticles. *Open Chem.* 16:556–570.

Ahmad, K. S., and Jaffri, S. B. 2018b. Carpogenic ZnO nanoparticles: Amplified nanophotocatalytic and antimicrobial action. *IET Nanobiotech.* 13:150–159.

Ahmad, K. S., Talat, M., Jaffri, S. B., and Shaheen, N. 2020. Innovatory role of nanomaterials as bio-tools for treatment of cancer. *Rev. Inorg. Chem.* 1:1–25.

Ahmadian, K., Jalilian, J., and Pirzad, A. 2021. Nano-fertilizers improved drought tolerance in wheat under deficit irrigation. *Agric. Water Manage.* 244:106544.

Ahmed, T., Noman, M., Manzoor, N., Shahid, M., Abdullah, M., Ali, L., and Li, B. 2021. Nanoparticle-based amelioration of drought stress and cadmium toxicity in rice via triggering the stress responsive genetic mechanisms and nutrient acquisition. *Ecotoxic. Envir. Safety.* 209:111829.

Ali, M., Jaffri, S. B., Ahmad, K. S., and Iqbal, S. 2019a. Sorptive Interactions of Fungicidal 2-(4'-Thiazolyl) Benzimidazole with Soils of Divergent Physicochemical Composition. *J. Int. Econ Envir. Geol.* 10:97–104.

Ali, S., Rizwan, M., Noureen, S., Anwar, S., Ali, B., Naveed, M., and Ahmad, P. 2019b. Combined use of biochar and zinc oxide nanoparticle foliar spray improved the plant growth and decreased the cadmium accumulation in rice (*Oryza sativa* L.) plant. *Envir. Sci. Poll. Res.* 26:11288–11299.

Amjad, I., Javaid, M., Ikhlaq, K., Gul, S., Jaffri, S. B., and Ahmad, K. S., 2019. Adsorption-desorption mechanism of synthesized benzimidazole based fungicide 2-(3'-Pyridyl) on selected soil minerals. *J. Int. Econ Envir. Geol.* 10:38–44.

Amjad, M., Iqbal, M. M., Abbas, G., Farooq, A. B. U., Naeem, M. A., Imran, M., and Jacobsen, S. E. 2021. Assessment of cadmium and lead tolerance potential of quinoa (*Chenopodium quinoa* Willd) and its implications for phytoremediation and human health. *Envir. Geochem. Health*:1–14.

Amna, Masood, S., Syed, J. H., Munis, M. F. H., and Chaudhary, H. J. 2015. Phyto-extraction of Nickel by Linum usitatissimum in Association with Glomus intraradices. *J. Int. Phytorem.* 17:981–987.

Bacaha, N., Shamas, R., Bakht, J., Rafi, A., and Farhatullah, G. A. 2015. Effect of heavy metal and EDTA application on plant growth and phyto-extraction potential of Sorghum (*Sorghum bicolor*). *J. Pak. Bot.* 47:1679–1684.

Bafana, A., Kumar, S. V., Temizel-Sekeryan, S., Dahoumane, S. A., Haselbach, L., and Jeffryes, C. S. 2018. Evaluating microwave-synthesized silver nanoparticles from silver nitrate with life cycle assessment techniques. *Sci. Total Envir.* 636:936–943.

Bashir, A., Rizwan, M., Ali, S., Adrees, M., ur Rehman, M. Z., and Qayyum, M. F. 2020. Effect of composted organic amendments and zinc oxide nanoparticles on growth and cadmium accumulation by wheat; a life cycle study. *Envir. Sci. Poll. Res.* 27:23926–23936.

Borges, J. A., Leon, M., Marturet, E., and Barrios, M. 2016. Phyto-stimulation in mulberry stakes (Morus alba L.) by plant extracts. *Bioagro.* 28(3):215–219.

Brereton, N. J. B., Gonzalez, E., Desjardins, D., Labrecque, M., and Pitre, F. E. 2020. Co-cropping with three phytoremediation crops influences rhizosphere microbiome community in contaminated soil. *Sci. Total Envir.* 711:135067.

Campinas, M., Viegas, R., Coelho, R., Lucas, H., and Rosa, M. J. 2021. Adsorption/coagulation/ceramic microfiltration for treating challenging waters for drinking water production. *Membranes.* 11:91.

Castro-Castellon, A. T., Hughes, J. M. R., Read, D. S., Azimi, Y., Chipps, M. J., and Hankins, N. P. 2021. The role of rhizofiltration and allelopathy on the removal of cyanobacteria in a continuous flow system. *Envir. Sci. Poll. Res.* 1–11.

Chen, R., Zhang, C., Zhao, Y., Huang, Y., and Liu, Z. 2018. Foliar application with nano-silicon reduced cadmium accumulation in grains by inhibiting cadmium translocation in rice plants. *Envir. Sci. Poll. Res.* 25:2361–2368.

Chiudioni, F., Marcheggiani, S., Puccinelli, C., and Mancini, L. 2021. Interaction between bacterial enteric pathogens and aquatic macrophytes. Can *Salmonella* be internalized in the plants used in phytoremediation processes?. *J. Int. Phytorem.* 23:18–25.

Cittrarasu, V., Kaliannan, D., Dharman, K., Maluventhen, V., Easwaran, M., Liu, W. C., and Arumugam, M. 2021. Green synthesis of selenium nanoparticles mediated from *Ceropegia bulbosa* Roxb extract and its cytotoxicity, antimicrobial, mosquitocidal and photocatalytic activities. *Sci. Rep.* 11:1–15.

Dai, C., Shen, H., Duan, Y., Liu, S., Zhou, F., Wu, D., and Tu, Y. J. 2019. TiO_2 and SiO_2 nanoparticles combined with surfactants mitigate the toxicity of Cd^{2+} to wheat seedlings. *Water Air Soil Poll.* 230:1–10.

Dal Ferro, N., De Mattia, C., Gandini, M. A., Maucieri, C., Stevanato, P., Squartini, A., and Borin, M. 2021. Green walls to treat kitchen greywater in urban areas: Performance from a pilot-scale experiment. *Sci. Total Envir.* 757:144189.

de Farias, V., Maranho, L. T., de Vasconcelos, E. C., da Silva Carvalho Filho, M. A., Lacerda, L. G., Azevedo, J. A. M., and Soccol, C. R. 2009. Phytodegradation potential of *Erythrina crista*-galli L., Fabaceae, in petroleum-contaminated soil. *Appl. Biochem. Biotech.* 157:10–22.

Eapen, S., Suseelan, K. N., Tivarekar, S., Kotwal, S. A., and Mitra, R. 2003. Potential for rhizofiltration of uranium using hairy root cultures of *Brassica juncea* and *Chenopodium amaranticolor. Envir. Res.* 91:127–133.

El-Gazzar, N., Almaary, K., Ismail, A., and Polizzi, G. 2020. Influence of *Funneliformis mosseae* enhanced with titanium dioxide nanoparticles (TiO_2NPs) on *Phaseolus vulgaris* L. under salinity stress. *PLoS ONE.* 15:e0235355.

El Gohary, E. G., Farag, S., El-Sayed, A., Khattab, R., and Mahmoud, D. 2021. Insecticidal activity and biochemical study of the clove oil (*Syzygium aromaticum*) nano-formulation on culex pipiens L.(Diptera: Culicidae). *J. Egypt. Aqua. Biol. Fish.* 25:227–239.

Galal, T. M., Eid, E. M., Dakhil, M. A., and Hassan, L. M. 2018. Bioaccumulation and rhizofiltration potential of *Pistia stratiotes* L. for mitigating water pollution in the Egyptian wetlands. *J. Int. Phytorem.* 20:440–447.

Gao, P., Cui, J., and Deng, Y. 2021. Direct regeneration of ion exchange resins with sulfate radical-based advanced oxidation for enabling a cyclic adsorption—regeneration treatment approach to aqueous perfluorooctanoic acid (PFOA). *J. Chem. Eng.* 405:126698.

Gordillo-Delgado, F., Zuluaga-Acosta, J., and Restrepo-Guerrero, G. 2020. Effect of the suspension of Ag-incorporated TiO_2 nanoparticles (Ag-TiO_2 NPs) on certain growth, physiology and phytotoxicity parameters in spinach seedlings. *PLoS ONE.* 15:e0244511.

Guha, T., Barman, S., Mukherjee, A., and Kundu, R. 2020. Nano-scale zero valent iron modulates Fe/Cd transporters and immobilizes soil Cd for production of Cd free rice. *Chemosphere.* 260:127533.

Hajabbasi, M. A., Zamani, J., Asadollahi, A., and Schulin, R. 2009. Using maize (*Zea mays* L.) and sewage sludge to remediate a petroleum contaminated calcareous soil. *Soil Sed. Contam.* 1–15.

Hao, Y., Ma, C., White, J. C., Adeel, M., Jiang, R., Zhao, Z., Rao, Y., Chen, G., Rui, Y., and Xing, B. 2020. Carbon-based nanomaterials alter the composition of the fungal endophyte community in rice (*Oryza sativa* L.). *Environ. Sci. Nano.* 7:2047–2060.

He, Y., Langenhoff, A. A., Sutton, N. B., Rijnaarts, H. H., Blokland, M. H., Chen, F., and Schröder, P. 2017. Metabolism of ibuprofen by *Phragmites australis*: Uptake and phytodegradation. *Envir. Sci. Tech.* 51: 4576–4584.

Hussain, A., Ali, S., Rizwan, M., ur Rehman, M. Z., Javed, M. R., Imran, M., and Nazir, R. 2018. Zinc oxide nanoparticles alter the wheat physiological response and reduce the cadmium uptake by plants. *Envir. Poll.* 242:1518–1526.

Hussain, A., Ali, S., Rizwan, M., ur Rehman, M. Z., Qayyum, M. F., Wang, H., and Rinklebe, J. 2019a. Responses of wheat (*Triticum aestivum*) plants grown in a Cd contaminated soil to the application of iron oxide nanoparticles. *Ecotoxic. Envir. Safety.* 173:156–164.

Hussain, A., Rizwan, M., Ali, Q., and Ali, S. 2019b. Seed priming with silicon nanoparticles improved the biomass and yield while reduced the oxidative stress and cadmium concentration in wheat grains. *Envir. Sci. Poll. Res.* 26:7579–7588.

Hussain, B., Lin, Q., Hamid, Y., Sanaullah, M., Di, L., Khan, M. B., and Yang, X. 2020. Foliage application of selenium and silicon nanoparticles alleviates Cd and Pb toxicity in rice (*Oryza sativa* L.). *Sci. Total Envir.* 712:136497.

Iftikhar, S., Saleem, M., Ahmad, K. S., and Jaffri, S. B. 2019. Synergistic mycoflora—natural farming mediated biofertilization and heavy metals decontamination of lithospheric compartment in a sustainable mode via *Helianthus annuus. J. Int. Envir. Sci. Tech.* 1–18.

Ignatius, A., Arunbabu, V., Neethu, J., and Ramasamy, E. V. 2014. Rhizofiltration of lead using an aromatic medicinal plant *Plectranthus amboinicus* cultured in a hydroponic nutrient film technique (NFT) system. *Envir. Sci. Poll. Res.* 21:13007–13016.

Ijaz, M., Aftab, M., Afsheen, S., and Iqbal, T. 2020a. Novel Au nano-grating for detection of water in various electrolytes. *Appl. Nanosci.* 10:4029–4036.

Ijaz, M., Zafar, M., Afsheen, S., and Iqbal, T. 2020b. A review on Ag-nanostructures for enhancement in shelf time of fruits. *J. Inorg. Organomet. Polymer. Mat.* 30:1475–1482.

Ijaz, M., Zafar, M., and Iqbal, T. 2020c. Green synthesis of silver nanoparticles by using various extracts: A review. *Inorg. Nano-Metal Chem.* 1–12.

Iqbal, T., Farooq, M., Afsheen, S., Abrar, M., Yousaf, M., and Ijaz, M. 2019. Cold plasma treatment and laser irradiation of *Triticum* spp. seeds for sterilization and germination. *J. Laser Appl.* 31:042013.

Iram, S., Ahmad, K. S., Noureen, S., and Jaffri, S. B. 2018. Utilization of wheat (*Triticum aestivum*) and Berseem (*Trifolium alexandrinum*) dry biomass for heavy metals biosorption. *Proceed. Pak. Acad. Sci. B.* 55:61–70.

Iram, S., Basri, R., Ahmad, K. S., and Jaffri, S. B. 2019. Mycological assisted phytoremediation enhancement of bioenergy crops *Zea mays* and *Helianthus annuus* in heavy metal contaminated lithospheric zone. *Soil Sed. Contam.* 28:411–430.

Iram, S., Iqbal, A., Ahmad, K. S., and Jaffri, S. B. 2020b. Congruously designed eco-curative integrated farming model designing and employment for sustainable encompassments. *Envir. Sci. Poll. Res.* 27:19543–19560.

Iram, S., Tariq, I., Ahmad, K. S., and Jaffri, S. B. 2020a. *Helianthus annuus* based biodiesel production from seed oil garnered from a phytoremediated terrain. *J. Int. Ambient Energy.* 1–9.

Ishtiaq, M., Iram, S., Ahmad, K. S., and Jaffri, S. B. 2020. Multi-functional bio-sorbents triggered sustainable detoxification of eco-contaminants besmirched hydrospheric swatches. *J. Int. Envir. Anal. Chem.* 1–16.

Jaffri, S. B., and Ahmad, K. S. 2018a. Neoteric environmental detoxification of organic pollutants and pathogenic microbes via green synthesized ZnO nanoparticles. *Envir. Tech.* 1–20.

Jaffri, S. B., and Ahmad, K. S. 2018b. *Prunus cerasifera* Ehrh. fabricated ZnO nano falcates and its photocatalytic and dose dependent in vitro bio-activity: Photodegradation and antimicrobial potential of biogenic ZnO nano falcates. *Open Chem.* 16:141–154.

Jaffri, S. B., and Ahmad, K. S. 2018c. Phytofunctionalized silver nanoparticles: Green biomaterial for biomedical and environmental applications. *Rev. Inorg. Chem.* 38:127–149.

Jaffri, S. B., and Ahmad, K. S. 2018d. Augmented photocatalytic, antibacterial and antifungal activity of prunosynthetic silver nanoparticles. *Artific. Cells Nanomed. Biotech.* 46:127–137.

Jaffri, S. B., and Ahmad, K. S. 2020a. Biomimetic detoxifier *Prunus cerasifera* Ehrh. silver nanoparticles: Innate green bullets for morbific pathogens and persistent pollutants. *Envir. Sci. Poll. Res.* 1–17.

Jaffri, S. B., and Ahmad, K. S. 2020b. Interfacial engineering revolutionizers: Perovskite nanocrystals and quantum dots accentuated performance enhancement in perovskite solar cells. *Critic. Rev. Solid State Mat. Sci.* 1–29.

Jaffri, S. B., Ahmad, K. S., Thebo, K. H., and Rehman, F. 2020. Sustainability consolidation via employment of biomimetic ecomaterials with an accentuated photo-catalytic potential: Emerging progressions. *Rev. Inorg. Chem.* 1:1–10.

Jaffri, S. B., Ahmad, K. S., ul Saba, M., Sher, G., and Sharif, M. 2019a. Sorptive interactions evaluation of benomyl metabolites mecarzole with the varyingly selected minerals. *J. Int. Econ. Envir. Geol.* 10: 50–54.

Jaffri, S. B., Nosheen, A., Iftikhar, S., and Ahmad, K. S. 2019b. Pedospheric environmental forensics aspects. In *Trends of Environmental Forensics in Pakistan* (pp. 39–59). Academic Press.

Jaskulak, M., Grobelak, A., Grosser, A., and Vandenbulcke, F. 2019. Gene expression, DNA damage and other stress markers in *Sinapis alba* L. exposed to heavy metals with special reference to sewage sludge application on contaminated sites. *Ecotoxic. Envir. Safety.* 181:508–517.

Kagalkar, A. N., Jadhav, M. U., Bapat, V. A., and Govindwar, S. P. 2011. Phytodegradation of the triphenylmethane dye Malachite Green mediated by cell suspension cultures of *Blumea malcolmii* Hook. *Biores. Tech.* 102:10312–10318.

Kah, M., Tufenkji, N., and White, J. C. 2019. Nano-enabled strategies to enhance crop nutrition and protection. *Nat. Nanotechnol.* 14:532–540.

Kasaishi, W., Oyamada, J., Morisada, S., Ohto, K., and Kawakita, H. 2021. Effects of precipitated size of water-soluble amide-containing polymers and pore size of filters on recovery of Pd nanoparticles dispersed in acetone solution including colloidal polymer. *Solvent Ext. Ion Exchange.* 1–16.

Kaur, L., Gadgil, K., and Sharma, S. 2013. Comparative study of natural phyto-extraction and induced phyto-extraction of lead using mustard plant (*Brassica juncea* arawali). *Int. J. Bioassay.* 2:352–357.

Khan, Z. S., Rizwan, M., Hafeez, M., Ali, S., Javed, M. R., and Adrees, M. 2019a. The accumulation of cadmium in wheat (*Triticum aestivum*) as influenced by zinc oxide nanoparticles and soil moisture conditions. *Envir. Sci. Poll. Res.* 26:19859–19870.

Khan, M. I., Dildar, S., Iqbal, T., Shakil, M., Tahir, M. B., Rafique, M., and Ijaz, M. 2019b. *In vivo* study of gold-nanoparticles using different extracts for kidney, liver function and photocatalytic application. *Chem. Rep.* 1:36–42.

Khanlarian, M., Roshanfar, M., Rashchi, F., and Motesharezadeh, B. 2020. Phyto-extraction of zinc, lead, nickel, and cadmium from zinc leach residue by a halophyte: *Salicornia europaea*. *Ecol. Eng.* 148:105797.

Konate, A., He, X., Zhang, Z., Ma, Y., Zhang, P., Alugongo, G. M., and Rui, Y. 2017. Magnetic (Fe_3O_4) nanoparticles reduce heavy metals uptake and mitigate their toxicity in wheat seedling. *Sustain.* 9:790.

Kumar, D., Singh, B., and Sharma, Y. C. 2017. Bioenergy and phytoremediation potential of Millettia pinnata. In *Phytoremediation Potential of Bioenergy Plants* (pp. 169–188). Springer, Singapore.

Kumar, M., Kumar, V., and Prasad, R. (Eds.). 2020. *Phyto-Microbiome in Stress Regulation*. Springer, New York.

Kumar Savani, A., Bhattacharyya, A., Boro, R. C., Dinesh, K., and Jc, N. S. 2021. Exemplifying endophytes of banana (*Musa paradisiaca*) for their potential role in growth stimulation and management of *Fusarium oxysporum* f. sp cubense causing panama wilt. *Folia Microbiol.* 1–14.

Kwong, R. M., Sagliocco, J. L., Harms, N. E., and Nachtrieb, J. G. 2018. Impacts of a pre-dispersal seed predator on achene production in the aquatic macrophyte, *Sagittaria platyphylla*. In *Proceedings of the 21st Australasian Weeds Conference, Sydney, Australia* (pp. 9–12). Available from: https://www.semanticscholar.org/paper/Impacts-of-a-pre-dispersal-seed-predator-on-achene-Kwong-Sagliocco/2092f66c71f3dbc563dd9439292b61e56cff0d03

Latef, A. A. H. A., Alhmad, M. F. A., and Abdelfattah, K. E. 2017. The possible roles of priming with ZnO nanoparticles in mitigation of salinity stress in lupine (*Lupinus termis*) plants. *J. Plant Growth Regul.* 36:60–70.

Lee, M., and Yang, M. 2010. Rhizofiltration using sunflower (*Helianthus annuus* L.) and bean (*Phaseolus vulgaris* L. var. vulgaris) to remediate uranium contaminated groundwater. *J. Hazard. Mat.* 173:589–596.

Lee, S., Xu, H., Rice, S. A., Chong, T. H., and Oh, H. S. 2021. Development of a quorum quenching-column to control biofouling in reverse osmosis water treatment processes. *J. Ind. Eng. Chem.* 94:188–194.

Li, J., Hu, J., Ma, C., Wang, Y., Wu, C., Huang, J., and Xing, B. 2016. Uptake, translocation and physiological effects of magnetic iron oxide (γ-Fe_2O_3) nanoparticles in corn (*Zea mays* L.). *Chemosphere.* 159:326–334.

Li, M., Adeel, M., Peng, Z., and Yukui, R. 2020. Physiological impacts of zero valent iron, Fe_3O_4 and Fe_2O_3 nanoparticles in rice plants and their potential as Fe fertilizers. *Environ. Pollut.* 1–10.

Li, Y., Liang, L., Li, W., Ashraf, U., Ma, L., Tang, X., and Mo, Z. 2020a. ZnO Nanoparticle-based seed priming modulates early growth and enhances physio-biochemical and metabolic profiles of fragrant rice against cadmium toxicity. *Res. Square.* 1–27.

Liu, C., Wei, L., Zhang, S., Xu, X., and Li, F. 2014. Effects of nanoscale silica sol foliar application on arsenic uptake, distribution and oxidative damage defense in rice (*Oryza sativa* L.) under arsenic stress. *RSC Adv.* 4:57227–57234.

Liu, J., Simms, M., Song, S., King, R. S., and Cobb, G. P. 2018. Physiological effects of copper oxide nanoparticles and arsenic on the growth and life cycle of rice (*Oryza sativa* japonica 'Koshihikari'). *Envir. Sci. Tech.* 52:13728–13737.

Liu, Y., Wu, T., White, J. C., and Lin, D. 2021. A new strategy using nanoscale zero-valent iron to simultaneously promote remediation and safe crop production in contaminated soil. *Nat. Nanotech.* 1–9.

López-Luna, J., Silva-Silva, M. J., Martinez-Vargas, S., Mijangos-Ricardez, O. F., González-Chávez, M. C., Solís-Domínguez, F. A., and Cuevas-Díaz, M. C. 2016. Magnetite nanoparticle (NP) uptake by wheat plants and its effect on cadmium and chromium toxicological behavior. *Sci. Total Envir.* 565: 941–950.

Ma, X., and Wang, X. 2018. Impact of engineered nanoparticles on the phytoextraction of environmental pollutants. In Ansari, A. A., editor. *Phytoremediation* (pp. 403–414). Switzerland: Springer Nature.

Madyiwa, S., Chimbari, M. J., Schutte, C. F., and Nyamangara, J. 2003. Greenhouse studies on the phytoextraction capacity of *Cynodon nlemfuensis* for lead and cadmium under irrigation with treated wastewater. *Phys. Chem. Earth. A.B.C.* 28:859–867.

Martins, F., Machado, S., Albergaria, T., and Delerue-Matos, C. 2017. LCA applied to nano scale zero valent iron synthesis. *J. Int. Life Cycle Assess.* 22:707–714.

Mohanty, C., Satpathy, S. S., and Mohanty, S. 2021. An eco-friendly approach for the eradication of heavy metal contaminants by nano-bioremediation. In *Recent Advancements in Bioremediation of Metal Contaminants* (pp. 220–236). IGI Global, New York.

Moradi, B., Maivan, H. Z., Hashtroudi, M. S., Sorahinobar, M., and Rohloff, J. 2021. Physiological responses and phytoremediation capability of *Avicennia marina* to oil contamination. *Acta Physiol. Plant.* 43:1–12.

Mousavi, S. M., Motesharezadeh, B., Hosseini, H. M., Alikhani, H., and Zolfaghari, A. A. 2018. Root-induced changes of Zn and Pb dynamics in the rhizosphere of sunflower with different plant growth promoting treatments in a heavily contaminated soil. *Ecotoxic. Envir. Safety*. 147:206–216.

Muthunarayanan, V., Santhiya, M., Swabna, V., and Geetha, A. 2011. Phytodegradation of textile dyes by water hyacinth (*Eichhornia crassipes*) from aqueous dye solutions. *J. Int. Envir. Sci.* 1:1702.

Naeem, H., Ahmad, K. S., and Jaffri, S. B. 2020. Biotechnological tools based lithospheric management of toxic Pyrethroid pesticides: A critical evaluation. *J. Int. Envir. Anal. Chem.* 1–24.

Naruka, A. K., Suganya, S., Kumar, P. S., Amit, C., Ankita, K., Bhatt, D., and Kumar, M. A. 2021. Kinetic modelling of high turbid water flocculation using native and surface functionalized coagulants prepared from shed-leaves of *Avicennia marina* plants. *Chemosphere.* 129894.

Nejad, Z. D., Jung, M. C., and Kim, K. H. 2018. Remediation of soils contaminated with heavy metals with an emphasis on immobilization technology. *Envir. Geochem. Health.* 40:927–953.

O'Connor, D., and Hou, D. 2018. Targeting cleanups towards a more sustainable future. *Envir. Sci. Proc. Impacts.* 20:266–269.

Okaiyeto, K., Hoppe, H., and Okoh, A. I. 2021. Plant-based synthesis of silver nanoparticles using aqueous leaf extract of *Salvia officinalis*: Characterization and its antiplasmodial activity. *J. Cluster Sci.* 32:101–109.

Omran, B. A. 2021. Facing lethal impacts of industrialization via green and sustainable microbial removal of hazardous pollutants and nanobioremediation. *Removal Emerg. Contam. Microb. Proc.* 133–160.

Osama, R., Awad, H. M., Zha, S., Meng, F., and Tawfik, A. 2021. Greenhouse gases emissions from duckweed pond system treating polyester resin wastewater containing 1, 4-dioxane and heavy metals. *Ecotoxic. Envir. Safety.* 207:111253.

Park, K. Y., Choi, S. Y., Ahn, S. K., and Kweon, J. H. 2021. Disinfection by-product formation potential of algogenic organic matter from *Microcystis aeruginosa*: Effects of growth phases and powdered activated carbon adsorption. *J. Hazard. Mat.* 408:124864.

Pati, P., McGinnis, S., and Vikesland, P. J. 2014. Life cycle assessment of "green" nanoparticle synthesis methods. *Envir. Eng. Sci.* 31:410–420.

Ramasamy, G., Rajkumar, P. K., and Narayanan, M. 2021. Generation of energy from salinity gradients using capacitive reverse electro dialysis: A review. *Envir. Sci. Poll. Res.* 1–10.

Rizwan, M., Ali, S., Ali, B., Adrees, M., Arshad, M., Hussain, A., and Waris, A. A. 2019d. Zinc and iron oxide nanoparticles improved the plant growth and reduced the oxidative stress and cadmium concentration in wheat. *Chemosphere.* 214:269–277.

Rizwan, M., Ali, S., ur Rehman, M. Z., Adrees, M., Arshad, M., Qayyum, M. F., and Imran, M. 2019b. Alleviation of cadmium accumulation in maize (*Zea mays* L.) by foliar spray of zinc oxide nanoparticles and biochar to contaminated soil. *Envir. Poll.* 248:358–367.

Rizwan, M., Ali, S., ur Rehman, M. Z., Malik, S., Adrees, M., Qayyum, M. F., and Ahmad, P. 2019c. Effect of foliar applications of silicon and titanium dioxide nanoparticles on growth, oxidative stress, and cadmium accumulation by rice (*Oryza sativa*). *Acta Physiol. Plant.* 41(3), 1–12.

Rizwan, M., Noureen, S., Ali, S., Anwar, S., ur Rehman, M. Z., Qayyum, M. F., and Hussain, A. 2019a. Influence of biochar amendment and foliar application of iron oxide nanoparticles on growth, photosynthesis, and cadmium accumulation in rice biomass. *J. Soils Sed.* 19:3749–3759.

Romeh, A. A., and Saber, R. A. I. 2020. Green nano-phytoremediation and solubility improving agents for the remediation of chlorfenapyr contaminated soil and water. *J. Envir. Manage.* 260:110104.

Roshanfar, M., Khanlarian, M., Rashchi, F., and Motesharezadeh, B. 2020. Phyto-extraction of zinc, lead, nickel, and cadmium from a zinc leach residue. *J. Clean. Prod.* 266:121539.

Rossi, L., Sharifan, H., Zhang, W., Schwab, A. P., and Ma, X. 2018. Mutual effects and in planta accumulation of co-existing cerium oxide nanoparticles and cadmium in hydroponically grown soybean (*Glycine max* (L.) Merr.). *Envir. Sci. Nano.* 5:150–157.

Rui, M., Ma, C., Tang, X., Yang, J., Jiang, F., Pan, Y., Xiang, Z., Hao, Y., Rui, Y., and Cao, W. 2017. Phytotoxicity of silver nanoparticles to peanut (Arachis hypogaea L.): Physiological responses and food safety. *ACS Sustain. Chem. Eng.* 5:6557–6567.

Sebastian, A., Nangia, A., and Prasad, M. N. V. 2019. Cadmium and sodium adsorption properties of magnetite nanoparticles synthesized from *Hevea brasiliensis* Muell. Arg. bark: Relevance in amelioration of metal stress in rice. *J. Hazard. Mat.* 371:261–272.

Shaheen, I., Ahmad, K. S., Jaffri, S. B., and Ali, D. 2021. Biomimetic [MoO_3@ZnO] semiconducting nanocomposites: Chemo-proportional fabrication, characterization and energy storage potential exploration. *Renew. Energy.* 167:568–579.

Shaheen, I., Ahmad, K. S., Jaffri, S. B., Zahra, T., and Azhar, S. 2016. Evaluating the adsorption and desorption behavior of triasulfuron as a function of soil physico-chemical characteristics. *Soil Envir.* 35:99–105.

Shang, H., Guv, H., Ma, C., Li, C., Chefetz, B., Polubesova, T., and Xing, B. 2019. Aggregation, dissolution, and toxicity of CuO nanoparticles on the rhizosphere as affected by maize (*Zea mays* L.) root exudates. *Abstracts the 16th International Phytotechnology Conference—Phytotechnologies for Food Safety and Environmental Health*, September 23–27, 2019. Changsha, China, p. 167.

Shazia, I., Ahmad, K. S., and Jaffri, S. B. 2018. Mycodriven enhancement and inherent phytoremediation potential exploration of plants for lithospheric remediation. *Sydowia* 70:141–153.

Shrivastava, M., Srivastav, A., Gandhi, S., Rao, S., Roychoudhury, A., Kumar, A., and Singh, S. D. 2019. Monitoring of engineered nanoparticles in soil-plant system: A review. *Envir. Nanotech. Monit. Manage.* 11:100218.

Tahir, M. B., Iram, S., Ahmad, K. S., and Jaffri, S. B. 2020a. Developmental abnormality caused by *Fusarium mangiferae* in mango fruit explored via molecular characterization. *Biologia.* 75:465–473.

Tahir, M. B., Malik, M. F., Ahmed, A., Nawaz, T., Ijaz, M., Min, H. S., and Siddeeg, S. M. 2020b. Semiconductor based nanomaterials for harvesting green hydrogen energy under solar light irradiation. *J. Int. Envir. Anal. Chem.* 1–17.

Uribe-López, M. C., Hidalgo-López, M. C., López-González, R., Frías-Márquez, D. M., Núñez-Nogueira, G., Hernández-Castillo, D., and Alvarez-Lemus, M. A. 2021. Photocatalytic activity of ZnO nanoparticles and the role of the synthesis method on their physical and chemical properties. *J. Photochem. Photobiol. A.* 404:112866.

Usman, M., Farooq, M., Wakeel, A., Nawaz, A., Cheema, S. A., Rehman, H. U., Ashraf, I., and Sanaullah, M. 2020. Nanotechnology in agriculture: Current status, challenges and future opportunities. *Sci. Total Environ.* 721:137778.

Venkatachalam, P., Jayaraj, M., Manikandan, R., Geetha, N., Rene, E. R., Sharma, N. C., and Sahi, S. V. 2017. Zinc oxide nanoparticles (ZnONPs) alleviate heavy metal-induced toxicity *in Leucaena leucocephala* seedlings: A physiochemical analysis. *Plant Physiol. Biochem.* 110:59–69.

Verma, V., Pandey, N. K., Gupta, P., Singh, K., and Singh, P. 2021. Humidity sensing enhancement and structural evolution of tungsten doped zinc oxide nano-sensors fabricated through co-precipitation synthesis. *J. Mat. Sci. Mat. Elect.* 1–10.

Vidal, J., Báez, M. E., and Salazar, R. 2021. Electro-kinetic washing of a soil contaminated with quinclorac and subsequent electro-oxidation of wash water. *Sci. Total Envir.* 761:143204.

Waani, S. P. T., Irum, S., Gul, I., Yaqoob, K., Khalid, M. U., Ali, M. A., and Arshad, M. 2021. TiO_2 nanoparticles dose, application method and phosphorous levels influence genotoxicity in Rice (*Oryza sativa* L.), soil enzymatic activities and plant growth. *Ecotoxic. Envir. Safety.* 213:111977.

Wang, S., Wang, F., and Gao, S. 2015. Foliar application with nano-silicon alleviates Cd toxicity in rice seedlings. *Envir. Sci. Poll. Res.* 22:2837–2845.

Wang, Y., Liu, Y., Zhan, W., Niu, L., Zou, X., Zhang, C., and Ruan, X. 2020a. A field experiment on stabilization of Cd in contaminated soils by surface-modified nano-silica (SMNS) and its phyto-availability to corn and wheat. *J. Soils Sed.* 20:91–98.

Wang, Y., Liu, Y., Zhan, W., Zheng, K., Lian, M., Zhang, C., and Li, T. 2020. Long-term stabilization of Cd in agricultural soil using mercapto-functionalized nano-silica (MPTS/nano-silica): A three-year field study. *Ecotoxic. Envir. Safety.* 197:110600.

Yadav, B. K., Siebel, M. A., and van Bruggen, J. J. 2011. Rhizofiltration of a heavy metal (lead) containing wastewater using the wetland plant *Carex pendula*. *Clean Soil Air Water.* 39:467–474.

Yadav, K. K., Gupta, N., Kumar, A., Reece, L. M., Singh, N., Rezania, S., and Khan, S. A. 2018. Mechanistic understanding and holistic approach of phytoremediation: A review on application and future prospects. *Ecol. Eng.* 120:274–298.

Yadav, K. K., Singh, J. K., Gupta, N., and Kumar, V. 2017. A review of nanobioremediation technologies for environmental cleanup: A novel biological approach. *J. Mater. Environ. Sci.* 8:740–757.

Yan, L., Van Le, Q., Sonne, C., Yang, Y., Yang, H., Gu, H., and Peng, W. 2021. Phytoremediation of radionuclides in soil, sediments and water. *J. Hazard. Mat.* 407:124771.

Ye, S., Zeng, G., Wu, H., Zhang, C., Dai, J., Liang, J., and Zhang, C. 2017. Biological technologies for the remediation of co-contaminated soil. *Critic. Rev. Biotech.* 37:1062–1076.

Zahra, T., Ahmad, K. S., Shaheen, I., Azhar, S., and Jaffri, S. B. 2017. Determining the adsorption and desorption behavior of thiabendazole fungicide for five different agricultural soils. *Soil Envir.* 36:1–10.

Zazouli, M. A., Mahdavi, Y., Bazrafshan, E., and Balarak, D. 2014. Phytodegradation potential of bisphenolA from aqueous solution by *Azolla filiculoides*. *J. Envir. Health Sci. Eng.* 12:1–5.

Zhang, P. W., Huang, Y. Z., Fan, C., and Chang, T. K. 2020. Application of waste lemon extract to toxic metal removal through gravitational soil flushing and composting stabilization. *Sustain.* 12:5751.

Zhang, R., Zhang, N., and Fang, Z. 2018. *In situ* remediation of hexavalent chromium contaminated soil by CMC-stabilized nanoscale zero-valent iron composited with biochar. *Water Sci. Tech.* 77:1622–1631.

Zhou, P., Adeel, M., Shakoor, N., Guo, M., Hao, Y., Azeem, I., and Rui, Y. 2021. Application of nanoparticles alleviates heavy metals stress and promotes plant growth: An overview. *Nanomat.* 11:26.

11 Iron Nanoparticles for Nano-Phytoremediation

Misbah Naz, A. M. Shackira, Mohammad Sarraf, Nair G Sarath, Sarah Bouzroud, Akshaya Prakash Chengatt, and Xiaorong Fan

11.1 INTRODUCTION

The size of iron nanoparticles (NPs) varies, ranging from 1 to 100 nm. They are super paramagnetic with copper oxide (CuO) and nickel oxide (NiO) (Darr et al. 2017). Two main forms of NPs are available: magnetite and an oxidized form of maghemite (Shah et al. 2015). Because of their super paramagnetism, they have been widely used in biomedical imaging, heavy electrons, catalysis and important chemical reactions (Khan, Saeed, and Khan 2019). Studies on zebrafish animal models have shown that iron oxide NPs are cytotoxic, and their stable properties help them reach a higher level of the ecosystem (Singh et al. 2021). In plants, they mainly affect the normal function of cells by destroying and blocking the aquaporins present on the cell membrane (Malhotra et al. 2020). Zhu, Tian, and Cai (2012) used pumpkin as their toxicity model to show how NPs transfer within plant tissues. Their transmission electron microscope (TEM) image is one of the earliest recorded descriptions of NPs entering the inside of root cells (from the root hair to the epidermis). It was found that the development of Arabidopsis is negatively affected by these NPs, and the root length and leaf number of exposed plants are reduced (Trujillo-Reyes et al. 2014).

According to the TEM image, regardless of the concentration of the iron precursor, about 90% of the NPs prepared with Vibrio flexneri have a diameter from 5 to 25 nm (Siddiqi, Husen, and Rao 2018). Mineralogical analysis showed that vZVI NPs are formed by 24% of iron oxide and 76% of alpha iron. The centered body of these NPs displays a cubic crystal shape. Fourier transform infrared spectroscopy revealed that the presence of -OH and C=C groups of mulberry extract might be implicated in iron NPs' establishment and stabilization (Murgueitio et al. 2018). Based on this finding, iron NPs are now produced by using mulberry berry extract (*vaccinium floribundum*) vZVI (Ossai et al. 2020). Generally, as shown by the chromatographic analysis of total petroleum hydrocarbon (TPH) products, iron NPs are used to remove them successfully. For instance, it has been demonstrated that water and soil treatments with vZVI particles allowed the removal of more than 80% of petroleum hydrocarbons. However, the presence of iron in these nano-sized particles seems to strongly impact this treatment (Kuppusamy et al. 2016). The great ability of pollutant removal (for example, of TPHs) is linked to NPs' high surface area, upper reactivity and more importantly in-situ processing (Murgueitio et al. 2018). Therefore, compared with conventional bioremediation methods, this technology can be effectively used for the rapid soil and water remediation from petroleum hydrocarbons. Iron NPs can be produced to exhibit the ability to remove TPHs from soil and water samples was investigated by Alfadul (2006) using synthetic spherical nanoscale iron particles with a size ranging between 5 and 10 nm. vZVI NP application allowed the removal of more than 85% of TPHs found in water samples and up to 82% of the TPH content in soil samples (Hong et al. 2020). The results show that the addition of vZVI NPs accelerates the elimination of TPHs by creating a great reducing environment, which suggests that these NPs can be widely used to prompt the removal of TPHs from water and soil compartments (Ren et al. 2014).

DOI: 10.1201/9781003186298-11

Iron oxide (IO) NPs are composed of maghemite (γ-Fe2O3) and/or magnetite (Fe3O4) particles with a diameter ranging from 1 to 100 nanometers, which can be used for magnetic data storage, bio sensing, drug delivery, etc. (Banerjee et al. 2010). In NPs, the surface area-to-volume ratio increases significantly. IO and iron oxyhydroxide are widely distributed in nature and play important roles in many geological and biological processes (Laurent et al. 2010). They are used as iron (Fe) ore, pigments, catalysts and thermite and are present in hemoglobin. IO is an inexpensive and durable pigment in paints, coatings and colored concrete (Ali et al. 2016). Nanomaterials (NMs) can play a role in phytoremediation systems by directly removing pollutants, promoting plant growth and increasing plants' utilization of pollutants. Plant extraction is the most effective and recognized phytoremediation strategy to remediate contaminated soil (da Conceição Gomes et al. 2016).

11.2 SYNTHESIS OF IRON NANOPARTICLES (THE REDUCTION OF FE(II) OR FE(III) SALT WITH SODIUM BOROHYDRIDE IN AN AQUEOUS MEDIUM)

In the presence of solvent-containing carbonyl groups, researchers have used borohydride to reduce Fe ions to synthesize nano-sized Fe particles. When we change the content of the solvent, the synthesis and the form of Fe will also change accordingly (Samaddar et al. 2018). Each synthetic Fe particle sample was characterized by TEM and powder X-ray diffractometer (XRD). The content of zero-valent iron (ZVI) in the synthetic particles was measured to study the effect of the solvent ratio on the reduction of Fe particles in the aqueous organic solvent system (Saif, Tahir, and Chen 2016). With the addition of the solvent, the proportion of ZVI in the synthetic particles decreases, forming a well-dispersed Fe particle suspension. Their dispersibility was also characterized by a light scattering particle size analyzer, which allowed us to confirm the changes in the dispersion characteristics of wet synthesized Fe particles in a multi-component system (Bagheri and Julkapli 2016). NP synthesis technology has attracted attention in catalysts' technical application and in industrial and therapeutic applications. These size ranges are in the particles' nanostructured domain. Pure Fe NPs have attracted special attention due to their reactivity and relatively biologically inert nature (Nath and Banerjee 2013). Fe NPs are particularly used in tumoral disease treatment and as tool to perform drug delivery. Unfortunately, due to their reactivity, pure Fe NPs have been difficult to understand. This is because they accelerate the tendency to form oxides in the air, which is due to their increased surface area-to-volume ratio (Sajid et al. 2015). Synthetic methods use polyphenols or long Amine chains use air-stabilized Fe NPs with a diameter size range of about 2 to 10 nm, but apparently due to internal pressure, the crystallographic defects are transformed into the face-centered cubic (FCC) stage. It has been found that FCC crystals form icosahedral and decahedral shapes. Within this size, they are used as a catalyst for boron and carbon nanotube development (Quideau et al. 2011).

11.2.1 ADVANTAGES OF CATALYSTS

The growth of nanotubes can be detached from catalyst synthesis for the first time. In addition, the catalyst size can be selected in advance to grow nanotubes of a specific size (Smalley et al. 2004). That is, (1) NPs' size distribution has been identified for many synthesis procedures, (2) the appropriate size range of vZVI NPs has been determined for nanotube synthesis, (3) the internal pressure of NPs has been quantified (4) as has their FCC phase, (5) Fe NP shapes have been discovered (6) with an internal crystalline structure, (7) FCC crystal deformation is due to carbon in the octahedral position, (8) Fe NPs have high stability in the air and (9) the addition of Cu (low concentration) decreases Fe NPs' size (Thomas and Raja 2006). Compared with other solvents, the possibility of hydroxylation in water is higher. For Fe(III), the hydroxylation of ions can start in a pH range as low as 1 and as high as 5 and can occur in acid. Fe(II) is more resistant to hydrolysis and can be in an ionic form, with a pH of up to 7 to 9 (Que, Domaille, and Chang 2008). This results in the formation of IO NPs at a lower pH than otherwise. In addition, due to the loss of control over maintaining Fe ions, the size distribution of NPs is due to oxide contamination occurring at different rates. For this, a case buffer must be added. An example of such a buffer is citric acid, which can be obtained from a pH of 3 to about a pH of 6 (Guo and Barnard 2013).

11.2.2 Using Iron Nanoparticles to Promote the Phytoremediation of Contaminated Soil

Soil pollution is widely considered by the world's scientific community as an environmental issue that affects living organisms. Phytoremediation has emerged as a promising technology for soil decontamination from environmental pollutants and has been widely accepted. Various studies have been conducted to improve the efficiency of phytoremediation through various methods. Nanotechnology application offers a new way to improve phytoremediation efficiency and to contribute to soil decontamination (Liu, Xin, and Zhou 2018). This chapter reviews the latest advances in the use of NMs to promote the phytoremediation of contaminated soil. NMs can contribute to the phytoremediation system by prompting the direct removal of pollutants or through the stimulation of plant growth and the improvement of pollutant utilization by plants. Plant extraction is the most effective and recognized phytoremediation strategy for remediating contaminated soil (Wang et al. 2017). Nanoscale zero-valent iron (nZVI) is the most studied NM for phytoremediation due to its great soil and water remediation. Fullerene NPs can increase the use of pollutants by plants. Generally, the use of NMs to promote the phytoremediation of contaminated soil is an effective strategy, but it is still in the exploratory and experimental stage. More application case experience is needed, and the long-term performance of NMs in phytoremediation systems needs further study (Rizwan et al. 2014).

Phytoremediation consists of the use of plants for pollutant removal, degradation or containment in environmental media (soil, water and air). Phytoremediation techniques for soil pollution include plant volatilization, plant extraction, plant degradation, plant stabilization and rhizosphere degradation (Wenzel et al. 1999). Plant volatilization can be defined as the ability of plants to absorb pollutants and transform or degrade them into volatile forms. These volatile products are then translocated to leaves and evaporated. Arsenic and mercury are usually studied in the process of plant volatilization because of their great ability to transform to gaseous forms through plant organisms (Jeevanantham et al. 2019). With the development of nanotechnology, the repair and pollution reduction of NMs used in the environment have attracted more and more attention from environmental researchers. NMs are materials with nanoscale dimensions (1–100 nm) in at least one dimension (Mura et al. 2013). Extensive research has been conducted on carbon-based NMs and metal-based NMs (Bayatsarmadi et al. 2017). Engineering NMs in many areas of soil and soil remediation has successfully developed groundwater in Europe and the United States (Mueller et al. 2012). Recently, some studies have reported the application of NMs in the phytoremediation of contaminated soil. Incorporating NMs into conventional phytoremediation is a promising system. An NM-assisted phytoremediation system includes three main parts: plants, pollutants and NMs (El Wakeil, Alkahtani, and Gaafar 2017). NMs can increase phytoremediation efficiency through direct action on plants. They can also indirectly participate in the plant-pollutant interaction. Moreover, they affect the final repair efficiency. On this basis, the following section discusses how NMs can play a role in phytoremediation from the following three aspects: NMs directly remove pollutants, stimulate plant growth and increase the utilization of pollutants by plants (Rai et al. 2020).

11.3 NANOMATERIALS' FUNCTION IN THE PHYTOREMEDIATION SYSTEM

Nano-phytoremediation entails either employing reactive NPs for the degradation/transformation of various kinds of pollutants or aiding the plant to detoxify pollutant as an intermediate (Figures 11.1 and 11.2). A number of NPs have been found to be exceptional for the purpose of environmental cleanup, which include nanoscale zeolites, fullerenes, metal oxides, carbon nanotubes and fibers and bimetallic NPs (Patil et al. 2016). These NPs can remediate a number of pollutants such as organo chloride pesticides, toxic metal ions, chlorinated compounds, polychlorinated biphenyls and inorganic anions (Karn, Kuiken, and Otto 2009). NPs may either directly remove the pollutants from the soil/water or indirectly assist plants in the phytoremediation process.

11.3.1 Direct Removal of Pollutants

NPs can be effectively exploited for the direct removal of various types of pollutants (toxic metals, explosives, herbicides, pesticides and organic and inorganic contaminants) from contaminated water, air and soil. Extensive application of NPs in the environmental cleanup process is mainly due to their large surface area compared to their bulk, which provides them with a high rate of reactivity for the chemical/biological process of decontamination (Sozer and Kokini 2009). NPs can also adhere to the roots of higher plants and are further transported to the shoot, thereby enhancing the efficiency of phytoremediation (Srivastav et al. 2018).

NPs have the potential to directly degrade, reduce, absorb or adsorb contaminants in the environment, which is influenced mainly by not only the type of pollutant, concentration of pollutant and charge of pollutant but also the soil characteristics including the soil texture, pH and soil aggregates (Srivastav et al. 2018; Song et al. 2019). Among different types of NPs, nZVI is extensively used in the field of phytoremediation (Song et al. 2019). Hence, the chemistry of nZVI is known much more than other NPs. During its interaction with pollutants, nZVI is converted from Fe^0 to Fe^{2+} and subsequently oxidized to Fe^{3+} (Patil et al. 2016). Organic contaminants such as molinate, atrazine and chlorpyrifos can be effectively degraded by nZVI. nZVI also has the potential to reduce other pollutants including tetrachloroethylene (PCE), trichloroethylene (TCE) and cis-1, 2-dichloroethylene (c-DCE).

In addition to nZVI, nFe_3O_4 (magnetite NP) and Pd/Fe (bimetallic NP) are excellent in the degradation of organic pollutants such as lindane, pentachlorophenol, pyrene, polychlorinated

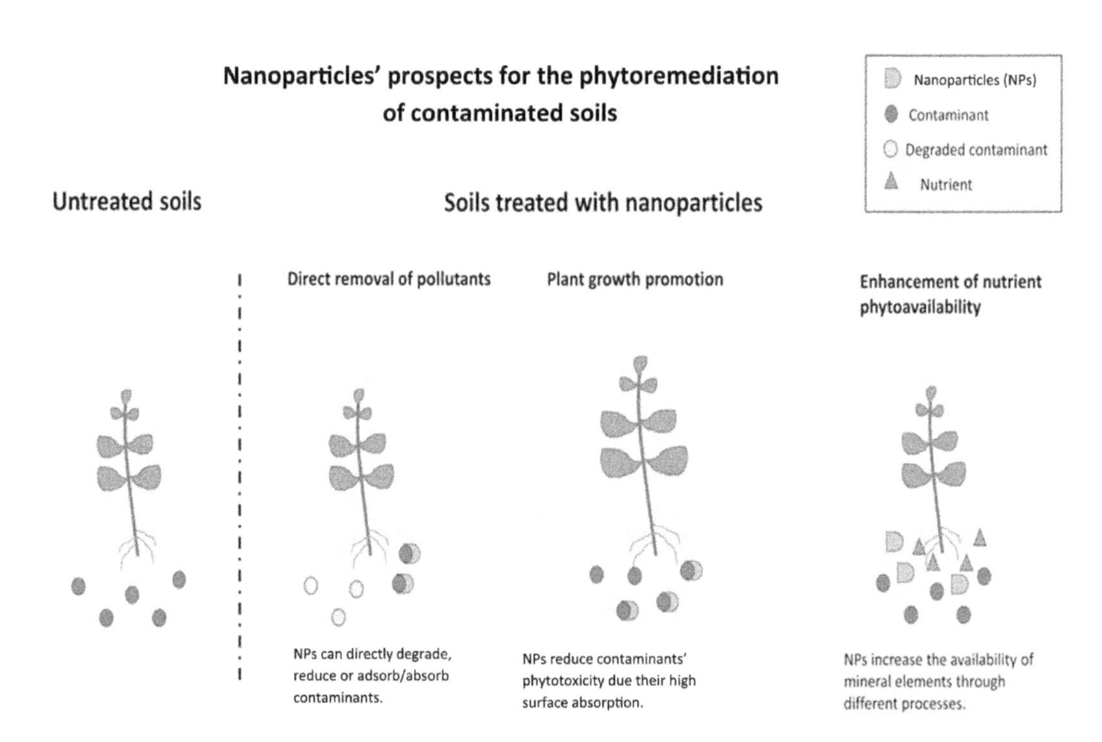

FIGURE 11.1 Nanoparticles' prospects for the phytoremediation of contaminated soils.

biphenyls (PCBs), 2, 4-dinitrotoluene and ibuprofen (Satapanajaru et al. 2008; Reddy et al. 2013; Singh et al. 2012). Moreover, organic contaminants, including DDT, diuron and pyrene, can be effectively degraded through photocatalyis and thermal destruction by TiO_2 NPs ($nTiO_2$) (Makarova et al. 2000). In addition, PEI-Cu NPs are efficient degraders of pesticides such as atrazine (Kalidhasan, Dror, and Berkowitz 2017).

The reclamation of polluted water bodies with toxic metal ions, explosives such as TNT and chlorinated compounds is also facilitated by NP-like bi-metallic NPs, for example, Cu/Fe, silver (Ag)/Fe and Ni/Fe (Koutsospyros et al. 2012; Nie et al. 2013; Yan et al. 2010). Engineered NPs including carbon nanotubes (CNTs) have also found extensive application in the water decontamination process because of their increased adsorption rate and potential to interact with the functional group of the target compounds. CNTs are excellent agents for the remediation of toxic metal ions such as Zn^{2+}, Cr^{3+} and Pb^{2+} (Savage and Diallo 2005; Liang et al. 2004; Patil et al. 2016).

11.3.2 Promoting Plant Growth

The growth and development of plants may be greatly influenced by various NPs, which may offer considerable benefits in the phytoremediation of different kinds of pollutants. The small size and large surface area-to-volume ratio provide them with increased affinity for the adsorption of pollutants, thereby enhancing the rate of phytoremediation to a greater extent (Mohammadi et al. 2020). In addition to their small size, factors including their physical and chemical structure, surface layer, reactivity, shape and NP concentration determine the efficiency of nano-phytoremediation (Ma et al. 2010; Khodakovskaya et al. 2012). The increased adsorption/absorption of pollutants through NMs may induce several physiological and biochemical alterations in plants so that they can cope with the stress caused by the pollutants. For example, ion NPs (Fe^0) enhance the growth of sunflower (*Helianthus annuus*) plants under chromium (Cr) stress by significantly boosting the activity of antioxidant enzymes, for example, SOD, POX, CAT and APX (Mohammadi et al. 2020). Nano-silicon application in *Oryza sativa* helps the plant tolerate heavy metal stress by regulating the accumulation of toxic ions such as cadmium (Cd), lead (Pb), zinc (Zn) and Cu (Wang, Wang, and Gao 2015).

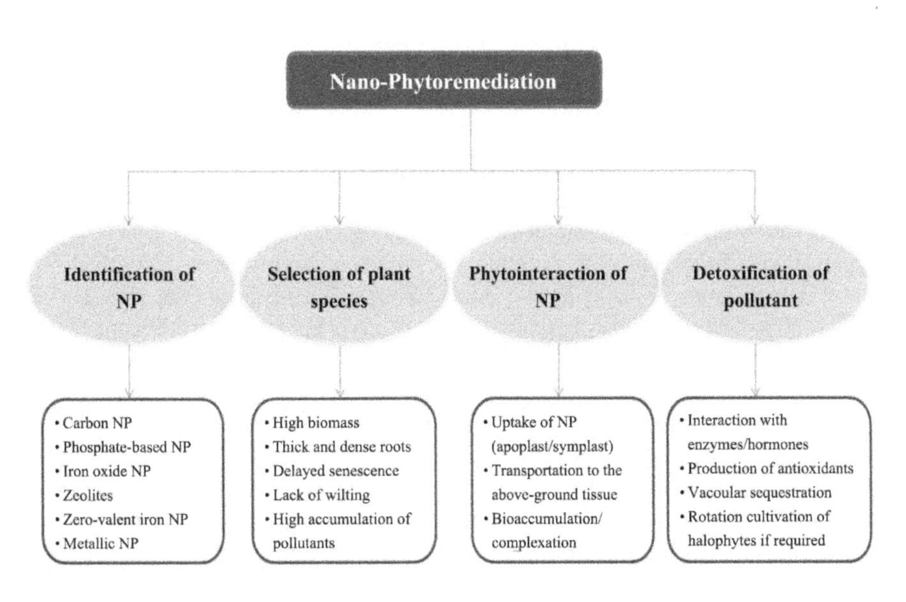

FIGURE 11.2 Nanomaterials' purpose in phytoremediation classification.

Likewise, nano-zinc oxide (ZnO) and nano-silicon dioxide (SiO_2) application in plants results in increased water uptake and the accumulation of osmolytes including proline, thereby improving the tolerance for environmental pollutants (Shalaby et al. 2016).

Recently, much attention has given to increasing plant productivity under stress conditions and can be achieved to some extent with engineered NPs. For example, NPs such as single-walled carbon nanotubes (SWNTs) have the potential to enhance the rate of electron transfer in chloroplasts during the light reaction of photosynthesis (Giraldo et al. 2014; Shang et al. 2019). In addition, SWNTs can suppress the production of reactive oxygen species (ROS) in the chloroplasts, thereby enhancing the quantum yield and efficiency of photosynthesis (Shang et al. 2019; Giraldo et al. 2014). The phytoremediation of salinized soils or desalination is greatly influenced by NPs through the modification of the photosynthetic process including the enlarged leaf surface, augmented pigment composition, increased rate of photosystem II (PSII) and better carbon dioxide (CO_2) assimilation and sub-stomatal CO_2 concentration, as reported in tomato and sunflower by the application of nano-SiO_2 and nano-$FeSO_4$, respectively (Haghighi, Afifipour, and Mozafarian 2012; Siddiqui and Al-Whaibi 2014; Torabian, Zahedi, and Khoshgoftar 2017; Rastogi et al. 2019). Moreover, NP applications in plants also alter the expression pattern of several stress genes, such as antioxidant enzyme-encoding genes, which further detoxifies the pollutant. For example, in *Arabidopsis thaliana*, nano-Ag application has induced the expression of genes including SOD, GPX and Cyt-P450-dependent oxidase as reported by Banerjee and Kole (2016).

11.3.3 Enhanced Phytoavailability

Soils are in general enriched with different elements of nutrients, minerals and metals, which will be taken up by plants for their growth and development through the root system. However, some nutrients, such as phosphorus (P), are not mobilized much through plant roots and need a supplemental mechanism for their absorption. NPs play a crucial role in this regard by enhancing the mobility of ions in the soil through various mechanisms such as adsorbing-desorbing, increasing the dissolution rate of ions, changing the soil pH and boosting the root exudates in the rhizosphere (Zahra et al. 2020). For example, P mobilization was increased in *Brassica napus* plants upon exposure to aluminum oxide (Al_2O_3) NPs under hydroponic culture conditions (Santner et al. 2012). Likewise, in *Lactuca sativa*, P phytoavailability was stimulated by approximately 56% after the application of TiO_2 NPs (Hanif et al. 2015). In general, TiO_2 NPs have been extensively used to enhance the P mobility in many plans, including wheat and cucumber (Rafique et al. 2018; Rafique et al. 2018; Servin et al. 2013). This is because Ti^{3+} possesses the features of cations that provide more sites for PO_4^{3-} absorption. This will induce a pH reduction in soil that will subsequently enable increased root exudation and finally increase P mobilization (Zahra et al. 2015; Zahra et al. 2020).

NPs modify the availability of mineral elements and other ions in the soil, thereby influencing the nutritional status of the plant. For example, the phytoavailability of minerals such as nitrogen (N), P and toxic metal ions of Zn and Pb is increased significantly when the soils are exposed to CeO_2 and TiO_2 NPs. The increased phytoavailability of Pb is attributed to the reduced soil pH and increased sorption sites for Pb^{2+} ions as induced by the CeO_2 NP. However, in the case of mineral N, the increased phytoavailability is due to the altered mineralization (i.e., nitrification) by the antimicrobial or catalytic properties of both NPs. Likewise, P is increasingly mobilized through the anti-microbial properties of NPs that modify the activity of P-solubilizing microbes. In the case of metal ion Zn, the increased phytoavailability accounts for the competition between positively charged ions and the NP of the same charge (Duncan and Owens 2019). In addition, NMs including fullerene are known to enhance the phytoavailability of various environmental pollutants as reported by Song et al. (2019).

TABLE 11.1
The Use of Nanoparticles to Increase Mineral Bioavailability for Plants

Mineral element	Nanoparticle	Plant species	References
Phosphorus (P)	Al_2O_3	*Brassica napus*	(Santner et al. 2012)
	TiO_2	*Lactuca sativa*	(Hanif et al. 2015)
	TiO_2	*Triticum aestivum*	(Rafique et al. 2018)
	TiO_2	*Cucumis sativus*	(Servin et al. 2013)
Nitrogen (N)	CeO_2	*Phaseolus vulgaris*	(Das et al. 2018)
	TiO_2	*Spirodela polyrrhiza*	(Movafeghi et al. 2018)
Zinc (Zn)	ZnO	*Gossypium hirsutum*	(Venkatachalam et al. 2017)
Lead (Pb)	TiO_2	*Oryza sativa*	(Cai et al. 2017)

11.4 ECOFRIENDLY APPLICATION OF NANOMATERIALS: NANO BIOREMEDIATION

11.4.1 Iron Oxide Nanoparticles

Nano bioremediation is the process of the elimination of both organic and inorganic environmental pollutants with the use of NPs formed by living organisms such as algae, fungi and bacteria (Yadav et al. 2017). That is, nano-phytoremediation refers to the use of NPs produced by or involving plants to eliminate organic or inorganic pollutants. Therefore, nano-phytoremediation involves the following strategies: nano-phytoextraction, nano-phytodegradation, nano-phytostabilization, nano-rhizofiltration and nano-phytovolatilization (Singh 2010). Classical phytoremediation, which is the process of using plants to eliminate contaminants from soil, has various limitations including its time-consuming nature and lower biomass of plants (Zhao et al. 2010). Thus, it becomes necessary to amend the phytoremediation process. The incorporation of nanotechnology into phytoremediation is a good option that augments phytoremediation's potential. Therefore, NPs can be used for the reclamation of water and soils polluted with toxic metals and other inorganic and organic substances (Singh 2010).

The most commonly used NPs for environmental cleanup are Fe NPs (Kanel et al. 2005; Shipley, Engates, and Guettner 2011). The electron-donating capacity of Fe makes it reduce, and this is its role in remediating a contaminated matrix. Its strong reducing property makes it an excellent choice to remove any contaminant that can be reduced (Mehndiratta et al. 2013). The increased absorption capability, surface modifiability, biocompatibility, increased reaction potential and high surface area of IO NMs are the characteristics that attract wide attention to this field of environmental restoration (Waychunas, Kim, and Banfield 2005; Komárek, Vaněk, and Ettler 2013; Martínez-Fernández et al. 2017).

The usage of IO NPs in the field of agriculture and environmental remediation has increased recently, which indicates their effectiveness and environmentally friendly nature. Pollutants such as Cr, PCBs and trichloroethylene (TCE) can be removed from soil using Fe NPs (Varanasi, Fullana, and Sidhu 2007; Zhang et al. 2011; Singh, Misra, and Singh 2012). In addition, various studies have underlined the scavenging nature of IO NPs for the toxic forms of heavy metals (Zhang et al. 2010; Lin, Lu, and Liu 2012; Tuutijärvi et al. 2012; Akhbarizadeh, Shayestefar, and Darezereshki 2013; Jiang et al. 2013; Roy and Bhattacharya 2012; Komárek et al. 2015; Chen and Li 2010; Nassar 2010; Xu et al. 2012; Chomoucka et al. 2010; Martínez-Fernández, Barroso, and Komárek 2016).

The amorphous form of IO NPs makes them the most suitable substances for land contaminated with toxic metals due to their large surface energies and considerable surface area-to-volume ratio

(Anjum et al. 2019). The added advantages of IO NPs are that they can be manipulated easily by applying an external magnetic field, they are recyclable and they are cheaper (Mulens-Arias, Rojas, and Barber 2020). In the absence of materials that coat their surface, magnetic IO particles are hydrophobic. Therefore, these particles cluster together, thus increasing their particle size. They then show strong magnetic interactions, which is the reason for their ferromagnetic behavior (Hamley 2003). Moreover, when each of these large particles comes into the vicinity of one another, each cluster falls into the magnetic field of one another. This leads to a further magnetization of particles (Tepper et al. 2003). This again enhances their aggregation properties (Gupta and Gupta 2005). A similar report by Xia et al. (2012) also supports the fact that due to the strong magnetic attraction among NPs, van der Waals forces and a high surface energy, the aggregation of IOs with bare surfaces occurs. Coating a shell on IO NPs makes them hydrophilic, and making them functionalized and compatible with the environment can prevent aggregation (Sheng-Nan et al. 2014). Usually, it is desired to coat IO NPs with a high-density substance to stabilize them. The addition of a polymer or a surfactant during the time of preparation helps in preventing the aggregation of a nanoscale particulate stabilizer. The adherence of these polymers occurs in a substrate-specific manner (Ghosh et al. 2011).

The synthesis of IO NPs can be carried out by using three methods, namely, physical, chemical and microbial methods. Among these, the chemical method is the easiest, fastest and cheapest. Yadav and Fulekar (2018), Zhao et al. (2020), Asab, Zereffa, and Abdo Seghne (2020), Kayani et al. (2014), Vallejos et al. (2016) and Zhu et al. (2013) synthesized IO NPs with the co-precipitation method, solvothermal method, microemulsion method, sol-gel method, chemical vapor deposition method and sonochemical method, respectively.

The characteristics of magnetic IO NPs vary with the procedure adopted for their synthesis. Magnetite (Fe_3O_4), maghemite (γ-Fe_2O_3) and hematite (α-Fe_2O_3) are the different forms of IO NPs. The superparamagnetic property and high saturation magnetization of magnetite and maghemite lead to their biomedical applications including tumor targeting and magnetic resonance imaging (Gupta and Gupta 2005; Atta et al. 2015; Li et al. 2015; Augustin et al. 2016; Karimzadeh, Dizaji, and Aghazadeh 2016; Yin et al. 2016). $Fe2^+$/$Fe3^+$ salts' double precipitation of hydrazinated Fe(II) oxalate decomposition by temperature, microemulsion, sol-gel syntheses, hydrolysis and pyrosol are the methods employed in magnetite and maghemite synthesis (Alibeigi and Vaezi 2008; Babay, Mhiri, and Toumi 2015; Jafari et al. 2015; Rane and Verenkar 2001; Chin and Yaacob 2007; Woo et al. 2003; Herrero et al. 1997).

Magnetite can be converted to maghemite through calcination in the air ($4Fe_3O_4 + O_2 \rightarrow 6\gamma\text{-}Fe_2O_3$). Magnetite has a spinal structure. Fe^{3+} occupies the tetrahedral sites and Fe^{3+} and Fe^{2+} occupy the octahedral sites of magnetite. Maghemite has the same structure as that of magnetite except that it has cationic vacancies in one-third of the octahedral sites (Laurent et al. 2008). Nanomaghemite can also be formed by the oxidation of nZVI (Zhang et al. 2011). Synthetic nanomaghemite (nFe_2O_3) is a good sorbent of metals and metalloids. Its magnetic property makes it easily separable and can therefore be recovered and reused (Komárek et al. 2015). SiO_2- (Hu, Wang, and Pan 2010), organic molecules- (Ozmen et al. 2010) or goethite- (Cundy, Hopkinson, and Whitby 2008) coated magnetite NPs have great capacity for the remediation of metal ions such as Cu^{2+}, Hg^{2+} and Pb^{2+}. The sorption capability of Cu^{2+} can be reduced by the clustering of IO NPs. This proves the efficiency of IO NPs in removing metal ions (Gilbert et al. 2009). This treatment, which can be called assisted natural remediation, can reduce the concentration of metals in leachates and water, thereby reducing microbial toxicity (Cundy, Hopkinson, and Whitby 2008). Nano bioremediation using IO NPs can decrease the economic burden since they can be recovered and reused (Dave and Chopda 2014).

11.4.2 Applications of Iron Nanoparticles

Currently, the applications of NPs in different areas of science are increasing considerably. This is because NPs' nanoscale size and other properties make them more efficient than other materials.

The ultra-nanosize ranging from 1 to 100 nm in diameter enables them to exhibit unique physical, chemical, optical, mechanical, electrical and magnetic properties. (Srivastava, Chaubey, and Ojha 2009; Chaturvedi, Dave, and Shah 2012).

NPs have more applications in the environmental, medical, agricultural, chemical, industrial, energy production, information and communication areas. Fe NPs also have various applications in various fields. These applications include the ecofriendly decontamination of pollutants from the environment—especially toxic metal absorbents—the degradation of ionic dyes and the magnetization of sediments/cycling of Fe of stratified marine environments. Moreover, they have very high antioxidant and anti-bacterial potential (Harshiny, Iswarya, and Matheswaran 2015). NPs can play an important role in the remediation process by directly removing pollutants from the environment (due to their large surface area-to-volume ratio) by increasing the phytoavailability of pollutants and indirectly promoting plant growth. This remediation technique with the help of NPs is generally termed nano-remediation.

The application of nZVI has been widely used for groundwater remediation and has been successful in the remediation of chlorinated solvents, Cr and other toxic compounds. The strong reduction potential of zero-valent Fe makes it a more suitable agent for the removal of chlorinated solvents. It converts the toxic form to nontoxic forms such as ethene and ethane (Macé et al. 2006). NPs used in phytoremediation technology increase the efficacy of a plant to extract toxic metals, and they improve plant growth. The adsorption of toxic metals is improved in NP-assisted phytoremediation. The application of zero-valent Fe plays a significant role in the removal of toxic metals Cd and Pb. The phytoextraction potential of *Boehmeria nivea* and *Lolium perenne* showed an improved antioxidant system and became more tolerant under heavy metal stress (Song et al. 2019). Ryegrass has been used for the remediation of toxic Pb from the soil. Huang et al. (2018) studied the application of nZVI in the phytoremediation of Pb. It effectively increased the Pb accumulation in plants even at a low concentration of nZVI.

The overuse of pesticides on agricultural fields potentially increases the arsenic in soil, which is the most toxic heavy metal chiefly present in pesticides, herbicides, phosphate fertilizers and wood preservatives. The decontamination of arsenic-affected soil with plants is sustainable and ecofriendly. Plants absorb these toxic metals from the soil and remediate arsenic pollution. The phytostabilization of arsenic by sunflower was effectively improved in nZVI-treated plants over the control (Vítková, Puschenreiter, and Komárek 2018). Gao and Zhou (2013) studied the potential use of nZVI with the plant *Impatiens balsamina*. The results revealed that the combination (plant with nZVI) made it more efficient in the remediation of e-waste-contaminated soil than without nZVI. Hussain et al. (2019) reported the potential of IO NPs (Fe NPs) in alleviating Cd stress in wheat plants. IO NPs (20 ppm) diminish the effects of Cd toxicity through antioxidant pathways and result in improved growth, higher efficiency in photosynthesis and decreased Cd absorption. Similarly, Cd(II) and Cr(VI) toxicity in wheat plants was reduced by applying citrate-coated magnetite NPs (López-Luna et al. 2016). The PCB- and trinitrotoluene-contaminated soil was remediated with *Panicum maximum* and *Helianthus annuus*, and its phytoremediation potential was increased more in nZVI plants than the control. Thus, it changed the texture of the contaminated soil and made it more suitable for the growth of other plants (Jiamjitrpanich et al. 2012).

11.5 USING NANOMATERIALS TO FACILITATE PHYTOREMEDIATION

Phytoremediation is a sustainable approach of environmental cleanup in which pollutants are effectively tackled in an ecofriendly manner with the help of plants. To date, a number of plants have been identified as promising candidates for phytoremediation. However, in some cases, the decontamination process may not be satisfactorily fulfilled because of various problems, for example, the limited bioavailability of pollutants, lack of an efficient uptake mechanism, etc. These can be overcome by the employment of an intermediate agent in the remediation process, and for this purpose, NPs and engineered NMs offer legitimate solutions since NPs are harmless to plants. The following sections discuss the role of NPs in the phytoremediation process and how NPs are involved.

TABLE 11.2
Examples of the Application of Iron Nanoparticles in the Phytoremediation of Contaminated Soils

Iron nanoparticle type	Type of pollutant	Plant species	Action	References
Zero-valent iron nanoparticles	Cadmium (Cd)	*Boehmeria nivea*	Plant growth promotion Increased Cd absorption by plant tissues	(Gong et al. 2021)
		Salix alba	Increased the bioconcentration factor of Cd and plant growth	(Mokarram-Kashtiban et al. 2019)
		Helianthus annuus L.	Cd immobilization	Vítková, Puschenreiter, and Komárek (2018)
	Lead (Pb)	*Celosia argentea*	Reduction of Pb bioavailability	(Emina, Okuo, and Anegbe 2017)
		Kochia scoparia	Enhancement of Pb uptake and accumulation in plant	(Daryabeigi Zand et al. 2020)
		Lolium perenne	Increased Pb uptake and accumulation by plant	(Huang et al. 2018)
		Helianthus annuus L.	Cd immobilization	Vítková, Puschenreiter, and Komárek (2018)
	Zinc (Zn)	*Helianthus annuus L.*	Zn immobilization	Vítková, Puschenreiter, and Komárek (2018)
	Arsenic (As)	*Helianthus annuus L.*	As immobilization	Vítková, Puschenreiter, and Komárek (2018)
	Polychlorinated biphenyl (PCB)	*Impatiens balsamina*	Increased PCB accumulation by plant	(Gao and Zhou 2013)
	2,4,6-trinitrotoluene (TNT)	*Panicum maximum*	TNT degradation and removal	(Jiamjitrpanich et al. 2012)
FeNPs	Cadmium (Cd)	*Triticum aestivum*	Cd toxicity mitigation	(Hussain et al. 2019)
Magnetite NPs	Chromium (Cr) and Cadmium (Cd)	*Triticum aestivum*	Cd and Cr toxicity alleviation	(López-Luna et al. 2016)

11.5.1 How Nanomaterials Can Function in the Phytoremediation System

Many NMs have been widely employed in the remediation of environmental pollutants during recent years due to the ecofriendly attributes of nano-phytoremediation technology. The unique feature of NPs that makes them ideal for this technique include their large surface area-to-volume ratio when it is converted from a bulk to a nanoscale particle. Due to this particular quality, NMs can absorb/adsorb a large number of materials (pollutants) with which they come into contact in their surroundings, thereby increasing the reactivity (Rizwan et al. 2014; Tyagi et al. 2017). Thus, NMs make suitable agents for the environmental cleanup of various pollutants. However, for successful remediation with NPs, a number of criteria have to be fulfilled. The most important one is the selection of the proper NPs, which must be based on the type of pollutants and the nature of the plants with which they will interact.

11.5.2 Selection of an Ideal Nanoparticle for Phytoremediation

In recent years, a variety of NPs have been found to be excellent in removing certain pollutants, which reinforces the fact that selection of the proper NPs for remediation is crucial in defining the efficiency of the process. For example, ZnO and Fe NPs are excellent for the removal of toxic metal ions, whereas gold (Au) and Ag NPs are exceptional for removing organic/inorganic pollutants (Yadav et al. 2017; Verma, Roy, and Bharadvaja 2021). In addition, the stability of NPs is of great significance since it may negatively affect the process of remediation; sometimes, there is a chance that the NP itself becomes a pollutant (Srivastav et al. 2018). Hence, before selecting an NP for phytoremediation, it has to be ensured that the selected NP is appropriate for the kind of pollutants to be removed, and it must be harmless and nontoxic for the plant.

11.5.3 Selection of the Proper Plant Species for Remediation

It is equally important to choose the right plant species for remediation since each plant differs in its metabolism, and accordingly, the interaction with the NP also differs, which is critical for defining the efficiency. For any type of pollutant, before selecting the plants, it must be assured that the plant is fast growing with a dense root system and can produce a large biomass. Well-developed and expansive roots are crucial since they provide a large surface area for the pollutants to absorb/adsorb so as to initiate the decontamination process. Another important aspect for consideration is the tolerance level of plants towards the contaminants. This obviously must be good if they are hyperaccumulators of pollutants and can carry out their growth and development process without any sign of toxicity (Verma, Roy, and Bharadvaja 2021). Moreover, if the accumulated contaminants are mainly translocated aboveground, then it will be more convenient to handle the biomass after successful remediation. It is more ideal when the plant selected is non-consumable for humans and other animals so that the chance to enter and contaminate the food chain can be avoided (Srivastav et al. 2018). In addition, in the recent approaches towards nano-phytoremediation, the generation of transgenic crops with a high potential to accumulate and detoxify contaminants is more focused. For this, plants that are easily manipulated through genetic engineering techniques are highly recommended since they can be easily incorporated with stress-resistant genes that will, in turn, enhance the rate of detoxification.

11.5.4 Phytointeraction with the Nanoparticle

The proper interaction between plants and the selected NP is very important as this will increase the reactivity between them, which will subsequently improve the efficiency of nano-phytoremediation. The size, shape and chemical nature of the NP are critical since these factors will influence the uptake of NPs by the plant (Romeh and Saber 2020). Among these features, size is the rate-determining feature since the uptake will be limited if the size of the NP is larger. An increased size of the NP will also negatively affect the transportation of the NP within the plant. Hence, the penetration of the NP through the plant root is the first and crucial step in the process of nano-phytoremediation and thereby defines the efficiency of the technique (Schwab et al. 2016). Immediately after entering the root, the NP can move inward either through the apoplastic pathway (through cell walls) or through the symplastic pathway (through cytoplasm) (Sattelmacher 2001; Roberts and Oparka 2003; Srivastav et al. 2018). In addition to the roots, NPs can directly enter plants through structures such as the stomata, cuticle and trichomes (Khan et al. 2019).

The uptake and distribution of NPs within plants may be affected by several internal and external factors, which also influence the rate of nano-phytoremediation. External features involve the atmospheric parameters including the temperature, moisture content, soil pH and amount of organic matter in the soil. One major internal feature is the anatomy of the roots and the chemical nature of the enzymes and other compounds in the plant. In addition to phytoenzymes, the nature of the plant growth regulators also becomes critical for the effective reactivity of NPs in the detoxification process. Moreover, the bioavailability of NPs is also considered because it will delimit the uptake of NPs from the soil; it is suggested that coating the NPs with polymers or similar compounds will increase the bioavailability (Srivastav et al. 2018). NPs employ several strategies for the decontamination/detoxification of pollutants within plants. These may include mechanisms such as adsorption, absorption, redox reactions, degradation, precipitation and transformation. However, the exact reaction between the NPs and the plant components that results in detoxification has yet to be revealed.

11.6 CONCLUSION

Nanotechnology can provide effective solutions to many pollution-related problems, such as heavy metal pollution, oil pollution, adverse effects of chemical pollutants and soil pollution. Nano bioremediation is one of the promising technologies to detoxify various harmful pollutants in the environment. The unique application of NPs makes them very useful in various fields, especially in environmental remediation. The combination of NP biosynthesis and repair is a new hope for a clean environment. A sustainable future large number of NPs (Au, Ag, Cu, Fe and Zn) are obtained from living organisms, including bacteria, fungi and plant extracts. The biosynthesis of NPs is very simple and cost-effective, minimizes the use of harmful chemicals and takes less time. Many potential environmental pollutants, such as heavy metals and organic molecules such as toxic dyes, have already been removed from polluted areas by NPs. Considering that they have great potential to reduce environmental pollution, researchers are most interested in finding novel and effective methods to biosynthesize various NPs.

REFERENCES

Akhbarizadeh, Razegheh, Mohammad Reza Shayestefar, and Esmaeel Darezereshki. 2013. "Competitive removal of metals from wastewater by maghemite nanoparticles: a comparison between simulated wastewater and AMD." *Mine Water and the Environment*:1–7.

Alfadul, Sulaiman Mohammed. 2006. *Using Magnetic Extractants for Removal of Pollutants from Water via Magnetic Filtration*. Oklahoma State University.

Ali, Attarad, Muhammad Zia Hira Zafar, Ihsan ul Haq, Abdul Rehman Phull, Joham Sarfraz Ali, and Altaf Hussain. 2016. "Synthesis, characterization, applications, and challenges of iron oxide nanoparticles." *Nanotechnology, Science and Applications* 9:49.

Alibeigi, Samaneh, and Mohammad Reza Vaezi. 2008. "Phase transformation of iron oxide nanoparticles by varying the molar ratio of Fe2+: Fe3+." *Chemical Engineering & Technology: Industrial Chemistry-Plant Equipment-Process Engineering-Biotechnology* 31 (11):1591–1596.

Anjum, Muzammil, R Miandad, Muhammad Waqas, F Gehany, and MA Barakat. 2019. "Remediation of wastewater using various nano-materials." *Arabian Journal of Chemistry* 12 (8):4897–4919.

Asab, Goshu, Enyew Amare Zereffa, and Teshome Abdo Seghne. 2020. "Synthesis of silica-coated fe3o4 nanoparticles by microemulsion method: characterization and evaluation of antimicrobial activity." *International Journal of Biomaterials* 2020.

Atta, Aly H, Mosad A El-ghamry, Adel Hamzaoui, and Moamen S Refat. 2015. "Synthesis and spectroscopic investigations of iron oxide nano-particles for biomedical applications in the treatment of cancer cells." *Journal of Molecular Structure* 1086:246–254.

Augustin, Ewa, Bartłomiej Czubek, Anna M Nowicka, Agata Kowalczyk, Zbigniew Stojek, and Zofia Mazerska. 2016. "Improved cytotoxicity and preserved level of cell death induced in colon cancer cells by doxorubicin after its conjugation with iron-oxide magnetic nanoparticles." *Toxicology in Vitro* 33:45–53.

Babay, Salem, Tahar Mhiri, and Mouhamed Toumi. 2015. "Synthesis, structural and spectroscopic characterizations of maghemite γ-Fe2O3 prepared by one-step coprecipitation route." *Journal of Molecular Structure* 1085:286–293.

Bagheri, Samira, and Nurhidayatullaili Muhd Julkapli. 2016. "Modified iron oxide nanomaterials: functionalization and application." *Journal of Magnetism and Magnetic Materials* 416:117–133.

Banerjee, Joydeep, and Chittaranjan Kole. 2016. "Plant nanotechnology: an overview on concepts, strategies, and tools." *Plant Nanotechnology*:1–14.

Banerjee, Reshmi, Yelena Katsenovich, Leonel Lagos, M McIintosh, Xueji Zhang, and C-Z Li. 2010. "Nanomedicine: magnetic nanoparticles and their biomedical applications." *Current Medicinal Chemistry* 17 (27):3120–3141.

Bayatsarmadi, Bita, Yao Zheng, Anthony Vasileff, and Shi-Zhang Qiao. 2017. "Recent advances in atomic metal doping of carbon-based nanomaterials for energy conversion." *Small* 13 (21):1700191.

Cai, Fei, Xinyi Wu, Haiyun Zhang, Xiaofang Shen, Meng Zhang, Weixiao Chen, Qian Gao, Jason C White, Shu Tao, and Xilong Wang. 2017. "Impact of TiO2 nanoparticles on lead uptake and bioaccumulation in rice (Oryza sativa L.)." *NanoImpact* 5:101–108.

Chaturvedi, Shalini, Pragnesh N Dave, and NK Shah. 2012. "Applications of nano-catalyst in new era." *Journal of Saudi Chemical Society* 16 (3):307–325.

Chen, Yen-Hua, and Fu-An Li. 2010. "Kinetic study on removal of copper (II) using goethite and hematite nano-photocatalysts." *Journal of Colloid and Interface Science* 347 (2):277–281.

Chin, Ang Bee, and Iskandar Idris Yaacob. 2007. "Synthesis and characterization of magnetic iron oxide nanoparticles via w/o microemulsion and Massart's procedure." *Journal of Materials Processing Technology* 191 (1–3):235–237.

Chomoucka, Jana, Jana Drbohlavova, Dalibor Huska, Vojtech Adam, Rene Kizek, and Jaromir Hubalek. 2010. "Magnetic nanoparticles and targeted drug delivering." *Pharmacological Research* 62 (2):144–149.

Cundy, Andrew B, Laurence Hopkinson, and Raymond LD Whitby. 2008. "Use of iron-based technologies in contaminated land and groundwater remediation: a review." *Science of the Total Environment* 400 (1–3):42–51.

da Conceição Gomes, Maria Angélica, Rachel Ann Hauser-Davis, Adriane Nunes de Souza, and Angela Pierre Vitória. 2016. "Metal phytoremediation: general strategies, genetically modified plants and applications in metal nanoparticle contamination." *Ecotoxicology and Environmental Safety* 134:133–147.

Darr, Jawwad A, Jingyi Zhang, Neel M Makwana, and Xiaole Weng. 2017. "Continuous hydrothermal synthesis of inorganic nanoparticles: applications and future directions." *Chemical Reviews* 117 (17):11125–11238.

Daryabeigi Zand, Ali, Alireza Mikaeili Tabrizi, Ali Daryabeigi Zand, and Alireza Mikaeili Tabrizi. 2020. "Effect of zero-valent iron nanoparticles on the phytoextraction ability of Kochia scoparia and its response in Pb contaminated soil." *Environmental Engineering Research* 26 (4).

Das, Pallabi, Shaswat Barua, Shuvasree Sarkar, Niranjan Karak, Pradip Bhattacharyya, Nadeem Raza, Ki-Hyun Kim, and Satya Sundar Bhattacharya. 2018. "Plant extract—mediated green silver nanoparticles: efficacy as soil conditioner and plant growth promoter." *Journal of Hazardous Materials* 346:62–72.

Dave, Pragnesh N, and Lakhan V Chopda. 2014. "Application of iron oxide nanomaterials for the removal of heavy metals." *Journal of Nanotechnology* 2014.

Duncan, Elliott, and Gary Owens. 2019. "Metal oxide nanomaterials used to remediate heavy metal contaminated soils have strong effects on nutrient and trace element phytoavailability." *Science of the Total Environment* 678:430–437.

El Wakeil, Nabil, Saad Alkahtani, and Nawal Gaafar. 2017. "Is nanotechnology a promising field for insect pest control in IPM programs?" In *New Pesticides and Soil Sensors*, 273–309. Elsevier.

Emina, Anwuli, James Okuo, and Bala Anegbe. 2017. "Pb uptake by quail grass (Celosia argentea) grown on Pb—acid battery soil treated with starch stabilized zero-valent iron nanoparticles." *Ife Journal of Science* 19 (2):283–291.

Gao, Yuan Yuan, and Qi Xing Zhou. 2013. "Application of nanoscale zero valent iron combined with Impatiens balsamina to remediation of e-waste contaminated soils." *Advanced Materials Research* 790:73–76.

Ghosh, Saikat, Wei Jiang, Julian D McClements, and Baoshan Xing. 2011. "Colloidal stability of magnetic iron oxide nanoparticles: influence of natural organic matter and synthetic polyelectrolytes." *Langmuir* 27 (13):8036–8043.

Gilbert, Benjamin, Reyn K Ono, Kristen A Ching, and Christopher S Kim. 2009. "The effects of nanoparticle aggregation processes on aggregate structure and metal uptake." *Journal of Colloid and Interface Science* 339 (2):285–295.

Giraldo, Juan Pablo, Markita P Landry, Sean M Faltermeier, Thomas P McNicholas, Nicole M Iverson, Ardemis A Boghossian, Nigel F Reuel, Andrew J Hilmer, Fatih Sen, and Jacqueline A Brew. 2014. "Plant nanobionics approach to augment photosynthesis and biochemical sensing." *Nature materials* 13 (4):400–408.

Gong, Xiaomin, Danlian Huang, Yunguo Liu, Dongsheng Zou, Xi Hu, Lu Zhou, Zhibin Wu, Yang Yang, and Zhihua Xiao. 2021. "Nanoscale zerovalent iron, carbon nanotubes and biochar facilitated the phytoremediation of cadmium contaminated sediments by changing cadmium fractions, sediments properties and bacterial community structure." *Ecotoxicology and Environmental Safety* 208:111510.

Guo, Haibo, and Amanda S Barnard. 2013. "Naturally occurring iron oxide nanoparticles: morphology, surface chemistry and environmental stability." *Journal of Materials Chemistry A* 1 (1):27–42.

Gupta, Ajay Kumar, and Mona Gupta. 2005. "Synthesis and surface engineering of iron oxide nanoparticles for biomedical applications." *biomaterials* 26 (18):3995–4021.

Haghighi, Maryam, Zahra Afifipour, and Maryam Mozafarian. 2012. "The effect of N-Si on tomato seed germination under salinity levels." *Journal of Biological & Environmental Sciences* 6 (16):87–90.

Hamley, IW. 2003. "Nanotechnology with soft materials." *Angewandte Chemie International Edition* 42 (15):1692–1712.

Hanif, Hina Umber, Muhammad Arshad, Muhammad Arif Ali, Niaz Ahmed, and Ishtiaq Ahmed Qazi. 2015. "Phyto-availability of phosphorus to Lactuca sativa in response to soil applied TiO2 nanoparticles." *Pakistan Journal of Agricultural Sciences* 52 (1):177–182.

Harshiny, Muthukumar, Chandrasekaran Nivedhini Iswarya, and Manickam Matheswaran. 2015. "Biogenic synthesis of iron nanoparticles using Amaranthus dubius leaf extract as a reducing agent." *Powder Technology* 286:744–749.

Herrero, E, MV Cabanas, M Vallet-Regi, JL Martinez, and JM Gonzalez-Calbet. 1997. "Influence of synthesis conditions on the γ-Fe2O3 properties." *Solid State Ionics* 101:213–219.

Hong, Jingping, Bo Wang, Guiqin Xiao, Ning Wang, Yuhua Zhang, Andrei Y Khodakov, and Jinlin Li. 2020. "Tuning the metal—support interaction and enhancing the stability of titania-supported cobalt fischer—tropsch catalysts via carbon nitride coating." *ACS Catalysis* 10 (10):5554–5566.

Hu, Haibo, Zhenghua Wang, and Ling Pan. 2010. "Synthesis of monodisperse Fe3O4@ silica core—shell microspheres and their application for removal of heavy metal ions from water." *Journal of Alloys and Compounds* 492 (1–2):656–661.

Huang, Danlian, Xiang Qin, Zhiwei Peng, Yunguo Liu, Xiaomin Gong, Guangming Zeng, Chao Huang, Min Cheng, Wenjing Xue, and Xi Wang. 2018. "Nanoscale zero-valent iron assisted phytoremediation of Pb in sediment: impacts on metal accumulation and antioxidative system of Lolium perenne." *Ecotoxicology and Environmental Safety* 153:229–237.

Hussain, Afzal, Shafaqat Ali, Muhammad Rizwan, Muhammad Zia ur Rehman, Muhammad Farooq Qayyum, Hailong Wang, and Jörg Rinklebe. 2019. "Responses of wheat (Triticum aestivum) plants grown in a Cd contaminated soil to the application of iron oxide nanoparticles." *Ecotoxicology and Environmental Safety* 173:156–164.

Jafari, A, S Farjami Shayesteh, M Salouti, and K Boustani. 2015. "Effect of annealing temperature on magnetic phase transition in Fe3O4 nanoparticles." *Journal of Magnetism and Magnetic Materials* 379:305–312.

Jeevanantham, S, A Saravanan, RV Hemavathy, P Senthil Kumar, PR Yaashikaa, and D Yuvaraj. 2019. "Removal of toxic pollutants from water environment by phytoremediation: a survey on application and future prospects." *Environmental Technology & Innovation* 13:264–276.

Jiamjitrpanich, Waraporn, Preeda Parkpian, Chongrak Polprasert, and Rachain Kosanlavit. 2012. "Enhanced phytoremediation efficiency of TNT-contaminated soil by nanoscale zero valent iron." *2nd International Conference on Environment and Industrial Innovation IPCBEE*. Available from: https://www.researchgate.net/publication/229090834_Enhanced_Phytoremediation_Efficiency_of_TNT-Contaminated_Soil_by_Nanoscale_Zero_Valent_Iron

Jiang, Wenjun, Miguel Pelaez, Dionysios D Dionysiou, Mohammad H Entezari, Dimitra Tsoutsou, and Kevin O'Shea. 2013. "Chromium (VI) removal by maghemite nanoparticles." *Chemical Engineering Journal* 222:527–533.

Kalidhasan, Sethu, Ishai Dror, and Brian Berkowitz. 2017. "Atrazine degradation through PEI-copper nanoparticles deposited onto montmorillonite and sand." *Scientific Reports* 7 (1):1–13.

Kanel, Sushil Raj, Bruce Manning, Laurent Charlet, and Heechul Choi. 2005. "Removal of arsenic (III) from groundwater by nanoscale zero-valent iron." *Environmental Science & Technology* 39 (5):1291–1298.

Karimzadeh, Isa, Hamid Rezagholipour Dizaji, and Mustafa Aghazadeh. 2016. "Development of a facile and effective electrochemical strategy for preparation of iron oxides (Fe3O4 and γ-Fe2O3) nanoparticles from aqueous and ethanol mediums and in situ PVC coating of Fe3O4 superparamagnetic nanoparticles for biomedical applications." *Journal of Magnetism and Magnetic Materials* 416:81–88.

Karn, Barbara, Todd Kuiken, and Martha Otto. 2009. "Nanotechnology and in situ remediation: a review of the benefits and potential risks." *Environmental Health Perspectives* 117 (12):1813–1831.

Kayani, Zohra Nazir, Sana Arshad, Saira Riaz, and Shahzad Naseem. 2014. "Synthesis of iron oxide nanoparticles by sol—gel technique and their characterization." *IEEE Transactions on Magnetics* 50 (8):1–4.

Khan, Ibrahim, Khalid Saeed, and Idrees Khan. 2019. "Nanoparticles: properties, applications and toxicities." *Arabian Journal of Chemistry* 12 (7):908–931.

Khan, Mujeebur Rahman, Vojtech Adam, Tanveer Fatima Rizvi, Baohong Zhang, Faheem Ahamad, Izabela Jośko, Ye Zhu, Mingying Yang, and Chuanbin Mao. 2019. "Nanoparticle—plant interactions: two-way traffic." *Small* 15 (37):1901794.

Khodakovskaya, Mariya V, Kanishka De Silva, Alexandru S Biris, Enkeleda Dervishi, and Hector Villagarcia. 2012. "Carbon nanotubes induce growth enhancement of tobacco cells." *ACS Nano* 6 (3):2128–2135.

Komárek, Michael, Carla M Koretsky, Krishna J Stephen, Daniel S Alessi, and Vladislav Chrastný. 2015. "Competitive adsorption of Cd (II), Cr (VI), and Pb (II) onto nanomaghemite: a spectroscopic and modeling approach." *Environmental Science & Technology* 49 (21):12851–12859.

Komárek, Michael, Aleš Vaněk, and Vojtěch Ettler. 2013. "Chemical stabilization of metals and arsenic in contaminated soils using oxides—a review." *Environmental Pollution* 172:9–22.

Koutsospyros, Agamemnon, Julius Pavlov, Jacqueline Fawcett, David Strickland, Benjamin Smolinski, and Washington Braida. 2012. "Degradation of high energetic and insensitive munitions compounds by Fe/Cu bimetal reduction." *Journal of Hazardous Materials* 219:75–81.

Kuppusamy, Saranya, Thavamani Palanisami, Mallavarapu Megharaj, Kadiyala Venkateswarlu, and Ravi Naidu. 2016. "Ex-situ remediation technologies for environmental pollutants: a critical perspective." *Reviews of Environmental Contamination and Toxicology* 236:117–192.

Laurent, Sophie, Jean-Luc Bridot, Luce Vander Elst, and Robert N Muller. 2010. "Magnetic iron oxide nanoparticles for biomedical applications." *Future Medicinal Chemistry* 2 (3):427–449.

Laurent, Sophie, Delphine Forge, Marc Port, Alain Roch, Caroline Robic, Luce Vander Elst, and Robert N Muller. 2008. "Magnetic iron oxide nanoparticles: synthesis, stabilization, vectorization, physicochemical characterizations, and biological applications." *Chemical Reviews* 108 (6):2064–2110.

Li, Huan, Kai Yan, Yalei Shang, Lochan Shrestha, Rufang Liao, Fang Liu, Penghui Li, Haibo Xu, Zushun Xu, and Paul K Chu. 2015. "Folate-bovine serum albumin functionalized polymeric micelles loaded with superparamagnetic iron oxide nanoparticles for tumor targeting and magnetic resonance imaging." *Acta Biomaterialia* 15:117–126.

Liang, Pei, Yan Liu, Li Guo, Jing Zeng, and Hanbing Lu. 2004. "Multiwalled carbon nanotubes as solid-phase extraction adsorbent for the preconcentration of trace metal ions and their determination by inductively coupled plasma atomic emission spectrometry." *Journal of Analytical Atomic Spectrometry* 19 (11):1489–1492.

Lin, Sen, Diannan Lu, and Zheng Liu. 2012. "Removal of arsenic contaminants with magnetic γ-Fe2O3 nanoparticles." *Chemical Engineering Journal* 211:46–52.

Liu, Jianv, Xin Xin, and Qixing Zhou. 2018. "Phytoremediation of contaminated soils using ornamental plants." *Environmental Reviews* 26 (1):43–54.

López-Luna, J, MJ Silva-Silva, S Martinez-Vargas, OF Mijangos-Ricardez, MC González-Chávez, FA Solís-Domínguez, and MC Cuevas-Díaz. 2016. "Magnetite nanoparticle (NP) uptake by wheat plants and its effect on cadmium and chromium toxicological behavior." *Science of the Total Environment* 565:941–950.

Ma, Xingmao, Jane Geiser-Lee, Yang Deng, and Andrei Kolmakov. 2010. "Interactions between engineered nanoparticles (ENPs) and plants: phytotoxicity, uptake and accumulation." *Science of the Total Environment* 408 (16):3053–3061.

Macé, Christian, Steve Desrocher, Florin Gheorghiu, Allen Kane, Michael Pupeza, Miroslav Cernik, Petr Kvapil, Ramesh Venkatakrishnan, and Wei-xian Zhang. 2006. "Nanotechnology and groundwater remediation: a step forward in technology understanding." *Remediation Journal: The Journal of Environmental Cleanup Costs, Technologies & Techniques* 16 (2):23–33.

Makarova, Olga V, Tijana Rajh, Marion C Thurnauer, Amy Martin, Patricia A Kemme, and Donald Cropek. 2000. "Surface modification of TiO2 nanoparticles for photochemical reduction of nitrobenzene." *Environmental Science & Technology* 34 (22):4797–4803.

Malhotra, Nemi, Jiann-Shing Lee, Rhenz Alfred D Liman, Johnsy Margotte S Ruallo, Oliver B Villaflores, Tzong-Rong Ger, and Chung-Der Hsiao. 2020. "Potential toxicity of iron oxide magnetic nanoparticles: a review." *Molecules* 25 (14):3159.

Martínez-Fernández, Domingo, Didac Barroso, and Michael Komárek. 2016. "Root water transport of Helianthus annuus L. under iron oxide nanoparticle exposure." *Environmental Science and Pollution Research* 23 (2):1732–1741.

Martínez-Fernández, Domingo, Martina Vítková, Zuzana Michálková, and Michael Komárek. 2017. "Engineered nanomaterials for phytoremediation of metal/metalloid-contaminated soils: implications for plant physiology." In *Phytoremediation*, 369–403. Springer.

Mehndiratta, Poorva, Arushi Jain, Sudha Srivastava, and Nidhi Gupta. 2013. "Environmental pollution and nanotechnology." *Environment and Pollution* 2 (2):49.

Mohammadi, Hamid, Ali Reza Amani-Ghadim, Amir Abbas Matin, and Mansour Ghorbanpour. 2020. "Fe 0 nanoparticles improve physiological and antioxidative attributes of sunflower (Helianthus annuus) plants grown in soil spiked with hexavalent chromium." *3 Biotech* 10 (1):1–11.

Mokarram-Kashtiban, Sahar, Seyed Mohsen Hosseini, Masoud Tabari Kouchaksaraei, and Habibollah Younesi. 2019. "The impact of nanoparticles zero-valent iron (nZVI) and rhizosphere microorganisms on the phytoremediation ability of white willow and its response." *Environmental Science and Pollution Research* 26 (11):10776–10789.

Movafeghi, Ali, Alireza Khataee, Mahboubeh Abedi, Roshanak Tarrahi, Mohammadreza Dadpour, and Fatemeh Vafaei. 2018. "Effects of TiO2 nanoparticles on the aquatic plant Spirodela polyrrhiza: evaluation of growth parameters, pigment contents and antioxidant enzyme activities." *Journal of Environmental Sciences* 64:130–138.

Mueller, Nicole C, Jürgen Braun, Johannes Bruns, Miroslav Černík, Peter Rissing, David Rickerby, and Bernd Nowack. 2012. "Application of nanoscale zero valent iron (NZVI) for groundwater remediation in Europe." *Environmental Science and Pollution Research* 19 (2):550–558.

Mulens-Arias, Vladimir, José Manuel Rojas, and Domingo F Barber. 2020. "The intrinsic biological identities of iron oxide nanoparticles and their coatings: unexplored territory for combinatorial therapies." *Nanomaterials* 10 (5):837.

Mura, Stefania, Giovanna Seddaiu, Fabio Bacchini, Pier Paolo Roggero, and Gian Franco Greppi. 2013. "Advances of nanotechnology in agro-environmental studies." *Italian Journal of Agronomy*:e18–e18.

Murgueitio, Erika, Luis Cumbal, Mayra Abril, Andrés Izquierdo, Alexis Debut, and Oscar Tinoco. 2018. "Green synthesis of iron nanoparticles: application on the removal of petroleum oil from contaminated water and soils." *Journal of Nanotechnology* 2018.

Nassar, Nashaat N. 2010. "Rapid removal and recovery of Pb (II) from wastewater by magnetic nanoadsorbents." *Journal of Hazardous Materials* 184 (1–3):538–546.

Nath, Debjani, and Pratyusha Banerjee. 2013. "Green nanotechnology—a new hope for medical biology." *Environmental Toxicology and Pharmacology* 36 (3):997–1014.

Nie, Xiaoqin, Jianguo Liu, Xianwei Zeng, and Dongbei Yue. 2013. "Rapid degradation of hexachlorobenzene by micron Ag/Fe bimetal particles." *Journal of Environmental Sciences* 25 (3):473–478.

Ossai, Innocent Chukwunonso, Aziz Ahmed, Auwalu Hassan, and Fauziah Shahul Hamid. 2020. "Remediation of soil and water contaminated with petroleum hydrocarbon: a review." *Environmental Technology & Innovation* 17:100526.

Ozmen, Mustafa, Keziban Can, Gulsin Arslan, Ali Tor, Yunus Cengeloglu, and Mustafa Ersoz. 2010. "Adsorption of Cu (II) from aqueous solution by using modified Fe3O4 magnetic nanoparticles." *Desalination* 254 (1–3):162–169.

Patil, Sayali S, Utkarsha U Shedbalkar, Adam Truskewycz, Balu A Chopade, and Andrew S Ball. 2016. "Nanoparticles for environmental clean-up: a review of potential risks and emerging solutions." *Environmental Technology & Innovation* 5:10–21.

Que, Emily L, Dylan W Domaille, and Christopher J Chang. 2008. "Metals in neurobiology: probing their chemistry and biology with molecular imaging." *Chemical Reviews* 108 (5):1517–1549.

Quideau, Stephane, Denis Deffieux, Céline Douat-Casassus, and Laurent Pouysegu. 2011. "Plant polyphenols: chemical properties, biological activities, and synthesis." *Angewandte Chemie International Edition* 50 (3):586–621.

Rafique, Rafia, Zahra Zahra, Nasar Virk, Muhammad Shahid, Eric Pinelli, Jean Kallerhoff, Tae Jung Park, and Muhammad Arshad. 2018. "Data on rhizosphere pH, phosphorus uptake and wheat growth responses upon TiO2 nanoparticles application." *Data in Brief* 17:890–896.

Rafique, Rafia, Zahra Zahra, Nasar Virk, Muhammad Shahid, Eric Pinelli, Tae Jung Park, Jean Kallerhoff, and Muhammad Arshad. 2018. "Dose-dependent physiological responses of Triticum aestivum L. to soil applied TiO2 nanoparticles: alterations in chlorophyll content, H2O2 production, and genotoxicity." *Agriculture, Ecosystems & Environment* 255:95–101.

Rai, Prabhat Kumar, Ki-Hyun Kim, Sang Soo Lee, and Jin-Hong Lee. 2020. "Molecular mechanisms in phytoremediation of environmental contaminants and prospects of engineered transgenic plants/microbes." *Science of the Total Environment* 705:135858.

Rane, KS, and VMS Verenkar. 2001. "Synthesis of ferrite grade γ-Fe 2 O 3." *Bulletin of Materials Science* 24 (1):39–45.

Rastogi, Anshu, Durgesh Kumar Tripathi, Saurabh Yadav, Devendra Kumar Chauhan, Marek Živčák, Mansour Ghorbanpour, Nabil Ibrahim El-Sheery, and Marian Brestic. 2019. "Application of silicon nanoparticles in agriculture." *3 Biotech* 9 (3):1–11.

Reddy, A, V Madhavi, K Gangadhara Reddy, and G Madhavi. 2013. "Remediation of chlorpyrifos-contaminated soils by laboratory-synthesized zero-valent nano iron particles: effect of pH and aluminium salts." *Journal of Chemistry* 2013.

Ren, Lixia, Hongwei Lu, Li He, and Yimei Zhang. 2014. "Enhanced electrokinetic technologies with oxidization—reduction for organically-contaminated soil remediation." *Chemical Engineering Journal* 247:111–124.

Rizwan, Md, Man Singh, Chanchal K Mitra, and Roshan K Morve. 2014. "Ecofriendly application of nanomaterials: nanobioremediation." *Journal of Nanoparticles* 2014.

Roberts, Alison G, and K J Oparka. 2003. "Plasmodesmata and the control of symplastic transport." *Plant, Cell and Environment* 26:103–124.

Romeh, Ahmed Ali, and Refaat Ahmed Ibrahim Saber. 2020. "Green nano-phytoremediation and solubility improving agents for the remediation of chlorfenapyr contaminated soil and water." *Journal of Environmental Management* 260:110104.

Roy, Arup, and Jayanta Bhattacharya. 2012. "Removal of Cu (II), Zn (II) and Pb (II) from water using microwave-assisted synthesized maghemite nanotubes." *Chemical Engineering Journal* 211:493–500.

Saif, Sadia, Arifa Tahir, and Yongsheng Chen. 2016. "Green synthesis of iron nanoparticles and their environmental applications and implications." *Nanomaterials* 6 (11):209.

Sajid, Muhammad, Muhammad Ilyas, Chanbasha Basheer, Madiha Tariq, Muhammad Daud, Nadeem Baig, and Farrukh Shehzad. 2015. "Impact of nanoparticles on human and environment: review of toxicity factors, exposures, control strategies, and future prospects." *Environmental Science and Pollution Research* 22 (6):4122–4143.

Samaddar, Pallabi, Yong Sik Ok, Ki-Hyun Kim, Eilhann E Kwon, and Daniel CW Tsang. 2018. "Synthesis of nanomaterials from various wastes and their new age applications." *Journal of Cleaner Production* 197:1190–1209.

Santner, Jakob, Erik Smolders, Walter W Wenzel, and Fien Degryse. 2012. "First observation of diffusion-limited plant root phosphorus uptake from nutrient solution." *Plant, Cell & Environment* 35 (9):1558–1566.

Satapanajaru, T, P Anurakpongsatorn, P Pengthamkeerati, and H Boparai. 2008. "Remediation of atrazine-contaminated soil and water by nano zerovalent iron." *Water, Air, and Soil Pollution* 192 (1):349–359.

Sattelmacher, Burkhard. 2001. "The apoplast and its significance for plant mineral nutrition." *New Phytologist* 149 (2):167–192.

Savage, Nora, and Mamadou S Diallo. 2005. "Nanomaterials and water purification: opportunities and challenges." *Journal of Nanoparticle research* 7 (4):331–342.

Schwab, Fabienne, Guangshu Zhai, Meaghan Kern, Amalia Turner, Jerald L Schnoor, and Mark R Wiesner. 2016. "Barriers, pathways and processes for uptake, translocation and accumulation of nanomaterials in plants—Critical review." *Nanotoxicology* 10 (3):257–278.

Servin, Alia D, Maria Isabel Morales, Hiram Castillo-Michel, Jose Angel Hernandez-Viezcas, Berenice Munoz, Lijuan Zhao, Jose E Nunez, Jose R Peralta-Videa, and Jorge L Gardea-Torresdey. 2013. "Synchrotron verification of TiO2 accumulation in cucumber fruit: a possible pathway of TiO2 nanoparticle transfer from soil into the food chain." *Environmental Science & Technology* 47 (20):11592–11598.

Shah, Rhythm R, Todd P Davis, Amanda L Glover, David E Nikles, and Christopher S Brazel. 2015. "Impact of magnetic field parameters and iron oxide nanoparticle properties on heat generation for use in magnetic hyperthermia." *Journal of Magnetism and Magnetic Materials* 387:96–106.

Shalaby, Tarek A, Yousry Bayoumi, Neama Abdalla, Hussein Taha, Tarek Alshaal, Said Shehata, Megahed Amer, Éva Domokos-Szabolcsy, and Hassan El-Ramady. 2016. "Nanoparticles, soils, plants and sustainable agriculture." In *Nanoscience in Food and Agriculture 1*, 283–312. Springer.

Shang, Yifen, Md Hasan, Golam Jalal Ahammed, Mengqi Li, Hanqin Yin, and Jie Zhou. 2019. "Applications of nanotechnology in plant growth and crop protection: a review." *Molecules* 24 (14):2558.

Sheng-Nan, Sun, Wei Chao, Zhu Zan-Zan, Hou Yang-Long, Subbu S Venkatraman, and Xu Zhi-Chuan. 2014. "Magnetic iron oxide nanoparticles: synthesis and surface coating techniques for biomedical applications." *Chinese Physics B* 23 (3):037503.

Shipley, Heather J, Karen E Engates, and Allison M Guettner. 2011. "Study of iron oxide nanoparticles in soil for remediation of arsenic." *Journal of Nanoparticle Research* 13 (6):2387–2397.

Siddiqi, Khwaja Salahuddin, Azamal Husen, and Rifaqat AK Rao. 2018. "A review on biosynthesis of silver nanoparticles and their biocidal properties." *Journal of Nanobiotechnology* 16 (1):1–28.

Siddiqui, Manzer H, and Mohamed H Al-Whaibi. 2014. "Role of nano-SiO2 in germination of tomato (Lycopersicum esculentum seeds Mill.)." *Saudi Journal of Biological Sciences* 21 (1):13–17.

Singh, Brajesh Kumar. 2010. "Exploring microbial diversity for biotechnology: the way forward." *Trends in Biotechnology* 28 (3):111–116.

Singh, Nisha, Jacky Bhagat, Ekta Tiwari, Nitin Khandelwal, Gopala Krishna Darbha, and SK Shyama. 2021. "Metal oxide nanoparticles and polycyclic aromatic hydrocarbons alter nanoplastic's stability and toxicity to zebrafish." *Journal of Hazardous Materials* 407:124382.

Singh, Ritu, Virendra Misra, Mohana Krishna Reddy Mudiam, Lalit Kumar Singh Chauhan, and Rana Pratap Singh. 2012. "Degradation of γ-HCH spiked soil using stabilized Pd/Fe0 bimetallic nanoparticles: pathways, kinetics and effect of reaction conditions." *Journal of Hazardous Materials* 237:355–364.

Singh, Ritu, Virendra Misra, and Rana Pratap Singh. 2012. "Removal of Cr (VI) by nanoscale zero-valent iron (nZVI) from soil contaminated with tannery wastes." *Bulletin of Environmental Contamination and Toxicology* 88 (2):210–214.

Smalley, Richard E, Jason H Hafner, Daniel T Colbert, and Ken Smith. 2004. *Catalytic Growth of Single-Wall Carbon Nanotubes from Metal Particles*. Google Patents.

Song, Biao, Piao Xu, Ming Chen, Wangwang Tang, Guangming Zeng, Jilai Gong, Peng Zhang, and Shujing Ye. 2019. "Using nanomaterials to facilitate the phytoremediation of contaminated soil." *Critical Reviews in Environmental Science and Technology* 49 (9):791–824.

Sozer, Nesli, and Jozef L Kokini. 2009. "Nanotechnology and its applications in the food sector." *Trends in Biotechnology* 27 (2):82–89.

Srivastav, Akansha, Krishna Kumar Yadav, Sunita Yadav, Neha Gupta, Jitendra Kumar Singh, Ravi Katiyar, and Vinit Kumar. 2018. "Nano-phytoremediation of pollutants from contaminated soil environment: current scenario and future prospects." In *Phytoremediation*, 383–401. Springer.

Srivastava, Manish, S Chaubey, and Animesh K Ojha. 2009. "Investigation on size dependent structural and magnetic behavior of nickel ferrite nanoparticles prepared by sol—gel and hydrothermal methods." *Materials Chemistry and Physics* 118 (1):174–180.

Tepper, T, F Ilievski, CA Ross, TR Zaman, RJ Ram, SY Sung, and BJH Stadler. 2003. "Magneto-optical properties of iron oxide films." *Journal of Applied Physics* 93 (10):6948–6950.

Thomas, John Meurig, and Robert Raja. 2006. "The advantages and future potential of single-site heterogeneous catalysts." *Topics in Catalysis* 40 (1):3–17.

Torabian, Shahram, Morteza Zahedi, and Amir Hossein Khoshgoftar. 2017. "Effects of foliar spray of nanoparticles of FeSO4 on the growth and ion content of sunflower under saline condition." *Journal of Plant Nutrition* 40 (5):615–623.

Trujillo-Reyes, J, S Majumdar, CE Botez, JR Peralta-Videa, and JL Gardea-Torresdey. 2014. "Exposure studies of core—shell Fe/Fe3O4 and Cu/CuO NPs to lettuce (Lactuca sativa) plants: are they a potential physiological and nutritional hazard?" *Journal of Hazardous Materials* 267:255–263.

Tuutijärvi, T, Riku Vahala, M Sillanpää, and Guohua Chen. 2012. "Maghemite nanoparticles for As (V) removal: desorption characteristics and adsorbent recovery." *Environmental Technology* 33 (16):1927–1936.

Tyagi, I, VK Gupta, H Sadegh, R Shahryari Ghoshekandi, and AS Hamdy Makhlouf. 2017. "Nanoparticles as adsorbent; a positive approach for removal of noxious metal ions: a review." *Science Technology and Development* 34 (3):195–214.

Vallejos, Stella, Francesco Di Maggio, Tahira Shujah, and Chris Blackman. 2016. "Chemical vapour deposition of gas sensitive metal oxides." *Chemosensors* 4 (1):4.

Varanasi, Patanjali, Andres Fullana, and Sukh Sidhu. 2007. "Remediation of PCB contaminated soils using iron nano-particles." *Chemosphere* 66 (6):1031–1038.

Venkatachalam, P, N Priyanka, K Manikandan, I Ganeshbabu, P Indiraarulselvi, N Geetha, K Muralikrishna, RC Bhattacharya, M Tiwari, and N Sharma. 2017. "Enhanced plant growth promoting role of phycomolecules coated zinc oxide nanoparticles with P supplementation in cotton (Gossypium hirsutum L.)." *Plant Physiology and Biochemistry* 110:118–127.

Verma, Ayushi, Arpita Roy, and Navneeta Bharadvaja. 2021. "Remediation of heavy metals using nanophytoremediation." In *Advanced Oxidation Processes for Effluent Treatment Plants*, 273–296. Elsevier.

Vítková, Martina, Markus Puschenreiter, and Michael Komárek. 2018. "Effect of nano zero-valent iron application on As, Cd, Pb, and Zn availability in the rhizosphere of metal (loid) contaminated soils." *Chemosphere* 200:217–226.

Wang, Li, Bin Ji, Yuehua Hu, Runqing Liu, and Wei Sun. 2017. "A review on in situ phytoremediation of mine tailings." *Chemosphere* 184:594–600.

Wang, Shihua, Fayuan Wang, and Shuangcheng Gao. 2015. "Foliar application with nano-silicon alleviates Cd toxicity in rice seedlings." *Environmental Science and Pollution Research* 22 (4):2837–2845.

Waychunas, Glenn A, Christopher S Kim, and Jillian F Banfield. 2005. "Nanoparticulate iron oxide minerals in soils and sediments: unique properties and contaminant scavenging mechanisms." *Journal of Nanoparticle Research* 7 (4):409–433.

Wenzel, Walter W, Domy C Adriano, David Salt, and Robert Smith. 1999. "Phytoremediation: a plant—microbe-based remediation system." *Bioremediation of Contaminated Soils* 37:457–508.

Woo, Kyoungja, Ho Jin Lee, J-P Ahn, and Yong Sung Park. 2003. "Sol—gel mediated synthesis of Fe2O3 nanorods." *Advanced Materials* 15 (20):1761–1764.

Xia, Tian, Jingping Wang, Chunli Wu, Fuchang Meng, Zhan Shi, Jie Lian, Jing Feng, and Jian Meng. 2012. "Novel complex-coprecipitation route to form high quality triethanolamine-coated Fe 3 O 4 nanocrystals: their high saturation magnetizations and excellent water treatment properties." *CrystEngComm* 14 (18):5741–5744.

Xu, Piao, Guang Ming Zeng, Dan Lian Huang, Chong Ling Feng, Shuang Hu, Mei Hua Zhao, Cui Lai, Zhen Wei, Chao Huang, and Geng Xin Xie. 2012. "Use of iron oxide nanomaterials in wastewater treatment: a review." *Science of the Total Environment* 424:1–10.

Yadav, KK, JK Singh, N Gupta, and VJJMES Kumar. 2017. "A review of nanobioremediation technologies for environmental cleanup: a novel biological approach." *Journal of Materials and Environmental Science* 8 (2):740–757.

Yadav, Virendra Kumar, and MH Fulekar. 2018. "Biogenic synthesis of maghemite nanoparticles (γ-Fe2O3) using Tridax leaf extract and its application for removal of fly ash heavy metals (Pb, Cd)." *Materials Today: Proceedings* 5 (9):20704–20710.

Yan, Weile, Andrew A Herzing, Xiao-qin Li, Christopher J Kiely, and Wei-xian Zhang. 2010. "Structural evolution of Pd-doped nanoscale zero-valent iron (nZVI) in aqueous media and implications for particle aging and reactivity." *Environmental Science & Technology* 44 (11):4288–4294.

Yin, Perry T, Shreyas Shah, Nicholas J Pasquale, Olga B Garbuzenko, Tamara Minko, and Ki-Bum Lee. 2016. "Stem cell-based gene therapy activated using magnetic hyperthermia to enhance the treatment of cancer." *Biomaterials* 81:46–57.

Zahra, Zahra, Muhammad Arshad, Muhammad Arif Ali, Muhammad Qudrat Ullah Farooqi, and Hyung Kyoon Choi. 2020. "Phosphorus phytoavailability upon nanoparticle application." In *Sustainable Agriculture Reviews 41*, 41–61. Springer.

Zahra, Zahra, Muhammad Arshad, Rafia Rafique, Arshad Mahmood, Amir Habib, Ishtiaq A Qazi, and Saud A Khan. 2015. "Metallic nanoparticle (TiO2 and Fe3O4) application modifies rhizosphere phosphorus availability and uptake by Lactuca sativa." *Journal of Agricultural and Food Chemistry* 63 (31):6876–6882.

Zhang, Man, Feng He, Dongye Zhao, and Xiaodi Hao. 2011. "Degradation of soil-sorbed trichloroethylene by stabilized zero valent iron nanoparticles: effects of sorption, surfactants, and natural organic matter." *Water Research* 45 (7):2401–2414.

Zhang, MeiYi, Yu Wang, DongYe Zhao, and Gang Pan. 2010. "Immobilization of arsenic in soils by stabilized nanoscale zero-valent iron, iron sulfide (FeS), and magnetite (Fe 3 O 4) particles." *Chinese Science Bulletin* 55 (4):365–372.

Zhao, Shengzhe, Xujiang Yu, Yuna Qian, Wei Chen, and Jianliang Shen. 2020. "Multifunctional magnetic iron oxide nanoparticles: an advanced platform for cancer theranostics." *Theranostics* 10 (14):6278.

Zhao, Zhongqiu, Meizhu Xi, Guangyu Jiang, Xiaona Liu, Zhongke Bai, and Yizong Huang. 2010. "Effects of IDSA, EDDS and EDTA on heavy metals accumulation in hydroponically grown maize (Zea mays, L.)." *Journal of Hazardous Materials* 181 (1–3):455–459.

Zhu, Shenmin, Jingjing Guo, Junping Dong, Zhaowen Cui, Tao Lu, Chenglin Zhu, Di Zhang, and Jun Ma. 2013. "Sonochemical fabrication of Fe3O4 nanoparticles on reduced graphene oxide for biosensors." *Ultrasonics Sonochemistry* 20 (3):872–880.

Zhu, Xiaoshan, Shengyan Tian, and Zhonghua Cai. 2012. "Toxicity assessment of iron oxide nanoparticles in zebrafish (Danio rerio) early life stages." *PLoS ONE* 7 (9):e46286.

12 Silver Nanoparticles for Nano-Phytoremediation

Recent Advancements and the Potential of Nano Silver-Consolidated Phytoremediation in Ecospheric Decontamination

Shaan Bibi Jaffri, Khuram Shahzad Ahmad, and Asma Jabeen

12.1 INTRODUCTION

Human civilization has been innovating modern approaches via enhanced urbanization and industrialization. Converting manual methods into completely mechanized forms is often seen as good progress. However, the negative impacts on the environment exerted due to such urban sprawl and unregulated industrialization, in addition to the chemicals of modern agricultural methods, cannot be overlooked (Ijaz et al. 2020c; Jaffri and Ahmad 2018d). In fact, the environment and other floral, faunal and microbial species have been undergoing the worst kinds of impacts in human history in terms of disruption to the overall health of the ecological compartments. In this regard, all compartments are equally disturbed by the contaminants produced by anthropogenic activities including the lithosphere, hydrosphere and atmosphere. The contamination of all compartments has its own implications, but pedospheric matrices and the hydrosphere are particularly prone to pollution since they act as a sink for different contaminants (Zahra et al. 2017).

A variety of pollutants that are released from industrial activities, for example, solvents, dyes and pharmaceuticals, find their way into wastewater streams. In a similar manner, soils are a reception ground for the contaminants associated with agricultural practices. For instance, the agrochemicals sprayed on a target crop are received by the target in a very meagre percentage, while the rest of the chemical is received by the immediate ground. In this way, the persistent organic pollutants (POPs), polyaromatic hydrocarbons (PAHs) and benzene-based products, including phenanthrene, heavy metals, pesticides and fertilizers, become a part of the pedospheric matrices and undergo several biochemical procedures (Ali et al. 2019; Amjad et al. 2019). The persistence and stability associated with different types of contaminants pose a serious threat to the achievement of the Sustainable Development Goals. In addition to such contaminants, soils and water bodies are also reception media for various types of nanomaterials released because of anthropogenic activities. The pedospheric compartment has been known especially as a primary sink for the aggregation of nanoparticles (NPs) in environmental settings (Ma et al. 2018) and in soils exceeding the hydrospheric and atmospheric compartments in accumulating nanomaterials (Nair 2018).

In the face of such contamination, the sustainable remediation of the pedospheric matrices and hydrosphere has become an urgent matter to protect the environment from further deterioration and maintain its integrity (Shaheen et al. 2016). A myriad of physical, chemical and biological methods have been devised, and research is ongoing in each domain to develop the best possible solution

DOI: 10.1201/9781003186298-12

that is acceptable in terms of monetary considerations, environmental quality and simpler operation and that is not marked by complicated operational procedures requiring labor-intensive work. More specifically, the cleanup of the hydrosphere is performed through different techniques, for example, precipitation–filtration (Estay et al. 2021), ion exchange (Mabrouk et al. 2021), flocculation (Naruka et al. 2021), reverse–osmosis (Jafari et al. 2021), coagulation (Campinas et al. 2021), electro-dialysis (Ramasamy et al. 2021), carbon adsorption (Xu et al. 2021), chelate-flushing (Omran 2021), and oxidation–reduction (Lukashev et al. 2021). In the case of soils, the extraction of contaminants is often conducted through soil flushing (Senevirathna et al. 2021), solidification (Tang et al. 2021), electro-kinetic treatment (Ali et al. 2019), stabilization (Bazarbekova et al. 2021), and vitrification (Shu et al. 2021). The ongoing research in this area related to physicochemical and biological methods suggests further investigation since none of these methods has completely fulfilled the sustainability criteria. In fact, in most cases, the site to be restored is deprived of its intrinsic value due to the disruption of the soil inherent structure; sometimes, the chemical treatments might remove the target pollutant but end up releasing even more synthetic substance into the environment.

Considering the unsatisfactory results associated with physicochemical methods, the utilization of nature's tools for the remediation of the environmental compartments has become a sustainable, facile, economic and environmentally friendly option (Shazia et al. 2018; Iram et al. 2018; Iftikhar et al. 2019). In this type of remediation, the pedospheric matrices and hydrospheric regions are detoxified by utilizing different plants that have the potential of taking up pollutants in their bodies and cleaning the environment where they are grown (Yan et al. 2021). This phyto-technology referred to as phytoremediation has been gaining interest in the last four decades because of its advantageous features and public acceptance on a global scale, in addition to its aesthetic appeal. Phytoremediation is based on the utilization of the metabolic system of hyper-accumulator plants to remove different types of contaminants and nutrients from the immediate environment where they are grown (Sangsuwan and Prapagdee 2021). The contaminants are then stored in the bodies of such phytoremediant plants obtainable upon harvesting. The selection of the plant species aimed at the cleanup of soils and water bodies needs to be very meticulous to derive the maximum results. For instance, the hyper-accumulator plants should ideally possess larger root systems since a greater length of the roots in close proximity to the contaminated site will lead to greater contaminant uptake due to enhanced interaction. Moreover, to ensure the effective detoxification of the lithosphere and hydrosphere, the hyper-accumulator plants need to be aggressive in the accumulation of the heavy metals, nutrients and other contaminants (Khalid et al. 2021; Han et al. 2021). Furthermore, the rate of reproduction in such plant species is also an integral factor in determining the overall success of the phytoremediation system. If the plants rapidly reproduce, then this will be representative of self-sustenance and will directly reduce the costs incurred for the remediation procedure. The phytoremediation process is comparatively dynamic and is aimed at the removal, transferal, stabilization, metabolization and detoxification of contaminants by employing specialized plants and has several advantages including long-term applicability (Nedjimi 2021; Rono et al. 2021).

Nanotechnology and nanoscience have been revolutionizing human life since their inception. A great deal of nanomaterials are being produced on a daily basis and applied in practical life to derive maximum benefits. The application of different types of nanomaterials have encompassed all fields such as biomedicine, photovoltaics, textiles, cosmetics, food packaging and agricultural products, including nano-fertilizers and nano-pesticides (Ahmadian et al. 2021; Morab et al. 2021). Nanomaterials are referred to as a unique class of materials that have at least one dimension less than 100 nm (Krishna et al. 2018), and they are engineered to obtain desired results depending upon the type of physicochemical and biological application in which they are being employed. A great variety of engineered nanomaterials have been fabricated through different routes, and they are often classified broadly into carbon-based, polymeric, metal/metal oxide-based, semiconductors and ceramic nanomaterials (Jaffri and Ahmad 2018a, 2018b; Ijaz et al. 2020a). Such nanomaterials exist in a myriad of morphologies, for example, nanospindles, nanofibers, nanofalcates, nanotubes, quantum dots and nanocrystals (Ijaz et al. 2020b; Tahir et al. 2020). The type of application usually takes into account the physicochemical features of any NPs such as the morphological aspects, size ranges, type, functionalization and any type of surficial coating on it (Iqbal et al. 2019; Afsheen et al. 2020). Previous studies are supportive of the repetitive utilization of silver (Ag) and gold (Au) NPs in various industrial, medical and agricultural applications by the dint of their auspicious features (Khan et al. 2019).

Among many different NPs, Ag NPs are studied in various applications in the agriculture, communications, biomedical and energy fields, among others. In this way, there are fair chances of the release of these Ag NPs into the environmental compartments. Interestingly, 25% of the consumer products using nano-range materials are based on the utilization of Ag NPs (Vance et al. 2015). Therefore, these NPs need to be studied comprehensively to gain a better understanding of them in the environment and agricultural systems. Moreover, plants express varied physiological and molecular reactions upon exposure to Ag NPs compared with the salt of silver nitrate (Noori et al. 2020). In terms of biological applications, Ag NPs have been known for their antimicrobial (Jaffri and Ahmad 2018c), antifungal (Ahmad and Jaffri 2018b) and antiviral (Ismail et al. 2021) actions against various species. Additionally, they are also employed in different domains of radiotherapy and wound healing (Sears et al. 2021), food packaging, water disinfection, and as nano-pesticides for crop improvement. The field of agri-nanotechnology is revolutionizing agricultural products by utilizing Ag NPs to improve the pedospheric quality and plant development and maturation (Fraceto et al. 2016; White and Gardea-Torresdey 2018). Previous studies conducted with Ag NPs expressed their profound impacts on the growth of plants by affecting plants' physiology, biochemistry, metabolism, taking up of water and other nutrients, rate of photosynthesis, rate of respiration and transpiration, stress inflicted due to oxidative reactions, generation of reactive oxygen species (ROS) and antioxidant reactions. Notably, Ag might exhibit varied responses in different environments. For example, Ag NPs in the nanoscales have proved to be more effective and reactive inside bio-systems than their bulk form because of their minute size ranges and comparatively larger surficial area (Dobias and Bernier-Latmani 2013). Furthermore, Ag NPs can undergo various types of oxidative reactions inside the pedospheric matrices that can lead to the emission of Ag ions, which are characterized by their higher availability and interaction than those of Ag NPs (Castro-Mayorga et al. 2017).

The miraculous features of nanotechnology, especially Ag NPs and phytoremediation, can be employed inside one consolidated system and used for the remediation of environmental contamination occurring in the pedospheric matrices and hydrosphere. The concept of nanotools-enabled phytoremediation is a great candidate for ecological detoxification from different pollutants and is an emerging concept. However, the analysis of the literature in this regard shows that it is a highly understudied and underapplied domain despite the vast potential associated with it. A large number of review articles have been produced on the miraculous functionality of Ag NPs (Das et al. 2020; Saim et al. 2021; Dawadi et al. 2021) and phytoremediation (Durand et al. 2021; Nguyen et al. 2021; Rathika et al. 2021) in an isolated manner. However, no review article has been produced that reports the potential and utilization of Ag NPs for environmental detoxification in collaboration with the process of phytoremediation. Considering the auspicious role of nanotechnology combined with phytoremediation, the current chapter summarizes for the first time the studies performed with Ag NPs coupled with different hyper-accumulators to detoxify the pedospheric and hydrospheric compartments. The chapter highlights the significant role of the consolidation of nanotechnology and phytoremediation, the remarkable features associated with the utilization of Ag NPs and the synthetic routes towards Ag NPs' fabrication. The chapter also emphasizes the role of Ag NPs in environmental cleanup by coupling Ag NPs with the concept of phytoremediation. More specifically, current and previous studies are compared to gain a better comprehension of the interaction developed between Ag NPs and phytoremediant plants in a nanotools-enabled phytoremediation system. Finally, the chapter discusses the challenges of and recommendations for using an Ag NP-coupled phytoremediation system (ACPS) in the practical sense and the reservations associated with it.

12.2 METHODS

12.2.1 Search Approach

This chapter is based on a comprehensive study of the role of Ag NPs in environmental detoxification coupled with the concept of phytoremediation. Data were collected in a detailed manner through strict selection from a variety of the databases, namely, Web of Science, Google Scholar, Science Direct, Cochrane, Embase, ProQuest, and PubMed. Different types of research-based articles were included by considering peer-reviewed research articles, short communications, reviews, perspectives, different book chapters and reports produced by different organizations. An effort was made to consider only research items published by authentic authors such as Elsevier, Taylor & Francis, Springer, Springer Nature, De Gruyter, IET, etc. In terms of the timescale, all of the articles produced that address Ag coupled with phytoremediation were considered, but since the topic is quite understudied, other facts and figures researched were chosen from publications produced from 2010–2021. Special emphasis was placed on articles produced in the last three years. The search criteria in different types of electronic search engines was varied by feeding them terms such as "Environmental pollution", "Soil pollution", "Heavy metals in soils", "Phytoremediation", "Hyper-accumulators", "Nano-remediation", "Nano-phytoremediation", "Ag NPs", "Synthesis of Ag NPs", "Ag NPs in phytoremediation", "Ag NPs environmental remediation", etc.

12.3 ECOLOGICAL INTEGRITY: THE ROLE OF NANOTECHNOLOGY AND PHYTOREMEDIATION

Environmental integrity in the current era is seriously threatened due to the uncontrolled release of contaminants into the lithospheric and hydrospheric regions. Therefore, sustainable efforts are required immediately to protect the environment from further deterioration. The physico-chemical methods employed for environmental remediation are characterized by various limitations in terms of the monetary considerations, operational complexity and damage to the environment. In the conventional way of remediating the pedospheric zone, the target soils are cleaned either through the in-situ mode or they are excavated and moved to the remediation site or landfill for disposal without treatment. In such an excavation mode, the pollution source is not actually treated but is shifted from one place to another, and the problem of the contamination of such pedospheric swatches remains consistent. To add to the issue, such treatment can be labelled as a source of spreading further environmental threats by the transferal of the contaminants to the lower soil profiles and to other compartments in close proximity in an immediate or slower manner depending upon different factors (Naeem et al. 2020; Tahir et al. 2020a). In another remediation strategy known as soil washing, the contaminated pedospheric samples are remediated through a costly washing method that subsequently leads to the production of residues rich in metallic content. Such residues are laden with metals and need further processing, which incurs the most cost. Furthermore, this type of remediation is also associated with the rendering of the treated land unsuitable for the growth of plants as this land is converted to barren land with no biological organisms carrying out activities (Iram et al. 2019; Ishtiaq et al. 2020; Iram et al. 2020). Such concerns and inherent pedospheric micro-floral and micro-faunal disruption discourage the scientific community from adopting these methods on a practical scale.

Phytoremediation is a sustainable alternative to such limitations since it uses nature's tools and solar irradiance for the cleanup of environmental compartments in an easier and simpler way (Iram et al. 2020a). The word phytoremediation is derived from the Greek *phyto*, which refers to the plant, and *remedium* is Latin and refers to the concept of balancing or remediation. Phytoremediation is advantageous over other remediation modes in terms of the exclusive and discerning uptake potential of the root systems of phytoremediant plants. Furthermore, phytoremediation is also known for the application of naturally occurring processes together with other procedures concerning the degradation of pollutants, bioaccumulation and translocation (Parihar et al. 2021; Viana et al. 2021).

Therefore, the application of phytoremediation for the cleanup of the hydrospheric and pedospheric matrices is a good choice to reduce higher amounts of contaminants. The success of this concept can be realized from the fact that phytoremediation only started about a few decades ago, but as a green alternative, previous studies have identified over 400 plants as hyper-accumulators. In recent years, the inherent detoxification potential of hyper-accumulators has further been augmented by engineering them with different microbial species. Genetically engineered phytoremediant plants are known for their increased potential of metal decontamination and survival under stressful circumstances (Shah and Daverey 2021; Rai et al. 2021). For instance, the utilization of the endophytic rhizospheric microbial species to modify phytoremediated plants has indicated an increase in the extent of the environmental cleanup (Durand et al. 2021; Lin et al. 2021; Sun et al. 2021).

NPs of different metals and substances are usually aggregates of atoms or molecules. NPs have their origin in natural reserves, that is, they are produced by volcanic eruptions, lunar dust, composites of different minerals, etc. They can also be incidentally released from several anthropogenic activities, such as engineered NPs and combustive processes (Ahmad et al. 2020). Some studies support the facts regarding the toxicity of NPs to living organisms, but notably, the toxicity is exhibited only in the tested organisms, which points towards further investigation in this regard (Haghighat et al. 2021; Chang et al. 2021). Nevertheless, multitudes of studies have approved the positive role of NPs in several domains (Said et al. 2021; Mohammed et al. 2021). In fact, the concept of nano-phytoremediation in the present era has shown greater potential for environmental remediation by exerting the synergistic impacts of the nano-range of NPs and inherent potential of hyper-accumulator plants. For instance, the treatment of plants with titanium dioxide (TiO_2) NPs not only resulted in an increase in the plants' dry weight, chlorophyll generation and rate of photosynthesis but also enhanced the activity of some specific enzymes. In addition, no toxic impacts were seen (Zhang et al. 2019). In other reports, an increment was observed in the activity of Rubisco carboxylase upon treatment with TiO_2 NPs (Gao et al. 2006). In addition to TiO_2 NPs, aluminum (Al) NPs were also evaluated in *Phaseolus vulgaris* and *Lolium perenne* plants, and the results indicated no adverse impacts on the growth of the target plants (Doshi et al. 2008).

12.4 AGRICULTURAL SIGNIFICANCE AND SYNTHETIC ROUTES OF AG NPS

Among all NPs, Ag NPs have received special emphasis from the scientific community because of their physicochemical and biological characteristics. Such features have contributed to the remarkable catalytic and photocatalytic potential of Ag NPs, in addition to their astounding antimicrobial impacts, employed in a range of nano-biotechnological investigations (Jaffri and Ahmad 2020a, 2020b). The route of administration for Ag NPs can be through foliar application from where they ideally enter the stomatal openings. Ag NPs usually reside in trichome bases followed by their transportation to other plant tissues (Tripathi et al. 2021; Li et al. 2020). In some cases, the accumulation of NPs can occur in the photosynthetic surficial region, leading to the heating of the foliar region, and subsequently end up in a gaseous exchange. This gaseous exchange is attributable to the obstruction in the stomata and thus causes alterations in the physiological and cellular functionality of the plants (Zhang et al. 2019; Lalau et al. 2020). The rate and extent of accumulation of NPs vary in the case of different types of plants, which can be associated with the alleviated ionic potential and sinking potential of different plants. This, in turn, is associated with the existence of the polyphenolic content and other types of heterocyclic compounds. When Ag NPs were applied on zucchini plants being held inside a hydroponic solution during an investigation of their impacts on the emergence of the seeds and growth and maturation of the roots, the results indicated the absence of any kind of negative impact (Almutairi and Alharbi 2015).

The previous data dealing with the investigation of Ag NPs coupled with phytoremediation are very limited. This identifies this domain as still in its infancy, with all reports published in approximately the last 10 years. However, the promising results are what makes this field interesting and worth investigating. Ag NPs possess remarkable surface plasmon resonance (SPR), lower electro thermal resistance and antimicrobial potential. With the passage of time, when the utilization of Ag NPs was augmented in different fields, new routes also emerged in terms of their synthesis. Many procedures have been tested for the fabrication of Ag NPs, namely, physical, chemical, thermal and spray pyrolysis (Emil Kaya et al. 2020), mechanical ball milling and biogenic and electrochemical approaches (Hoang et al. 2021). Different types of synthetic routes produce Ag NPs with variable particle size ranges, distributions, stability indices and other considerations. Generally, two famous approaches are adopted for the synthesis of Ag NPs. They are the "top to bottom" method or the "bottom to top" method. In the latter method, Ag NPs are synthesized through physicochemical and biological methods through the self-assemblage of the atoms to newer nuclei, which subsequently grow into a particle that has a nano-range size. In the top to bottom method, material in bulk form is broken down through a number of lithographic methods, for example, thermal/laser ablation, grinding, sputtering and milling, into finer particles that have alleviated sizes (Beyene et al. 2017). Chemical reduction is one of the most frequently used bottom to top approaches in the synthesis of Ag NPs. In this method, Ag NPs are synthesized by reacting metal salt with chemically reducing agents, such as, N,N-dimethyl formamide (DMF), sodium borohydride ($NaBH_4$), Tollen's reagent and ascorbate (Kamaraj and Vivekanand 2020). In addition, Ag NPs have been stabilized with capping agents for their use in long-term applications.

There has been increasing concern regarding the utilization of chemically synthesized Ag NPs and their employment in environmental remediation and other applications due to some extent of toxicity associated with them. However, this challenge is easily met by adopting the green route for synthesizing Ag NPs with nature's tools, specifically, plants, microbes, etc., as a source of reducing substances in a sustainable and facile manner without using any chemical reducing agents (Shaheen et al. 2021). The concept of green chemistry and the biogenic synthetic route have been gaining attention in the last two decades. The main goal of the green route is the cost-effective synthesis of Ag NPs while ensuring the elimination of the chemicals used for the capping or reduction of the NPs. A number of reducing agents from plants have been derived and used for the synthesis of Ag NPs, for example, Ixora brachypoda (Bhat et al. 2021), Rehmannia glutinosa (Yong et al. 2021), Grewia lasiocarpa (Akwu et al. 2021), Spondia mombins (Asomie et al. 2021) and Tabebuia rosea (Dugganaboyana et al. 2021). Different parts of plants have different phytochemicals that can be used for the reduction cum stabilization of the NPs. Green synthesized Ag NPs can a be a sustainable alternative to chemically synthesized Ag NPs, thus their coupling with the concept of phytoremediation for environmental detoxification with no toxicity objections (Saim et al. 2021). In fact, scant studies have been published in this regard that express the utilization of green Ag NP-enabled phytoremediation. Although many other applications with Ag NPs are frequently reported, the less-studied green Ag NPs in association with phytoremediation must spark an interest in the scientific community. This opens great room for investigating the use of such Ag NPs.

12.5 AG NP-ENABLED SUSTAINABLE PHYTOREMEDIATION

Agri-nanotechnology-based products have been used frequently in the improvement of agricultural production and yield. Ag NPs represent auspicious NPs that have been extensively used in agriculture in the form of stimulators for plant growth (Tripathi et al. 2021), fungicidal and bactericidal agents to prevent the pathogenicity in plants incurred due to pathogenic fungi or bacteria (Ahmad and Jaffri 2018a) and agents specialized in the enhancement of fruit ripening (Ijaz et al. 2020b). With an increment in the utilization of Ag NPs, there is also an increasing chance of their inevitable release into ecological settings during fabrication and modification to form different products, in addition to handling or recycling. Meanwhile, a large number of studies have been reported for Ag NPs in terms of environmental remediation. For instance, Ag NPs have been used for the photocatalytic degradation of methylene blue (Wali et al. 2019), norfloxacin (Kanagamani et al. 2019), nitrophenol (4-NP) (Samuel et al. 2020) and paraoxon-ethyl (Zhao et al. 2020). However, it is quite interesting that only a limited number of studies can be found that report the utilization of Ag NPs in environmental remediation combined with phytoremediation (Verma et al. 2021). The scarce studies conducted in this regard are usually based on the Ag NP-enabled phytoremediation of different pollutants including chlorfenapyr, fipronil, molinate, heavy metals, atrazine and chlorpyrifos (Cao et al. 2020). Furthermore, Ag NP-enabled phytoremediation has also been employed to combat various plant diseases. The following section elucidates the studies performed with Ag NPs and phytoremediation.

Chlorfenapyr-contaminated pedospheric and hydrospheric matrices have been successfully detoxified by the integration of green NPs, agents specialized for solubility improvement and *Plantago major*-based phytoremediation. The experimental results supported the faster and efficacious removal of Chlorfenapyr from the soil and water samples by the collaborative use of green NPs, specifically, F—Fe^0, Ip—Ag^0 and Br—Ag^0, and *Plantago major*, which was supported over an activated charcoal medium. The remediation of Chlorfenapyr was performed in a sustainable and cost-effective manner through by utilizing nano-enabled phytoremediation (Romeh and Saber 2020). In another report by Romeh (2018), another toxic pollutant was degraded by utilizing nanotools-enabled phytoremediation. In a nanotools-enabled phytoremediation consortia, Ag NPs were added to different plants, namely, *Brassica* Ag NPs, *Plantago* Ag NPs, *Ipomoea* Ag NPs and *Camellia* Ag NPs, and used for the detoxification of flooded pedospheric and hydrospheric samples contaminated with Fipronil. The 6-day treatment resulted in an effective detoxification of the samples, and degraded products of Fipronil were obtained, such as fipronil amide and fipronil desulfenyl, expressing the candidacy of such Ag NP-enabled phytoremediation consortia in environmental cleanup processes (Romeh 2018). There has been increasing interest in the collaborative treatment of Ag NPs and plant growth-promoting rhizobacteria (PGPR) aimed at the alleviation of heavy metals including manganese (Mn), nickel (Ni), cobalt (Co) and lead (Pb) in the shoots of the maize plant. The synergistic utilization of Ag NPs and PGPR improved the growth of the maize plants by the induction of gibberellic acid (GA) and abscisic acid (ABA) phytohormone generation. Such phytohormones assist in enduring stress and augmenting the uptake of nutrients (Khan and Bano 2016b).

The treatment of municipal wastewater with Ag NPs and PGPR is also found to enhance the area and length of the root, in addition to effectively countering different types of stresses, for example, oxidative and osmotic stress (Khan and Bano 2016a). Ag NPs are inherently equipped for the immobilization or adsorption of heavy metals via alleviation in their motion and absorption. Furthermore, Ag NPs have the potential to attenuate the oxidative harms inflicted on plants through heavy metal absorption. Treatment of *M. oleifera* with Ag NPs and subsequent exposure to Cd and Pb expressed their enhanced ability of scavenging free radicals and repressing the physiological forbearance towards the stress. The addition of Ag NPs enhanced the photosynthetic pigment, polyphenolic content and the tolerance index of the growth and antioxidant activities. Such results indicate the value of Ag NPs in agro-systems to mitigate the adverse impacts of heavy metals on agriculture crop production, in addition to the sustainable bioremediation of the contaminated sites (Azeez et al. 2019). Ag NPs were utilized in association with *Phragmites australis* for the phytoremediation of contaminated sites, and impacts on the rhizospheric microbial community were observed (Fernandes et al. 2017).

Ag NPs have also been employed for environmental remediation in terms of different plant diseases. For instance, Ag NPs have been used in collaboration with cyanobacterium *Calothrix elenkinii (Ce)* for biocontrol against *A. alternata*-attacked tomato plants. The Ag NP-Ce consortia that were employed on the leaves of the tomato plant expressed remarkable augmentation of the foliar chlorophyll, carotenoids and polyphenol oxidase. The synergistic impacts of Ag NP-Ce consortia on the biocontrol of *A. alternata*-attacked tomato plants in an agro-climatic region signifies a sustainable approach (Mahawar et al. 2020). In another report based on the citrus disease known as Huanglongbing (HLB), regularly referred to as "yellow dragon disease", Ag NPs were applied on the foliar regions of the citrus to eradicate the bacteria-causing HLB. Compared to other methods, the Ag NP-based remediation produced 3–60 times better results through foliar sprinkling (Stephano-Hornedo et al. 2020). A study conducted with roses applied Ag NPs and Co NPs to combat various negative impacts including yellowing and the abscission of foliar regions, and there was an increase in the quality of the roses (Ha et al. 2020).

12.6 CHALLENGES AND CONCLUSION

Ag NPs' beneficial collaboration with phytoremediation is indicated by the effective results obtained against a variety of pollutants and heavy metals and by its ability to combat different diseases. However, the major challenge lies in the further research required to evaluate the toxicity caused by Ag NPs to different types of species. There is a considerable inhomogeneity in this regard if the existing literature is considered. This is actually further complicated by the variety of Ag NPs synthesized through different routes, and the variation in the size ranges, morphologies, surficial chemistry, etc. further adds to the puzzle. Therefore, the future directions need to closely address these issues and employ Ag NPs in close association with phytoremediation to derive the maximum benefits. In this regard, the toxicity-based objections also make practical application a challenging matter. However, the previous studies performed with biogenic NPs and the advancements made in the field of green chemistry are suggestive of producing sustainable, facile and effective results that can be employed for the remediation of the environment (Figure 12.1). In addition to the different concerns associated with the utilization of Ag NPs, some challenges are also associated with the type of the hyper-accumulators used. Some of the major disputes in this regard are listed as follows.

- Phytoremediation is often a time-consuming process compared to other detoxification methods. However, it can be adopted by developing a synergy between hyper-accumulator plants and nanomaterials.
- There are issues of lesser biomass and a slower rate of growth in the case of most hyper-accumulator plants. Therefore, the choice of the plant selected for phytoremediation needs to be faster in growth so that it can act as a self-sustaining system that does not incur further costs.
- The pedospheric and hydrospheric regions contain different types of pollutants and contaminants. However, their bioavailability is limited, which is attributable to the slower motion of the metal fragments of the metallic ions in soils.
- In the changing global scenario, fewer plants are effective in the detoxification of pollutants. Even their efficacy can be adversely affected if they are attacked by pests and other phyto-pathogens.
- The currently developed method of nanotools-enabled phytoremediation is ineffective in detoxifying lower soil profiles and groundwater.

TABLE 12.1
Different Types of Plants Used for the Phytoremediation of Toxic Pollutants

Plant	Common name	Family	Tested against	Reference
Lantana camara L	Big-sage	Verbenaceae	Cd	Liu et al. (2019)
Sedum plumbizincicola	Stonecrops	Crassulaceae	Cd	Liu et al. (2017)
Sedum alfredii	Leaf blade	Crassulaceae	Cd	Tao et al. (2020)
Pteris vittata	Chinese brake	Pteridaceae	As	da Silva et al. (2018)
Sedum alfredii	Leaf blade	Crassulaceae	Cd	Ma et al. (2020b)
Blepharidium guatemalense	Irayol blanco	Rubiaceae	Ni	Gutiérrez et al. (2021)
Stanleya pinnata	Desert princesplume	Brassicaceae/ Cruciferae	Se	Wang et al. (2018)
Abelmoschus manihot	Sunset muskmallow	Malvaceae	Cd	Wu et al. (2018)
Pteris vittata	Chinese brake	Pteridaceae	As, Cd, Zn	Abid et al. (2019)
Noccaea caerulescens	Alpine Pennycress	Brassicaceae	Ni, Zn, Fe, Mn, Cu, Co	Tang et al. (2019)
Microsorum pteropus	Java fern	Polypodiaceae	Cd	Lan et al. (2019)

TABLE 12.2
Different Synthetic Routes Adopted for the Synthesis of Ag NPs Used in Different Applications

Method	Synthetic details	Size of NPs (nm)	Morphology	Application	Reference
Green synthesis	Endophytic *Streptomyces laurentii*	7–28	Spherical	Medical textile industry	Eid et al. (2020)
Green synthesis	*Crataegus pentagyna*	25–45	Spherical	Degradation of dyes and antibacterial action	Ebrahimzadeh et al. (2020)
Green synthesis	*Lysiloma acapulcensis*	1.2–62	Quasi-spherical	Antimicrobial action	Garibo et al. (2020)
Green synthesis	Seaweed *Spyridia filamentosa*	20–30	Spherical	Cytotoxic activity against MCF-7 cells	Valarmathi et al. (2020)
Green synthesis	Onion (O), tomato (T), acacia catechu (C) alone and mixed OTC extracts	5 and 100	Spherical	Photocatalytic action against methyl orange (MO), methyl red (MR) and Congo red (CR)	Chand et al. (2020)
Green synthesis	*Pseudoduganella eburnea* MAHUQ-39	8–24	Spherical	Biomedical applications	Huq (2020)
Green synthesis	*Gomphrena globosa*	15.64, 19.44 and 22.16	Spherical triangle	Antibacterial activity	Tamilarasi and Meen (2020)
Green synthesis	*Curcuma longa*	15–40	Spherical	Wound healing activity against fibroblast cells and antimicrobial applications	Maghimaa and Alharbi (2020)
Green synthesis	*Diospyros lotus*	27	Cubic	Photocatalytic applications	Yasmin et al. (2020)
Green synthesis	*Camellia sinensis*	10 and 20	Spherical	Antibacterial ceramic applications	Göl et al. (2020)
Green synthesis	Endophytic *Actinomycetes*	13–40	Spherical	Textile fabrics coating	Salem et al. (2020)
Green synthesis	*Solanum surattense*	24	Spherical	Environmental and biomedical applications	Mani et al. (2020)
Chemical synthesis	Tri-sodium citrate functionalization	8 ± 1.1	-	Hg(II) detection and antimicrobial activity	Kokilavani et al. (2020)
Chemical synthesis	Starch modification	9 ± 1.3 nm	Spherical	Mercuric ion detection	Janani et al. (2020)
Chemical synthesis	Polyethylene glycol functionalization	11 ± 1.4	Spherical	Hg(II) detection	Kokilavani et al. (2020)
Chemical synthesis	Chitosan-biguanidine coating	25 ± 5	Quasi-spherical	Human lung cancer	Ma et al. (2020a)
Chemical synthesis	Chitosan and different Mw-CDP	7 and 5	Sphere and triangle shapes	Biofilm preparation	Affes et al. (2020)
Electrochemical synthesis	Polyaniline nanowires	115–135	Wires	Electrochemical DNA Sensor	Tran et al. (2020)
Electrochemical synthesis	Polyethylenimine-Encapsulation	~100 and 500	Non-uniform	Catalytic reduction of H_2O_2	Kim and Park (2020)

Photochemical synthesis	Thioxanthone-Anthracene (TX-A)	45–53	Spherical	Nanocomposite films	Mutlu et al. (2021)
Flame spray pyrolysis	Malondialdehyde/MDA equivalents	-	-	Cytotoxic and oxidative impacts in hemocytes of *Mytilus galloprovincialis*	Efthimiou et al. (2021)
Ultrasonic-spray-pyrolysis (USP)	Water solution of $AgNO_3$	250–700	Spherical	Size and size-distribution evaluation	Emil Kaya et al. (2020)
Hydrothermal synthesis	Ag NPs/strontium-titanate-perovskite	35–40	Cubic	Photocatalytic production of hydrogen	Ramos-Sanchez et al. (2020)
Bio-mechanochemical synthesis	Eggshell membrane and origanum plant	35	Irregular	Antibacterial activity	Baláž et al. (2017)
Green solid-state mechanochemical approach	*Pseudevernia furfuracea* and *Lobaria pulmonaria*	28.9 ± 2.5 and 23.8 ± 3.4	Spherical	Antibacterial activity	Goga et al. (2021)
Mechanochemical synthesis	*Thymus serpyllum* L., *Sambucus nigra* L. and *Thymus vulgaris* L.	< 10	Grains/ clusters/ agglomerates	Antibacterial activity	Kováčová et al. (2020)
Double arc discharge	15 and 30A of discharge currents	23	Spherical	Antibacterial activity	El-Khatib et al. (2020)
DC arc-discharge	125 V electrical DC voltage	73	Variable	Efficiency enhancement in polymer solar cells	Wongrat et al. (2019)
γ-irradiation assisted synthesis	Graphene oxide sheets	< 13.9	Oligomeric clusters	Trace recognition of dyes	Yue et al. (2017)
Microwave-assisted synthesis	Carboxymethylated gum kondagogu	9±2	Spherical	Catalytic, antibacterial and antioxidant potential	Seku et al. (2018)
Laser-assisted synthesis	Silica aerogel	~2–4	Near-spherical or marginally prolate	Supercritical deposition technique	Arakcheev et al. (2017)

FIGURE 12.1 Different types of plant species used for the biogenic synthesis of Ag NPs utilized in a myriad of applications: (a) *Prunus cerasifera* Ehrh, (b) *Carica papaya*, (c) *Ocimum sanctum, (d) Salvia officinalis, (e) Cocos nucifera L, (f) Parthenium hysterophorus, (g) Eriobotrya japonica, (h) Cinnamonum cassia, (i) Citrus sinensis, (j) Areca catechu, (k) Ocimum tenuiflorum and (l) Crataegus ambigua*

CONFLICTS OF INTEREST

None

FUNDING

None

REFERENCES

Abid, R., Manzoor, M., De Oliveira, L. M., da Silva, E., Rathinasabapathi, B., Rensing, C., and Ma, L. Q. 2019. Interactive effects of As, Cd and Zn on their uptake and oxidative stress in As-hyperaccumulator *Pteris vittata. Envir. Poll.* 248:756–762.

Affes, S., Maalej, H., Aranaz, I., Kchaou, H., Acosta, N., Heras, Á., and Nasri, M. 2020. Controlled size green synthesis of bioactive silver nanoparticles assisted by chitosan and its derivatives and their application in biofilm preparation. *Carb. Polymer*. 236:116063.

Afsheen, S., Naseer, H., Iqbal, T., Abrar, M., Bashir, A., and Ijaz, M. 2020. Synthesis and characterization of metal sulphide nanoparticles to investigate the effect of nanoparticles on germination of soybean and wheat seeds. *Mat. Chem. Phys*. 252:123216.

Ahmad, K. S., and Jaffri, S. B. 2018a. Phytosynthetic Ag doped ZnO nanoparticles: Semiconducting green remediators: Photocatalytic and antimicrobial potential of green nanoparticles. *Open Chem*. 16:556–570.

Ahmad, K. S., and Jaffri, S. B. 2018b. Carpogenic ZnO nanoparticles: Amplified nanophotocatalytic and antimicrobial action. *IET Nanobio*. 13:150–159.

Ahmad, K. S., Talat, M., Jaffri, S. B., and Shaheen, N. 2020. Innovatory role of nanomaterials as bio-tools for treatment of cancer. *Rev. Inorg. Chem*. 1:1–10.

Ahmadian, K., Jalilian, J., and Pirzad, A. 2021. Nano-fertilizers improved drought tolerance in wheat under deficit irrigation. *Agric. Water Manage*. 244:106544.

Akwu, N. A., Naidoo, Y., Singh, M., Nundkumar, N., Daniels, A., and Lin, J. 2021. Two temperatures biogenic synthesis of silver nanoparticles from *Grewia lasiocarpa* E. Mey. ex Harv. leaf and stem bark extracts: Characterization and applications. *BioNanoSci.* 1–17.

Ali, M., Jaffri, S. B., Ahmad, K. S., and Iqbal, S. 2019. Sorptive interactions of fungicidal 2-(4'-Thiazolyl) benzimidazole with soils of divergent physicochemical composition. *J. Int. Econ. Envir. Geol.* 10:97–104.

Almutairi, Z. M., and Alharbi, A. 2015. Effect of silver nanoparticles on seed germination of crop plants. *J. Adv. Agric.* 4:283–288.

Amjad, I., Javaid, M., Ikhlaq, K., Gul, S., Jaffri, S. B., and Ahmad, K. S. 2019. Adsorption-desorption mechanism of synthesized benzimidazole based fungicide 2-(3'-Pyridyl) on selected soil minerals. *J. Int. Econ. Envir. Geol.* 10:38–44.

Arakcheev, V., Bagratashvili, V., Bekin, A., Khmelenin, D., Minaev, N., Morozov, V., and Rybaltovsky, A. 2017. Laser assisted synthesis of silver nanoparticles in silica aerogel by supercritical deposition technique. *J. Supercritic. Fluid.* 127:176–181.

Asomie, J., Aina, A., Owolo, O., Olukanni, O., Okojie, D., Aina, F., and Feyisara Banji, A. 2021. Biogenic synthesis and characterization of Silver nanoparticles from seed extract of *Spondia mombins* and screening of its antibacterial activity. *J. Int. Nano Dimen.* 1:1–10.

Azeez, L., Adejumo, A. L., Lateef, A., Adebisi, S. A., Adetoro, R. O., Adewuyi, S. O., and Olaoye, S. 2019. Zero-valent silver nanoparticles attenuate Cd and Pb toxicities on *Moringa oleifera* via immobilization and induction of phytochemicals. *Plant Physiol. Biochem.* 139:283–292.

Baláž, M., Daneu, N., Balážová, Ľ., Dutková, E., Tkáčiková, Ľ., Briančin, J., and Baláž, P. 2017. Biomechanochemical synthesis of silver nanoparticles with antibacterial activity. *Adv. Powder Tech.* 28: 3307–3312.

Bazarbekova, A., Shon, C. S., Kissambinova, A., Kim, J. R., Zhang, D., and Moon, S. W. (2021). Potential of limestone powder to improve the stabilization of sulfate-contained saline soil. In *IOP Conference Series: Materials Science and Engineering* (Vol. 1040, No. 1, p. 012016). IOP Publishing.

Beyene, H. D., Werkneh, A. A., Bezabh, H. K., and Ambaye, T. G. 2017. Synthesis paradigm and applications of silver nanoparticles (AgNPs), a review. *Sust. Mat. Tech.* 13:18–23.

Bhat, M., Chakraborty, B., Kumar, R. S., Almansour, A. I., Arumugam, N., Kotresha, D., and Nayaka, S. 2021. Biogenic synthesis, characterization and antimicrobial activity of *Ixora brachypoda* (DC) leaf extract mediated silver nanoparticles. *J. King Saud Univ. Sci.* 33:101296.

Campinas, M., Viegas, R., Coelho, R., Lucas, H., and Rosa, M. J. 2021. Adsorption/coagulation/ceramic microfiltration for treating challenging waters for drinking water production. *Membranes* 11:91.

Cao, J., Feng, Y., Lin, X., and Wang, J. 2020. A beneficial role of arbuscular mycorrhizal fungi in influencing the effects of silver nanoparticles on plant-microbe systems in a soil matrix. *Envir. Sci. Poll. Res.* 1–15.

Castro-Mayorga, J. L., Randazzo, W., Fabra, M. J., Lagaron, J. M., Aznar, R., and Sanchez, G. 2017. Antiviral properties of silver nanoparticles against norovirus surrogates and their efficacy in coated polyhydroxyalkanoates systems. *LWT Food Sci Technol.* 79:503–510.

Chand, K., Cao, D., Fouad, D. E., Shah, A. H., Dayo, A. Q., Zhu, K., and Dong, S. 2020. Green synthesis, characterization and photocatalytic application of silver nanoparticles synthesized by various plant extracts. *J. Arab. Chem.* 13:8248–8261.

Chang, X., Wang, X., Li, J., Shang, M., Niu, S., Zhang, W., and Xue, Y. 2021. Silver nanoparticles induced cytotoxicity in HT22 cells through autophagy and apoptosis via PI3K/AKT/mTOR signaling pathway. *Ecotoxic. Envir. Safety.* 208:111696.

Das, C. A., Kumar, V. G., Dhas, T. S., Karthick, V., Govindaraju, K., Joselin, J. M., and Baalamurugan, J. 2020. Antibacterial activity of silver nanoparticles (biosynthesis): A short review on recent advances. *Biocatal. Agric. Biotech.* 101593.

da Silva, E. B., Lessl, J. T., Wilkie, A. C., Liu, X., Liu, Y., and Ma, L. Q. 2018. Arsenic removal by As-hyperaccumulator *Pteris vittata* from two contaminated soils: A 5-year study. *Chemosphere* 206:736–741.

Dawadi, S., Katuwal, S., Gupta, A., Lamichhane, U., Thapa, R., Jaisi, S., and Parajuli, N. 2021. Current research on silver nanoparticles: Synthesis, characterization, and applications. *J. Nanomat.* 1:1–20.

Dobias, J., and Bernier-Latmani, R. 2013. Silver release from silver nanoparticles in natural waters. *Environ. Sci. Technol.* 47(9):4140–4146.

Doshi, R., Braida, W., Christodoulatos, C., Wazne, M., and O'Connor, G. 2008. Nano-aluminum: Transport through sand columns and environmental effects on plants and soil communities. *Envir. Res.* 106:296–303.

Dugganaboyana, G. K., Jayendra, R., Narayan, A., and Konasur, M. S. 2021. A novel green biogenic synthesis of silver nanoparticles using *Tabebuia rosea* (Bertol.) DC fruit extract and its antioxidant and antibacterial potential. *J. Int. Pharma. Sci. Nanotech.* 14:5323–5333.

Durand, A., Leglize, P., and Benizri, E. 2021. Are endophytes essential partners for plants and what are the prospects for metal phytoremediation? *Plant and Soil.* 1–30.

Ebrahimzadeh, M. A., Naghizadeh, A., Amiri, O., Shirzadi-Ahodashti, M., and Mortazavi-Derazkola, S. 2020. Green and facile synthesis of Ag nanoparticles using *Crataegus pentagyna* fruit extract (CP-AgNPs) for organic pollution dyes degradation and antibacterial application. *Bioorg. Chem.* 94:103425.

Efthimiou, I., Kalamaras, G., Papavasileiou, K., Anastasi-Papathanasi, N., Georgiou, Y., Dailianis, S., and Vlastos, D. 2021. ZnO, Ag and ZnO-Ag nanoparticles exhibit differential modes of toxic and oxidative action in hemocytes of mussel *Mytilus galloprovincialis*. *Sci. Total Envir.* 144699.

Eid, A. M., Fouda, A., Niedbała, G., Hassan, S. E. D., Salem, S. S., Abdo, A. M., and Shaheen, T. I. 2020. Endophytic Streptomyces laurentii mediated green synthesis of Ag-NPs with antibacterial and anticancer properties for developing functional textile fabric properties. *Antibiotic.* 9:641.

El-Khatib, A. M., Doma, A. S., Abo-Zaid, G. A., Badawi, M. S., Mohamed, M. M., and Mohamed, A. S. 2020. Antibacterial activity of some nanoparticles prepared by double arc discharge method. *Nano-Struct. Nano-Obj.* 23:100473.

Emil Kaya, E., Kaya, O., Alkan, G., Gürmen, S., Stopic, S., and Friedrich, B. 2020. New proposal for size and size-distribution evaluation of nanoparticles synthesized via ultrasonic spray pyrolysis using search algorithm based on image-processing technique. *Mat.* 13:38.

Estay, H., Ruby-Figueroa, R., Gim-Krumm, M., Seriche, G., Quilaqueo, M., Díaz-Quezada, S., and Barros, L. 2021. Changing the conventional clarification method in metal sulfide precipitation by a membrane-based filtration process. *J. Mat. Res. Tech.* 1–10.

Fernandes, J. P., Mucha, A. P., Francisco, T., Gomes, C. R., and Almeida, C. M. R. 2017. Silver nanoparticles uptake by salt marsh plants—Implications for phytoremediation processes and effects in microbial community dynamics. *Marine Poll. Bull.* 119:176–183.

Fraceto, L. F., Grillo, R., de Medeiros, G. A., Scognamiglio, V., Rea, G., and Bartolucci, C. 2016. Nanotechnology in agriculture: Which innovation potential does it have? *Front. Environ. Sci.* 4:1–20.

Gao, F., Hong, F., Liu, C., Zheng, L., Su, M., Wu, X., and Yang, P. 2006. Mechanism of nano-anatase TiO_2 on promoting photosynthetic carbon reaction of spinach. *Biol. Trace Element Res.* 111:239–253.

Garibo, D., Borbón-Nuñez, H. A., de León, J. N. D., Mendoza, E. G., Estrada, I., Toledano-Magaña, Y., and Susarrey-Arce, A. 2020. Green synthesis of silver nanoparticles using *Lysiloma acapulcensis* exhibit high-antimicrobial activity. *Sci. Rep.* 10:1–11.

Goga, M., Baláž, M., Daneu, N., Elečko, J., Tkáčiková, Ľ., Marcinčinová, M., and Bačkor, M. 2021. Biological activity of selected lichens and lichen-based Ag nanoparticles prepared by a green solid-state mechanochemical approach. *Mat. Sci. Eng. C.* 119:111640.

Göl, F., Aygün, A., Seyrankaya, A., Gür, T., Yenikaya, C., and Şen, F. 2020. Green synthesis and characterization of *Camellia sinensis* mediated silver nanoparticles for antibacterial ceramic applications. *Mat. Chem. Phy.* 250:123037.

Gutiérrez, D. M. N., Pollard, A. J., van der Ent, A., Cathelineau, M., Pons, M. N., Sánchez, J. A. C., and Echevarria, G. 2021. *Blepharidium guatemalense*, an obligate nickel hyperaccumulator plant from non-ultramafic soils in Mexico. *Chemoecol.* 1–19.

Ha, N. T. M., Do, C. M., Hoang, T. T., Dai Ngo, N., and Nhut, D. T. 2020. The effect of cobalt and silver nanoparticles on overcoming leaf abscission and enhanced growth of rose (*Rosa hybrida* L. 'Baby Love') plantlets cultured in vitro. *Plant Cell Tissue Organ Cult.* 1–13.

Haghighat, F., Kim, Y., Sourinejad, I., Yu, I. J., and Johari, S. A. 2021. Titanium dioxide nanoparticles affect the toxicity of silver nanoparticles in common carp (*Cyprinus carpio*). *Chemosphere* 262:127805.

Han, L., Yang, G., Qin, Y., Wang, H., Cao, M., and Luo, J. 2021. Impact of O_3 on the phytoremediation effect of *Celosia argentea* in decontaminating Cd. *Chemosphere* 266:128940.

Hoang, V. T., Dinh, N. X., Pham, T. N., Hoang, T. V., Tuan, P. A., Huy, T. Q., and Le, A. T. 2021. Scalable electrochemical synthesis of novel biogenic silver nanoparticles and its application to high-sensitive detection of 4-nitrophenol in aqueous system. *Adv. Polymer Tech.* 1:1–26.

Huq, M. 2020. Green synthesis of silver nanoparticles using *Pseudoduganella eburnea* MAHUQ-39 and their antimicrobial mechanisms investigation against drug resistant human pathogens. *J. Int. Mol. Sci.* 21:1510.

Iftikhar, S., Saleem, M., Ahmad, K. S., and Jaffri, S. B. 2019. Synergistic mycoflora—natural farming mediated biofertilization and heavy metals decontamination of lithospheric compartment in a sustainable mode via *Helianthus annuus*. *J. Int. Envir. Sci. Tech.* 1–18.

Ijaz, M., Aftab, M., Afsheen, S., and Iqbal, T. 2020c. Novel Au nano-grating for detection of water in various electrolytes. *Appl. Nanosci.* 10:4029–4036.

Ijaz, M., Zafar, M., Afsheen, S., and Iqbal, T. 2020b. A review on Ag-nanostructures for enhancement in shelf time of fruits. *J. Inorg. Organ. Polymer. Mat.* 30:1475–1482.

Ijaz, M., Zafar, M., and Iqbal, T. 2020a. Green synthesis of silver nanoparticles by using various extracts: A review. *Inorg. Nano-Metal Chem.* 1–12.

Iqbal, T., Farooq, M., Afsheen, S., Abrar, M., Yousaf, M., and Ijaz, M. 2019. Cold plasma treatment and laser irradiation of *Triticum* spp. seeds for sterilization and germination. *J. Laser Appl.* 31:042013.

Iram, S., Ahmad, K. S., Noureen, S., and Jaffri, S. B. 2018. Utilization of wheat (*Triticum aestivum*) and Berseem (*Trifolium alexandrinum*) dry biomass for heavy metals biosorption. *Proceed. Pak. Acad. Sci. B.* 55:61–70.

Iram, S., Basri, R., Ahmad, K. S., and Jaffri, S. B. 2019. Mycological assisted phytoremediation enhancement of bioenergy crops *Zea mays* and *Helianthus annuus* in heavy metal contaminated lithospheric zone. *Soil Sed. Contam.* 28:411–430.

Iram, S., Iqbal, A., Ahmad, K. S., and Jaffri, S. B. 2020. Congruously designed eco-curative integrated farming model designing and employment for sustainable encompassments. *Envir. Sci. Poll. Res.* 27: 19543–19560.

Iram, S., Tariq, I., Ahmad, K. S., and Jaffri, S. B. 2020a. *Helianthus annuus* based biodiesel production from seed oil garnered from a phytoremediated terrain. *J. Int. Ambient Energy* 1–9.

Ishtiaq, M., Iram, S., Ahmad, K. S., and Jaffri, S. B. 2020. Multi-functional bio-sorbents triggered sustainable detoxification of eco-contaminants besmirched hydrospheric swatches. *J. Int. Envir. Anal. Chem.* 1–16.

Ismail, G. A., El-Sheekh, M. M., Samy, R. M., and Gheda, S. F. 2021. Antimicrobial, antioxidant, and antiviral activities of biosynthesized silver nanoparticles by phycobiliprotein crude extract of the cyanobacteria *Spirulina platensis* and *Nostoc linckia*. *BioNanoSci.* 1–16.

Jafari, M., Vanoppen, M., van Agtmaal, J. M. C., Cornelissen, E. R., Vrouwenvelder, J. S., Verliefde, A., and Picioreanu, C. 2021. Cost of fouling in full-scale reverse osmosis and nanofiltration installations in the Netherlands. *Desal.* 500:114865.

Jaffri, S. B., and Ahmad, K. S. 2018a. Neoteric environmental detoxification of organic pollutants and pathogenic microbes via green synthesized ZnO nanoparticles. *Envir. Tech.* 1:1–10.

Jaffri, S. B., and Ahmad, K. S. 2018b. *Prunus cerasifera* Ehrh. fabricated ZnO nano falcates and its photocatalytic and dose dependent in vitro bio-activity: Photodegradation and antimicrobial potential of biogenic ZnO nano falcates. *Open Chem.* 16:141–154.

Jaffri, S. B., and Ahmad, K. S. 2018c. Phytofunctionalized silver nanoparticles: Green biomaterial for biomedical and environmental applications. *Rev. Inorg. Chem.* 38:127–149.

Jaffri, S. B., and Ahmad, K. S. 2018d. Augmented photocatalytic, antibacterial and antifungal activity of prunosynthetic silver nanoparticles. *Artific. Cell Nanomed. Biotech.* 46:127–137.

Jaffri, S. B., and Ahmad, K. S. 2020a. Biomimetic detoxifier *Prunus cerasifera* Ehrh. silver nanoparticles: Innate green bullets for morbific pathogens and persistent pollutants. *Envir. Sci. Poll. Res.* 1–17.

Jaffri, S. B., and Ahmad, K. S. 2020b. Interfacial engineering revolutionizers: Perovskite nanocrystals and quantum dots accentuated performance enhancement in perovskite solar cells. *Critic. Rev. Solid State Mat. Sci.* 1–29.

Janani, B., Syed, A., Raju, L. L., Marraiki, N., Elgorban, A. M., Zaghloul, N. S., and Khan, S. S. 2020. Highly selective and effective environmental mercuric ion detection method based on starch modified Ag NPs in presence of glycine. *Optic. Comm.* 465:125564.

Kamaraj, P., and Vivekanand, P. A. 2020. Review on bio-synthesized silver nanoparticles and their antimicrobial applications. *J. Malaya Mat.* 2:4301–4308.

Kanagamani, K., Muthukrishnan, P., Shankar, K., Kathiresan, A., Barabadi, H., and Saravanan, M. 2019. Antimicrobial, cytotoxicity and photocatalytic degradation of norfloxacin using *Kleinia grandiflora* mediated silver nanoparticles. *J. Clust Sci.* 30:1415–1424.

Khalid, M., Saeed, U. R., Hassani, D., Hayat, K., Pei, Z., and Nan, H. 2021. Advances in fungal-assisted phytoremediation of heavy metals: A review. *Pedosphere* 31:475–495.

Khan, M. I., Dildar, S., Iqbal, T., Shakil, M., Tahir, M. B., Rafique, M., and Ijaz, M. 2019. In vivo study of gold-nanoparticles using different extracts for kidney, liver function and photocatalytic application. *Chem. Rep.* 1:36–42.

Khan, N., and Bano, A. 2016a. Role of plant growth promoting rhizobacteria and Ag-nano particle in the bioremediation of heavy metals and maize growth under municipal wastewater irrigation. *J. Int. Phytorem.* 18:211–221.

Khan, N., and Bano, A. 2016b. Modulation of phytoremediation and plant growth by the treatment with PGPR, Ag nanoparticle and untreated municipal wastewater. *J. Int. Phytorem.* 18:1258–1269.

Kim, K. T., and Park, D. S. 2020. Simple electrochemical synthesis of polyethylenimine-encapsulated Ag nanoparticles from solid AgCl applied in catalytic reduction of H_2O_2. *Catalyst.* 10:1416.

Kokilavani, S., Syed, A., Raju, L. L., Al-Rashed, S., Elgorban, A. M., Thomas, A. M., and Sudheerkhan, S. 2020. Citrate functionalized Ag NPs-polyethylene glycol nanocomposite for the sensitive and selective detection of mercury (II) ion, photocatalytic and antimicrobial applications. *Physica E.* 124:114335.

Kokilavani, S., Syed, A., Thomas, A. M., Marraiki, N., Al-Rashed, S., Elgorban, A. M., and Khan, S. S. 2020. Polyethylene glycol functionalised Ag NPs based optical probe for the selective and sensitive detection of Hg (II). *J. Mol. Liquid.* 307:112978.

Kováčová, M., Daneu, N., Tkáčiková, Ľ., Búreš, R., Dutková, E., Stahorský, M., and Baláž, M. 2020. Sustainable one-step solid-state synthesis of antibacterially active silver nanoparticles using mechanochemistry. *Nanomat.* 10:2119.

Krishna, V. D., Wu, K., Su, D., Cheeran, M. C. J., Wang, J. P., Perez, A. 2018. Nanotechnology: Review of concepts and potential application of sensing platforms in food safety. *Food Microbiol.* 75:47–54.

Lalau, C. M., Simioni, C., Vicentini, D. S., Ouriques, L. C., Mohedano, R. A., Puerari, R. C., and Matias, W. G. 2020. Toxicological effects of AgNPs on duckweed (*Landoltia punctata*). *Sci. Total Envir.* 710:136318.

Lan, X. Y., Yan, Y. Y., Yang, B., Li, X. Y., and Xu, F. L. 2019. Subcellular distribution of cadmium in a novel potential aquatic hyperaccumulator—*Microsorum pteropus*. *Envir. Poll.* 248:1020–1027.

Li, W. Q., Qing, T., Li, C. C., Li, F., Ge, F., Fei, J. J., and Peijnenburg, W. J. 2020. Integration of subcellular partitioning and chemical forms to understand silver nanoparticles toxicity to lettuce (*Lactuca sativa* L.) under different exposure pathways. *Chemosphere* 258:127349.

Lin, H., Liu, C., Li, B., and Dong, Y. 2021. *Trifolium repens* L. regulated phytoremediation of heavy metal contaminated soil by promoting soil enzyme activities and beneficial rhizosphere associated microorganisms. *J. Hazard. Mat.* 402:123829.

Liu, H., Zhao, H., Wu, L., Liu, A., Zhao, F. J., and Xu, W. 2017. Heavy metal ATPase 3 (HMA3) confers cadmium hypertolerance on the cadmium/zinc hyperaccumulator *Sedum plumbizincicola*. *New Phytol.* 215:687–698.

Liu, S., Ali, S., Yang, R., Tao, J., and Ren, B. 2019. A newly discovered Cd-hyperaccumulator *Lantana camara* L. *J. Hazard. Mat.* 371:233–242.

Lukashev, P. E., Sidorov, M. I., Stavrovskiy, M. E., Lazarev, S. A., Ragutkin, A. V., Korolev, E. I., and Gazanova, N. S. 2021. Development of the mathematical model of catalytic oxidation-reduction reaction kinetics for wastewater treatment technology using vortex layer devices. In *IOP Conference Series: Materials Science and Engineering* (Vol. 1027, No. 1, p. 012017). IOP Publishing.

Ma, C., White, J. C., Zhao, J., Zhao, Q., and Xing, B. 2018. Uptake of engineered nanoparticles by food crops: Characterization, mechanisms, and implications. *Annu. Rev. Food Sci. Technol.* 9: 129–153.

Ma, D., Han, T., Karimian, M., Abbasi, N., Ghaneialvar, H., and Zangeneh, A. 2020a. Immobilized Ag NPs on chitosan-biguanidine coated magnetic nanoparticles for synthesis of propargylamines and treatment of human lung cancer. *J. Int. Biol. Macromol.* 165:767–775.

Ma, L., Wu, Y., Wang, Q., and Feng, Y. 2020b. The endophytic bacterium relieved healthy risk of pakchoi intercropped with hyperaccumulator in the cadmium polluted greenhouse vegetable field. *Envir. Poll.* 264:114796.

Mabrouk, W., Lafi, R., Fauvarque, J. F., Hafiane, A., and Sollogoub, C. 2021. New ion exchange membrane derived from sulfochlorated polyether sulfone for electrodialysis desalination of brackish water. *Polymers. Adv. Tech.* 32:304–314.

Maghimaa, M., and Alharbi, S. A. 2020. Green synthesis of silver nanoparticles from *Curcuma longa* L. and coating on the cotton fabrics for antimicrobial applications and wound healing activity. *J. Photochem. Photobiol. B.* 204:111806.

Mahawar, H., Prasanna, R., Gogoi, R., Singh, S. B., Chawla, G., and Kumar, A. 2020. Synergistic effects of silver nanoparticles augmented *Calothrix elenkinii* for enhanced biocontrol efficacy against *Alternaria blight* challenged tomato plants. *3 Biotech.* 10:1–10.

Mani, M., Chang, J. H., Gandhi, A. D., Vizhi, D. K., Pavithra, S., Mohanraj, K., and Kumaresan, S. 2020. Environmental and biomedical applications of AgNPs synthesized using the aqueous extract of *Solanum surattense* leaf. *Inorg. Chem. Comm.* 121:108228.

Mohammed, S. S. S., Lawrance, A. V., Sampath, S., Sunderam, V., and Madhavan, Y. 2021. Facile green synthesis of silver nanoparticles from sprouted Zingiberaceae species: Spectral characterisation and its potential biological applications. *Mat. Tech.* 1:1–14.

Morab, P. N., Sumanth Kumar, G. V., and Akshay, K. 2021. Foliar nutrition of nano-fertilizers: A smart way to increase the growth and productivity of crops. *J. Pharmacog. Phytochem.* 10:1325–1330.

Mutlu, S., Metin, E., Yuksel, S. A., Bayrak, U., Nuhoglu, C., and Arsu, N. 2021. In-situ photochemical synthesis and dielectric properties of nanocomposite thin films containing Au, Ag and MnO nanoparticles. *J. Europ. Polymer.* 144:110238.

Naeem, H., Ahmad, K. S., and Jaffri, S. B. 2020. Biotechnological tools based lithospheric management of toxic Pyrethroid pesticides: A critical evaluation. *J. Int. Envir. Anal. Chem.* 1–24.

Nair, R. 2018. Plant response strategies to engineered metal oxide nanoparticles: A review. In *Phytotoxicity of Nanoparticles*. Edited by M. Faisal, Q. Saquib, A. A. Alatar, and A. A. Al-Khedhairy (pp. 377–393). Springer International Publishing.

Naruka, A. K., Suganya, S., Kumar, P. S., Amit, C., Ankita, K., Bhatt, D., and Kumar, M. A. 2021. Kinetic modelling of high turbid water flocculation using native and surface functionalized coagulants prepared from shed-leaves of *Avicennia marina* plants. *Chemosphere.* 129894.

Nedjimi, B. 2021. Phytoremediation: A sustainable environmental technology for heavy metals decontamination. *SN Appl. Sci.* 3:1–19.

Nguyen, T. Q., Sesin, V., Kisiala, A., and Emery, R. N. 2021. Phytohormonal roles in plant responses to heavy metal stress: Implications for using macrophytes in phytoremediation of aquatic ecosystems. *Envir. Toxicol. Chem.* 40:7–22.

Noori, A., Donnelly, T., Colbert, J., Cai, W., Newman, L. A., and White, J. C. 2020. Exposure of tomato (*Lycopersicon esculentum*) to silver nanoparticles and silver nitrate: Physiological and molecular response. *J. Int. Phytorem.* 22:40–51.

Omran, B. A. 2021. Facing lethal impacts of industrialization via green and sustainable microbial removal of hazardous pollutants and nanobioremediation. *Removal Emerg. Contam. Microb. Proc.* 133–160.

Parihar, J. K., Parihar, P. K., Pakade, Y. B., and Katnoria, J. K. 2021. Bioaccumulation potential of indigenous plants for heavy metal phytoremediation in rural areas of Shaheed Bhagat Singh Nagar, Punjab (India). *Envir. Sci. Poll. Res.* 28:2426–2442.

Rai, K. K., Pandey, N., Meena, R. P., and Rai, S. P. 2021. Biotechnological strategies for enhancing heavy metal tolerance in neglected and underutilized legume crops: A comprehensive review. *Ecotoxic. Envir. Safety* 208:111750.

Ramasamy, G., Rajkumar, P. K., and Narayanan, M. 2021. Generation of energy from salinity gradients using capacitive reverse electro dialysis: A review. *Envir. Sci. Poll. Res.* 1–10.

Ramos-Sanchez, J. E., Camposeco, R., Lee, S. W., and Rodriguez-Gonzalez, V. 2020. Sustainable synthesis of AgNPs/strontium-titanate-perovskite-like catalysts for the photocatalytic production of hydrogen. *Catal. Today* 341:112–119.

Rathika, R., Srinivasan, P., Alkahtani, J., Al-Humaid, L. A., Alwahibi, M. S., Mythili, R., and Selvankumar, T. 2021. Influence of biochar and EDTA on enhanced phytoremediation of lead contaminated soil by *Brassica juncea. Chemosphere* 271:129513.

Romeh, A. A. A. 2018. Green silver nanoparticles for enhancing the phytoremediation of soil and water contaminated by fipronil and degradation products. *Water Air Soil Poll.* 229:1–13.

Romeh, A. A. A., and Saber, R. A. I. 2020. Green nano-phytoremediation and solubility improving agents for the remediation of chlorfenapyr contaminated soil and water. *J. Envir. Manage.* 260:110104.

Rono, J. K., Le Wang, L., Wu, X. C., Cao, H. W., Zhao, Y. N., Khan, I. U., and Yang, Z. M. 2021. Identification of a new function of metallothionein-like gene OsMT1e for cadmium detoxification and potential phytoremediation. *Chemosphere* 265:129136.

Said, M. M., Rehan, M., El-Sheikh, S. M., Zahran, M. K., Abdel-Aziz, M. S., Bechelany, M., and Barhoum, A. 2021. Multifunctional hydroxyapatite/silver nanoparticles/cotton gauze for antimicrobial and biomedical applications. *Nanomat.* 11:429.

Saim, A. K., Kumah, F. N., and Oppong, M. N. 2021. Extracellular and intracellular synthesis of gold and silver nanoparticles by living plants: A review. *Nanotech. Envir. Eng.* 6:1–11.

Salem, S. S., El-Belely, E. F., Niedbała, G., Alnoman, M. M., Hassan, S. E. D., Eid, A. M., and Fouda, A. 2020. Bactericidal and in-vitro cytotoxic efficacy of silver nanoparticles (Ag-NPs) fabricated by endophytic actinomycetes and their use as coating for the textile fabrics. *Nanomat.* 10:2082.

Samuel, M. S., Jose, S., Selvarajan, E., Mathimani, T., and Pugazhendhi, A. 2020. Biosynthesized silver nanoparticles using Bacillus amyloliquefaciens; Application for cytotoxicity effect on A549 cell line and photocatalytic degradation of p-nitrophenol. *J. Photochem. Photobiol. B*. 202:111642.

Sangsuwan, P., and Prapagdee, B. 2021. Cadmium phytoremediation performance of two species of Chlorophytum and enhancing their potentials by cadmium-resistant bacteria. *Envir. Tech. Innov*. 21: 101311.

Sears, J., Swanner, J., Fahrenholtz, C. D., Snyder, C., Rohde, M., Levi-Polyachenko, N., and Singh, R. 2021. Combined photothermal and ionizing radiation sensitization of triple-negative breast cancer using triangular silver nanoparticles. *J. Int. Nanomed*. 16:851.

Seku, K., Gangapuram, B. R., Pejjai, B., Kadimpati, K. K., and Golla, N. 2018. Microwave-assisted synthesis of silver nanoparticles and their application in catalytic, antibacterial and antioxidant activities. *J. Nanostruct. Chem*. 8:179–188.

Senevirathna, S. T. M. L. D., Mahinroosta, R., Li, M., and KrishnaPillai, K. 2021. In situ soil flushing to remediate confined soil contaminated with PFOS-an innovative solution for emerging environmental issue. *Chemosphere* 262:127606.

Shah, V., and Daverey, A. 2021. Effects of sophorolipids augmentation on the plant growth and phytoremediation of heavy metal contaminated soil. *J. Clean. Prod*. 280:124406.

Shaheen, I., Ahmad, K. S., Jaffri, S. B., and Ali, D. 2021. Biomimetic [MoO_3@ZnO] semiconducting nanocomposites: Chemo-proportional fabrication, characterization and energy storage potential exploration. *Renew. Energy* 167:568–579.

Shaheen, I., Ahmad, K. S., Jaffri, S. B., Zahra, T., and Azhar, S. 2016. Evaluating the adsorption and desorption behavior of triasulfuron as a function of soil physico-chemical characteristics. *Soil Envir*. 35:99–105.

Shazia, I., Ahmad, K. S., and Jaffri, S. B. 2018. Mycodriven enhancement and inherent phytoremediation potential exploration of plants for lithospheric remediation. *Sydowia* 70:141–153.

Shu, X., Huang, W., Shi, K., Chen, S., Zhang, S., Li, B., and Lu, X. 2021. Microwave vitrification of simulated radioactively contaminated soil: Mechanism and performance. *J. Solid State Chem*. 293:121757.

Stephano-Hornedo, J. L., Torres-Gutiérrez, O., Toledano-Magaña, Y., Gradilla-Martínez, I., Pestryakov, A., Sánchez-González, A., and Bogdanchikova, N. 2020. Argovit™ silver nanoparticles to fight Huanglongbing disease in Mexican limes (*Citrus aurantifolia* Swingle). *RSC Adv*. 10:6146–6155.

Sun, X., Song, B., Xu, R., Zhang, M., Gao, P., Lin, H., and Sun, W. 2021. Root-associated (rhizosphere and endosphere) microbiomes of the *Miscanthus sinensis* and their response to the heavy metal contamination. *J. Envir. Sci*. 104:387–398.

Tahir, M. B., Iram, S., Ahmad, K. S., and Jaffri, S. B. 2020a. Developmental abnormality caused by *Fusarium mangiferae* in mango fruit explored via molecular characterization. *Biologia*. 75:465–473.

Tahir, M. B., Malik, M. F., Ahmed, A., Nawaz, T., Ijaz, M., Min, H. S., and Siddeeg, S. M. 2020. Semiconductor based nanomaterials for harvesting green hydrogen energy under solar light irradiation. *J. Int. Envir. Anal. Chem*. 1–17.

Tamilarasi, P., and Meena, P. 2020. Green synthesis of silver nanoparticles (Ag NPs) using *Gomphrena globosa* (Globe amaranth) leaf extract and their characterization. *Mat. Today Proceed*. 33:2209–2216.

Tang, H., Shu, X., Huang, W., Miao, Y., Shi, M., Chen, S., and Lu, X. 2021. Rapid solidification of Sr-contaminated soil by consecutive microwave sintering: Mechanism and stability evaluation. *J. Hazard. Mat*. 407:124761.

Tang, Y. T., Sterckeman, T., Echevarria, G., Morel, J. L., and Qiu, R. L. 2019. Effects of the interactions between nickel and other trace metals on their accumulation in the hyperaccumulator *Noccaea caerulescens*. *Envir. Exp. Bot*. 158:73–79.

Tao, Q., Zhao, J., Li, J., Liu, Y., Luo, J., Yuan, S., and Wang, C. 2020. Unique root exudate tartaric acid enhanced cadmium mobilization and uptake in Cd-hyperaccumulator *Sedum alfredii*. *J. Hazard. Mat*. 383:121177.

Tran, L. T., Tran, H. V., Dang, H. T. M., Huynh, C. D., and Mai, T. A. 2020. Silver nanoparticles decorated polyaniline nanowires-based electrochemical DNA sensor: Two-step electrochemical synthesis. *J. Electrochem. Soc*. 167:087508.

Tripathi, D., Rai, K. K., and Pandey-Rai, S. 2021. Impact of green synthesized WcAgNPs on in-vitro plant regeneration and with anolides production by inducing key biosynthetic genes in *Withania coagulans*. *Plant Cell Rep*. 40:283–299.

Valarmathi, N., Ameen, F., Almansob, A., Kumar, P., Arunprakash, S., and Govarthanan, M. 2020. Utilization of marine seaweed *Spyridia filamentosa* for silver nanoparticles synthesis and its clinical applications. *Mat. Lett*. 263:127244.

Vance, M. E., Kuiken, T., Vejerano, E. P., McGinnis, S. P., Hochella, M. F., Jr., Rejeski, D., and Hull, M. S. 2015. Nanotechnology in the real world: Redeveloping the nanomaterial consumer products inventory. *Beilstein J. Nanotechnol.*6:1769–1780.

Verma, A., Roy, A., and Bharadvaja, N. 2021. Remediation of heavy metals using nanophytoremediation. In *Advanced Oxidation Processes for Effluent Treatment Plants* (pp. 273–296). Elsevier.

Viana, D. G., Pires, F. R., Ferreira, A. D., Egreja Filho, F. B., de Carvalho, C. F. M., Bonomo, R., and Martins, L. F. 2021. Effect of planting density of the macrophyte consortium of *Typha domingensis* and *Eleocharis acutangula* on phytoremediation of barium from a flooded contaminated soil. *Chemosphere* 262:127869.

Wali, L. A., Alwan, A. M., Dheyab, A. B., and Hashim, D. A. 2019. Excellent fabrication of Pd-Ag NPs/PSi photocatalyst based on bimetallic nanoparticles for improving methylene blue photocatalytic degradation. *Optik* 179:708–717.

Wang, J., Cappa, J. J., Harris, J. P., Edger, P. P., Zhou, W., Pires, J. C., and Pilon-Smits, E. A. 2018. Transcriptome-wide comparison of selenium hyperaccumulator and nonaccumulator Stanleya species provides new insight into key processes mediating the hyperaccumulation syndrome. *Plant Biotech. J.* 16:1582–1594.

White, J. C., and Gardea-Torresdey, J. 2018. Achieving food security through the very small. *Nature Nanotechnol.* 13(8):627.

Wongrat, E., Wongkrajang, S., Chuejetton, A., Bhoomanee, C., and Choopun, S. 2019. Rapid synthesis of Au, Ag and Cu nanoparticles by DC arc-discharge for efficiency enhancement in polymer solar cells. *Mat. Res. Innov.* 23:66–72.

Wu, M., Luo, Q., Zhao, Y., Long, Y., Liu, S., and Pan, Y. 2018. Physiological and biochemical mechanisms preventing Cd toxicity in the new hyperaccumulator *Abelmoschus manihot. J. Plant Growth Regul.* 37:709–718.

Xu, L., Zhang, M., Wang, Y., and Wei, F. 2021. Highly effective adsorption of antibiotics from water by hierarchically porous carbon: Effect of nanoporous geometry. *Envir. Poll.* 116591.

Yan, L., Van Le, Q., Sonne, C., Yang, Y., Yang, H., Gu, H., and Peng, W. 2021. Phytoremediation of radionuclides in soil, sediments and water. *J. Hazard. Mat.* 407:124771.

Yasmin, S., Nouren, S., Bhatti, H. N., Iqbal, D. N., Iftikhar, S., Majeed, J., and Rizvi, H. 2020. Green synthesis, characterization and photocatalytic applications of silver nanoparticles using Diospyros lotus. *Green Proc. Synth.* 9:87–96.

Yong, D. W., Lieu, Z. Z., Cao, X., Yong, X. E., Wong, J. Z., Cheong, Y. S., and Chin, W. S. 2021. Biogenic synthesis of silver nanoparticles with high antimicrobial and catalytic activities using Sheng Di Huang (*Rehmannia glutinosa*). *Chem. An Asian J.* 16:237–246.

Yue, Y., Zhou, B., Shi, J., Chen, C., Li, N., Xu, Z., and Fu, H. 2017. γ-Irradiation assisted synthesis of graphene oxide sheets supported Ag nanoparticles with single crystalline structure and parabolic distribution from interlamellar limitation. *Appl. Surf. Sci.* 403:282–293.

Zahra, T., Ahmad, K. S., Shaheen, I., Azhar, S., and Jaffri, S. B. 2017. Determining the adsorption and desorption behavior of thiabendazole fungicide for five different agricultural soils. *Soil Envir.* 36:1–10.

Zhang, W., He, J., Zhang, L., He, S. Y., Ryser, E. T., and Li, H. 2019. Stomata facilitated sorption of silver nanoparticles by *Arabidopsis thaliana. Geophys. Res. Abstr.* 21:1. Available from: https://meetingorganizer.copernicus.org/EGU2019/EGU2019-15061.pdf

Zhao, L., Deng, C., Xue, S., Liu, H., Hao, L., and Zhu, M. 2020. Multifunctional g-C3N4/Ag NPs intercalated GO composite membrane for SERS detection and photocatalytic degradation of paraoxon-ethyl. *J. Chem. Eng.* 402:126223.

13 Magnetic Nanoparticles for Nano-Phytoremediation

Mohamed Nouri

13.1 INTRODUCTION

The pollution of the environment is the greatest challenge of our time and has created serious human costs. Pollution is mainly the result of the rapid progress of urbanization and industrialization, changes in urban lifestyle, variations of industrial strategy, the economic evolution mode and residents' consumption style (Fayiga et al. 2018; Mohammadi et al. 2020; Ukaogo et al. 2020). The constant upsurge in industrial waste and urban effluent affects ecosystems and human health worldwide by polluting the soil, water and air. The remediation of environmental matrices has become a vital interest with the intense increase in environmental pollution in recent decades.

Recently, to remediate the destroyed environment that has different impurity types, abundant remediation strategies have been developed, including coagulation/flocculation, reverse osmosis, photocatalysis, filtration, chemical precipitation, ion exchange, electro-chemical methods, electrodialysis, advanced oxidation, electrode ionization, reduction, biological decomposition and adsorption (Kaur and Roy 2020; Peng and Guo 2020; Singh et al. 2021). Adsorption is one of the main methods to remove venomous pollutants from water and wastewater. Therefore, various adsorbents with different components have been synthesized, especially low-cost adsorbents, such as agricultural waste (Chakraborty et al. 2020; Nouri 2021). Notably, conventional approaches encounter many difficulties when producing concentrated torrents; they also require high energy, have high costs, engender enormous amounts of residue, are low in process and instigate consequential contamination (Kaur and Roy 2020; Peng and Guo 2020).

Environmentally friendly remediation strategies are an alternative to restrict biodiversity destruction. Phytoremediation is a well-liked method of biological remediation. Compared to other methods, phytoremediation is more economical and simple, has a pleasing appearance, large adaptability and public acceptance and causes only slight destruction to the soil structure (Song et al. 2019). Nevertheless, phytoremediation often takes a prolonged time (many years), and its treatment is controlled by climate conditions, pollutant toxicity and soil quality. To evolve the effects of phytoremediation, agronomic management, chemical additive treatment, rhizosphere microbial inoculation, genetic engineering and other measures have been taken (Song et al. 2019). The extension of nanotechnology has brought inventive ideas for the phytoremediation of polluted ecosystems. Of course, researchers hope to conjoin the benefits of both approaches in a novel remediation technology for prospective pollutants (Zhu et al. 2019). The latest investigations have confirmed that it is practicable to employ nanomaterials and plants to clean up various pollutants in the ecosystem milieus (Liu et al. 2020b; Mohammadi et al. 2020; Romeh 2020; Romeh and Saber 2020; Hidalgo et al. 2020).

In addition to their encouraging effect on contaminant remediation, nanoparticles (NPs) possibly will react with living and nonliving constituents, both positively and negatively; for this reason, countless attempts have been made to assess the synergistic influence of the conjoined use of NP and phytoremediation techniques and to reveal their biological, chemical and physical exchanges in water or soil (Zand and Tabrizi 2020; Youssef et al. 2020).

To date, there are no harmonious decisions as to whether the joint technologies are advantageous in enhancing the efficiency of pollutant elimination and whether the combination of phytoremediation

DOI: 10.1201/9781003186298-13

and NPs to eliminate contaminants has not been extensively reported, this is why the principal purpose of this chapter is on the functioning of NPs applied in nano-phytoremediation. The interactions between pollutants, plants and MNPs are discussed. In addition, the challenges associated with the use of MNPs in nano-phytoremediation are considered to identify future research needs. Given that the most used and effective MNPs are iron MNPs (Fe-MNPS) (Singh et al. 2020), this chapter has focused on iron MNPs for nano-phytoremediation.

13.2 NANOPARTICLES

NPs are aggregates of atoms or molecules that vary in dimension from 1 to 100 nm, which can radically change their physicochemical properties compared to bulk materials (Martínez-Fernández and Komárek 2016). They are predominantly partitioned into two categories, namely, inorganic and organic NPs. The organic group includes carbon (C) NPs, while the inorganic category involves magnetic NPs (MNPs), noble metal NPs and semiconductor NPs, such as TiO_2 and ZnO (Figure 13.1).

Several NPs are liberated into soil and other environments through the utilization and discarding of nanomaterial products. It is approximated that around 5500, 3000, 550, 300, 55, 55, 55, 55, 0.6 and 0.6 tons of SiO_2, TiO_2, ZnO, CNT, FeO_x, Ag, AlO_x, CeO_x, quantum dot and fullerenes, respectively, are engendered worldwide each year (Piccinno et al. 2012). It is estimated that 0.1 to 2% of all NPs formed are freed into nature throughout the manufacturing process (Keller et al. 2013). About 63–91% of NPs are deposited in landfills, and 8–28%, 0.4%–7% and 0.1%–1.5% were assessed to be in soil, water forms, and air, respectively (Keller et al. 2013).

In addition, NPs' annual production involving SiO_2, TiO_2, ZnO_2, FeO_2, AlO_2 and CeO_2 has been approximated at 270,000 tons (Kumar et al. 2019). SiO_2, TiO_2, FeO_x, ZnO and Al_2O_3 are the most abundant NPs discharged into soil and landfills (Khan 2020a). The fate of NPs in soil and other milieus differs depending on their intrinsic properties and environmental properties (Klaine et al. 2008; Saleh 2020). The development of nanotechnology guides the continuous delivery of NPs into nature, resulting in the NPs piling up in organisms (e.g., plants).

It is estimated that the world's largest emissions of NPs are in soils, landfill sites and aquatic and air environments (Keller et al. 2013). Scientists all over the globe have been studying the impacts of numerous NP types on the ecosystem, but the outcomes are inconclusive as to the comparative suitability and harm of NPs.

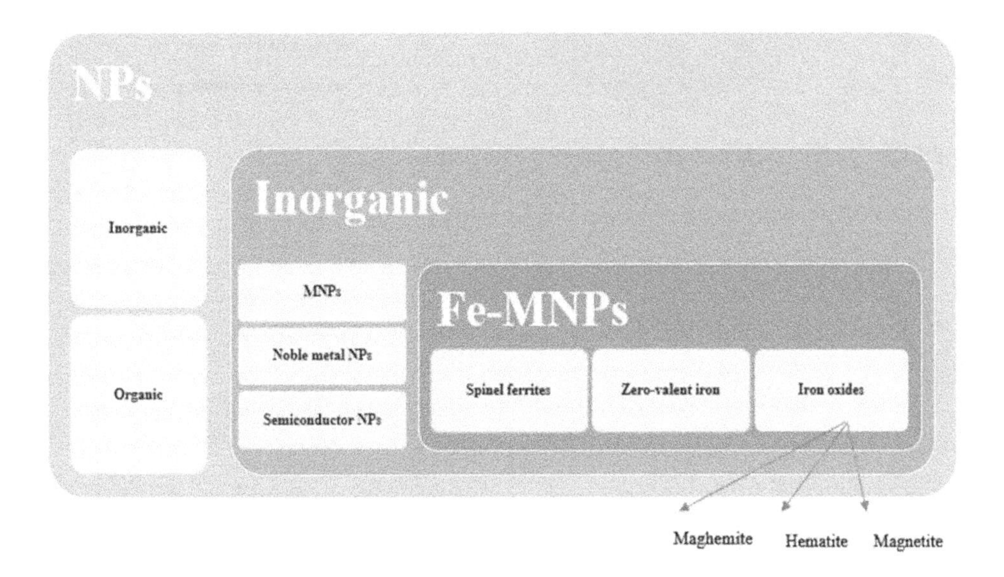

FIGURE 13.1 Main NP classes.

Metal NPs are utilized to cleanse pollutants. However, care should be exercised to prevent utilizing an NP that could poison nature. In this regard, Fe-MNPs are an opportunity (Souza et al. 2020). Remember, there are many factors that affect the performance of NPs (Figure 13.2) (Hasan et al. 2020; OECD 2010).

13.3 MAGNETIC NPS

MNPs are favored by researchers because of their various properties (Figure 13.3) that render them more attractive in diverse applications (Singh et al. 2020; Yogalakshmi et al. 2020).

MNPs consist of magnetic ingredients including iron (Fe), nickel (Ni) and Cobalt (Co). Because of minimal coercivity and great magnetization, they promote magnetic division for reutilization (Tombuloglu et al. 2021). They may reach surroundings such as landfill sites, sewage sludge and wastewater throughout their manufacture or after utilization. In this context, they may amass in soil and water (López-Luna et al. 2018). Nevertheless, investigations on their probable ecotoxicological and toxicological influences are incomplete and have yielded divisive results based on the type of plants and NPs employed and the conditions of manipulation (Tombuloglu et al. 2021). However, MNPs have gained a lot of consideration due to their green and durable chemical properties (Khan et al. 2020b).

Fe-MNPs have been widely studied in several environmental treatments. Fe-MNPs are the greatest usual metal-reducing agents, which are extensively employed to remove dyes, phenols and oil (Singh et al. 2020; Yogalakshmi et al. 2020).

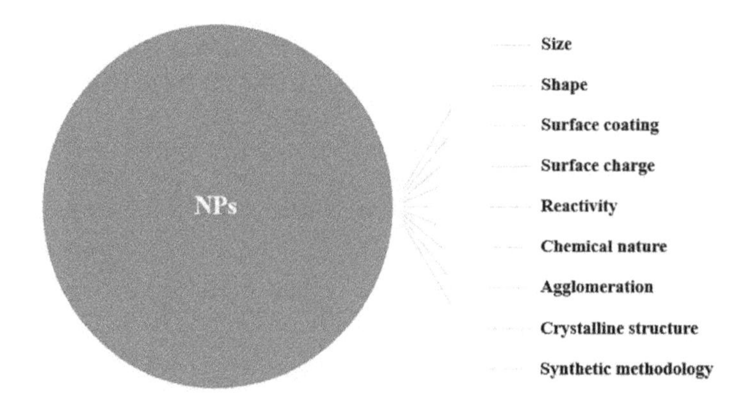

FIGURE 13.2 The principal factors influencing NP properties.

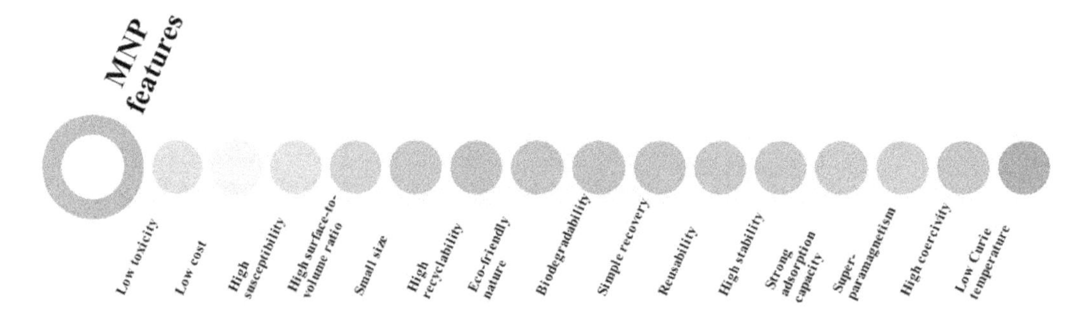

FIGURE 13.3 Main MNP features.

Fe-MNPs are good candidates for immobilizing contaminants in soils, water sources and sediments, and since they are naturally omnipresent, they have the promise to be a sustainable and economical technology used to develop adsorbents to perform remediation (Gillispie et al. 2019).

Fe-MNPs (less than 20 nm) have unique magnetic properties (such as super-paramagnetism). Superparamagnetic αFe_3O_4, αFe_2O_3, γFe_3O_4 and γFe_2O_3 NPs have attracted significant attention due to their exclusive mesoporous, physical and chemical properties (Ahmadi 2020). Moreover, zero-valent iron NPs (nZVI) are largely employed in the remediation of wastewater, soil and groundwater because of their high reactivity and particular redox potential (Zhang et al. 2020). However, there are numerous investigations on the influence of Fe-MNPs on flora; some studies highlight their beneficial impacts, and others report their toxic effects (Brasili et al. 2020; Pizarro et al. 2021; Yang et al. 2020).

13.4 NPS AND GREEN PLANTS

NPs can radically modify their physicochemical characteristics from bulk-sized particles and touch sundry components of the environment, including water, soil and air. NPs can react with their milieu, and plants are an important part of ecosystems. The four key factors affecting the contact between NPs and plants can be categorized into NPs, plants, contaminants and media (Figure 13.4).

The interaction between green plants and NPs is an intelligent system because it (1) controls the liberation of compounds, (2) reduces the environmental loss of products and (3) reduces plant toxicity (Fincheira et al. 2020). Moreover, NPs are a natural abiotic substance produced by plants under environmental stress, which vitally assists in the cleanup of contaminated air, water and soil (Khan 2020b).

Clearly, the successful application of NPs depends on the balance between their advantages and risks. In addition to the NP quantity and species of organisms, the pathway of exposure also crucially contributes to the generation of toxicity. NPs can stimulate plants growing at petite levels, whereas they become risky at superior amounts (Brasili et al. 2020). In addition, many NPs did not

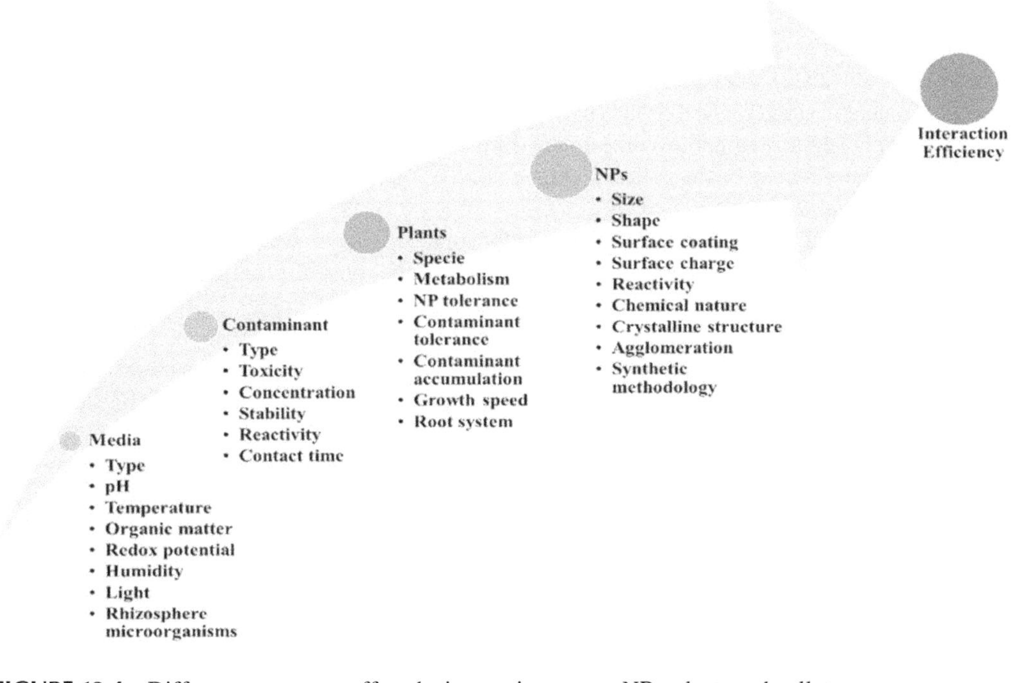

FIGURE 13.4 Different parameters affect the interaction among NPs, plants and pollutants.

disturb the germination of *Oryza sativa* and *zea mays* seeds, but the root dimension was considerably restrained (Liu et al. 2020a). However, the fully non-toxic employment of NPs cannot exist even if no poisonous influence has been detected (Liu et al. 2020a).

Accordingly, and as reported by Hasan et al.'s (2020) synthetic method, investigations of NPs' particle shape and size, dose and toxicity are essential before they are applied in the field.

The effects of NPs on plants depend on their properties (Figures 13.3 and 13.4) and plant species. Therefore, it is helpful to know whether plants can extract NPs from the original surrounding and whether NPs can be transferred to other organs of plants according to the NP characteristics. These data are pertinent in, for example, the treatment of plant waste, the harvesting of aboveground plant parts after media purification or in agricultural procedures of crops for human and animal consumption. For example, knowing the dissemination of NPs absorbed by diverse plant organs is helpful to develop safe harvesting technology and prevent the combine from being exposed to risky phyto-accumulated NPs. The distribution of NPs in plant organs may also affect the inclusion of NPs into the diet of animals and/or humans who consume NPs.

13.5 FE-MNPS AND GREEN PLANTS

Fe is an indispensable nutrient for plants, beneficial in the right dose but harmful in excess. Preceding studies have revealed that Fe-MNPs have beneficial or phytotoxic effects on plants. Clear effects on crop performances were discovered in previous experiments (Table 13.1).

13.5.1 $NF_{E2}O_3$

The appropriate concentration of nFe_2O_3 could upgrade the physiological function, improve environmental stress resistance, increase seed germination and develop seedling development of *Citrullus lanatus* (Li et al. 2020a), *Zea mays* (Li et al. 2016; Youssef et al. 2020), *Coriandrum satinvum* (Fahad et al. 2020), *Cicer arietinum* (Irum et al. 2020), *Arachis hypogaea* (Rui et al. 2016) and *Triticum aestivum* (Adrees et al. 2020) (Table 13.1).

Moreover, nFe_2O_3 can reduce the toxicity of cadmium $(Cd)^{2+}$ and the accumulation of Cd^{2+} in plant tissues by increasing the growth and iron content of maize seedlings. Furthermore, nFe_2O_3 could attenuate Cd^{2+} toxicity by involving Cd^{2+} detoxification mechanisms (Youssef et al. 2020).

Adrees et al. (2020) showed that nFe_2O_3 improved wheat physiology and growth and diminished Cd content in sundry wheat tissues under drought stress, including grains. Moreover, nFe_2O_3's restricted levels augmented photosynthesis but diminished oxidative stress. In general, the correct level of nFe_2O_3 treatment in Cd-polluted soils could be a successful way to produce crops more safely (Adrees et al. 2020).

However, Lu et al. (2020) confirmed that nFe_2O_3 applied to leaves was stored in the seedlings of wheat via stomata and then transferred from the leaves to the roots. Furthermore, nFe_2O_3 in leaves provoked a Fenton-like reaction, produced severe $^{\bullet}OH$, destroyed the redox balance and instigated oxidative stress. In addition, excessive $^{\bullet}OH$ can accelerate chlorophyll degradation, lead to the diminution of the chlorophyll amount and photosynthesis, and thus inhibit biomass synthesis (Lu et al. 2020). The results are of great significance to assess the effects of atmospheric Fe-MNPs on plants (Lu et al. 2020).

nFe_2O_3 at 4 mg Kg^{-1} decreased *A. thaliana* biomass and the chlorophyll content (Marusenko et al. 2013). Overall, 100 mg L^{-1} γFe_2O_3 NPs remarkably augmented MDA synthesis and diminished the chlorophyll level and activity of the root (Hu et al. 2017). Moreover, αFe_2O_3 NP treatment considerably lowered germination and the chlorophyll level. αFe_2O_3 NPs also caused damage to the root cell membranes and altered the morphology of the root, showing that they were barley phytotoxic (Tombuloglu et al. 2020a).

In another study, αFe_2O_3 and γFe_2O_3 had no considerable effect on watermelon fruit growth (Li et al. 2020a). Compared with the control, 400 ppm of γFe_2O_3 NP usage diminished the fresh

TABLE 13.1
The Fe-MNP Effects on Different Plant Species

MNPs		Size (nm)	Plant Species	Concentration	Observed Effect	References
Fe_2O_3	nFe_2O_3	5	*Glycine max*	15, 30 and 60 mg/pot	Chlorophyll level, root development and plant biomass indices increased	(Yang et al. 2020)
	γFe_2O_3	6	*Glycine max*	50, 100, 250, 500, 1,000 or 2,000 mg L^{-1}	Enhanced photosynthetic parameters and root elongation	(Alidoust and Isoda 2013)
	γFe_2O_3	6	*Oryza sativa*	50, 100, 250, 500, 1,000 or 2,000 mg L^{-1}	Enhanced root elongation	(Alidoust and Isoda 2014)
	γFe_2O_3	20	*Citrus maxima*	20, 50 and 100 mg L^{-1}	50 mg L^{-1} γFe_2O_3 significantly improved root activity by 23.8% and the chlorophyll level by 23.2% compared to the control; γFe_2O_3 had no influence on the expression of the AHA gene; 50 mg L^{-1} of γFe_2O_3 and Fe^{3+} considerably augmented the expression intensities of the gene and congruently had an advanced activity of ferric reductase	(Hu et al. 2017)
	γFe_2O_3	<50	*Arachis hypogaea*	2, 10, 50, 250, and 1,000 mg kg^{-1}	Root length, plant height, chlorophyll content and biomass increased	(Rui et al. 2016)
	nFe_2O_3	50–100	*Triticum aestivum*	25, 50 and 100 mg kg^{-1}	Improved photosynthesis performance and Fe contents and decreased Cd contents in tissues; NPs reduced oxidative stress in the shoots	(Adrees et al. 2020)
	nFe_2O_3	32.0–35.78	*Zea mays*	500, 1,000, 2,000, 4,000 and 8,000 mg kg^{-1}	Fe_2O_3 NPs proved dual effects: 500 mg kg^{-1} NPs notably promoted maize growing; 4,000 and 8,000 mg kg^{-1} NPs notably promoted the peroxidation of lipids	(Youssef et al. 2020)
	αFe_2O_3	50±0.2	*Cicer arietinum*	1, 5, 10, 15 and 20 mg L^{-1}	Could improve the development factors of chickpea	(Irum et al. 2020)
	αFe_2O_3	<50	*Arabidopsis thaliana*	4 mg kg^{-1}	Chlorophyll level and biomass decreased	(Marusenko et al. 2013)
	Fe_2O_3	10.2±2.6	*Mycorrhizal clover*	0.032, 0.32 and 3.2 mg kg^{-1}	Biomass reduced	(Feng et al. 2013)
	γFe_2O_3	20–100	*Helianthus annuus*	50 and 100 mg L^{-1}	The employment of 50 mg L^{-1} considerably downgraded the hydraulic conductivity of the roots by up to 57% and by 26% at 100 mg L^{-1} but had no influence on the shoot or root length or on plant biomass and did not stimulate oxidative stress	(Martínez-Fernández et al. 2016)
	γFe_2O_3	7–13	*Oryza sativa*	2, 20 and 200 mg L^{-1}	Had a significant impact on the production of abscisic acid and indole acetic acid hormones; however, provoked inhibition of the root phytohormone	(Gui et al. 2015)
	γFe_2O_3	20	*Citrus maxima*	20, 50 and 100 mg L^{-1}	100 mg L^{-1} amplified MDA synthesis and root activity and reduced the chlorophyll amount	(Hu et al. 2017)

(Continued)

TABLE 13.1 (*Continued*)
The Fe-MNP Effects on Different Plant Species

MNPs		Size (nm)	Plant Species	Concentration	Observed Effect	References
	αFe_2O_3	<50	*Arabidopsis thaliana*	4 mg kg^{-1}	Chlorophyll amount and biomass reduced	(Marusenko et al. 2013)
	γFe_2O_3	<50	*Glycine max*	250, 500, 750, 1,000, 1,250 and 1,500 mg L^{-1}	Lignin-induced growth inhibition and stem growth-inhibition	(Cunha et al. 2018)
	nFe_2O_3	20–30	*Oryza sativa*	50, 250 and 500 mg L^{-1}	Fe_2O_3 NPs did not improve plant growth at 50 mg L^{-1} and caused phytotoxicity at 500 mg L^{-1} under Fe deficiency	(Li et al. 2020b)
	nFe_2O_3	-	*Triticum aestivum*	0, 60 and 180 mg per plant	Initiated oxidative stress and diminished chlorophyll content, photosynthesis and biomass	(Lu et al. 2020)
	αFe_2O_3	14	*Hordeum vulgare*	50, 100, 200 and 400 mg L^{-1}	Considerably diminished germination and pigmentation, caused damage to the root cell membrane and changed the root morphology	(Tombuloglu, et al. 2020a)
	nFe_2O_3	20–100	*Solanum lycopersicum*	50 and 100 mg L^{-1}	100 mg L^{-1} lowered the root hydraulic	(Martínez-Fernández and Komárek 2016)
	αFe_2O_3	60	*Citrullus lanatus*	100, 200 and 400 ppm	Had no significant effect on plant development and physiological activities	(Li et al. 2020a)
	γFe_2O_3	30	*Citrullus lanatus*	100, 200 and 400 ppm		(Li et al. 2020a)
Fe_3O_4	*n*Fe_3O_4	25	*Lolium perrene* and *Cucurbita mixta*	30, 100 and 500 mg L^{-1}	No Fe absorption, amplified root length, but engendered oxidative stress	(Wang et al. 2011)
	nFe_3O_4	20	*Oryza sativa*	50, 250 and 500 mg L^{-1}	A low dose (50 mg L^{-1}) upgraded growth under Fe deficiency, and a high concentration (500 mg L^{-1}) caused phytotoxicity	(Li et al. 2020b)
	n$CoFe_2O_4$	11.4	*Hordeum vulgare*	125, 250, 500 and 1,000 mg L^{-1}	Tb-doped $CoFe_2O_4$ NPs were absorbed by the roots, moved to the leaves and contributed to the increase of plant biomass	(Tombuloglu et al. 2021)
	n$CoFe_2O_4$	13.5	*Hordeum vulgare*	125, 250, 500 and 1,000 mg L^{-1}	Lower doses (125 to 500 mg L^{-1}) promoted growth, but higher doses (> 500 mg L^{-1}) restrained growth	(Tombuloglu, et al. 2020b)
	n$CoFe_2O_4$	25	*Medicago falcata*	1, 2 and 4 mg L^{-1}	Significantly increased root development, chlorophyll a fluorescence and miRNA expression and provoked genotoxicity and genome instability	(Kokina et al. 2020)
	n$CoFe_2O_4$	25	*Eruca sativa*	1, 2 and 4 mg L^{-1}	NPs had little genotoxicity, promoted the growth of seedlings, and upgraded the resistance of plants to environmental stress	(Plaksenkova et al. 2019)

	nFe_3O_4	20	*Nicotiana benthamiana*	100 µg mL^{-1}	*n*Fe_3O_4 foliar deposition increased dehydrated and fresh plant masses, stimulated antioxidants, and increased salicylic acid production, salicylic acid-sensitive PR gene expression and resistance of the plant to the *tobacco mosaic* virus	(Cai et al. 2020)
	nFe_3O_4	60–80	*Oryza Sativa*	5, 10 and 15 ppm	Reduced arsenic stress and promoted plant growth	(Khan et al. 2020a)
	nFe_3O_4	5–10	*Coriandrum satinvum*	10 mg L^{-1}	Reduced the Cd and Pb absorption and promoted plant growth	(Fahad et al. 2020)
	nFe_3O_4	<50	*Arabidopsis thaliana*	400, 2,000 and 4,000 mg L^{-1}	Diminished the root size	(Lee et al. 2010)
	n$CoFe_2O_4$	2.24–3.44	*Triticum aestivum*	500, 1,000, 2,000, 4,000 and 8,000 mg kg^{-1}	NPs were absorbed and distributed by wheat seedlings, photosynthetic pigments decreased and wheat seedling stress increased	(López—Luna et al. 2018)
	nFe_3O_4	25	*Linum usitatissimum*	0.5, 1 and 1.5 mg L^{-1}	A low concentration of iron oxide NPs provoked genotoxicity	(Kokina et al. 2017)
	nFe_3O_4	6.7	*Cucumis sativus*	50, 500 and 2,000 mg L^{-1}	At 50 mg L^{-1}, *n*Fe_3O_4 reduced the biomass and enzyme activity, but they increased significantly at a higher concentration (2000 mg L^{-1})	(Konate et al. 2018)
	nFe_3O_4	<100	*Solanum lycopersicum*	1 g/100 mL	*n*Fe_3O_4 could clog the plant's channels and prevent the absorption of water and nutrients; NPs have a negative impact on plant evolution and increase oxidative stress	(Pizarro et al. 2021)
	nFe_3O_4	50–60	*Lactuca sativa*	10 and 20 mg L^{-1}	No influence on the physiological indices	(Trujillo-Reyes et al. 2014)
	nFe_3O_4	12.50 ± 4.10	*Helianthus annuus*	50 or 500 ppm	NPs had no significant influence on seed germination and seedling growth; in most cases, the germination parameters decreased	(Kornarzyński et al. 2020)
	nFe_3O_4	7	*Cucumis sativus* and *Lactuca sativa*	116 µg mL^{-1}	No seed germination toxicity was observed	(Barrena et al. 2009)
	nFe_3O_4	50	*Vigna radiata*	300, 600 and 1,000 mg L^{-1}	Usually engendered no important phytotoxicity at ≤ 1,000 mg L^{-1}	(Sun et al. 2020)
	nFe_3O_4	27.23	*Prosopis juliflora*	100, 200, 300 and 400 mg kg^{-1}	A low content considerably augmented the shoot and root length, but when the concentration of NPs exceeded 200 mg kg^{-1}, the root and shoot length and the chlorophyll level decreased	(Kumari and Khan 2018)
*n*ZVI	*n*ZVI	-	*Typha latifolia*	25, 50, 200, 500 and 1,000 mg L^{-1}	Augmented plant development at small doses; produced toxicity at an advanced dose	(Ma et al. 2013)

(Continued)

TABLE 13.1 (*Continued*)
The Fe-MNP Effects on Different Plant Species

MNPs		Size (nm)	Plant Species	Concentration	Observed Effect	References
	*n*ZVI	-	*Oriza sativa*	10, 20, 40, 80 and 160 mg L^{-1}	Increased the root and shoot length, photosynthetic pigments and biomass	(Guha et al. 2018)
	*n*ZVI	20 and 100	*Salix alba*	0, 150 and 300 mg kg^{-1}	Small nZVI concentration significantly augmented root elongation, the leaf area and the BCF of Cd; however, elevated nZVI concentration decreased seedling growth and the BCF of Cu and Pb	(Mokarram-Kashtiban et al. 2019)
	*n*ZVI	54 ± 1	*Arabidopsis thaliana*	500 mg kg^{-1}	Augmented photosynthesis and the iron uptake in roots and leaves; consequently, biomass increased by 38%	(Yoon et al. 2019)
	*n*ZVI	20	*Oryza sativa*	50, 250 and 500 mg L^{-1}	A low dose improved rice growth under Fe deficiency and produced phytotoxicity at 500 mg L^{-1}	(Li et al. 2020b)
	*n*ZVI	26	*Oryza sativa*	50, 100, 150 and 200 mg L^{-1}	nZVI is beneficial in small amounts and phytotoxic at a high concentration	(Guha et al. 2020b)
	*n*ZVI	-	*Agrostis capillaris* and *Festuca rubra*	25, 50 and 100 mg L^{-1}	No negative consequences on seed germination but increased the shoot growth and reduced the inhibition rate of the elongation of *F. rubra* seedlings; nevertheless, nZVI treatments in hydroponics media had no influence on *F. rubra*, and *A. capillaris* acquired more biomass and longer roots	(Teodoro et al. 2020)
	*n*ZVI	33.8±3.59	*Oryza sativa*	50, 100 and 200 mg L^{-1}	Increased antioxidant activities and phyto-chelatin production, decreased Fe/Cd transporters and the bioavailable Cd amount and immobilized Cd in soil, which consequently improved crop yield and avoided Cd amassing in grains	(Guha et al. 2020a)
	*n*ZVI	2.25–36	*Solanum lycopersicum*	5, 50, 100 and 1,000 mg L^{-1}	A small nZVI dose (5 mg L^{-1}) encouraged seed germination, root and hypocotyl development, leaf chlorophyll and carotenoids contents; however, high nZVI doses (100 and 1,000 mg L^{-1}) interfered with the germination of tomato seeds	(Brasili et al. 2020)
	*n*ZVI	50	*Oryza sativa*	50, 100, 250, 500, 750 and 1000 mg L^{-1}	Significantly inhibited chlorophyll biosynthesis	(Zhang et al. 2020)
	*n*ZVI	< 50	*Solanum lycopersicum*	50 and 100 mg L^{-1}	No effect was detected	(Martínez-Fernández and Komárek 2016)
	*n*ZVI	60	*Nasturtium officinale*	0.5%, 2%, 5% and 10%	At a 2% dose, the effect was obvious, and there was no adverse effect on the soil parameters; in fact, soil phytotoxicity was diminished and can be detected at the greatest amount	(Baragaño et al. 2020a)
FeOOH	*aFeOOH*	2.7	*Nasturtium officinale*	0.2%, 1%, 2% and 5%	Good consequences attained at 0.2%; however, soil phytotoxicity was augmented at advanced amounts	(Baragaño et al. 2020a)

mass but augmented the water content. Different concentrations of γFe_2O_3 NPs had great effects on the nutritional characteristics of watermelon fruit and could diminish the contents of sugar and total amino acids (Li et al. 2020a). In addition, diverse concentrations of αFe_2O_3 and γFe_2O_3 can significantly increase the content of vitamin C in fruits (Li et al. 2020a).

13.5.2 nFe_3O_4

Some reports have indicated that nFe_3O_4 may have a negative, positive and/or no effects on seed germination, root elongation and crop development (Cai et al. 2020; Li et al. 2020b; Pizarro et al. 2021; Sun et al. 2020; Wang et al. 2011) (Table 13.1).

nFe_3O_4 may lead to severe alterations in crops, leading to genotoxicity (Kokina et al. 2017; Kokina et al. 2020). Kokina et al. (2020) showed that Fe_3O_4 NPs synchronously increase the expression of miR159c, stimulate genome instability and provoke genotoxicity in yellow medick.

Konate et al. (2018) reported that after 21 days of treatment with different nFe_3O_4 concentrations, more oxidative stress and a series of antioxidant enzyme activities were induced. Moreover, biomass restraint and oxidative stress were attained at small nFe_3O_4 concentrations (50 mg/L) (Konate et al. 2018). Likewise, Pizarro et al. (2021) reported that Fe_3O_4 NPs have a negative impact on growth and increase oxidative stress in plants. Fe_3O_4 NPs could also clog the plant's channels and prevent water and nutrient uptake (Pizarro et al. 2021). Conforming to Wang et al. (2011), Fe_3O_4 NPs were not transferred from the roots to the leaves in ryegrass and pumpkin but could provoke more oxidative stress in the two plant species.

In contrast, Plaksenkova et al. (2019) revealed that Fe_3O_4 NPs had a beneficial influence on the quality and yield of rocket seedlings and low genotoxicity and could improve the plants' environmental stress resistance. In addition, Fe_3O_4 NPs' foliar deposition increased the fresh and dry plant masses, activated the antioxidants of the plant and increased salicylic acid synthesis and the expression of salicylic acid-sensitive PR genes, thereby increasing plant resistance against the tobacco mosaic virus (Cai et al. 2020). Furthermore, a lower concentration of Fe_3O_4 NPs notably inhibits the arsenic level and improves plant growth, while a higher concentration does not. It seems that Fe_3O_4 NPs have different effects according to their dosage (Khan et al. 2020a). Similarly, nFe_3O_4 application has significantly augmented the length of the shoots and roots of *Prosopis juliflora*. However, at greater than 200 mg Kg^{-1} of NPs, the shoot and root elongation and chlorophyll content are lowered (Kumari and Khan 2018).

A new study showed a substantial diminution in the germination rate with CoFe2O4 NPs (up 37% at 1,000 mg L^{-1}); nonetheless, notable growth (up 38–65%) and biomass rise (up 72–133%) were revealed after 3 weeks of contact with $CoFe_2O_4$ NPs (Tombuloglu et al. 2021).

Finally, Kornarzyński et al. (2020) concluded that no considerable effects of nFe_3O_4 and the magnetic field on seedling growth and seed germination were demonstrated. In this regard, Barrena et al. (2009) concluded that 116 μg mL^{-1} of Fe_3O_4 NPs has no toxicity for *Cucumis sativus* and *Lactuca sativa* seed germination. In addition, another study proved that 2,000 mg L^{-1} of nFe_3O_4 did not have a significant influence on the augmentation of shoot and root elongation of wheat seedlings (Konate et al. 2017).

13.5.3 nZVI

nZVI is the most typically studied NP and is widely employed in environmental clean-up. It has been proved to be an operational adsorbent, catalyst and reductant and can be utilized for a variety of pollutants, such as metals, halogen organic composites, dyes and pharmaceuticals (Brasili et al. 2020; Irum et al. 2020; Liu et al. 2020b; Zand et al. 2020a).

In this regard, low nZVI concentrations can encourage plant development by increasing the leaf area and the root length of the seedling (Mokarram-Kashtiban et al. 2019).

nZVI exercised promoting influences exclusively at small doses on *Typha latifolia* (Ma et al. 2013), *Oriza sativa* (Guha et al. 2018, 2020a, 2020b; Li et al. 2020b), *Salix alba* (Mokarram-Kashtiban et al. 2019), *Arabidopsis thaliana* (Yoon et al. 2019), *Agrostis capillaris*, *Solanum lycopersicum* (Brasili et al. 2020) and *Festuca rubra* (Teodoro et al. 2020). However, excessive nZVI has caused damage to many plants such as hybrid poplars, cattail (Ma et al. 2013) and *Oryza sativa* at elevated concentrations (Zhang et al. 2020). Meanwhile, Martínez-Fernández and Komárek (2016) reported that no effect was detected after *Solanum lycopersicum* treatment with nZVI. In contrast with Martínez-Fernández and Komárek's (2016) results, Brasili et al. (2020) concluded that *Solanum lycopersicum* treatment with nZVI (100 and 1,000 mg L^{-1}) negatively influenced the germination of tomato seeds.

Using ZVI NPs at a small dose (50 mg L^{-1}) augmented chlorophyll content by 30.7%. They also improved plant stress as confirmed by decreased oxidative stress and reduced contents of phytohormones such as indole-3-acetic acid and gibberellin. Small-dose ZVI NP applications lead to a greater stock of Fe in plants and regulate the expression of YSL15 and IRT1 (Li et al. 2020b).

The utilization of 50 and 250 mg L^{-1} of nZVI attenuated the oxidative stress induced by Quinclorac. Nevertheless, 750 mg Kg^{-1} of nZVI severely destroyed the seedlings of *Oryza sativa*, perhaps due to the lack of active iron (Zhang et al. 2020).

It is emphasized that, at *Citrus maxima* harvest, the level of Fe deficiency was classed in the following succeeding order: Fe-free treatment > Fe^{3+} > α-Fe_2O_3, γ-Fe_2O_3 NPs > Fe_3O_4 NPs. This partly determined that nFe_3O_4 was less sensitive to the uptake and use by plants than other iron oxide NPs (Li et al. 2018).

The controversy in the published findings may be principally related to the plant species, NPs or culture medium utilized in diverse investigations. Therefore, more experiments are needed to completely comprehend the influences of Fe-based NPs so that they can be safely utilized in agriculture and environmental remediation (Li et al. 2020b).

Accordingly, some Fe-MNPs have an encouraging influence on plant growth, which is advantageous to the nano-phytoremediation of environmental pollutants. Nevertheless, some Fe-MNPs also have poisonous consequences on plants, as exposed in Figure 13.5.

13.6 FE-MNPS AND ECOSYSTEM REMEDIATION

In terms of environmental remediation and resource conservation, NPs are considered cost-effective, efficient and environmentally friendly replacements to current processing materials (Nwadinigwe and Ugwu 2019). The beneficial impacts of NPs on living communities, ecosystem functions and exchanges beyond ecosystem boundaries deserve exclusive consideration (Ahmad et al. 2019). The magnetic quality of MNPs has significantly influenced the physical characteristics of water pollutants and contributes to water purification (Shahi et al. 2021). Moreover, MNPs have been revealed to be efficient in the degradation of several types of dyes from textile effluents (Yogalakshmi et al. 2020) and in the clean-up of metal-polluted soil, sediment and water (Latif et al. 2020).

The combination of biosorption technologies and magnetic separation can effectively achieve the adsorption of metal ions, which has the advantages of being an environmentally friendly, low cost and flexible operation (Khan et al. 2020b). As a suitable adsorbent, MNPs have the following advantages: (1) a significant amount of particles can be formed by a simple and convenient technique; (2) important adsorption capacity; (3) low toxicity and an exceptional magnetic force; and (4) metal adsorbents are easy to separate from the wastewater treated by an external magnetic field (Khan et al. 2020b).

Fe-based NPs such as ZVI, Fe_2O_3, Fe_3O_4 and FeOOH NPs have been utilized for numerous usages in environmental fields. For example, nZVI is employed to clean up polluted soil and water. In this context, nZVI can simultaneously reduce the availability of As, Pb, Zn, Se, Cr, Cu, Cd, Sb, Ag, Hg and Ni in polluted soil (Baragaño et al. 2020c; Galdames et al. 2020; Slijepčević et al. 2021). The interaction between metals and nZVI can be recapitulated as oxidation/reoxidation, reduction, adsorption, coprecipitation and precipitation (Xue et al. 2018). Moreover, nZVI usually has been utilized in the remediation of organic pollutants (PAHs, perchlorate, nitrate, TCE, bromoform,

FIGURE 13.5 Benefits and toxic effects of Fe-MNPs on plants.

ibuprofen, DDT, etc.) (Galdames et al. 2020). Iron oxide has similarly been investigated as a substitute for nZVI, while the majority of studies have concentrated on water remediation (Singh et al. 2020). nFe_3O_4 and nFe_2O_3 have excellent potential to be utilized in water treatment because of their super-paramagnetic and adsorption attributes (Li et al. 2020b; Singh et al. 2021).

nFe_3O_4 has been revealed to be more effective as an arsenic (As) adsorbent, and iron oxide NPs are very effective in eliminating As from aqueous milieus due to the strong affinity of iron oxides for As (Pizarro et al. 2021). Moreover, Fe_3O_4 NPs can extract Cd, Ni, Cr, Pb, Cu, Zn and Fe from water and soil samples within a 20-min time interval, and the removal efficiency for metal ions is between 50 and 95% (Singh et al. 2021). Furthermore, nFe_3O_4 decreased the As, polyaromatic hydrocarbon (PAH) and total petroleum hydrocarbon (TPH) contents in polluted soils (Baragaño et al. 2020b).

In addition, due to their enormous specific surface area, unique super-paramagnetism and easy separation, MNPs are widely employed as enzyme or microbial immobilizing vectors in wastewater treatment (Shahi et al. 2021; Yogalakshmi et al. 2020). Iron oxide NPs are high potential wastewater remediation agents due to their chemical inertness, high adsorption capacity and super-paramagnetism. Fe_3O_4 NPs can encourage the synthesis and activity of degrading enzymes. They are likewise responsible for improving membrane permeability. Consequently, nFe_3O_4 is the most investigated immobilization vector in bioremediation (Shahi et al. 2021).

Iron oxide NPs have been applied in several environmental sectors, such as aquatic nano-remediation, due to their exclusive super-paramagnetic and nano-specific properties. However, iron oxide NPs provoke excessive Fe bioaccumulation in the whole soft part of *Biomphalaria glabrata* tissue. Chronic exposure to iron oxide NPs increased the behavior harms of snails and reduced fertility, while fertility reduction and mortality were detected after exposure to iron oxide NPs at 15.6 mg L^{-1} (Caixeta et al. 2021). Numerous morphological modifications and cardiotoxicity in zebrafish embryos and larvae have also been identified (Pereira et al. 2020). In addition, Lacalle et al. (2020) revealed that, different from the existence of organic change, nano-remediation with nZVI was not a useable choice for Cr(VI)-polluted soils under the experimental conditions. Kamran et al. (2020) concluded that nFe_2O_3 is poisonous to soil microorganisms and disturbs their role, that is, the mineralization of nitrogen (N) and carbon.

The toxicity of NPs may vary due to their nature and behavior that is different from that of bulk materials. This necessitates an understanding of how NPs spread and influence aquatic and terrestrial ecosystems. To investigate the NP toxicity and/or benefit, applicable analytic methods are needed to assess the NP amount that exists in water or other environments (Saleh 2020).

Finally, NPs can be employed directly to remove organic contaminants by adsorption, chemical modification and immobilization of the vectors of enzymes or microbes, and they can also serve as a facilitator in the phytoremediation and/or bioremediation of pollutants. Nevertheless, the human health and environmental dangers related to NPs must be assessed before employment for real usages, specifically in cases of the direct use of these elements by humans (Saleh 2020).

13.7 PHYTOREMEDIATION

Phytoremediation employs green plants to degrade, remove, contain or stabilize contaminants in environmental milieus (soil, water and air). Phytoremediation strategies applied to polluted ecosystems involve phytoextraction, phytovolatilization, phytostabilization, rhizodegradation and phytodegradation (Figure 13.6). It is recorded that phytoextraction is the most useful and acknowledged phytoremediation approach for reducing environmental pollution. Moreover, phytoremediation

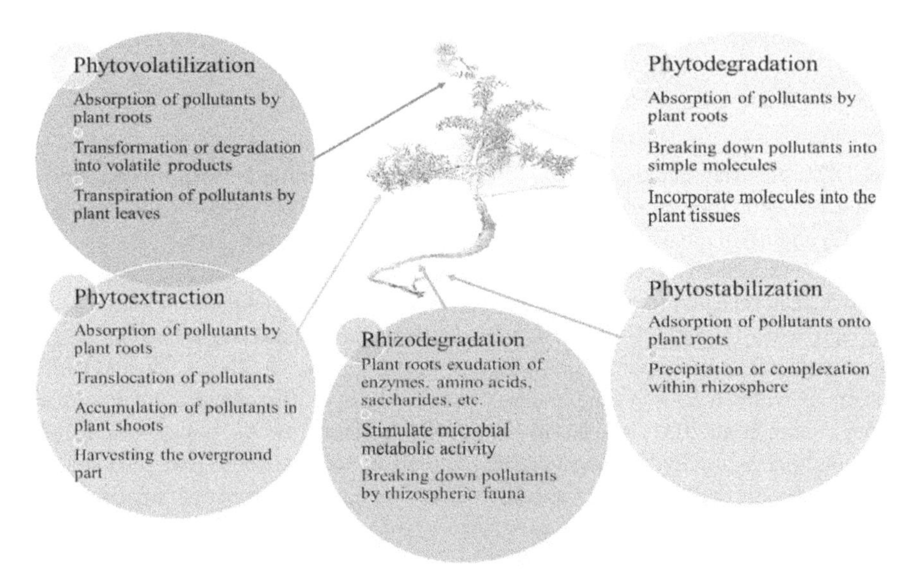

FIGURE 13.6 Phytoremediation technologies to reduce soil pollution. The plant is *Echinops spinosus*, a Zn phytoextraction candidate (Nouri et al. 2013).

Source: Adapted from Song et al. (2019)

technology has attracted increasingly more consideration because of its economical, environmental protection and self-propulsion advantages.

Phytoremediation involves a series of biochemical procedures that incorporate (1) systems of tolerance to probable contaminants; (2) the generation of phyto-chelators for transport; (3) microbial-plant exchanges; and (4) plant absorption (Ebrahimbabaie et al. 2020).

Phytoremediation is a promising technology, but its effect is influenced by many features, including the plant species, rhizoplane microorganisms, climate conditions, pollutant quantities, pollutant bioavailability and environmental matrix conditions (Zand et al. 2020a).

Due to the slow growth of hyperaccumulator plants, a limited treatment depth and long repair cycle, traditional phytoremediation methods are not economical for large-scale application (Souza et al. 2020). Therefore, nano-phytoremediation could be a successful method to solve this problem.

13.7.1 Phytoremediation of Fe-MNPs

Plant uptake and accumulation of NPs are the fundamental factors in the potential effect of nano-phytoremediation. NPs' uptake by plants is influenced by many features as recapitulated in Figure 13.4. Therefore, sometimes contradictory results will appear regarding NP plant uptake (Ebrahimbabaie et al. 2020).

The processes of NP uptake, accumulation and translocation in plants are influenced by (1) the stability and type of NPs, (2) the plant physiology and cell structure and (3) the interaction between NPs and the media (Fincheira et al. 2020). It is highlighted that the detailed absorption, accumulation and translocation of NPs are yet undiscovered (Fincheira et al. 2020). Plants have special obstructions to control the access of NPs; cell wall constituents can regulate the solubility and fluidity of NPs according to their characteristics (Fincheira et al. 2020).

Notably, rhizosphere, rhizoplane and/or endophytic microorganisms could modulate physiochemical interactions with NPs. In this context, various examinations have revealed the ability of bacterial and fungal species to tolerate, accumulate and/or detoxify metals to control the performance of NP-contaminants. In the meantime, fungi can create a robust association with plants by controlling the amount of metal NPs inside the plant (Fincheira et al. 2020).

Bystrzejewska-Piotrowska et al. (2012) studied the accumulation aptitude of nFe_3O_4 by *L. sativum* and *P. sativum*. These species have stored > 90% of nfe_3O_4 in the roots due to their high adsorption (approximately 7.8 g Kg^{-1} for *L. sativum* and 39.5 g Kg^{-1} for *P. sativum*).

Zhu et al. (2008) examined the consequence of nFe_3O_4 on pumpkins (*Cucurbita maxima*) and found that nFe_3O_4 was transported and accumulated in pumpkin tissue. However, Wang et al. (2011) revealed that nFe_3O_4 was not stored in ryegrass and pumpkin. It is worth remarking that Zhu et al. (2008) indicated that diverse plants have different responses to NPs. Bystrzejewska-Piotrowska et al. (2012) believed that the Fe_3O_4 accumulation in growth liquid and other media is dose-dependent.

13.8 NANO-PHYTOREMEDIATION

Nanotechnology can modernize numerous environmental restoration and water management technologies (Romeh 2020; Souri et al. 2020). The environmental employment of NPs has attracted specific consideration in recent years. These NPs have outstanding physicochemical features such as a large surface area and petite size (Figure 13.4), which improves their performance compared to their bulk counterparts (Hasan et al. 2020). Recently, there has been increasing attention on the progress of bioharmoniously engineered NPs with various physiochemical traits for different types of technological applications (Souri et al. 2020).

Nanotechnology is the science of creating and using "small" constituent parts with proportions of 10^{-9} m (Martínez-Fernández and Komárek 2016), characterized by the utilization of extremely small fabricated particles (NPs). Phytoremediation, in contrast, is an aspect of biological remediation that employs green plants to eliminate, transport, stabilize and/or degrade pollutants in soil, water and sediments. Nano-phytoremediation is a combination of nanotechnology and phytoremediation for the cleaning of contaminated milieus (Jiamjitrpanich et al. 2012; Nwadinigwe and Ugwu 2019; Pillai and Kottekottil 2016; Romeh and Saber 2020; Hidalgo et al. 2020).

As noted, the phytoremediation method is termed a clean and green eco-friendly method to clean up contaminated ecosystems, but often, it takes a long time to operate (Zhu et al. 2019). Therefore, grouping phytoremediation and nanotechnology (nano-phytoremediation) has attracted the consideration of numerous researchers for environmental management.

As described by Nwadinigwe and Ugwu (2019), numerous procedures including nano-phytodegradation, nano-phytoaccumulation and nano-phytostabilization can take place simultaneously in one plant. The grouping of nanotechnology and phytoremediation has been revealed to be more operative than NP-free soil (Jiamjitrpanich et al. 2012). It is crucial to guarantee that the by-products of poisonous pollutants that are degraded in the nano-phytoremediation procedure do not appear in the food chain or groundwater.

13.8.1 Fe-MNP-Based Nano-Phytoremediation

The use of NPs to sustain the phytoremediation of polluted soils is slowly attracting worldwide attention (Song et al. 2019). MNPs have an imperative task in the nano-phytoremediation approach. They can directly remove organic and inorganic pollutants (Figure 13.7), upgrade plant progress and improve the phytoavailability of pollutants (that is, increase the plants' pollutant uptake) (Song et al. 2019). Moreover, NPs can significantly regulate the molecular and genetic characteristics of hyperaccumulator plants, promote nano-phytoremediation (Rai et al. 2020) and in this way, increase the effectiveness of nano-phytoremediation in contaminated ecosystems. These effects depend on a number of factors (Figure 13.8). Therefore, nano-phytoremediation is a complex process in which plant physiology, NP transformation, pollutant reaction and media factors interact to guide the process (Figure 13.8). The recent studies of MNPs to upgrade the nano-phytoremediation of contaminated ecosystems are reviewed in Table 13.2.

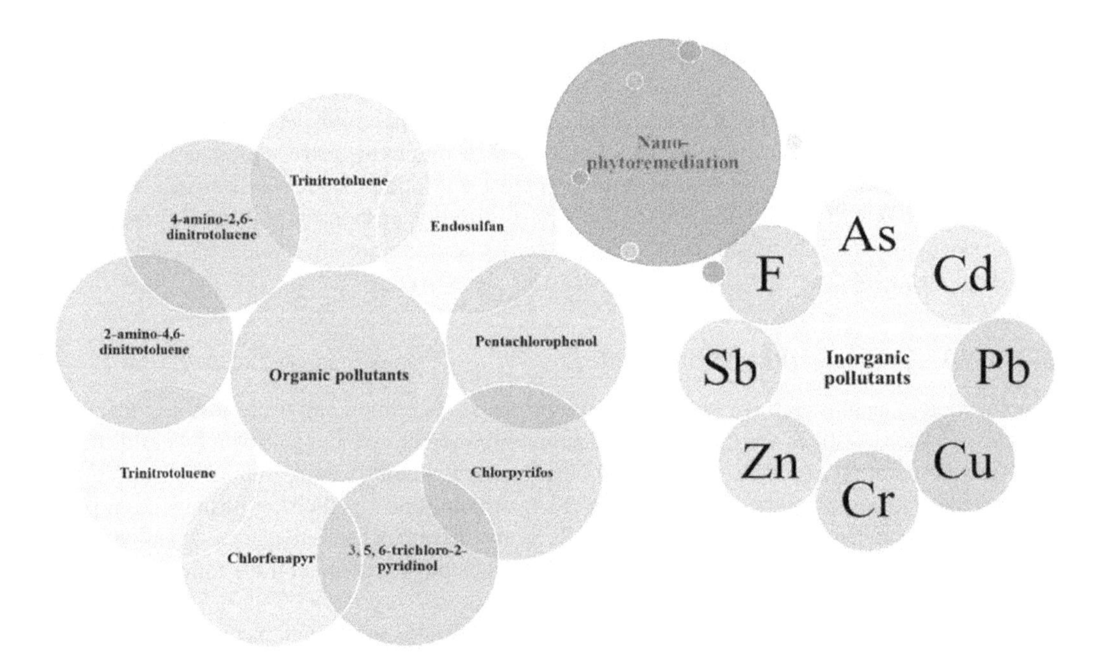

FIGURE 13.7 Recapitulation of the key pollutants remediated with Fe-MNP-based nano-phytoremediation.

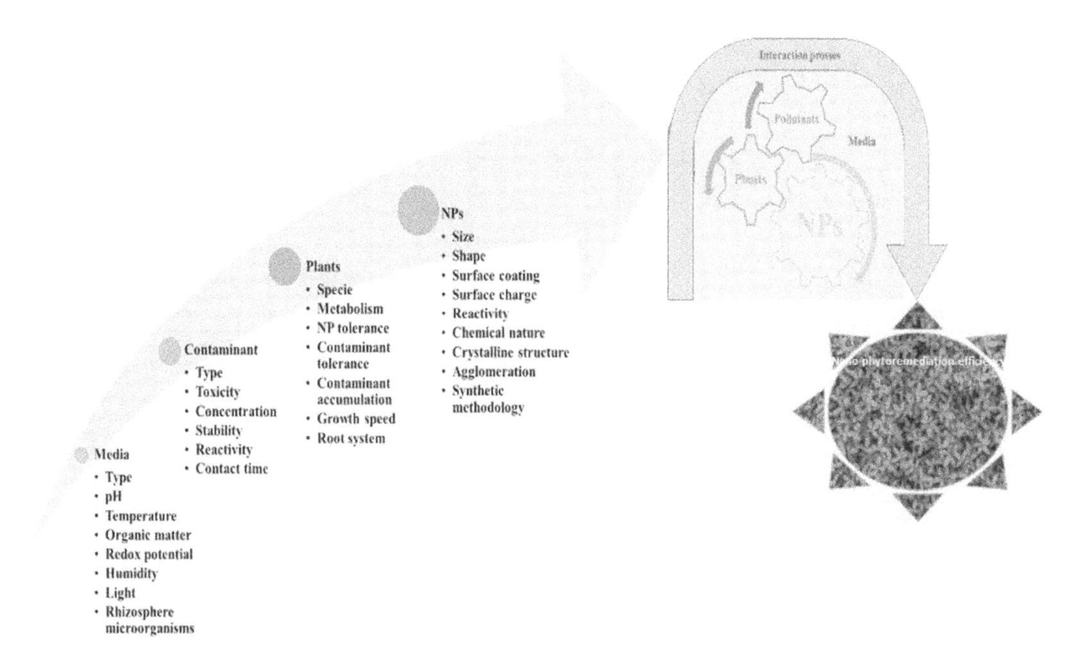

FIGURE 13.8 Parameters affecting pollutant nano-phytoremediation.

Certain studies have demonstrated the role of Fe-MNPs in improving nano-phytoremediation, and as shown in Table 13.2, nZVIs are widely used. In this regard, nZVI particles have been found to effectively accelerate the nano-phytoremediation of TNT (Jiamjitrpanich et al. 2013; Jiamjitrpanich et al. 2012), enriched soil with endosulfan (Pillai and Kottekottil 2016) and heavy metals (Vítková et al. 2018; Hidalgo et al. 2020; Gong et al. 2021). According to the research results, the most commonly identified Fe-MNP-based nano-phytoremediation strategies include the following.

13.8.1.1 Fe-MNP-Based Nano-Phytoextraction

The employment of reasonable levels of nZVI (100–500 mg Kg^{-1}) improved the aptitude to accumulate Pb in the roots (142.22 ± 4.47 mg Kg^{-1}) and shoots (49.78 ± 5.87 mg Kg^{-1}) of *K. scoparia*, judged against the control treatment. In contrast, advanced amounts of nZVI have shown Pb transfer suppression, chlorophyll content decrease, inhibitory impacts on plant growth and, thus, the efficiency of Pb remediation (Zand and Tabrizi 2020). In this context, minor amounts of utilized nZVI increased plant growth, while elevated levels produced nanotoxicity (Ma et al. 2013; Guha et al. 2020b).

In antimony (Sb)-polluted soil, the combined application of nZVI and plant growth-promoting bacteria (PGPR) stimulated the development of *T. repens*. The accumulation of Sb was greater in the aerial parts than in the roots of *T. repens*. The maximum accumulation of Sb in *T. repens* was 3,896.4 µg/Kg (Zand et al. 2020a). However, the usage of 1,000 mg/kg of nZVI has harmful consequences on plant evolution and phytoremediation capacity. Therefore, the joint employment of nZVI and PGPR can diminish the need for nZVI to remediate metalloid-contaminated soil (Zand et al. 2020a).

Additionally, the accumulation and transportation of nFe_3O_4 and F in plant tissues have been confirmed (Kumari and Khan 2018). In this respect, the application of nFe_3O_4 at 200 mg Kg^{-1} significantly augmented the stem and root length of *Prosopis juliflora* and encouraged the accumulation of F in stems and roots of up to 34.64 and 28.43 mg Kg^{-1}, respectively (Kumari and Khan 2018). Thus, the utilization of nFe_3O_4 up to 200 mg Kg^{-1} could improve the fluoride absorption efficiency of *Prosopis juliflora* (Kumari and Khan 2018).

TABLE 13.2
Current Examples of Fe-MNP-based Nano-Phytoremediation

Nano-Phytoremediation Strategy	Fe-Magnetic NPs	Plant Species	Pollutants	Media	Accumulation Efficiency (%) or Capacity (mg/Kg)	References
Nano-Phytoextraction	nZVI	*Trifolium repens*	Sb	Soil	3896.4 μg/Kg	(Zand et al. 2020a)
	nZVI	*Trifolium repens*	Cd	Soil	718.6 μg/Kg	(Zand et al. 2020b)
	nZVI	*Lolium perenne*	Pb	Soil	1175.4 μg per pot	(Huang et al. 2018)
	nZVI	*Boehmeria nivea*	Cd	Soil	Cd levels in the leaves, stems and roots improved by 31–73%, 29–52% and 16–50%, respectively	(Gong et al. 2017)
	nZVI	*Salix alba*	Pb, Cu and Cd	Soil	The elevated nZVI amount had no encouraging influence on the heavy metal uptake, and a small concentration of nZVI improved the root Cd store and the BCF level of Cd in the seedlings	(Mokarram-Kashtiban et al. 2019)
	nZVI	*Kochia scoparia*	Pb	Soil	857.18 μg per pot obtained with 500 mg Kg^{-1} of nZVI	(Zand and Tabrizi 2020)
	nZVI	*Avicennia Germinans*	Cd, Pb and As	Soil	The nZVI and plant reduced the As concentration by 85.19%, Cd by 50.30% and Pb by 62.49%	(Hidalgo et al. 2020)
	nZVI	*Boehmeria nivea*	Cd	Sediments	13%	(Gong et al. 2021)
	nZVI	*Plantago major*	TCP	Water	TCP notably amassed in *P. major* leaves and roots	(Romeh 2020)
	nFe3O4@ GSH	*Isatis cappadocica*	As	Water	566 mg kg-1 DW (59%)	(Souri et al. 2020)
	nFe3O4	*Prosopis juliflora*	F	Soil	Accumulation in roots and shoots up to 34.64 and 28.43 mg Kg^{-1}, respectively	(Kumari and Khan 2018)
Nano-Phytostabilization	nZVI	*Helianthus annuus*	As, Cd, Pb and Zn	Soil	Cd, Zn, Pb and As levels in roots and shoots reduced by 50–60%	(Vítková et al. 2018)
	nZVI	*Lolium perenne*	As, Cd, Pb and Zn	Soil	Cd, Zn, Pb and As levels in roots and shoots reduced by 50–60%	(Vítková et al. 2018)
	nZVI	*Helianthus annuus*	Cr	Soil	The Cr dose in roots and shoots diminished by 34.85–73.67% and 57.08–85.64%, respectively	(Mohammadi et al. 2020)
	nZVI	*Hordeum vulgare*	As	Soil	Diminished As availability and uptake, amplified plant development and augmented biomass	(Gil-Díaz et al. 2016)
	nZVI	*Panicum maximum*	TNT, 2-ADNT and 4-ADNT	Soil	TNT, 2-ADNT and 4-ADNT absorbed by roots in nZVI-added soil	(Jiamjitrpanich et al. 2013)

Nano-Phytodegradation	nZVI	*Panicum maximum*	Trinitrotoluene	Soil	Improved the removal competence of trinitrotoluene from 85.7% to 100% in 4 months	(Jiamjitrpanich et al. 2012)
	nZVI	*Alpinia calcarata*	Endosulfan	Soil	Removal rates were increased from 81.2% to 100%	(Pillai and Kottekottil 2016)
	nZVI	*Ocimum sanctum*	Endosulfan	Soil	Removal rates were amplified from 20.76% to 76.28%	(Pillai and Kottekottil 2016)
	nZVI	*Cymbopogon citratus*	Endosulfan	Soil	Removal rates were augmented from 65.08% to 86.16%	(Pillai and Kottekottil 2016)
	nZVI	*Oryza sativa*	PCP	Soil	100 mg nZVI Kg^{-1} improved grain yield, diminished the grain PCP level and augmented the soil PCP removal rate	(Liu et al. 2020b)
	nZVI	*Plantago major*	Chlorpyrifos	Water	Chlorpyrifos removed at 81.69%	(Romeh 2020)
	nZVI	*Plantago major*	Chlorfenapyr	Water	Chlorfenapyr was reduced by 93.74%	(Romeh and Saber 2020)
	nZVI	*Plantago major*	Chlorfenapyr	Soil	The degradation percentages were boosted by 71.22%	(Romeh and Saber 2020)

Gong et al. (2021) confirmed that 100 mg Kg^{-1} nZVI improved Cd accumulation by 13% and decreased dry biomass slightly but not significantly; other amounts of nZVI significantly downgraded the dry biomass. The existence of Fe and $Fe(OH)_x$ on the exterior of nZVI might react with Cd (establishing composites such as FeOCdOH), which could enhance the reducible amount of Cd in polluted sediments (Gong et al. 2021). nZVI usage can promote the phytoremediation of a Cd-polluted environment by modifying Cd portions, sediment characteristics and structure of the bacterial community (Gong et al. 2021).

Gong et al. (2017) showed that nZVI (100 mg Kg^{-1}) encouraged the accumulation and translocation of Cd in *Boehmeria nivea* seedlings. In addition, 100 mg Kg^{-1} nZVI could alleviate the oxidative damage of *Boehmeria nivea* under Cd stress, while 500 and 1,000 mg Kg^{-1} nZVI restrained plant development and provoked oxidative damage (Gong et al. 2017). Therefore, a low amount of nZVI can enhance the phytoremediation efficiency of metal-contaminated sites.

Huang et al. (2018) proved that the overall Pb levels in *Lolium perenne* with the usage of small amounts of nZVI (<500 mg Kg^{-1}) were higher than those without nZVI. Nevertheless, the total content of Pb in *Lolium perenne* diminished at elevated levels of nZVI (>1,000 mg Kg^{-1}). Moreover, small levels of nZVI (<500 mg Kg^{-1}) can effectively promote the growth of *Lolium perenne*, encourage the absorption and transportation of Pb, and improve the stabilization of Pb in sediments. This shows that nZVI-aided phytoremediation is an encouraging technique for the remediation of Pb-polluted sediments (Huang et al. 2018). Therefore, 100 mg Kg^{-1} was the optimal dose of nZVI to aid in the phytoremediation of lead-contaminated sediments (Huang et al. 2018).

The usage of nZVI notably augmented the accumulating competence of *Trifolium repens* for Cd (Zand et al. 2020b). The accumulation of Cd in *Trifolium repens* roots was significantly higher than the shoots, indicating that the roots of *Trifolium repens* were the preferred storage organ for Cd (Zand et al. 2020b).

Moreover, nZVI's interaction with *Avicennia Germinans* is an effective way to boost the bioavailability of Pb^{2+} and Cd^{2+} and allow plants to translocate them well to their aboveground parts (Hidalgo et al. 2020).

Furthermore, Mokarram-Kashtiban et al. (2019) reported that an elevated amount of nZVI had no encouraging consequence on heavy metal absorption and bioconcentration factor (BCF) for *Salix alba* and that it instigated stress and diminished development of *Salix alba*. Nevertheless, a small amount of nZVI had no damaging influence on plant development or the biochemical and physiological parameters and augmented the root measurement and leaf surface according to the plant root Cd accumulation and the Cd BCF level in seedlings. They also showed that a small quantity of nZVI could advance phytoremediation efficiency, and the use of rhizosphere microorganisms could diminish the stress of nZVI on plants, even at a high dose.

Souri et al. (2020) indicated the high efficiency of nFe3O4@GSH as a new As nano-phytoremediation promoter for *Isatis cappadocica*. They also proved that the advantageous consequences of nFe3O4@GSH on As nano-phytoremediation are due to the improvement of the activities of numerous enzyme and non-enzyme antioxidants (Souri et al. 2020).

13.8.1.2 Fe-MNP-Based Nano-Phytostabilization

The utilization of immobilization agents in general, including nZVI, can reduce the solubility and bioavailability of contaminants (Baragaño et al. 2020a). This is an important factor in nano-phytostabilization and in establishing a vegetation covering that can lead to ecological restoration. Increased levels of nZVI in the growing medium augmented the uptake of Fe and Mg and diminished the Fe translocation factor for *Festuca rubra* and *Agrostis capillaris* (Teodoro et al. 2020). The stabilization of inorganic pollutants can be deemed as one of the greatest practical and feasible substitutes for the restoration and protection of seriously polluted soils (Baragaño et al. 2020c).

Adrees et al. (2020) reported that the appropriate amount of nFe_2O_3 in Cd-contaminated soil immobilizes Cd in soil, decreases Cd bioavailability, diminishes Cd translocation to aerial tissues and grains and therefore reduces the Cd content in grains.

In general, the three promising procedures for the diminution of Cd stress by Fe-MNPs are (i) reducing Cd accumulation in roots and stems, (ii) diminishing oxidative stress caused by Cd accumulation and (iii) controlling nutrient uptake, which should lead to oxidative protection (Rahmatizadeh et al. 2019). Generally, NPs can affect the organismal, cellular or molecular level of plants in many ways by (1) disrupting physiological processes, (2) releasing particulate forms and ions, (3) inducing oxidative stress and (4) impairing or altering plant gene expression (Liu et al. 2020a).

Mohammadi et al. (2020) proved that with an nZVI concentration increase, the Cr absorption rate of plants decreased, which significantly enhanced the morphological and physiological characteristics of plants. nZVI treatment decreased the bioaccumulation factor (BAF) and transport factor (TF) in the roots and shoots, but with an increase in Cr(VI) concentrations, the BAF and TF decreased significantly. In addition, Cr stress increased the antioxidant enzyme activity, and adding nZVI further augmented this activity, thus improving the growth features (Mohammadi et al. 2020). Furthermore, nZVI could improve sunflower plant properties under Cr contamination by diminishing the absorption of Cr and increasing the activities of intracellular detoxification enzymes (Mohammadi et al. 2020).

Vítková et al. (2018) stated that the employment of nZVI may benefit the stabilization of As in the rhizosphere of sunflower. They added nZVI to polluted soil; after a 5-week growing period, the content of As in the soil declined by more than 80%, and the accumulation of As in the shoots and roots diminished by 24% and 47%, respectively (Vítková et al. 2018). Thus, the utilization of nZVI stabilizes metals in the soil and decreases the absorption of As, Cd, Pb and zinc (Zn) in *Helianthus annuus* and *Lolium perenne* (Vítková et al. 2018).

Gil-Díaz et al. (2016) confirmed the usefulness of nZVI in diminishing the availability of As in heavily As-polluted soil. In this regard, the inferior availability of As in soils treated with 10% nZVI encouraged the growth of barley and reduced the uptake of As (Gil-Díaz et al. 2016).

Jiamjitrpanich et al. (2013) indicated that the nano-phytostabiliziation of TNT and its metabolites in polluted soil resulted in advanced accumulation of TNT and its metabolites in the roots.

Nano-phytostabilization is an effective restoration method that employs species that can accumulate or precipitate pollutants in the root zone to establish vegetation cover, thus reducing pollutant mobility and bioavailability (Lebrun et al. 2019; Sigua et al. 2019). Therefore, species that simultaneously tolerate high levels of contamination and have a large fasciculate root organization, a fast-growing organization, long longevity, simple maintenance, appropriate contamination tolerance and an underground metal store are suitable applicants for nano-phytostabilization (Sigua et al. 2019). The interaction among the plant species, the designated immobilizer and consequently, the pollutant will determine the remediation procedure. The encouraging results of these interactions in soil remediation contribute to successful nano-phytostabilization. Therefore, understanding such interactions has recently attracted a lot of consideration (Teodoro et al. 2020).

Bidabadi (2020) proved that the utilization of Fe-MNPs notably enhanced plant development and encouraged plant protection tools against heavy metal toxicity. However, they had an undesirable effect on the phytoremediation features of *S. splendens* by lessening the heavy metal accumulation in plant tissues. Therefore, the utilization of Fe-MNPs is not appropriate to support the phytoremediation aptitude of *S. splendens* (Bidabadi 2020).

13.8.1.3 Fe-MNP-Based Nano-Phytodegradation

nZVI increased personal care product (PCP) degradation and facilitated its rhizosphere microbial degradation (Liu et al. 2020b). Liu et al. (2020b) revealed that the cooperation of nZVI and rice cultivation offers a broader path of PCP degradation. Additionally, root iron plate formation was encouraged by adding 100 mg of nZVI Kg^{-1} to PCP-polluted soil, which upgraded the quality and yield of the rice. Co-treatment with nZVI and rice agriculture has synergistically incorporated the dehalogenation capabilities of nZVI and rhizosphere microorganisms to realize soil clean-up (Liu et al. 2020b).

Romeh (2020) verified that the utilization of nZVI as a green nanotechnology and *P. major* as a candidate for the phytoremediation approach plays a key role in the elimination of chlorpyrifos from water with an important reduction in the toxic degradation product (TCP). Moreover, the grouping of nano-phytoremediation technology, green nanotechnology and solubility enhancement agents is of great significance for the clean-up of chlorfenapyr-polluted soil and water (Romeh and Saber 2020). The consequence of nZVI may be that chlorpyrifos degrades to TCP, which is stored in the leaves and roots of *P. major* (Romeh 2020).

The nZVI and enzyme degradation mechanism of endosulfan seems to be hydrolysis and continuous dehalogenation. Moreover, the adding of nZVI could promote the reductive dechlorination of endosulfan; thus, only a small amount of endosulfan was accumulated in these plants (*Cymbopogon citratus*, *Ocimum sanctum* and *Alpinia calcarata*) (Pillai and Kottekottil 2016). Researchers found that that nano-phytoremediation technology could remove endosulfan from soil rapidly and effectively, and *A. calcarata* and nZVI could accelerate the removal of endosulfan from the soil in 30 days (Pillai and Kottekottil 2016).

In the combination of *Plantago major* and nZVI, the degradation of chlorfenapyr increased from the soil (Romeh and Saber 2020). Likewise, the rapid and efficient elimination of chlorfenapyr from water was revealed using *Plantago major* and nZVI (Romeh and Ibrahim Saber 2020). Romeh and Saber (2020) confirmed that the effect of nano-phytoremediation was better than phytoremediation for cleaning-up chlorfenapyr-polluted water. Similarly, endosulfan nano-phytoremediation was more effective than endosulfan phytoremediation within 7 days (Pillai and Kottekottil 2016). Consequently, green nanotechnology based on nZVI and phytoremediation serves an imperative function in pesticide removal from the ecosystem.

Jiamjitrpanich et al. (2012) verified that the degradation and elimination competence of TNT in polluted soil was better by nano-phytoremediation than by nano-remediation and phytoremediation. The highest elimination competence of nano-phytoremediation was attained in soil with a TNT/nZVI ratio of 1/10 that is treated with *P. maximum* (Jiamjitrpanich et al. 2012).

Overall, Fe-MNPs appear to be a promising option for strategies to mobilize and immobilize soil nano-remediation when joined with other techniques such as phytoremediation (Baragaño et al. 2020c). When using NPs in nano-phytoremediation or agriculture, it is crucial to consider the serious consequences on plants under actual conditions (Pizarro et al. 2021).

Nano-phytoremediation is a novel practice to clean up polluted soils, water and sediments. To date, the manipulation of nZVI in soils with plants has been chiefly limited due to stresses produced in plants by these NPs (Huang et al. 2018; Mokarram-Kashtiban et al. 2019), and the data on the clean-up of polluted soils under the joint influence of plants and nZVI are still weak.

Indeed, the application of elevated amounts of certain NPs for clean-up applications has higher research interest because of their unclear consequences on the environment and likely lethal impacts on animals, microorganisms and plants; contradictory results have been published in the literature as discussed in section 13.4. An ideal plant for removing environmental pollutants should have a variability of the features shown in Figure 13.9.

Currently, nano-phytoremediation strategies are still in the exploration and experimental stage, but their employment provides a new way to advance the efficiency of nano-phytoremediation. Figure 13.10 illustrates the advantages and disadvantages of nano-phytoremediation.

13.9 NANO-PHYTOREMEDIATION APPLICATIONS

13.9.1 Nano-Phytoremediation of Pollutants in Soil

Soil contamination by poisonous substances is a recurring problem instigated by environmental catastrophes, industrialization, petroleum combustion and urbanization, which participate in soil contamination (Souza et al. 2020). Therefore, the clean-up of poisons in the soil is still a critical

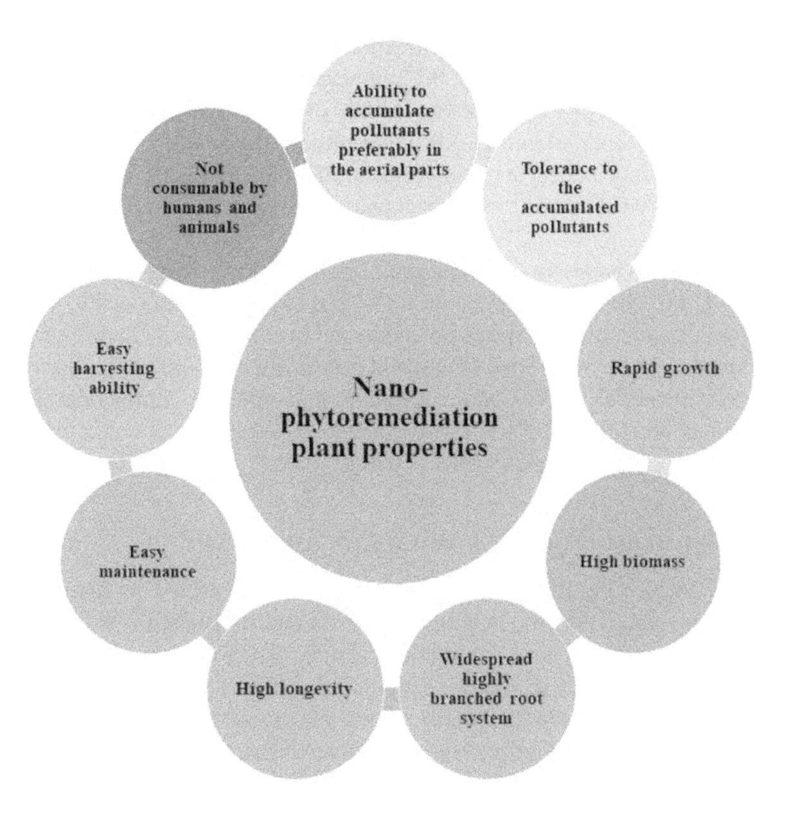

FIGURE 13.9 Ideal nano-phytoremediation plant features.

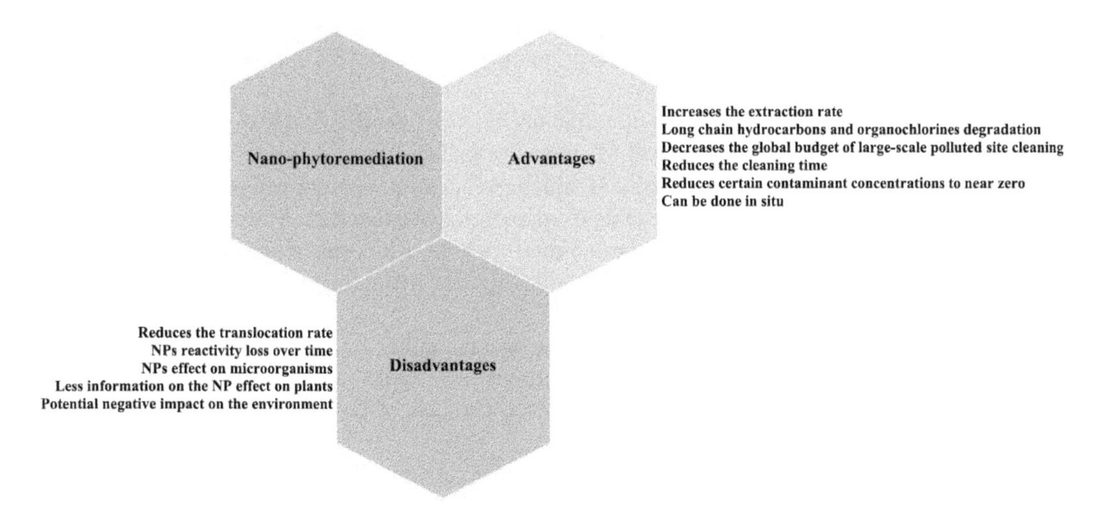

FIGURE 13.10 Nano-phytoremediation advantages and disadvantages.

subject as polluted soils could disturb the environment, human health and agricultural security (Nouri and Haddioui 2016a; Nouri and Haddioui 2016b; Nouri et al. 2017).

NPs employed for the nano-phytoremediation of polluted topsoil should be non-toxic to vegetations and their rhizosphere and rhizosplane microorganisms, and can be combined with pollutants to make them bioavailable to improve the phytoextraction procedure. Sundry NPs formed by topsoil microorganisms and plants also encourage the synthesis of phytohormones, resulting in a superior biomass and good extraction of soil pollutants to plant organs (Khan 2020b).

The joint application of nZVI and PGPR can diminish the need for nZVI in the remediation of metalloid-contaminated soil. The intelligent application of plants (*Trifolium repens*) combined with NPs and PGPR for antimony removal from soil has broad application prospects (Zand et al. 2020a).

The content of Cd in the tissues of *Trifolium repens* improved with an increase in the nZVI amount. Moreover, the combined treatment of nZVI and biochar is beneficial to the phytoremediation of topsoil metals and has broad usage prospects in topsoil clean-up (Zand et al. 2020b).

nZVI participates with an imperative function in soil in-situ clean-up by fixing Cd and enhancing vegetable growth (Guha et al. 2020a). Nevertheless, to guarantee the commercial use of nZVI, it is necessary to conduct repetitive field trials and exhaustive investigations on the impact of nZVI on topsoil and microflora ecosystems (Guha et al. 2020a).

13.9.2 Nano-Phytoremediation for Water Contamination

Regardless of the source, the majority of released NPs are expected to finally arrive to water milieus. Finding and quantifying the quantity of NPs in natural situations is difficult because utilized technologies and methods have not yet been standardized. Certain aquatic vegetation's aptitude to absorb and store metals to remove metal-contaminated water and sediments have been verified. Some vegetation has been distinguished that can absorb NPs via roots or leaves.

It is imperative to comprehend the exchanges between NPs and aquatic vegetation and to evaluate the probable influences on metal absorption and phytoremediation methods (Ebrahimbabaie et al. 2020).

Once arriving in aquatic and terrestrial settings, NPs are subjected to several biological, chemical and physical variations. NPs can undergo agglomeration, dissolution, adsorption, redox effects, deposition and further procedures, which frequently occur simultaneously. The diverse modifications of NPs depend on their native characteristics (Figures 13.4 and 13.8). In addition, NPs are very sensitive to the chemical conditions of the reception milieu (Figures 13.4 and 13.8) (Ebrahimbabaie et al. 2020).

The absorption, storage or utilization of pollutants by plants are the vital features to improve the quality of water and preserve the survival of aquatic organisms. Due to the strong absorption and accumulation potential of numerous forms of aquatic vegetation, phytoremediation has become a hopeful technology to remove surplus metals from aquatic environments. Moreover, several types of aquatic vegetation fix metals to the root and the root area. Rhizosphere and rhizoplane microorganisms also participate with imperative functions in the phytoremediation of aquatic locations (Ebrahimbabaie et al. 2020).

The capacity of aquatic vegetation to eliminate metals, either in a dissolved form or as NPs, depends on the characteristics of the roots, genotypes (e.g., rapid reproduction) and physiology. If vegetation is to be utilized for phytoremediation, then all of these factors must be counted (Ebrahimbabaie et al. 2020).

Constructed wetlands can utilize the aptitude of aquatic vegetation to fix and/or absorb NPs. So far, the elimination of NPs in constructed wetlands has not attracted enough consideration (Ebrahimbabaie et al. 2020).

Bao et al. (2019) assessed the removal efficiency of silver NPs by a vertical movement wetland technique. The results revealed that the removal rate of water-soluble NPs by a constructed wetland planted with *P. australis* was 78.53%, that of an unplanted system was 40.96% and that of plants was 37.57% (Bao et al. 2019). However, *P. australis* has not shown great effectiveness in removing cerium NPs (Hu et al. 2018), as these plants remove only 17.9% of the NPs (principally stored in the roots). A maximum number of NPs are stored in the biofilm of the system, and their harmfulness to vegetation and microorganisms is very obvious (Hu et al. 2018).

Huang et al. (2019) also revealed the effective elimination of silver NPs (about 96%) by planting *Iris pseudacorus* in a vertical movement wetland and detected the diminution of COD (83%), total N (61%), NH_3 (42%) and total phosphorus (P) (70%). However, there are few investigations on the elimination of NPs in wetlands and the role of vegetation in this process. The available findings are only separate studies that do not normally include the interaction with other environmental factors (Ebrahimbabaie et al. 2020).

The application of nFe3O4@GSH (200 mg/L) considerably upgraded *Isatis cappadocica* development in aqueous media and lessened the poisonous traces in As-treated plants (Souri et al. 2020).

13.10 CHALLENGES AND FUTURE PROSPECTS OF NANO-PHYTOREMEDIATION

With the progression of nanotechnology and bioremediation technology, the usage of NPs for phytoremediation is an innovative idea. Despite the many benefits of nano-phytoremediation, there are still several challenges in its application to pollutant remediation that require special attention in future studies, including NPs and nano-phytoremediation challenges (Figure 13.11).

Consequently, to achieve effective nano-phytoremediation under different environmental treatment conditions, it is obligatory to conduct a large number of studies on promising NPs and plant species. Most recent research was conducted in closed environments (laboratory or greenhouse), and the results are sometimes unreliable or even contradictory. Therefore, it is mandatory to perform further studies under normal environmental conditions and in the field to better comprehend the encouraging and undesirable effects of Fe-MNPs on plants, biodiversity and human life. In addition, the literature on the regeneration of Fe-MNPs is very limited, so it is crucial to study the regeneration process for environmental sustainable development.

Moreover, it is necessary to further study the responses of NPs to diverse vegetable species, contaminants, media types and climatic conditions in plant protection systems. Therefore, to diminish the potential risk of Fe-MNPs in the soil-plant system, further study is required on the fate, migration and action mechanism of Fe-MNPs in the soil-plant system. Additionally, it is obligatory to investigate the harmfulness of Fe-MNPs and their absorption by plants over a

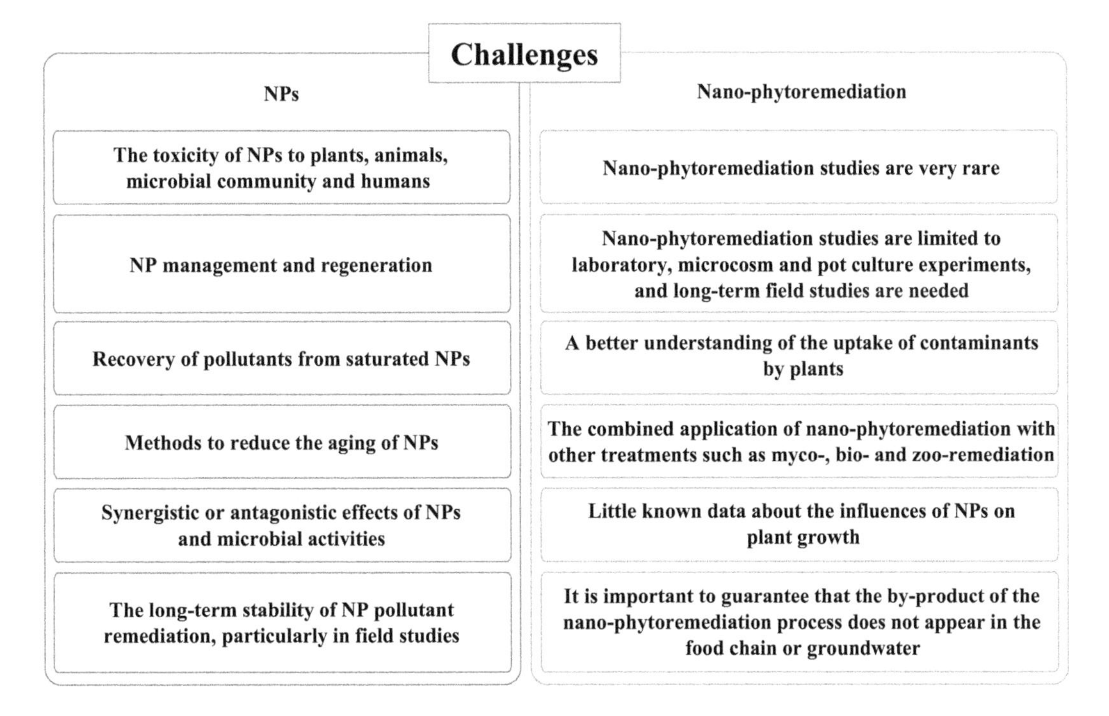

FIGURE 13.11 NPs and nano-phytoremediation challenges.

long time. More consideration should be given to whole life cycle experiments. Consequently, it is crucial to examine the trophic migration of Fe-MNPs in the food chain. Finally, the efficient production of NPs necessitates more care to amplify intelligent nano-materials for environmental purification.

13.11 CONCLUSION

This chapter is helpful to progress knowledge concerning nano-phytoremediation. The collected literature clearly shows that Fe-MNPs can remove pollutants, are easily absorbed by plants, promote plant development and advance the phytoavailability of pollutants; thus, FeMNPs improve the nano-phytoremediation efficiency of polluted environmental matrices.

Low concentrations of Fe-MNP can modify plant carotene and chlorophyll, intensify plant protein content, progress the plant photosynthesis rate, encourage growth, improve soil richness and diminish the availability of metals and toxicity to vegetation. However, high amounts of Fe-MNPs have poisonous effects on plants, which are mainly manifested as germination and growth inhibition, leaf toxicity, oxidative stress promotion and induced genotoxicity.

Accordingly, nano-phytoremediation technology that combines NPs with plants has broad application prospects in environmental remediation, and its development direction should be quickly found in large-scale application. To accomplish this, additional investigation is required, and the long-term performance of nano-phytoremediation systems needs to be further studied.

REFERENCES

Adrees, Muhammad, Zahra Saeed Khan, Shafaqat Ali, Muhammad Hafeez, Sofia Khalid, Muhammad Zia ur Rehman, Afzal Hussain, Khalid Hussain, Shahzad Ali Shahid Chatha, and Muhammad Rizwan. 2020. 'Simultaneous Mitigation of Cadmium and Drought Stress in Wheat by Soil Application of Iron Nanoparticles'. *Chemosphere* 238 (January). Elsevier Ltd. doi:10.1016/j.chemosphere.2019.124681.

Ahmad, Bilal, Abbu Zaid, Hassan Jaleel, M. Masroor A. Khan, and Mansour Ghorbanpour. 2019. 'Nanotechnology for Phytoremediation of Heavy Metals: Mechanisms of Nanomaterial-Mediated Alleviation of Toxic Metals'. In *Advances in Phytonanotechnology*, 315–327. Elsevier. doi:10.1016/b978-0-12-815322-2.00014-6.

Ahmadi, Mazaher. 2020. 'Iron Oxide Nanoparticles for Delivery Purposes'. In *Nanoengineered Biomaterials for Advanced Drug Delivery*, 373–393. Elsevier. doi:10.1016/b978-0-08-102985-5.00016-4.

Alidoust, Darioush, and Akihiro Isoda. 2013. 'Effect of ΓFe2O3 Nanoparticles on Photosynthetic Characteristic of Soybean (Glycine Max (L.) Merr.): Foliar Spray Versus Soil Amendment'. *Acta Physiologiae Plantarum* 35 (12). Springer: 3365–3375. doi:10.1007/s11738-013-1369-8.

Alidoust, Darioush, and Akihiro Isoda. 2014. 'Phytotoxicity Assessment of γ-Fe2O3 Nanoparticles on Root Elongation and Growth of Rice Plant'. *Environmental Earth Sciences* 71 (12). Springer Verlag: 5173–5182. doi:10.1007/s12665-013-2920-z.

Bao, Shaopan, Lei Liang, Jiaolong Huang, Xiawei Liu, Wei Tang, Jia Yi, and Tao Fang. 2019. 'Removal and Fate of Silver Nanoparticles in Lab-Scale Vertical Flow Constructed Wetland'. *Chemosphere* 214 (January). Elsevier Ltd: 203–209. doi:10.1016/j.chemosphere.2018.09.110.

Baragaño, D., J. Alonso, J. R. Gallego, M. C. Lobo, and M. Gil-Díaz. 2020a. 'Zero Valent Iron and Goethite Nanoparticles as New Promising Remediation Techniques for As-Polluted Soils'. *Chemosphere* 238 (January). Elsevier Ltd: 124624. doi:10.1016/j.chemosphere.2019.124624.

Baragaño, D., J. Alonso, J. R. Gallego, M. C. Lobo, and M. Gil-Díaz. 2020b. 'Magnetite Nanoparticles for the Remediation of Soils Co-Contaminated with as and PAHs'. *Chemical Engineering Journal* 399 (November). Elsevier B.V.: 125809. doi:10.1016/j.cej.2020.125809.

Baragaño, Diego, Rubén Forján, Lorena Welte, and José Luis R. Gallego. 2020c. 'Nanoremediation of As and Metals Polluted Soils by Means of Graphene Oxide Nanoparticles'. *Scientific Reports* 10 (1). Nature Research: 1–10. doi:10.1038/s41598-020-58852-4.

Barrena, Raquel, Eudald Casals, Joan Colón, Xavier Font, Antoni Sánchez, and Víctor Puntes. 2009. 'Evaluation of the Ecotoxicity of Model Nanoparticles'. *Chemosphere* 75 (7). Pergamon: 850–857. doi:10.1016/j.chemosphere.2009.01.078.

Bidabadi, Siamak Shirani. 2020. 'The Role of Fe-Nano Particles in Scarlet Sage Responses to Heavy Metals Stress'. *International Journal of Phytoremediation* 22 (12). Bellwether Publishing, Ltd.: 1259–1268. doi:10.1080/15226514.2020.1759507.

Brasili, Elisa, Irene Bavasso, Valerio Petruccelli, Giorgio Vilardi, Alessio Valletta, Chiara Dal Bosco, Alessandra Gentili, Gabriella Pasqua, and Luca Di Palma. 2020. 'Remediation of Hexavalent Chromium Contaminated Water through Zero-Valent Iron Nanoparticles and Effects on Tomato Plant Growth Performance'. *Scientific Reports* 10 (1). Nature Research: 1–11. doi:10.1038/s41598-020-58639-7.

Bystrzejewska-Piotrowska, G., M Asztemborska, R. Stęborowski, H. Polkowska-Motrenko, B. Danko, and J. Ryniewicz. 2012. 'Application of Neutron Activation for Investigation of Fe3O4 Nanoparticles Accumulation by Plants'. *Nukleonika* 57.

Cai, Lin, Liuti Cai, Huanyu Jia, Changyun Liu, Daibin Wang, and Xianchao Sun. 2020. 'Foliar Exposure of Fe3O4 Nanoparticles on Nicotiana Benthamiana: Evidence for Nanoparticles Uptake, Plant Growth Promoter and Defense Response Elicitor against Plant Virus'. *Journal of Hazardous Materials* 393 (July). Elsevier B.V.: 122415. doi:10.1016/j.jhazmat.2020.122415.

Caixeta, Maxwell Batista, Paula Sampaio Araújo, Cândido Carvalho Rodrigues, Bruno Bastos Gonçalves, Olacir Alves Araújo, Giovanni Bonatti Bevilaqua, Guilherme Malafaia, Luciana Damacena Silva, and Thiago Lopes Rocha. 2021. 'Risk Assessment of Iron Oxide Nanoparticles in an Aquatic Ecosystem: A Case Study on Biomphalaria Glabrata'. *Journal of Hazardous Materials* 401 (January). Elsevier B.V.: 123398. doi:10.1016/j.jhazmat.2020.123398.

Chakraborty, Rupa, Anupama Asthana, Ajaya Kumar Singh, Bhawana Jain, and Abu Bin Hasan Susan. 2020. 'Adsorption of Heavy Metal Ions by Various Low-Cost Adsorbents: A Review'. *International Journal of Environmental Analytical Chemistry*. Taylor and Francis Ltd. doi:10.1080/03067319.2020.1722811.

Cunha Lopes, Tamires Letícia, Rita de Cássia Siqueira-Soares, Guilherme Henrique Gonçalves de Almeida, Gabriele Sauthier Romano de Melo, Gabriela Ellen Barreto, Dyoni Matias de Oliveira, Wanderley Dantas dos Santos, Osvaldo Ferrarese-Filho, and Rogério Marchiosi. 2018. 'Lignin-Induced Growth Inhibition in Soybean Exposed to Iron Oxide Nanoparticles'. *Chemosphere* 211 (November). Elsevier Ltd: 226–234. doi:10.1016/j.chemosphere.2018.07.143.

Ebrahimbabaie, Parisa, Weeradej Meeinkuirt, and John Pichtel. 2020. 'Phytoremediation of Engineered Nanoparticles Using Aquatic Plants: Mechanisms and Practical Feasibility'. *Journal of Environmental Sciences (China)*. Chinese Academy of Sciences. doi:10.1016/j.jes.2020.03.034.

Fahad, Aamna Balouch, Muhammad Hassan Agheem, Sohail Ahmed Memon, Abdul Rehman Baloch, Aqsa Tunio, Abdullah, et al. 2020. 'Efficient Mitigation of Cadmium and Lead Toxicity in Coriander Plant Utilizing Magnetite (Fe3O4) Nanofertilizer as Growth Regulator and Antimicrobial Agent'. *International Journal of Environmental Analytical Chemistry*. Taylor and Francis Ltd. doi:10.1080/03067319.2020.1776861.

Fayiga, Abioye O., Mabel O. Ipinmoroti, and Tait Chirenje. 2018. 'Environmental Pollution in Africa'. *Environment, Development and Sustainability* 20 (1). Springer Netherlands: 41–73. doi:10.1007/s10668-016-9894-4.

Feng, Youzhi, Xiangchao Cui, Shiying He, Ge Dong, Min Chen, Junhua Wang, and Xiangui Lin. 2013. 'The Role of Metal Nanoparticles in Influencing Arbuscular Mycorrhizal Fungi Effects on Plant Growth'. *Environmental Science and Technology* 47 (16). American Chemical Society: 9496–9504. doi:10.1021/es402109n.

Fincheira, Paola, Gonzalo Tortella, Nelson Duran, Amedea B. Seabra, and Olga Rubilar. 2020. 'Current Applications of Nanotechnology to Develop Plant Growth Inducer Agents as an Innovation Strategy'. *Critical Reviews in Biotechnology*. Taylor and Francis Ltd. doi:10.1080/07388551.2019.1681931.

Galdames, Alazne, Leire Ruiz-Rubio, Maider Orueta, Miguel Sánchez-Arzalluz, and José Luis Vilas-Vilela. 2020. 'Zero-Valent Iron Nanoparticles for Soil and Groundwater Remediation'. *International Journal of Environmental Research and Public Health* 17 (16). MDPI AG: 1–23. doi:10.3390/ijerph17165817.

Gil-Díaz, M., S. Diez-Pascual, A. González, J. Alonso, E. Rodríguez-Valdés, J. R. Gallego, and M. C. Lobo. 2016. 'A Nanoremediation Strategy for the Recovery of an As-Polluted Soil'. *Chemosphere* 149 (April). Elsevier Ltd: 137–145. doi:10.1016/j.chemosphere.2016.01.106.

Gillispie, Elizabeth C., Stephen E. Taylor, Nikolla P. Qafoku, and Michael F. Hochella Jr. 2019. 'Impact of Iron and Manganese Nano-Metal-Oxides on Contaminant Interaction and Fortification Potential in Agricultural Systems—A Review'. *Environmental Chemistry* 16 (6). CSIRO: 377. doi:10.1071/EN19063.

Gong, Xiaomin, Danlian Huang, Yunguo Liu, Guangming Zeng, Rongzhong Wang, Jia Wan, Chen Zhang, Min Cheng, Xiang Qin, and Wenjing Xue. 2017. 'Stabilized Nanoscale Zerovalent Iron Mediated Cadmium Accumulation and Oxidative Damage of Boehmeria Nivea (L.) Gaudich Cultivated in Cadmium Contaminated Sediments'. *Environmental Science and Technology* 51 (19). American Chemical Society: 11308–11316. doi:10.1021/acs.est.7b03164.

Gong, Xiaomin, Danlian Huang, Yunguo Liu, Dongsheng Zou, Xi Hu, Lu Zhou, Zhibin Wu, Yang Yang, and Zhihua Xiao. 2021. 'Nanoscale Zerovalent Iron, Carbon Nanotubes and Biochar Facilitated the Phytoremediation of Cadmium Contaminated Sediments by Changing Cadmium Fractions, Sediments Properties and Bacterial Community Structure'. *Ecotoxicology and Environmental Safety* 208 (January). Academic Press: 111510. doi:10.1016/j.ecoenv.2020.111510.

Guha, Titir, Sandip Barman, Amitava Mukherjee, and Rita Kundu. 2020a. 'Nano-Scale Zero Valent Iron Modulates Fe/Cd Transporters and Immobilizes Soil Cd for Production of Cd Free Rice'. *Chemosphere* 260 (December). Elsevier Ltd. doi:10.1016/j.chemosphere.2020.127533.

Guha, Titir, Geetha Gopal, Rohan Chatterjee, Amitava Mukherjee, and Rita Kundu. 2020b. 'Differential Growth and Metabolic Responses Induced by Nano-Scale Zero Valent Iron in Germinating Seeds and Seedlings of Oryza Sativa L. Cv. Swarna'. *Ecotoxicology and Environmental Safety* 204 (November). Academic Press: 111104. doi:10.1016/j.ecoenv.2020.111104.

Guha, Titir, K.V.G. Ravikumar, Amitava Mukherjee, Anita Mukherjee, and Rita Kundu. 2018. 'Nanopriming with Zero Valent Iron (NZVI) Enhances Germination and Growth in Aromatic Rice Cultivar (Oryza Sativa Cv. Gobindabhog L.)'. *Plant Physiology and Biochemistry* 127 (June). Elsevier Masson SAS: 403–413. doi:10.1016/j.plaphy.2018.04.014.

Gui, Xin, Yingqing Deng, Yukui Rui, Binbin Gao, Wenhe Luo, Shili Chen, Le Van Nhan, et al. 2015. 'Response Difference of Transgenic and Conventional Rice (Oryza Sativa) to Nanoparticles (ΓFe2O3)'. *Environmental Science and Pollution Research* 22 (22). Springer Verlag: 17716–17723. doi:10.1007/s11356-015-4976-7.

Hasan, Murtaza, Saira Rafique, Ayesha Zafar, Suraj Loomba, Rida Khan, Shahbaz Gul Hassan, Muhammad Waqas Khan, et al. 2020. 'Physiological and Anti-Oxidative Response of Biologically and Chemically Synthesized Iron Oxide: Zea Mays a Case Study'. *Heliyon* 6 (8). Elsevier Ltd: e04595. doi:10.1016/j.heliyon.2020.e04595.

Hidalgo, Keyla T. Soto, Pedro J. Carrión-Huertas, Richard T. Kinch, Luis E. Betancourt, and Carlos R. Cabrera. 2020. 'Phytonanoremediation by Avicennia Germinans (Black Mangrove) and Nano Zero Valent Iron for Heavy Metal Uptake from Cienaga Las Cucharillas Wetland Soils'. *Environmental Nanotechnology, Monitoring and Management* 14 (December). Elsevier B.V.: 100363. doi:10.1016/j.enmm.2020.100363.

Hu, Jing, Huiyuan Guo, Junli Li, Qiuliang Gan, Yunqiang Wang, and Baoshan Xing. 2017. 'Comparative Impacts of Iron Oxide Nanoparticles and Ferric Ions on the Growth of Citrus Maxima'. *Environmental Pollution* 221 (February). Elsevier Ltd: 199–208. doi:10.1016/j.envpol.2016.11.064.

Hu, Xuebin, Xiaobo Liu, Xiangyu Yang, Fucheng Guo, Xiaoxuan Su, and Yi Chen. 2018. 'Acute and Chronic Responses of Macrophyte and Microorganisms in Constructed Wetlands to Cerium Dioxide Nanoparticles: Implications for Wastewater Treatment'. *Chemical Engineering Journal* 348 (September). Elsevier B.V.: 35–45. doi:10.1016/j.cej.2018.04.189.

Huang, Danlian, Xiang Qin, Zhiwei Peng, Yunguo Liu, Xiaomin Gong, Guangming Zeng, Chao Huang, et al. 2018. 'Nanoscale Zero-Valent Iron Assisted Phytoremediation of Pb in Sediment: Impacts on Metal Accumulation and Antioxidative System of Lolium Perenne'. *Ecotoxicology and Environmental Safety* 153 (May). Academic Press: 229–237. doi:10.1016/j.ecoenv.2018.01.060.

Huang, Juan, Chunni Yan, Jialiang Liu, Wenzhu Guan, Rajendra Prasad Singh, Chong Cao, and Jun Xiao. 2019. 'Feasibility Study of Vertical Flow Constructed Wetland for Tertiary Treatment of Nanosilver Wastewater and Temporal-Spatial Distribution of Pollutants and Microbial Community'. *Journal of Environmental Management* 245 (September). Academic Press: 28–36. doi:10.1016/j.jenvman.2019.04.128.

Irum, Samra, Nyla Jabeen, Khawaja Shafique Ahmad, Saima Shafique, Talha Farooq Khan, Hina Gul, Sadaf Anwaar, Nuzhat Imam Shah, Ansar Mehmood, and Syed Zaheer Hussain. 2020. 'Biogenic Iron Oxide Nanoparticles Enhance Callogenesis and Regeneration Pattern of Recalcitrant Cicer Arietinum L.' *PLoS ONE* 15 (12 December). Public Library of Science. doi:10.1371/journal.pone.0242829.

Jiamjitrpanich, Waraporn, Preeda Parkpian, Chongrak Polprasert, and Rachain Kosanlavit. 2012. 'Enhanced Phytoremediation Efficiency of TNT-Contaminated Soil by Nanoscale Zero Valent Iron'. In *2nd International Conference on Environment and Industrial Innovation*, edited by IACSIT Press. Singapore.

Jiamjitrpanich, Waraporn, Preeda Parkpian, Chongrak Polprasert, and Rachain Kosanlavit. 2013. 'Trinitrotoluene and Its Metabolites in Shoots and Roots of Panicum Maximum in Nano-Phytoremediation'. *International Journal of Environmental Science and Development*. EJournal Publishing, 7–10. doi:10.7763/ijesd.2013.v4.293.

Kamran, Muhammad, Hifsa Ali, Muhammad Farhan Saeed, Hafiz Faiq Bakhat, Zeshan Hassan, Muhammad Tahir, Ghulam Abbas, Muhammad Asif Naeem, Muhammad Imtiaz Rashid, and Ghulam Mustafa Shah. 2020. 'Unraveling the Toxic Effects of Iron Oxide Nanoparticles on Nitrogen Cycling through

Manure-Soil-Plant Continuum'. *Ecotoxicology and Environmental Safety* 205 (December). Academic Press: 111099. doi:10.1016/j.ecoenv.2020.111099.

Kaur, Simran, and Arpita Roy. 2020. 'Bioremediation of Heavy Metals from Wastewater Using Nanomaterials'. In *Environment, Development and Sustainability*. Springer Science and Business Media B.V. doi:10.1007/s10668-020-01078-1.

Keller, Arturo A., Suzanne McFerran, Anastasiya Lazareva, and Sangwon Suh. 2013. 'Global Life Cycle Releases of Engineered Nanomaterials'. *Journal of Nanoparticle Research* 15 (6). Springer: 1–17. doi:10.1007/s11051-013-1692-4.

Khan, A. G. 2020b. 'Promises and Potential of in Situ Nano-Phytoremediation Strategy to Mycorrhizo-Remediate Heavy Metal Contaminated Soils Using Non-Food Bioenergy Crops (Vetiver Zizinoides & Cannabis Sativa)'. *International Journal of Phytoremediation*. Taylor and Francis Inc. doi:10.1080/15226514.2020.1774504.

Khan, Fahad Saleem Ahmed, Nabisab Mujawar Mubarak, Mohammad Khalid, Rashmi Walvekar, Ezzat Chan Abdullah, Shaukat A. Mazari, Sabzoi Nizamuddin, and Rama Rao Karri. 2020b. 'Magnetic Nanoadsorbents' Potential Route for Heavy Metals Removal—a Review'. *Environmental Science and Pollution Research* 27 (19). Springer: 24342–24356. doi:10.1007/s11356-020-08711-6.

Khan, Sehresh, Nazneen Akhtar, Shafiq Ur Rehman, Shaukat Shujah, Eui Shik Rha, and Muhammad Jamil. 2020a. 'Biosynthesized Iron Oxide Nanoparticles (Fe3O4 NPs) Mitigate Arsenic Toxicity in Rice Seedlings'. *Toxics* 9 (1). MDPI AG: 2. doi:10.3390/toxics9010002.

Khan, Shams Tabrez. 2020a. 'Interaction of Engineered Nanomaterials with Soil Microbiome and Plants: Their Impact on Plant and Soil Health'. In *Sustainable Agriculture Reviews*, 41:181–199. Springer, Cham. doi:10.1007/978-3-030-33996-8_10.

Klaine, Stephen J., Pedro J.J. Alvarez, Graeme E. Batley, Teresa F. Fernandes, Richard D. Handy, Delina Y. Lyon, Shaily Mahendra, Michael J. McLaughlin, and Jamie R. Lead. 2008. 'Nanomaterials in the Environment: Behavior, Fate, Bioavailability, and Effects'. In *Environmental Toxicology and Chemistry*. John Wiley & Sons, Ltd. doi:10.1897/08-090.1.

Kokina, Inese, Ilona Mickeviča, Inese Jahundoviča, Andrejs Ogurcovs, Marina Krasovska, Marija Jermaļonoka, Irena Mihailova, Edmunds Tamanis, and Vjačeslavs Gerbreders. 2017. 'Plant Explants Grown on Medium Supplemented with Fe3O4 Nanoparticles Have a Significant Increase in Embryogenesis'. *Journal of Nanomaterials* 2017. Hindawi Limited. doi:10.1155/2017/4587147.

Kokina, Inese, Ilona Plaksenkova, Marija Jermaļonoka, and Anastasija Petrova. 2020. 'Impact of Iron Oxide Nanoparticles on Yellow Medick (*Medicago Falcata* L.) Plants'. *Journal of Plant Interactions* 15 (1). Taylor and Francis Ltd.: 1–7. doi:10.1080/17429145.2019.1708489.

Konate, Alexandre, Xiao He, Zhiyong Zhang, Yuhui Ma, Peng Zhang, Gibson Alugongo, and Yukui Rui. 2017. 'Magnetic (Fe3O4) Nanoparticles Reduce Heavy Metals Uptake and Mitigate Their Toxicity in Wheat Seedling'. *Sustainability* 9 (5). MDPI AG: 790. doi:10.3390/su9050790.

Konate, Alexandre, Yaoyao Wang, Xiao He, Muhammd Adeel, Peng Zhang, Yuhui Ma, Yayun Ding, et al. 2018. 'Comparative Effects of Nano and Bulk-Fe3O4 on the Growth of Cucumber (Cucumis Sativus)'. *Ecotoxicology and Environmental Safety* 165 (December). Academic Press: 547–554. doi:10.1016/j.ecoenv.2018.09.053.

Kornarzyński, Krzysztof, Agnieszka Sujak, Grzegorz Czernel, and Dariusz Wiącek. 2020. 'Effect of Fe3O4 Nanoparticles on Germination of Seeds and Concentration of Elements in Helianthus Annuus L. under Constant Magnetic Field'. *Scientific Reports* 10 (1). Nature Research: 1–10. doi:10.1038/s41598–020–64849-w.

Kumar, A., K. Gupta, S. Dixit, K. Mishra, and S. Srivastava. 2019. 'A Review on Positive and Negative Impacts of Nanotechnology in Agriculture'. *International Journal of Environmental Science and Technology*. Center for Environmental and Energy Research and Studies. doi:10.1007/s13762-018-2119-7.

Kumari, Sonu, and Suphiya Khan. 2018. 'Effect of Fe3O4 NPs Application on Fluoride (F) Accumulation Efficiency of Prosopis Juliflora'. *Ecotoxicology and Environmental Safety* 166 (December). Academic Press: 419–426. doi:10.1016/j.ecoenv.2018.09.103.

Lacalle, Rafael G., Carlos Garbisu, and José M. Becerril. 2020. 'Effects of the Application of an Organic Amendment and Nanoscale Zero-Valent Iron Particles on Soil Cr(VI) Remediation'. *Environmental Science and Pollution Research* 27 (25). Springer: 31726–31736. doi:10.1007/s11356-020-09449-x.

Latif, Abdul, Di Sheng, Kai Sun, Youbin Si, Muhammad Azeem, Aown Abbas, and Muhammad Bilal. 2020. 'Remediation of Heavy Metals Polluted Environment Using Fe-Based Nanoparticles: Mechanisms, Influencing Factors, and Environmental Implications'. *Environmental Pollution*. Elsevier Ltd. doi:10.1016/j.envpol.2020.114728.

Lebrun, Manhattan, Florie Miard, Romain Nandillon, Gabriella S. Scippa, Sylvain Bourgerie, and Domenico Morabito. 2019. 'Biochar Effect Associated with Compost and Iron to Promote Pb and As Soil Stabilization and Salix Viminalis L. Growth'. *Chemosphere* 222 (May). Elsevier Ltd: 810–822. doi:10.1016/j.chemosphere.2019.01.188.

Lee, Chang Woo, Shaily Mahendra, Katherine Zodrow, Dong Li, Yu-Chang Tsai, Janet Braam, and Pedro J.J. Alvarez. 2010. 'Erratum: Developmental Phytotoxicity of Metal Oxide Nanoparticles to Arabidopsis Thaliana'. *Environmental Toxicology and Chemistry* 29 (6). John Wiley & Sons, Ltd: 1399–1399. doi:10.1002/etc.234.

Li, Junli, Jing Hu, Chuanxin Ma, Yunqiang Wang, Chan Wu, Jin Huang, and Baoshan Xing. 2016. 'Uptake, Translocation and Physiological Effects of Magnetic Iron Oxide (γ-Fe2O3) Nanoparticles in Corn (Zea Mays L.)'. *Chemosphere* 159 (September). Elsevier Ltd: 326–334. doi:10.1016/j.chemosphere.2016.05.083.

Li, Junli, Jing Hu, Lian Xiao, Yunqiang Wang, and Xilong Wang. 2018. 'Interaction Mechanisms between α-Fe2O3, γ-Fe2O3 and Fe3O4 Nanoparticles and Citrus Maxima Seedlings'. *Science of the Total Environment* 625 (June). Elsevier B.V.: 677–685. doi:10.1016/j.scitotenv.2017.12.276.

Li, Junli, Fengting Wan, Wenjing Guo, Jiali Huang, Zhaoyi Dai, Licong Yi, and Yunqiang Wang. 2020a. 'Influence of α- and γ-Fe2O3 Nanoparticles on Watermelon (Citrullus Lanatus) Physiology and Fruit Quality'. *Water, Air, and Soil Pollution* 231 (4). Springer: 1–12. doi:10.1007/s11270-020-04511-3.

Li, Mingshu, Peng Zhang, Muhammad Adeel, Zhiling Guo, Andrew J. Chetwynd, Chuanxin Ma, Tonghao Bai, Yi Hao, and Yukui Rui. 2020b. 'Physiological Impacts of Zero Valent Iron, Fe3O4 and Fe2O3 Nanoparticles in Rice Plants and Their Potential as Fe Fertilizers'. *Environmental Pollution* 269 (December). Elsevier BV: 116134. doi:10.1016/j.envpol.2020.116134.

Liu, Yang, Bo Pan, Hao Li, Di Lang, Qing Zhao, Di Zhang, Min Wu, Christian E. W. Steinberg, and Baoshan Xing. 2020a. 'Can the Properties of Engineered Nanoparticles Be Indicative of Their Functions and Effects in Plants?' *Ecotoxicology and Environmental Safety* 205 (December). Academic Press: 111128. doi:10.1016/j.ecoenv.2020.111128.

Liu, Yangzhi, Ting Wu, Jason C. White, and Daohui Lin. 2020b. 'A New Strategy Using Nanoscale Zero-Valent Iron to Simultaneously Promote Remediation and Safe Crop Production in Contaminated Soil'. *Nature Nanotechnology* (November). Nature Research: 1–9. doi:10.1038/s41565-020-00803-1.

López-Luna, J., Camacho-Martínez, M. M., Solís-Domínguez, F. A., González-Chávez, M. C., Carrillo-González, R., Martinez-Vargas, S., Mijangos-Ricardez, O. F., and Cuevas-Díaz, M. C. 2018. 'Toxicity Assessment of Cobalt Ferrite Nanoparticles on Wheat Plants'. *Journal of Toxicology and Environmental Health—Part A: Current Issues* 81 (14). Taylor and Francis Inc.: 604–619. doi:10.1080/15287394.2018.1469060.

Lu, Kun, Danlei Shen, Xiaokai Liu, Shipeng Dong, Xueping Jing, Wei Wu, Yang Tong, Shixiang Gao, and Liang Mao. 2020. 'Uptake of Iron Oxide Nanoparticles Inhibits the Photosynthesis of the Wheat after Foliar Exposure'. *Chemosphere* 259 (November). Elsevier Ltd: 127445. doi:10.1016/j.chemosphere.2020.127445.

Ma, Xingmao, Arun Gurung, and Yang Deng. 2013. 'Phytotoxicity and Uptake of Nanoscale Zero-Valent Iron (NZVI) by Two Plant Species'. *Science of the Total Environment* 443 (January). Elsevier: 844–849. doi:10.1016/j.scitotenv.2012.11.073.

Martínez-Fernández, Domingo, Didac Barroso, and Michael Komárek. 2016. 'Root Water Transport of Helianthus Annuus L. under Iron Oxide Nanoparticle Exposure'. *Environmental Science and Pollution Research* 23 (2). Springer Verlag: 1732–1741. doi:10.1007/s11356-015-5423-5.

Martínez-Fernández, Domingo, and Michael Komárek. 2016. 'Comparative Effects of Nanoscale Zero-Valent Iron (NZVI) and Fe2O3 Nanoparticles on Root Hydraulic Conductivity of Solanum Lycopersicum L.' *Environmental and Experimental Botany* 131 (November). Elsevier B.V.: 128–136. doi:10.1016/j.envexpbot.2016.07.010.

Marusenko, Yevgeniy, Jessie Shipp, George A. Hamilton, Jennifer L.L. Morgan, Michael Keebaugh, Hansina Hill, Arnab Dutta, et al. 2013. 'Bioavailability of Nanoparticulate Hematite to Arabidopsis Thaliana'. *Environmental Pollution* 174 (March). Elsevier: 150–156. doi:10.1016/j.envpol.2012.11.020.

Mohammadi, Hamid, Ali Reza Amani-Ghadim, Amir Abbas Matin, and Mansour Ghorbanpour. 2020. 'Fe0 Nanoparticles Improve Physiological and Antioxidative Attributes of Sunflower (Helianthus Annuus) Plants Grown in Soil Spiked with Hexavalent Chromium'. *3 Biotech* 10 (1). Springer: 1–11. doi:10.1007/s13205-019-2002-3.

Mokarram-Kashtiban, Sahar, Seyed Mohsen Hosseini, Masoud Tabari Kouchaksaraei, and Habibollah Younesi. 2019. 'The Impact of Nanoparticles Zero-Valent Iron (NZVI) and Rhizosphere Microorganisms on the Phytoremediation Ability of White Willow and Its Response'. *Environmental Science and Pollution Research* 26 (11). Springer Verlag: 10776–10789. doi:10.1007/s11356-019-04411-y.

Nouri, Mohamed. 2021. 'Potentials and Challenges of Date Pits as Alternative Environmental Clean-up Ingredients'. *Biomass Conversion and Biorefinery*, January. Springer Science and Business Media LLC, 1–28. doi:10.1007/s13399-020-01215-w.

Nouri, Mohamed, Fernando Gonçalves, Jausé Paulo Sousa, Jörg Jörg Römbke, Mohamed Ksibi, Ruth Pereira, and Abdelmajid Haddioui. 2013. 'Metal and Phosphorus Uptake by Spontaneous Vegetation in an Abandoned Iron Mine from a Semiarid Area in Center Morocco: Implications for Phytoextraction'. *Environmental Research, Engineering and Management* 64 (2): 59–71. doi:10.5755/j01.erem.64.2.3866.

Nouri, M., and A. Haddioui. 2016a. 'Human and Animal Health Risk Assessment of Metal Contamination in Soil and Plants from Ait Ammar Abandoned Iron Mine, Morocco'. *Environmental Monitoring and Assessment* 188 (1). doi:10.1007/s10661-015-5012-6.

Nouri, M., and Abdelmajid Haddioui. 2016b. 'Ases Ment of Metals Con Tamina Tion and Ecological Risk in Ait Ammar Abandoned Iron Mine Soil, Moroco'. *Ekologia Bratislava* 35 (1). doi:10.1515/eko-2016-0003.

Nouri, Mohamed, Taoufik El Rasafi, and Abdelmajid Haddioui. 2017. 'Levels of Metals in Soils of Ait Ammar Iron Mine, Morocco: Human Health Risks'. *Acta Chemica Iasi* 25 (2). Walter de Gruyter GmbH: 127–144. doi:10.1515/achi-2017-0012.

Nwadinigwe, Alfreda Ogochukwu, and Emmanuel Chibuzor Ugwu. 2019. 'Overview of Nano-Phytoremediation Applications'. In *Phytoremediation: Management of Environmental Contaminants*, 6:377–382. Springer International Publishing. doi:10.1007/978-3-319-99651-6_15.

OECD (Organisation for Economic Co-operation andDevelopment). 2010. 'List of Manufactured Nanomaterials and List of Endpoints for Phase One of the Sponsorship Programme for the Testing of Manufactured Nanomaterials: Revision; Series on the Safety of Manufactured Nanomaterials 27'. www.oecd.org/officialdocuments/publicdisplaydocumentpdf/?cote=env/jm/mono(2010)46&doclanguage=en.

Peng, Hao, and Jing Guo. 2020. 'Removal of Chromium from Wastewater by Membrane Filtration, Chemical Precipitation, Ion Exchange, Adsorption Electrocoagulation, Electrochemical Reduction, Electrodialysis, Electrodeionization, Photocatalysis and Nanotechnology: A Review'. *Environmental Chemistry Letters*. Springer Science and Business Media Deutschland GmbH. doi:10.1007/s10311-020-01058-x.

Pereira, Aryelle Canedo, Bruno Bastos Gonçalves, Rafaella da Silva Brito, Lucélia Gonçalves Vieira, Emília Celma de Oliveira Lima, and Thiago Lopes Rocha. 2020. 'Comparative Developmental Toxicity of Iron Oxide Nanoparticles and Ferric Chloride to Zebrafish (Danio Rerio) after Static and Semi-Static Exposure'. *Chemosphere* 254 (September). Elsevier Ltd: 126792. doi:10.1016/j.chemosphere.2020.126792.

Piccinno, Fabiano, Fadri Gottschalk, Stefan Seeger, and Bernd Nowack. 2012. 'Industrial Production Quantities and Uses of Ten Engineered Nanomaterials in Europe and the World'. *Journal of Nanoparticle Research*. Springer. doi:10.1007/s11051-012-1109-9.

Pillai, Harikumar P. S., and Jesitha Kottekottil. 2016. 'Nano-Phytotechnological Remediation of Endosulfan Using Zero Valent Iron Nanoparticles'. *Journal of Environmental Protection* 7 (5). Scientific Research Publishing, Inc,: 734–744. doi:10.4236/jep.2016.75066.

Pizarro, Carmen, Mauricio Escudey, Eliana Caroca, Carolina Pavez, and Gustavo E. Zúñiga. 2021. 'Evaluation of Zeolite, Nanomagnetite, and Nanomagnetite-Zeolite Composite Materials as Arsenic (V) Adsorbents in Hydroponic Tomato Cultures'. *Science of the Total Environment* 751 (January). Elsevier B.V. doi:10.1016/j.scitotenv.2020.141623.

Plaksenkova, Ilona, Marija Jermaļonoka, Linda Bankovska, Inese Gavarāne, Vjačeslavs Gerbreders, Eriks Sledevskis, Jānis Sniķeris, and Inese Kokina. 2019. 'Effects of Fe3O4 Nanoparticle Stress on the Growth and Development of Rocket Eruca Sativa'. *Journal of Nanomaterials* 2019. Hindawi Limited. doi:10.1155/2019/2678247.

Rahmatizadeh, Razieh, Seyyed Mohammad Javad Arvin, Rashid Jamei, Hossein Mozaffari, and Farkondeh Reza Nejhad. 2019. 'Response of Tomato Plants to Interaction Effects of Magnetic (Fe_3O_4) Nanoparticles and Cadmium Stress'. *Journal of Plant Interactions* 14 (1). Taylor and Francis Ltd.: 474–481. doi:10.1080/17429145.2019.1626922.

Rai, Prabhat Kumar, Ki Hyun Kim, Sang Soo Lee, and Jin Hong Lee. 2020. 'Molecular Mechanisms in Phytoremediation of Environmental Contaminants and Prospects of Engineered Transgenic Plants/Microbes'. *Science of the Total Environment* 705 (February). Elsevier B.V.: 135858. doi:10.1016/j.scitotenv.2019.135858.

Romeh, Ahmed Ali. 2020. 'Synergistic Effect of Ficus-Zero Valent Iron Supported on Adsorbents and Plantago Major for Chlorpyrifos Phytoremediation from Water'. *International Journal of Phytoremediation*. Taylor and Francis Inc. doi:10.1080/15226514.2020.1803201.

Romeh, Ahmed Ali, and Refaat Ahmed Ibrahim Saber. 2020. 'Green Nano-Phytoremediation and Solubility Improving Agents for the Remediation of Chlorfenapyr Contaminated Soil and Water'. *Journal of Environmental Management* 260 (April). Academic Press: 110104. doi:10.1016/j.jenvman.2020.110104.

Rui, Mengmeng, Chuanxin Ma, Yi Hao, Jing Guo, Yukui Rui, Xinlian Tang, Qi Zhao, et al. 2016. 'Iron Oxide Nanoparticles as a Potential Iron Fertilizer for Peanut (Arachis Hypogaea)'. *Frontiers in Plant Science* 7 (June2016). Frontiers Media S.A.: 815. doi:10.3389/fpls.2016.00815.

Saleh, Tawfik A. 2020. 'Trends in the Sample Preparation and Analysis of Nanomaterials as Environmental Contaminants'. *Trends in Environmental Analytical Chemistry*. Elsevier B.V. doi:10.1016/j.teac.2020.e00101.

Shahi, Mamta Patra, Priyanka Kumari, Deepika Mahobiya, and Sushil Kumar Shahi. 2021. 'Nano-Bioremediation of Environmental Contaminants: Applications, Challenges, and Future Prospects'. In *Bioremediation for Environmental Sustainability*, 83–98. Elsevier. doi:10.1016/b978-0-12-820318-7.00004-6.

Sigua, Gilbert C., Jeff M. Novak, Don W. Watts, Jim A. Ippolito, Thomas F. Ducey, Mark G. Johnson, and Kurt A. Spokas. 2019. 'Phytostabilization of Zn and Cd in Mine Soil Using Corn in Combination with Biochars and Manure-Based Compost'. *Environments* 6 (6). MDPI AG: 69. doi:10.3390/environments6060069.

Singh, Anita, Sudesh Chaudhary, and Brijnandan S. Dehiya. 2021. 'Fast Removal of Heavy Metals from Water and Soil Samples Using Magnetic Fe3O4 Nanoparticles'. *Environmental Science and Pollution Research* 28 (4). Springer Science and Business Media Deutschland GmbH: 3942–3952. doi:10.1007/s11356-020-10737-9.

Singh, Harpreet, Neha Bhardwaj, Shailendra Kumar Arya, and Madhu Khatri. 2020. 'Environmental Impacts of Oil Spills and Their Remediation by Magnetic Nanomaterials'. *Environmental Nanotechnology, Monitoring and Management*. Elsevier B.V. doi:10.1016/j.enmm.2020.100305.

Slijepčević, Nataša, Dragana Tomašević Pilipović, Đurđa Kerkez, Dejan Krčmar, Milena Bečelić-Tomin, Jelena Beljin, and Božo Dalmacija. 2021. 'A Cost Effective Method for Immobilization of Cu and Ni Polluted River Sediment with NZVI Synthesized from Leaf Extract'. *Chemosphere* 263 (January). Elsevier Ltd: 127816. doi:10.1016/j.chemosphere.2020.127816.

Song, Biao, Piao Xu, Ming Chen, Wangwang Tang, Guangming Zeng, Jilai Gong, Peng Zhang, and Shujing Ye. 2019. 'Using Nanomaterials to Facilitate the Phytoremediation of Contaminated Soil'. *Critical Reviews in Environmental Science and Technology* 49 (9). Taylor and Francis Inc.: 791–824. doi:10.1080/10643389.2018.1558891.

Souri, Zahra, Naser Karimi, Leila Norouzi, and Xingmao Ma. 2020. 'Elucidating the Physiological Mechanisms Underlying Enhanced Arsenic Hyperaccumulation by Glutathione Modified Superparamagnetic Iron Oxide Nanoparticles in Isatis Cappadocica'. *Ecotoxicology and Environmental Safety* 206 (December). Academic Press: 111336. doi:10.1016/j.ecoenv.2020.111336.

Souza, Lilian Rodrigues Rosa, Luiza Carolina Pomarolli, and Márcia Andreia Mesquita Silva da Veiga. 2020. 'From Classic Methodologies to Application of Nanomaterials for Soil Remediation: An Integrated View of Methods for Decontamination of Toxic Metal(Oid)S'. *Environmental Science and Pollution Research*. Springer. doi:10.1007/s11356-020-08032-8.

Sun, Yuhuan, Wenjie Wang, Fangyuan Zheng, Shuwu Zhang, Fayuan Wang, and Shaowen Liu. 2020. 'Phytotoxicity of Iron-Based Materials in Mung Bean: Seed Germination Tests'. *Chemosphere* 251 (July). Elsevier Ltd: 126432. doi:10.1016/j.chemosphere.2020.126432.

Teodoro, Manuel, Rafael Clemente, Ermengol Ferrer-Bustins, Domingo Martínez-Fernández, Maria Pilar Bernal, Martina Vítková, Petr Vítek, and Michael Komárek. 2020. 'Nanoscale Zero-Valent Iron Has Minimum Toxicological Risk on the Germination and Early Growth of Two Grass Species with Potential for Phytostabilization'. *Nanomaterials* 10 (8). MDPI AG: 1537. doi:10.3390/nano10081537.

Tombuloglu, Huseyin, Yassine Slimani, Thamer Marhoon AlShammari, Muhammed Bargouti, Mehmet Ozdemir, Guzin Tombuloglu, Sultan Akhtar, et al. 2020a. 'Uptake, Translocation, and Physiological Effects of Hematite (α-Fe2O3) Nanoparticles in Barley (Hordeum Vulgare L.)'. *Environmental Pollution* 266 (November). Elsevier Ltd: 115391. doi:10.1016/j.envpol.2020.115391.

Tombuloglu, Huseyin, Yassine Slimani, Thamer Marhoon AlShammari, Guzin Tombuloglu, Munirah A. Almessiere, Huseyin Sozeri, Abdulhadi Baykal, and Ismail Ercan. 2021. 'Delivery, Fate and Physiological Effect of Engineered Cobalt Ferrite Nanoparticles in Barley (Hordeum Vulgare L.)'. *Chemosphere* 265 (February). Elsevier Ltd. doi:10.1016/j.chemosphere.2020.129138.

Tombuloglu, Huseyin, Yassine Slimani, Guzin Tombuloglu, Thamer Alshammari, Munirah Almessiere, Ayşe Demir Korkmaz, Abdulhadi Baykal, and Anna Cristina S. Samia. 2020b. 'Engineered Magnetic Nanoparticles Enhance Chlorophyll Content and Growth of Barley through the Induction of Photosystem Genes'. *Environmental Science and Pollution Research* 27 (27). Springer: 34311–34321. doi:10.1007/s11356-020-09693-1.

Trujillo-Reyes, J., S. Majumdar, C. E. Botez, J. R. Peralta-Videa, and J. L. Gardea-Torresdey. 2014. 'Exposure Studies of Core-Shell Fe/Fe3O4 and Cu/CuO NPs to Lettuce (Lactuca Sativa) Plants: Are They a Potential Physiological and Nutritional Hazard?' *Journal of Hazardous Materials* 267 (February). Elsevier: 255–263. doi:10.1016/j.jhazmat.2013.11.067.

Ukaogo, Prince O., Ugochukwu Ewuzie, and Chibuzo V. Onwuka. 2020. 'Environmental Pollution: Causes, Effects, and the Remedies'. In *Microorganisms for Sustainable Environment and Health*, 419–429. Elsevier. doi:10.1016/b978-0-12-819001-2.00021-8.

Vítková, Martina, Markus Puschenreiter, and Michael Komárek. 2018. 'Effect of Nano Zero-Valent Iron Application on As, Cd, Pb, and Zn Availability in the Rhizosphere of Metal(Loid) Contaminated Soils'. *Chemosphere* 200 (June). Elsevier Ltd: 217–226. doi:10.1016/j.chemosphere.2018.02.118.

Wang, Huanhua, Xiaoming Kou, Zhiguo Pei, John Q. Xiao, Xiaoquan Shan, and Baoshan Xing. 2011. 'Physiological Effects of Magnetite (Fe3O4) Nanoparticles on Perennial Ryegrass (Lolium Perenne L.) and Pumpkin (Cucurbita Mixta) Plants'. *Nanotoxicology* 5 (1). Taylor & Francis: 30–42. doi:10.3109/17435390.2010.489206.

Xue, Wenjing, Danlian Huang, Guangming Zeng, Jia Wan, Min Cheng, Chen Zhang, Chanjuan Hu, and Jing Li. 2018. 'Performance and Toxicity Assessment of Nanoscale Zero Valent Iron Particles in the Remediation of Contaminated Soil: A Review'. *Chemosphere*. Elsevier Ltd. doi:10.1016/j.chemosphere.2018.07.118.

Yang, Xueling, Darioush Alidoust, and Chunyan Wang. 2020. 'Effects of Iron Oxide Nanoparticles on the Mineral Composition and Growth of Soybean (Glycine Max L.) Plants'. *Acta Physiologiae Plantarum* 42 (8). Springer: 1–11. doi:10.1007/s11738-020-03104-1.

Yogalakshmi, Kadapakkam Nandabalan, Anamika Das, Gini Rani, Vijay Jaswal, and Jatinder Singh Randhawa. 2020. 'Nano-Bioremediation: A New Age Technology for the Treatment of Dyes in Textile Effluents'. In *Bioremediation of Industrial Waste for Environmental Safety*, 313–347. Springer Singapore. doi:10.1007/978-981-13-1891-7_15.

Yoon, Hakwon, Yu Gyeong Kang, Yoon Seok Chang, and Jae Hwan Kim. 2019. 'Effects of Zerovalent Iron Nanoparticles on Photosynthesis and Biochemical Adaptation of Soil-Grown Arabidopsis Thaliana'. *Nanomaterials* 9 (11). MDPI AG. doi:10.3390/nano9111543.

Youssef, Ola A., Amel A. Tammam, Ranya F. El-Bakatoushi, Asmaa M. Alframawy, Mahmoud M. Emara, and Laila M. El-Sadek. 2020. 'Hematite Nanoparticles Influence Ultrastructure, Antioxidant Defenses, Gene Expression, and Alleviate Cadmium Toxicity in *Zea Mays*'. *Journal of Plant Interactions* 15 (1). Taylor and Francis Ltd.: 54–74. doi:10.1080/17429145.2020.1745307.

Zand, Ali Daryabeigi, and Alireza Mikaeili Tabrizi. 2020. 'Effect of Zero-Valent Iron Nanoparticles on the Phytoextraction Ability of Kochia Scoparia and Its Response in Pb Contaminated Soil'. *Environmental Engineering Research* 26 (4). Korean Society of Environmental Engineering: 200227. doi:10.4491/eer.2020.227.

Zand, Ali Daryabeigi, Alireza Mikaeili Tabrizi, and Azar Vaezi Heir. 2020a. 'The Influence of Association of Plant Growth-Promoting Rhizobacteria and Zero-Valent Iron Nanoparticles on Removal of Antimony from Soil by Trifolium Repens'. *Environmental Science and Pollution Research* 27 (34): 42815–42829. doi:10.1007/s11356-020-10252-x.

Zand, Ali Daryabeigi, Alireza Mikaeili Tabrizi, and Azar Vaezi Heir. 2020b. 'Incorporation of Biochar and Nanomaterials to Assist Remediation of Heavy Metals in Soil Using Plant Species'. *Environmental Technology and Innovation* 20 (November). Elsevier B.V.: 101134. doi:10.1016/j.eti.2020.101134.

Zhang, Ruyang, Xiaohan Bai, Jihai Shao, Anwei Chen, Haiyong Wu, and Si Luo. 2020. 'Effects of Zero-Valent Iron Nanoparticles and Quinclorac Coexposure on the Growth and Antioxidant System of Rice (Oryza Sativa L.)'. *Ecotoxicology and Environmental Safety* 203 (October). Academic Press: 111054. doi:10.1016/j.ecoenv.2020.111054.

Zhu, Hao, Jie Han, John Q. Xiao, and Yan Jin. 2008. 'Uptake, Translocation, and Accumulation of Manufactured Iron Oxide Nanoparticles by Pumpkin Plants'. *Journal of Environmental Monitoring* 10 (6): 713–717. doi:10.1039/b805998e.

Zhu, Yi, Fang Xu, Qin Liu, Ming Chen, Xianli Liu, Yanyan Wang, Yan Sun, and Lili Zhang. 2019. 'Nanomaterials and Plants: Positive Effects, Toxicity and the Remediation of Metal and Metalloid Pollution in Soil'. *Science of the Total Environment*. Elsevier B.V. doi:10.1016/j.scitotenv.2019.01.234.

14 Sustainability Aspects of Nanoremediation and Nanophytoremediation

Irshad Ahmad Bhat, Khushboo Guleria, Mudasir Fayaz, Jasfeeda Qadir, Roof-Ul-Qadir, and Prof. Zahoor A. Kaloo

14.1 INTRODUCTION

The environment needs continuous observation of the impurities that are added to it through different anthropogenic activities; this will, in turn, lead us to understand and help us to manage the threats posed to human health and the environment. Increases in the global temperature and environmental pollution level are some of the notable global defects of current times (Rogers 2006; Soler and Sánchez 2014; Pino et al. 2017). In addition, the scarcity of fresh water, especially in underdeveloped countries, is creating a life-threatening situation that needs to be addressed as quickly as possible (Qu et al. 2013; Thines et al. 2017). Nanotechnology, as one of the interdisciplinary areas of science, has generated tremendous potential and has received sufficient attention from researchers. Presently, nanotechnology is a swiftly evolving area of investment from which most countries are interested in obtaining commercial benefits. Nanotechnology can be summarized as technological innovation and development at the atomic and molecular levels that uses a measuring scale of one to one hundred nm in any dimension. Since the generally employed treatment methods have a number of limitations in decreasing the amount of pollutants exclusively in water and soil, this has ultimately resulted in their reduced use on a comprehensive scale and in long-term applications (Ifang et al. 2014). Employment of nanomaterial and microbes for the abstraction of environmental pollutants has received great impetus and consideration from the scientific community as it has one important feature: sustainability (Werkneh and Rene 2019). Moreover, the use of nanoparticles (NPs) has been thoroughly explored for their incredible potential in the remediation of water, air and soil contaminants (Boente et al. 2018; Guerra et al. 2018a, 2018b; Mueller and Nowack 2010; Tratnyek and Johnson 2006; Wang et al. 2018; Yunus et al. 2012). Nanoscience actually provides prospective solutions by allowing the investigation of materials at the nanoscale (Das et al. 2015).

After nanomaterial has been employed in the remediation process, it is very important to ensure that the utilized material does not act as a contaminant itself. Therefore, there is an urgent need for biodegradable material in this field of application. Using biodegradable nanomaterial will ensure that there is no production of waste after the remediation process, which will, in turn, increase consumer confidence and the recognition of a new greener/safer alternative for different types of environmental pollution remediation. Off-targeting results in low efficiency, and overcoming the target-specific capture of pollutants is a good option and especially appealing. Thus, considerable work has been published in this regard that considers not only the use of nanotechnology but also merging it with physical and chemical modifications on the surface of the materials to obtain engineered NPs that overcome the challenges involved in environmental remediation (Kamat and Meisel 2003). Although NPs have generated tremendous potential for environmental remediation, unfortunately, some NPs are inherently instable under different environmental conditions; therefore, the preparation method involved needs special formulation at the nanoscale. In addition, actions are

DOI: 10.1201/9781003186298-14

needed to prevent aggregation and increase stability. There are chances of toxicity because of the use of metallic NPs during environmental remediation, along with the creation of by-products and recovery costs form the site, and these factors may limit the use of NPs. This is why an intense understanding is needed of the material required for NP fabrication and performance optimization to have excellent NPs that can address environmental issues. However, there are many options to remediate pollutants such as absorption, adsorption, chemical reactions, photocatalysis and filtration, as summarized in Figure 14.1.

14.2 NANOREMEDIATION: A NEW AND EVOLVING TECHNOLOGY FOR THE ELIMINATION OF TOXIC WASTE PRODUCTS FROM THE ENVIRONMENT

The word remediate means problem solving, and nanoremediation elucidates the use of different NPs/nanomaterials and plants in a technique called nanophytoremediation that controls environmental contaminants (Khan 2020). Currently, nanoremediation is a favorable approach to contain pollution. Nanoremediation takes into consideration the applicability of NPs in the conversion and detoxification of environmental contaminants. The properties of nanomaterials that make them compatible for contaminant removal and that make pollutants less toxic are chemical reduction and catalysis. For nanoremediation in-situ, no groundwater is pumped for aboveground treatment, and no soil is transported to other locations for treatment and disposal (Otto et al. 2008). For the elimination of different contaminants from air, water and soil, various conventional methods are employed including absorption, adsorption, photocatalysis, chemical reactions and filtration [molecule]. However, nanoremediation has the prospective ability not only to decrease the overall resulting cost, time and lifting and disposal of contaminated soil offsite but also to bring the concentration

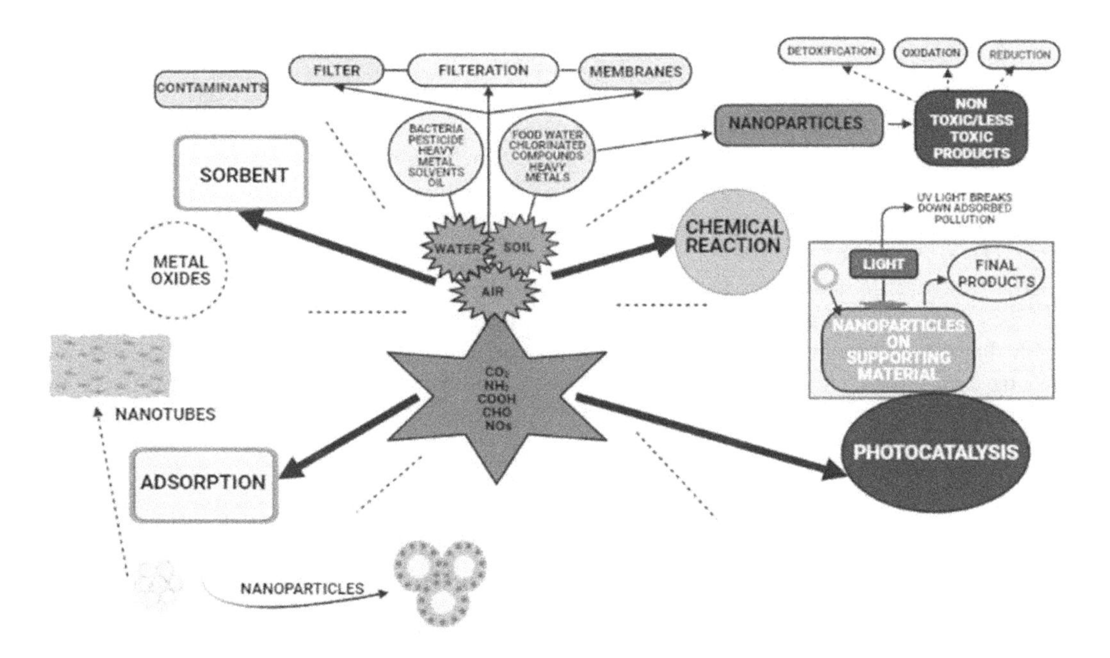

FIGURE 14.1 Application of environmental nanotechnology involving adsorption, chemical reaction, photocatalysis and filtration, primarily for the removal of pollutants form air, water and soil.

Source: *Reproduced with permission from Guerra, F.D., Attia, M., Whitehead, D., Alexis, F., 2018a. Nanotechnology for environmental remediation: materials and applications. Molecules 23 (7), 1760.*

of contaminants down to an almost level zero (Karn et al. 2009). The novel properties and enhanced efficacy of nanomaterials make them compatible for remediation processes because of their high surface area-to-volume ratio that ultimately provides them with higher reactivity. Nanomaterials are actually anticipated for in-situ applications as they have novel properties, namely, a very minute size and groundbreaking surface coatings that enable them to infuse very small spaces and remain suspended in groundwater. In addition, nano-sized particles travel farther than macro ones and attain a wider distribution (Tratnyek and Johnson 2006). It is very well-known that every new technology possesses both benefits and limitations. Similarly, nanoremediation has its own probable good and bad characteristics in effectively eradicating contaminated sites. In-situ applications of nanoremediation compared to other remediation processes provide us with very important benefits such as being cost-effective and faster than other methods to clean up polluted sites (Yan et al. 2006).

However, at the same time, there are growing concerns about the use of nanomaterials to clean up contaminated environments. At the field scale, nanoremediation is almost untested. Questions have been raised regarding not only the real-world usefulness of nanoremediation for large-scale field sites contaminated with different sorts of pollutants but also the way that these NPs behave in dissimilar environments, which differ in the soil porosity, temperature, hydrogeology and pH. One benefit of NPs is that they generally move farther than macroparticles. It has been generally observed in practice, however, that NPs currently employed in the remediation process do not actually move far from the injection site (Karn et al. 2009). Similarly, other questions involve the confrontational effects on the environment in general, probability of persistence/bioaccumulation and changes in microbial populations (Mueller and Nowack 2010; Grieger et al. 2010). Various kinds of NPs and nanomaterials can be used for the techniques summarized in Figure 14.2.

14.3 TYPES OF NANOPARTICLES

14.3.1 Nanomaterials of an Inorganic Nature, Based on Metals and Their Oxides

Various NPs are employed for remediating a number of pollutants typically from water, which includes heavy metals and chlorinated pollution-causing agents. Among these nanomaterials, metal and metal oxides have shown promising results as they are highly proficient adsorbents and possess swift kinetics (Santhosh et al. 2016). Nanomaterials are being increasingly used for environmental remediation because they have potential application towards both in-situ and ex-situ aqueous systems (Das et al. 2015).

The antibacterial, antiviral and antifungal capabilities of silver (Ag) NPs are well established, and they are thus useful as water decontaminators (Chou et al. 2005; Bosetti et al. 2002). Ag NPs of fewer than 10 nm have been found to be highly toxic to *E. coli* and *Pseudomonas aeruginosa*. Ag NPs have also proved to be effective against viruses as they prevent their binding to host cells by favorably binding to virus glycoproteins. The size and shape of Ag NPs have a great influence on

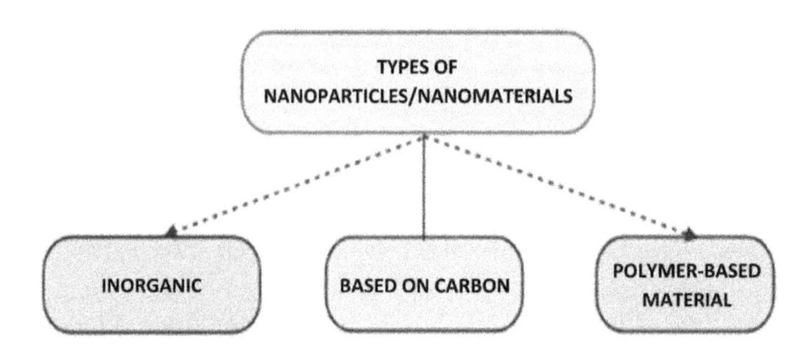

FIGURE 14.2 Types of nanoparticles.

the effect of antibacterial activity; particles of 11–23 nm have lower activity (Gogoi et al. 2006). In addition, triangular Ag nanomaterials have better antibacterial activity than rod- and sphere-shaped Ag NPs (Pal et al. 2007). Ag NPs, together with other metal oxides, have increased the efficiency of the resulting coupled NPs.

Titanium oxides are a metal-based nanomaterial that is frequently used for environmental remediation. They have been exclusively prepared for the study of waste treatment, air purification and as photocatalysts in treating water because they have cost-effective, non-toxic, photocatalytic, semi-conducting and energy-converting properties (Adesina 2004). Titanium dioxide (TiO_2) NPs are regularly used for their potential to remove organic pollutants and are activated by light. They also produce highly reactive oxidants such as hydroxyl radicals that assist in sterilizing fungi, bacteria, viruses and algae (Cho et al. 2005; Li et al. 2014). TiO_2 NPs alone have little photocatalytic potential; accordingly, they are doped with transition metals to enhance their performance. Investigations have been carried out with metal-doped TiO_2 NPs. The sol-gel electrospinning technique utilized by Park and Lee (2014) assigned TiO_2 nanofibers to be Ag-doped and were subsequently evaluated for their photocatalytic degradation of 2-chlorophenol underneath ultraviolet (UV) irradiation while the TiO_2 nanofibers were maintained as the control. From the results obtained, there was a clear indication that Ag-doped TiO_2 nanofibers showed enhanced photodegradation over the control. This significant increase was credited to including a sufficient number of Ag on the surface, which aids in effectively capturing photoinduced electrons, shifting these captured electrons to the surface of nanofibers where adsorbed oxygen is present and stretching the response array to the visible region. Inorganic compounds of titanium oxide known as titanates have also been found to be effective in environmental nanoremediation. Chen et al. (2013) assessed the hydrothermal method that fabricates acidic, basic and neutral titanate nanotubes and utilized these materials for the catalytic reduction of nitric oxide (NO) with ammonia. Titanate nanotubes were doped with manganese oxides to produce nanosheets, nanorods and nanotubes and were subsequently evaluated for the reduction of NO with ammonia; neutral (Mn/TNTs) titanates were found to possess better catalytic reduction potential than acidic and basic (Mn/TNTs) titanates, whereas basic (Mn/TNTs) titanates showed the least activity.

Likewise, bimetallic oxides have been explored for their nanoremediation properties. Rasalingam et al. (2014) utilized TiO_2-silicon dioxide (SiO_2) bimetallic nanomaterial for the photocatalytic degradation of methylene blue dye from water, and the results showed efficient photoactivity. Bimetallic oxide nanomaterials have tremendous potential to remove many pollutants as can be seen from Table 14.1.

Moreover, the adsorbents of a nano metallic nature are especially eye-catching as they can be retained and detached easily from treated water. The literature has extensively described the use of iron and iron oxide NPs for the removal of different heavy metals, for example, Ni2+, Cu2+, Co2+ and Cd2+ (Hooshyar et al. 2013; Ebrahim et al. 2016; Poguberović et al. 2016), and for the remediation of chlorinated organic solvents (Guo et al. 2017; Han and Yan 2016). However, there are some problems when using NPs of this class in environmental remediation. One of the greatest concerns is accumulation as this affects the reactivity of material and therefore diminishes the benefits of using nanoscale materials as a way to improve efficiency. In addition, working with NPs of metal and metal oxide causes possible toxicity. Furthermore, the cost of remediation technology is also one of the significant aspects to consider when choosing the NPs of this class as remediation material.

In examining the structure of iron NPs, they have a core comprising elemental iron (zerovalent iron) covered with a shell that has mixed valent oxides of iron, namely, Fe(II) and Fe(III). Figure 14.3 demonstrates the way in which iron NPs help in environmental remediation. Zerovalent iron present inside the core donates electrons, which are actually used for the reduction of heavy metals and chlorinated compounds. Furthermore, this shell part of the iron NP aids in pollutant removal, such as heavy metals, and has a higher standard reducing potential (E0) than the Fe2+/Fe couple (Li and Zhang 2006).

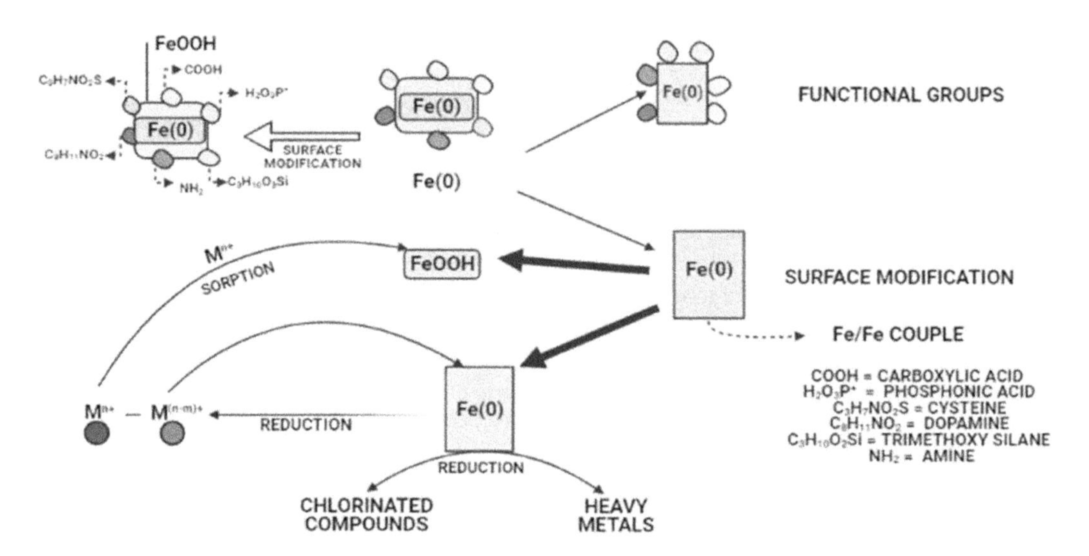

FIGURE 14.3 Degradation mechanism of chlorinated contaminants and heavy metals from aqueous systems using iron nanoparticles/nanomaterials.

14.3.2 Nanoparticles Based on Silica

Silica materials have considerable importance due to their flexibility for various applications including catalysis and adsorption. Furthermore, mesoporous silica nanomaterial has a number of benefits from an environmental remediation point of view viz-a-viz a high surface area, facile surface modification, large pore volumes and a tunable pore size (Tsai et al. 2016). Extensive studies have conveyed that utilising silica materials for environmental remediation in the gas phase shows promising results. The modification of mesoporous silica material has been also performed as shown in many publications (Brigante et al. 2016). Table 14.1 summarizes some of the reported works found in the literature that investigate the use of silica nanomaterials for the environmental remediation of different contaminants. Silica material has the presence of a hydroxyl group on the surface, which is further important for modification, adsorption of gas and wetting one of the surface phenomenon. Designing new adsorbents and catalysts involves the embedding of functional groups onto the pore walls (Huang et al. 2003).

The abstraction of carbon dioxide (CO_2) and hydrogen sulfide (H_2S) using an amine surface modified by silica xerogels from natural gas was attributed to the fact that there is a high disposal of amine groups on the surface of the silica materials, which subsequently resulted in increased efficiency towards the removal. It was found that 80% of the removal of CO_2 was obtained in the first 30 minutes, revealing a high rate of adsorption and large capacity. A similar outcome was achieved for the removal of H_2S as 80% adsorption took place in first 35 minutes (Huang et al. 2003). Subsequent investigations have revealed the efficacy of amine-modified aluminosilicates for the removal of CO_2 and other carbonyl compounds, along with ketones and aldehydes. CO_2 capture was made possible through the reversible adsorption of gaseous molecules onto the aminosilica material. Likewise, for the removal of aldehydes and ketones, imine or hemiaminal formation was required. For CO_2, the remediation recyclability and capacity of these silica-based mesoporous NPs were analyzed, and it was found that CO_2 gets adsorbed reversibly and that the material relics were stable even after 50 cycles of continuous adsorption-desorption. Subsequently these silica-based materials can be considered a good alternative to conventional CO_2 capture in that they are cost-effective, synthesis is easy and they possess enhanced performance and stability (Qi et al. 2011). Instead of subjecting post-treatment scaffold material to a functionalization technique, the amine functional group is integrated due to the polymerization of aziridine. However, one issue arises when incorporation is performed that actually confines the use of the material to pollutants especially reacting with amines. Moreover, this monomer of aziridine is difficult to care for without dedicated equipment.

Amine functionalized porous silica nanomaterial was used by Drese et al. (2011) for aldehyde reduction, such as formaldehyde, which is a low molecular weight compound. The results obtained showed 1.4 mmol·g–1 and 10.8 mmol·g–1 of formaldehyde retention on silica material that has primary and secondary amines, respectively, and an insignificant quantity for the tertiary amines. The results deducted from this experiment convey that primary and secondary amines are suitable for aldehyde remediation, which agrees with the covalent capture of the target contaminant by the creation of imine and hemiaminal intermediates. Remediation of some higher molecular weight and fewer volatility aldehydes was also carried out by Bollini et al. (2011). Unfortunately, the required time to achieve comparable performance for the target pollutants was more than 10 hours, which is more than enough when compared with the adsorption of formaldehyde (Nomura and Jones 2014). This lengthy reaction time has restrained the use of these materials in an industrial environment, which requires the swift remediation of pollutants.

The removal of dyes organic in nature from wastewater by utilising silica-based nanomaterial has also been reported. Tsai et al. (2016) investigated the functionalization of mesoporous silica with –COOH groups because carboxylic acid can form hydrogen bonds with different types of compounds, such as metal ions, dyes and pollutants. The group was able to successfully functionalize mesoporous silica SBA-16 with the tunable loadings of carboxylic acid groups. The study revealed that the interaction between the functional groups of carboxylic acid and the target pollutant for adsorption depends on a particular pH value, which was found to be reliable with the proposed hydrogen bonding capture model. For methylene blue at a pH of 9, maximum uptake was attained. In basic conditions, these materials may be effective for adsorption, but this has, in turn, led to the creation of limitations on the practical aspects of these materials. In addition, different metals ions have been removed including Cd2+, Co2+, Cu2+, Zn2+, Ni2+, Al2+, Cr3+, Pb2+, Hg2+ and U6+ (Wang et al. 2015; Hernández et al. 2012; Nakanishi et al. 2015; Repo et al. 2013; Arencibia et al. 2010; Rostamain et al. 2011). Figure 14.4 explains mesoporous silica nanomaterial and its characteristic surface properties that are particularly important for adsorption.

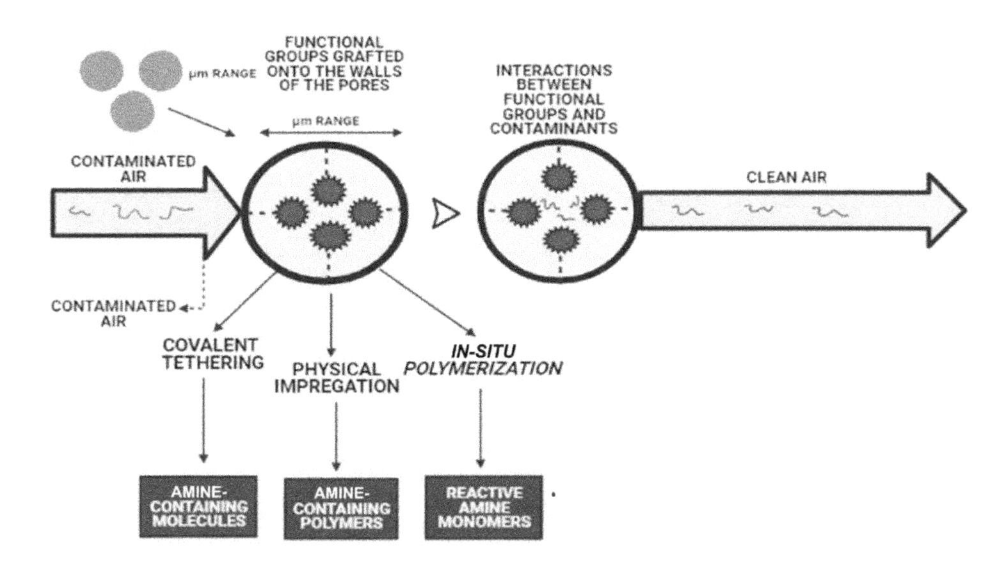

FIGURE 14.4 Mesoporous silica materials used for remediating the environment.

14.3.3 Carbon-Based Nanomaterials

Compared to nanomaterials based on metal, elemental carbon has exceptional physical, chemical and electronic properties due to its mutable hybridization state (Ren et al. 2011). This hybridization state has yielded many structural conformations such as fullerene C60, fullerene C540, single-walled nanotubes, multi-walled nanotubes and graphene (Mauter and Elimelech 2008). Considerable research has been carried out to determine the appropriateness of carbon nanotubes (CNTs) and graphene for remediation; the requirements of treating the surface and activating the carbon material are first involved. The adsorption properties of multi-walled and single-walled carbon tubes make them predominantly helpful for the degradation of organic and inorganic contaminants in the air and in aqueous systems (Ren et al. 2011; Theron et al. 2008; Di Paola et al. 2012; Kharisov et al. 2014). NPs based on carbon are used for environmental remediation through photocatalytic methods. Figure 14.5 demonstrates the photocatalytic approach for the remediation of environmental contaminants. Due to UV, photons get excited when they have energy more than or equivalent to the band gap of the nanotubes, promoting the creation of valence band holes (h+) and conduction band electrons (e−). These holes play an important role in the formation of radicals of hydroxyl, which ultimately lead to the oxidation of chlorinated organic compounds. Heavy metal contaminants are reduced by the electrons from superoxide radicals. Moreover, there is already published work that has indicated the use of graphene to fabricate photocatalytic nanocomposites (Liu et al. 2011; Yang et al. 2013a). Graphene in association with TiO_2 NPs has shown enhanced photocatalytic activity when matched to simple TiO_2 NPs because of a surge in conductivity (Zhang et al. 2010a).

14.3.4 Graphene Materials

Two forms of graphene such as modified and pristine have already been explored for environmental remediation uses. Pristine graphene was utilized in aqueous solution to adsorb fluoride, and the adsorption potential of fluoride by graphene was 35.59 mg/g at pH 7 and 298 K. Due to its increased efficiency and adsorption capacity, graphene is very effective in adsorbing fluoride from aqueous systems. Pristine graphene can be directly employed for environmental remediation; however, different environmental remediation techniques depend upon the exercise of modifying graphene to remediate different pollutants (Wang et al. 2013). An effective surface area can be

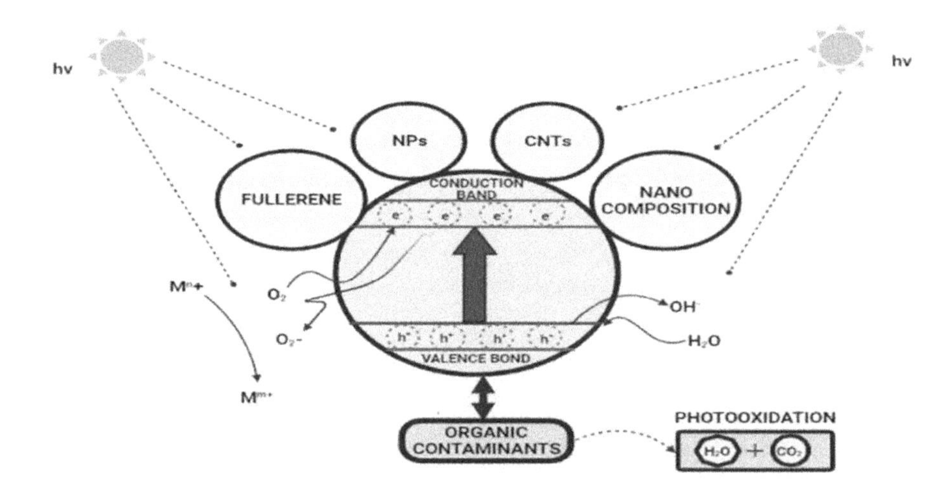

FIGURE 14.5 Mechanisms of photocatalytic degradation for metal and organic contaminants.

enhanced by preventing the accretion of graphene layers, thus creating modified graphene, which is more promising than pristine graphene (An and Jimmy 2011). One good example of modified graphene is graphene oxide that has been used for the environmental remediation of different pollutants, for example, sulphur oxide (SOx), hydrogen sulfide (H_2S), ammonia (NH_3), volatile organic compounds, heavy metals, pesticides and pharmaceuticals. In addition, graphene oxide (GO) has been reported to present high adsorption capacity for cationic metals. Nonetheless, the removal of anionic metals requires the modification of GO with organic or metal oxides (Wang et al. 2013).

14.3.5 Polymer-Based Nanomaterials/Nanoparticles

Because of their greater surface area-to-volume ratio, NPs possess enhanced performance and higher reactivity. However, some reasons can limit the use of these nanotechnologies due to a lack of functionality such as aggregation, lower stability and non-specificity. To increase the stability of nanoscale material, the best substitute is to use host material that will serve as a support or matrix for other NPs (Züttel et al. 2002). Polymers mostly find their use in detection and chemical contamination removal including iron, arsenic, manganese (Mn), nitrate, and heavy metals, different gaseous pollutants such as carbon monoxide, sulphur dioxide and NOx and pollutants that are organic in nature, for example, aromatic/aliphatic hydrocarbons, pharmaceuticals and a wide range of biological agents involving bacteria, parasites and viruses. To overcome the shortcomings of pristine NPs and to provide them with required characters, polymeric hosts include surfactants, emulsifiers, stabilizing agents and surface functionalized ligands.

Aromatic hydrocarbons have been degraded from the soil by amphiphilic polyurethane NPs, thereby validating the claim that organic NPs can be planned with anticipated properties (Tungittiplakorn et al. 2004). NPs' flexibility in the soil can be enhanced by making a hydrophilic surface, and a hydrophobic interior promotes an affinity towards organic contaminants. Phenanthrene pollutants have been removed from aquifer sand with almost an 80% recovery rate by polyurethane NPs. Amphiphilic polyurethane (APU) NPs' removal potential for phenanthrene improved by an increase in the size of the hydrophobic backbone. In addition, it was observed that by having more ionic groups on the hydrophobic backbone, the aggregation of particles was prevented because of polyvalent cations (Tungittiplakorn et al. 2004). For environmental remediation, these NPs could be helpful; however, no description on their biodegradability has given rise to serious questions about their fate after application.

Since many investigations have been conducted regarding the potential of NPs/nanomaterials, Table 14.1 provides a glimpse of the type of NPs and the applications achieved.

14.4 NANOREMEDIATION CHALLENGES INVOLVED

The increased reactivity due to a decrease in the particle size may affect a substance and make it harmful. Hence, the basic property that makes nanomaterials widely applicable also makes them a potential toxic substance. Accordingly, a complete understanding about the transport and mobility of nanomaterial use should be evolved, and their potential risks and toxicity should be keenly understood. The following sections will aid comprehension of the hardships faced by researchers during nanoremediation.

14.5 ECONOMICS OF NANOREMEDIATION

Nanotechnology has a cross-disciplinary nature, and its valuation for economic impact is rather complex because this technology is still in its early stages of development, innovation and growth. In addition, it is far more difficult to estimate its economic influences on the separation sciences as there are no indicators for this field. Nevertheless, nanotechnology will certainly leave its impact on specific disciplines in the near future. Additionally, once nanomaterial use has achieved its full

TABLE 14.1
Various Types of NPs/Nanomaterials and Their Role in Environmental Remediation

Nanoparticle/Nanomaterial	Applicability	Reference
Ag-mesoporous TiO_2	Degradation of Rhodamine B [RhB] and *E. coli.* from water	Zhigang xiong et al. (2011)
Metal-doped TiO_2	2-chlorophenol as the water contaminant degraded	Ju young park andin-hwa Lee (2014)
TiO_2 photocatalytic disinfection	Disinfectant of water from MS-2 phage and *E. coli.*	Min cho et al. (2005)
Ag-doped titania	Water contaminant *E. coli* removed	Hassan younas (2014)
CNTs	Removal of aerosols from the atmosphere	Yildiz and Bradford (2013)
Ag/Graphene oxide	Bactericidal-Staphylococcus aureus and *E. coli*	Qi Bao et al. (2011)
Titanate nanotubes	Catalytic reduction of nitric oxide	xiongbo chen et al. (2013)
TiO_2-SiO_2	Degradation of MB dyes, PO_4^{3-} and NO_3^-	Rasalingam et al. (2014)
Zerovalent iron/Sorbent	Ni^{2+} and Co^{2+} water contaminants removed	Hooshyar et al. (2013)
Iron oxide NPs	Ni^{2+}, Zn^{2+}, Cd^{2+} and Cu^{2+} metal pollutants cleared from water	Ebrahim et al. (2016)
Silver NPs	Disinfecting of water from *E. coli*	Gupta and Silver (1998)
TiO_2 nanopowder film	Degradation of aromatic hydrocarbons in polluted groundwater	Alizadeh Fard et al. (2013)
TiO_2 NPs	Reduced nutrient desorption and eradicating algal blooms from water	Da Silva et al. (2016)
Anatase TiO_2	Phenanthrene eliminated from the soil surface	Gu et al. (2012)
TiO_2 NPs	Biological nitrogen exclusion from wastewater	Li et al. (2014)
Zerovalent iron NPs	Removal of As(III) and Cr(VI) from aquatic ecosystems	Sofija S. Poguberović et al. (2016)
S-nZVI	Trichloroethene contaminant dechlorinated in water	Mengyu Guo et al. (2017)
Fe NPs	Degradation of 2,4-dichlorophenol in water	Yanlai Han and Weile Yan (2016)
Fe NPs as a sorbent	Ni(II) heavy metal reduced in water	Xiao-qin Li and Wei-Xian Zhang (2006)
Nanoscale bimetallic Fe/Pd	Hydrodechlorinaion of chlorinated ethanes (pollutants) in soil and water	Lien and Zhang (2005)
Nanoscale palladized bimetallic Fe	Monochloroacetic acid and 2,4-dichlorophenol dechlorinated in water	Chao Chen et al. (2008)
Bimetallic Fe/Pd	Degradation of lindane and chlorinated pollutant in water	Nagpal et al. (2010)
Ni/Fe bimetallic NPs	Nanoremediation of polybrominated diphenyl ethers present in soil	Yingying Xie et al. (2014)
Ni/Fe	Dilapidation of trichloroethylene in water	Yit Hong Tee et al. (2005)
Nanoscale Cu/Fe	HCB (hexachlorobenzene) POPs dechlorinated	Zhu et al. (2010)

(Continued)

TABLE 14.1 ***(Continued)***
Various Types of NPs/Nanomaterials and Their Role in Environmental Remediation

Nanoparticle/Nanomaterial	Applicability	Reference
Silver-doped TiO_2	2,4,6-trichlorophenol degraded from water	Rengaraj and Li (2006)
Nanosilver-decorated TiO_2	Capable of decomposing NOx and volatile organic compounds (VOC)—volatile air pollutants	Chutima Srisitthiratkul et al. (2011)
Cu-, Fe- and Ag-doped TiO_2	Nitrate removal from water	Duan et al. (2023)
NPs of silica made by adding salicylic acid and poly (propylene imine)	Removal of pyrene, phenanthrene, Pb2+, Hg2+ and Cd2+ from polluted water solutions	Mamadou S. Diallo et al. (2007)
Poly-amidoamine dendrimer-fused membrane entailing chitosan and a dendrimer	CO_2 separated from a feed gas mixture of CO_2/N_2 on porous substrates	Takayuki Koukestsu et al. (2007)
Chitosan-capped gold NPs	Aqueous solutions decontaminated from Zn2+ and Cu2+	Sugunan et al. (2005)
Al_2O_3/CNTs	Fluoride adsorbed from water	Yan-Hui Li et al. (2001)
Purified nanotubes	Zinc(II) adsorbed from water	Chungsying Lu and Huantsung Chiu (2006)
PHB-TiO_2	Organic dyes removed and wastewater disinfected	Saw-Peng Yew et al. (2006)
Amine-grafted silica xerogel	CO_2/H_2S removed from natural gas	Helen Y. Huang et al. (2003)
Porous silica	Adsorption of aldehydes from the air	Akihiro Nomura and Jones (2013)
Porous silica functionalized with amine organic species	Adsorption of CO_2 from simulated flue gas stream	Praveen Bollini et al. (2011)
Aminosilica	Removal of aldehydes and ketones from bio-oils	Jeffrey H. Drese et al. (2011)
Multimeric amine with aminosilicas	Adsorption of formaldehyde and VOCs	Nomura and Jones (2014)
Mesoporous silica; amino-functionalized	Heavy metal-like Pb^{2+} removed from wastewater	Shugo Wang et al. (2015)
Mesoporous silica; thiol-functionalized	Mercury adsorbed	Amaya Arencibia et al. (2010)
Mesoporous silica; thiol-functionalized	Hg^{2+}, Pb^{2+} and Cd^{2+} removed from water	Rahele Rostamian et al. (2011)
Graphene	Fluoride adsorption from water	Yanhui Li et al. (2011)
Ag NPs and graphene oxide	Bactericidal agent for water disinfection	Qi Bao et al. (2011)
Graphite oxide	Removal of ammonia	Mykola Seredych and Teresa J. Bandosz (2007)
nZVI	Degradation of DDT in soil	El-Temsah and Joner (2013)
(G/Ag NPs-MS) graphene/silver NPs with melamine	Disinfecting water from *E. coli* and Staphylococcus aureus	Can Hui Deng et al. (2017)
CdS-Graphene	Reduced heavy ions/metals (Cr(IV)) in water	Nan Zhang et al. (2013)
TiO_2-Graphene nanocomposite	Degradation of volatile aromatic pollutant—gaseous	Yanhui Zhang et al. (2010a)
ZnO-Graphene	Increased photocatalytic reduction of heavy metal-like Cr(VI)	Xinjuan Liu et al. (2011)

(Continued)

TABLE 14.1 *(Continued)*
Various Types of NPs/Nanomaterials and Their Role in Environmental Remediation

Nanoparticle/Nanomaterial	Applicability	Reference
Multimeric polyurethane NPs	Remediation of soil polluted with polynuclear aromatic hydrocarbons	Warapong Tungittiplakorn et al. (2004)
Dendrimers with ethylene diamine	Removal of heavy metal from water viz, Cu(11)	Diallo et al. (2005)
Cellulose nanocrystals with amine modification	VOCs removed from the atmosphere	Guerra et al. (2018b)
Metal carbon polymer nanocomposite	Degradation of hexavalent chromium ions from water	Prateek Khare et al. (2016)
Hydrogel nanocomposite	Methylene blue dye removed	Hemant Mittal et al. (2015)
Polymer-functionalized magnetic nanoparticles	Enhanced removal of heavy metals from water	Gei et al. (2012)
$Zn_{12}O_{12}$ nano cage	Adsorption of carbon disulphide from the atmosphere	Ghenaatian et al. (2013)
Bimetallic Ni/Fe NPs	Degradation of diphenyl ethers from soil	Xie et al. (2014)

potential, wide-ranging economic and societal advances will be attained. Once nanotechnological expansion occurs, this will create a way for policymakers and stakeholders to design and assess more confidently the economic forecasts and benefits of nanoremediation (Vijaya et al. 2018). It is hoped that the investments being made in nanotechnologies will prove to be worthwhile and not an economic burden in the future.

14.6 HEALTH AND SAFETY RISKS

The nanosize of the material has raised number of objections regarding its potential toxicity to the environment and exposure to mammals, because NPs will be absorbed in cell membranes as they can easily pass through the skin. Different characters of nanomaterial determine the rate of adsorption in cells such as their size, aggregation and sedimentation (Binderup et al. 2013). The possible avenues for NP exposure include oral, skin, nutrient supplements, skin care products and food coloring (Ma et al. 2015). Some studies have mentioned that during the early phases of life, nanomaterials act as allergens and may prompt an immune response at later stages of life, causing allergic inflammation (Sly and Schüepp 2011). Iron oxide NPs, which occur naturally, bind to different metals such as copper; because they can travel long distances downstream from mining sites, this indicates that nanomaterials' toxicity depends upon their transport (Hochella et al. 2005). There are observations about colloidal NPs' ability to move sorbed pollutants and transport them long distances, which affects the ecosystem. Hence, it is very easy to deduce that although nanomaterial may not be harmful to the environment, they can transport pollutants with them. Many researchers have found that aqueous systems contain mineral NPs. According to Villalobos et al. (2003), numerous processes of a natural and biological origin might create mineral NPs of metal oxides and sulphides. Nanomaterials could affect aquatic organisms as revealed by some toxicological studies (Gao et al. 2018). The results of nanomaterial exposure to neonates during the early phases of life have been thoroughly explored. Meldrum et al. (2017) assessed that nanomaterial exposure can cause different ailments, for example, asthma, diminished lung function and cough, without affecting neonates. Furthermore, different epidemiological investigations have revealed that nanomaterial exposure may affect the cardiovascular system, for example, an altered coagulation of blood, which may ultimately result in changes in cardiac functions (Araujo and Nel 2009; Quan et al. 2010; Liou et al. 2012). Particular NPs have a variety of applications, but numerous unknown NPs are added to the environment unintentionally, which may have serious negative results on human health. The disposal of plastics in a non-scientific way has enhanced the amount of plastic in the environment at an alarming rate, particularly in aquatic systems (Bouwmeester et al. 2015; Prata 2018). Because of the presence of nanoplastics in the ecosystem, these materials can far more easily enter the food chain. Moreover, from the food chain, these materials may find their way to human systems and ultimately affect human health through circulation.

14.7 FUTURE OF NANOREMEDIATION

The speed with which innovations and unique development take place in science and technology has extended the range of nanotechnology and its implications across multidisciplinary areas. Nanotechnological exploration has generated many practical applications in terms of economics, and in recent times, the economic relevance side has intensified, thereby extensively increasing nanotechnological research. Because of widespread application and the potential of nanotechnology, especially in environmental remediation, the fabrication of sensors and green synthesis is currently attaining impetus. Nanomaterials have a high efficiency that has actually led to the formulation of a broader range of products. Nanoscale material produced through synthetic and biological methods has paved the way for toxic metal remediation, an area of significance. By utilising pulsed laser deposition technology, microscopic structures have been coated with nanometre-thick particulates. The scope of commercial applications has extended by blending microelectronics with molecular biology. Altering the optical characteristics of NPs and enhancing the magnitude of their applications in several aspects of nanotechnology need to be followed up for proficient pollutant remediation. By putting nanotechnology to appropriate uses, a constant inter-relationship between energy and the environment can be established. The progress made in nanotechnology broadens its role in the protection of the environment and remediation. Nanoscale material demand from a global perspective has been continuously increasing. Despite their advantages, the environmental impact in terms of their toxicity, transportation, bioavailability and bioaccumulation has yet to be fully interpreted. Eliminating waste at the source through the aid of nanotechnology via environmentally friendly manufacturing aspects holds sufficient potential. Therefore, a complete analysis about the negative effects of nanomaterials on the environment and their production is to be given priority for the potential service of nanoscale material in the monitoring and judicious supervision of contaminants through remediation.

14.8 CONCLUSION

Continuous and swift efforts that are required to create sustainable methods for the management of toxic contaminants have led us to transform ideas into experimental designs and manufacture useful products with due convenience and cost-efficiency. Environmental remediation by nanotechnology is one tool under exploration globally. By remediating contaminants with a target-based approach, the potential of nanomaterials has enhanced their role in different fields, such as sensor fabrication, catalysis, remediation and green synthesis. Their extraordinary adsorption and reactive capacity prompted the utilization of nanomaterials in the early stages of green processing. A vision regarding the commercial aspect has been achieved by combining molecular biology with microelectronics. Concerns about discarding harmful metals can be managed with the production of synthetic nanomaterials, thereby enhancing the demand for these nanomaterials while raising issues about the negative effects on the environment and human health in the future. Hence, before fully considering nanotechnology and broadening its practicable possibilities, its toxicological effects on the environment in the long term need to be elucidated.

14.9 NANOPHYTOREMEDIATION

Phytoremediation takes into consideration the expenditure of plants to remove, reduce and degrade contaminants in different environmental media such as water, air, sediments and soil. One of the greatest concerns currently faced by the world is pollution that can be remediated with phytoremediation (Asante-Badu et al. 2020). Some of the compounds that can be remediated by utilizing phytoremediation include wastes of an organic origin, metals, metalloids, inorganic substances, petroleum hydrocarbons and chlorinated solvents (Misra and Misra 2019). The level of these contaminants is on the rise because of enhanced anthropogenic activities, for example, pesticide and herbicide use in agriculture, petrochemical usage in industry, spillage of fuel and solvents, processing of wood and defense activities such as chemical weapons and explosives (Tripathi et al. 2020). One important feature of phytoremediation is cost-effectiveness, which implicates plants in metabolizing contaminants in their tissues and thereby reducing these contaminants. The accumulation of contaminants from different environmental media is a natural ability of plants. Nanophytoremediation encompasses nanotechnology and phytoremediation together for environmental remediation. Nanotechnology enhances the effectiveness of phytoremediation (Song et al. 2019). The nanophytoremediation of pollutants is affected by different factors including the physical and chemical aspects of pollutant such as their solubility in water, molecular weight, bond type, hydrophobicity, toxicity, mobility and reactivity. Moreover, the characteristic features of the environment that affect chemical uptake include the temperature, the amount of moisture in the soil, pH, light, microorganisms, humidity and oxygen. In addition, different plant features influence chemical absorption viz-a-viz the root system involved, the type of plant, the potential to accumulate contaminants and a high biomass. The chemical configuration, size and type are prime features upon which the intake of NPs into the plant rely. The mobility and translocation of NPs from the roots inside the plant primarily depends upon the size of the nanomaterial. To determine how proficiently plants take in nanomaterials, the penetration process is one factor (Schwab et al. 2016). Inside plants roots, NPs move via apoplastic and symplastic transport (Sattelmacher 2001; Roberts 2003). Environmental situations serve an important function in this uptake (Burken and Schnoor 1996). Growth ingredients are affected by temperature and subsequently, the length of roots. The root structure varies both in natural conditions and in greenhouses. The accomplishments of phytoremediation, explicitly, phytoextraction, rest upon contaminant-specific hyperaccumulator plants. The NPs utilized in phytoremediation should possess some of the ideal characteristics, specifically, non-toxicity towards plants, potential binding with the pollutant, a significant increase in the plant growth regulators within the plant and enhanced germination and biomass production.

Nanophytoremediation generally utilizes two types of NPs/nanomaterials, namely, ones that occur naturally and others that are plant-assisted and genetically modified for cleaning polluted environments. Some selected NPs have expressively increased plant growth, and nanoaugmentation-enhanced phytoremediation has resulted in considerable increments in pollutant removal from the soil environment. Some NPs have also been recognized as promoting plant growth due to their potential to increase the growth hormones in plants and encourage the healthier intake of pollutants. In addition to the advantages already known of nanomaterials for soil remediation and other media of the environment, testing their potential in the field and accessing the toxicity of nanocarbon to plants and microbes in soil is an important aspect (Sajid et al. 2015).

The effectiveness of nanomaterials in the soil is influenced by the type of contaminant, amount present, size, charge, organic content and other physical/chemical properties of soil. Various techniques are involved in using NPs in the soil remediation process such as injection under pressure, recirculation, pressure pulse technology and hydraulics. The type of plant species that is used in the nanophytoremediation method for soil remediation is particularly grounded in the application of NPs and method of phytoremediation (e.g., phytostabilization, phytoextraction and phytodegradation). It is obvious that nanophytoremediation has impending potential for remediating and neutralizing pollutants of an organic nature in sediments. Contaminant removal takes less time than

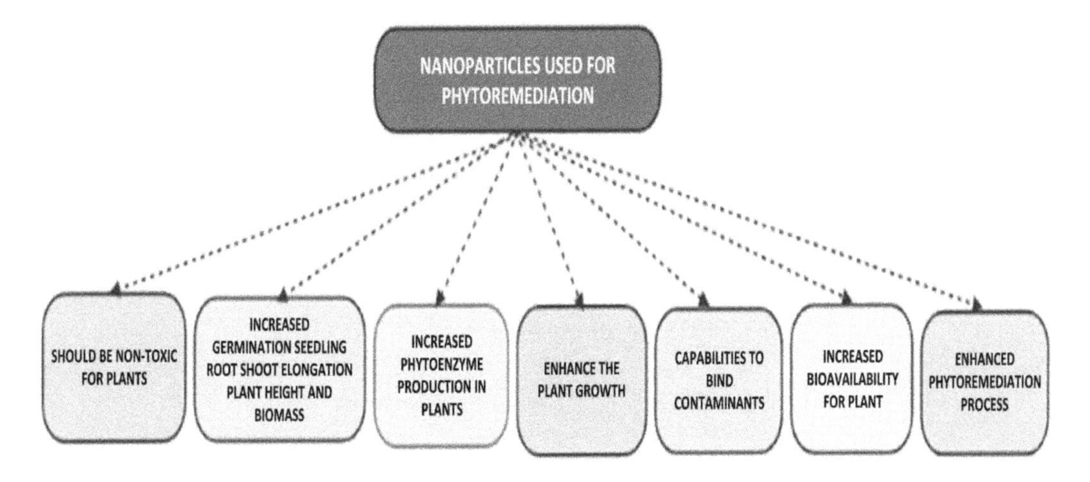

FIGURE 14.6 Desired properties of nanoparticles for nanophytoremediation.

normal, which is accelerated by the plant species. Significant pollutant removal results have been obtained from contaminated soil (Ma et al. 2010; Pillai and Kottekottil 2016; Souri et al. 2017). Table 14.2 offers a small glimpse of various plant species and different NPs utilized to remove contaminants.

14.9.1 Phytodegradation

In phytodegration, organic contaminants are taken up by plants for the purpose of decomposition (metabolized) or mineralization that is attained by the aid of different enzymes present inside plant cells. Some examples include nitro-aromatic compound decomposition by nitroreductase, pesticide decomposition by enzyme dehalogenase and aniline decomposition by volatile enzymes (Kumar et al. 2018). The tissues of *Armoracia rusticana* have been found to possess the potential to degrade benzophenone (Chen et al. 2016). The label biosphere green liver has been given to these plants. In addition to organic compounds, phytodegradation can remove or reduce insecticides, different herbicides, nutrients of an inorganic nature and chlorinated solvents (Kumar et al. 2019).

14.9.2 Phytostabilization (Phytoimmobilization)

To diminish the amount of pollutants from agricultural and natural environments, phytostabilization is of paramount importance; this is an in-situ technique in which toxic contaminants are integrated with the lignin root cell wall to make them less toxic (Khalid et al. 2017). Under the influence of exudates from roots, metal contaminants are precipitated and consequently move into the soil medium. One of the major pitfalls involved in this in-situ method is that the concentration of pollutants and their overpowering dispersion in soil cannot be avoided (Fan et al. 2017). It has been found by Lebrun et al. (2018) that the phytoimmobilization of polluted soil can be attained by shielding it with the type of vegetation that is resistant to toxic elements; this constrains soil erosion and the discharge of contaminants into groundwater. These substances immobilize pollutants by inhibiting their uptake and reducing their mobility in soil. The renovation of vegetation and remediation of contaminated areas of agricultural production can be achieved by these plants that have stabilizing potential (Eskander and Saleh 2017; Saha et al. 2017). Soil contaminated with inorganic pollutants such as arsenic (As), copper (Cu), lead (Pb), zinc (Zn), cadmium (Cd) and chromium (Cr) can be remediated through phytostabilization (Yadav et al. 2018).

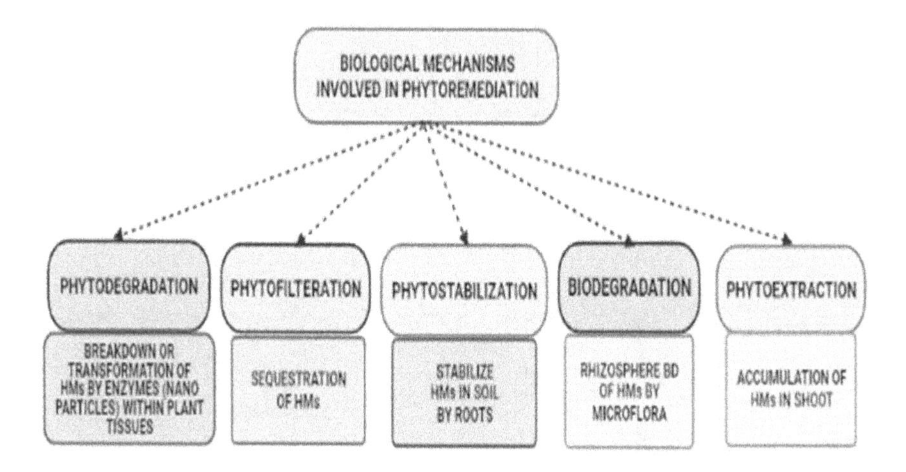

FIGURE 14.7 Methods of phytoremediation involved and their role.

14.9.3 Phytovolatilization

Some plants actually have the potential to engulf/absorb contaminants of metal, metalloids and organic and heavy metals and afterward, to release them through vaporization. Mercury (Hg), As and selenium (Se) are absorbed by the roots, converted into non-toxic forms and finally volatilized (Limmer and Burken 2016). Plants that have phytovolatilization potential as verified by different studies include *Stanleya pinnata*, *Astragalus bisulcatus*, and transgenic plants (with bacterial genes) such as Liriodendron tulipifera, Brassica napus, Arabidopsis thaliana, and Nicotiana tabacum (Ali et al. 2013; Mani and Kumar 2014). One of the drawbacks of this method is that harmful compounds released into the atmosphere may undergo precipitation and re-deposit back into the environment, which will give rise to other toxic substances (Nikolić and Stevović 2015).

14.9.4 Phytoextraction (Phytoaccumulation)

Contaminant absorption via roots succeeds through the translocation and accumulation in air particles. Insights provided by Parmar and Singh (2015) about phytoextraction encompass the removal by plants of pollutants from different environmental media such as the soil, groundwater and surface water. Phytoextraction is primarily associated with contaminated soil remediation from various metals, for example, Cd, nickel (Ni), Cu, Zn, and Pb, in addition to organic compounds. Hyperaccumulator plant species are favored for this method, and these plants have great tendency to absorb metals in certain portions (0.01–1% dry weight) (Parmar and Singh 2015). Hyperaccumulators rely on the use of sorption and cationic pumps to remove different contaminants including organic compounds, salts and metals by engrossing the water available to plants (Xiao et al. 2017). This process starts by planting metal-accumulating plants in polluted soils, which is integrated with vast agricultural practices. *Elsholtzia splendens*, *Alyssum bertolonii*, *Thlaspi caerulescens* and *Pteris vittata* are known examples of hyperaccumulating plants for Cu, Ni, Zn/Cd, and As, respectively (van der Ent et al. 2013; Xiao et al. 2017; Reeves et al. 2018).

14.9.5 Rhizodegradation/Phytostimulation

Due to the microbial activity of the rhizosphere (roots), organic compounds in the soil are broken down (Echereme et al. 2018). Roots associated with particular fungi and bacteria enhance the biodegradation of pollutants (Khalid et al. 2017). Some symbiotic free-living mycorrhizal fungi associated with plant roots are considerably more significant for the production and biochemical accessibility of nutrients including nitrogen, cobalt, phosphorus, potassium, sulphur, and calcium. Sarwar et al. (2017) found that certain bacterial species such as pseudomonas, bacillus and Actinobacter have enriched their adaptability to the environment and are resistant to Pb, Cd and Cu in the rhizosphere soil around the plant. Moreover, these plants have the inherent potential to release some of the biodegradable enzymes. Due to the changing nutrient conditions around the rhizosphere, microbial association is therefore heterogeneous in nature, but the species belonging to the genus *Pseudomonas* are predominantly associated with roots (Singh and Singh 2016). The death of plants should be considered, and agronomic methods must be used to reduce plant death by timely planting in the growing period, digging an adequate hole, and feasibly filling it with unpolluted plant soil to better ensure the survival of plants and a greater efficiency of phytoremediation.

14.9.6 Phytofiltration

Phytofiltration relies upon the absorption of precipitate pollutants, particularly radioactive elements and heavy metals, to remediate the environment through the plant root structure or submerged organs. In this method, plants kept in a hydroponic system perform rhizofiltration once wastewater passes through them or the pollutants remain in the plant structures that have the potential to absorb contaminants (Parmar and Singh 2015). Different metallic compounds can be remediated with phytofiltration such as Cu, Ni, vanadium, Cr, Pb and other radionuclides, namely, caesium, strontium and uranium. An investigation conducted by Cule et al. (2016) found that in the *Canabis indica* remediation for Pb in wastewater removal, efficiency was enhanced by 81%.

14.10 NANOMATERIAL-FACILITATED PHYTOREMEDIATION FOR THE REMOVAL OF HEAVY METALS

14.10.1 Lead (Pb)

Soil polluted with heavy metals is one of the most critical issues around the globe because it poses a serious threat to human health and food safety. Contaminated soil can be remediated from heavy metals with in-situ phytoremediation (Liang et al. 2017). The phytoremediation of contaminated soil can be increased by the application of NPs (Khan and Bano 2016). The two most commonly studied heavy metal pollutants are Pb and Cd as these are the most common at polluted sites. Generally for Pb removal from the soil, phytoextraction is considered (Ali et al. 2013). Because of its swift growth rate and high potential to tolerate Pb, cost-effective *Lolium perenne* is employed for phytoextraction. In addition, NPs have been found to be effective in promoting Pb phytoextraction by *Lolium perenne*. In another study conducted by Liang et al. (2017), the application of nanomaterials such as nano-hydroxyapatite in the Pb phytoextraction by *Lolium perenne* was determined, and it was found that efficiency increased from 16% to 31%. Huang et al. (2018) assessed different concentrations of nZVI particles on the phytoextraction of Pb by *Lolium perenne*. Concentrations of 100mg/kg nZVI caused the maximum amount of Pb to be phytoextracted.

14.10.2 Cadmium (Cd)

As a result of industrial processes, mining, electroplating and phosphate fertilizers, harmful metals such as Cd are added to the soil (Godt et al. 2006; Mahabadi et al. 2007). One of the primary phytoremediation methods is to use hyperaccumulators to extract Cd, but these plant species are confined in their number and capacity (Kirkham 2006). NPs have been found to enhance the phytoextraction of Cd from soil. In soyabean plants, utilising Tio_2 NPs on Cd accumulation showed a positive response (Singh and Lee 2016). The authors added TiO_2 NPs of 100, 200 and 300mg/kg to the soil and analyzed the accumulation and distribution of Cd in plants on the 60th day after sowing; with the assistance of TiO_2 NPs, the Cd accumulation in the shoots increased by approximately 1.9, 2.1 and 2.6 times, while the Cd accumulation in the roots increased by 2.5, 2.6 and 3.3 times, respectively. The maximum accumulation of Cd reached 1,534.7 mg/g with TiO_2 NPs of 300 mg/kg.

14.10.3 Arsenic (As)

As is highly harmful and has carcinogenic properties; it is added to the environment via the use of As-possessing pesticides, herbicides, phosphate fertilizers, wood preservatives and some operations related to industry (Singh et al. 2015). The phytoremediation techniques that have been found to be more fruitful for As reduction in soil are phytoextraction and phytostabilization. Three individual systems are used by plants to absorb As, namely, active uptake by the symplast, passive uptake by the apoplast and direct transcellular transport from the environment to the plant vascular system (Vithanage et al. 2012). Salicyclic acid NPs are effective in As phytoextraction (Souri et al. 2017).

The interaction between NPs and plants lead to various morphological and physiological changes; for example, the application of Ag NPs to *Zea mays* L. enhanced the amount of ABA and gibberellin, which aid the plant in enduring stress and enhance the nutrient uptake for enriched growth (Khan and Bano 2016). Similarly, in *Vicia faba* L., germination increased due to graphene oxide (Anjum et al. 2014), and the response of plants to NPs depends on the type, dose and the plant species involved (Dimkpa et al. 2012). Chemical/physical qualities including the size, shape, surface area, reactivity and amount of NPs determine their effectiveness, which considerably varies among plants (Ma et al. 2010; Khodakovskaya et al. 2012). Salicylic acid NPs are also considered to increase phytoremediation and plant growth under the influence of As stress (Souri et al. 2017). Pradhan et al. (2013) and Pradhan et al. (2014) assessed that Mn NP application increased not only the root and shoot length but also nitrogen uptake. Nano-fertilizers aid in the regulation of and release of nutrients into the soil system (Subramanian et al. 2015), which enhances the germination of seeds, the chlorophyll amount and soil fertility (Varma and Khanuja 2017) and promotes enzyme activity in plants (Yang et al. 2017).

A primary concern that must be prioritized while utilizing NPs for nanophytoremediation is that it must be ensured that they are not toxic to plants and to plants' rhizospheric micro-biota. Micro-biota has the potential to bind pollutants and afterwards, to make them bioavailable to complete the phytoextraction process (Khan 2020). The biological synthesis of engineered NPs is rapid by the plants and microbes whose roots grow at polluted sites (Singh et al. 2009). However, these engineered NPs need to be evaluated for their toxic effects on the environment and human health, and subsequently, this knowledge will be helpful in designing and employing NPs (Ma and Wang 2018). Due to inconsistencies found in the literature regarding nanotechnology, addressing NP toxicity is the need of the hour to understand their genotoxic potential (Singh et al. 2009). The biological synthesis of NP purification can be accomplished through their filtration with antioxidants to reduce heavy metals to their respective NPs such as Ag, gold and platinum (Yadav et al. 2017).

14.11 NANOREMEDIATION ASPECTS OF SUSTAINABILITY

The sustainability aspect of nanoremediation and nanophytoremediation can be broadly viewed as an all-inclusive approach where benefits from remediation in different spheres, for example, the environment, economic and social spheres, are maximized for all shareholders for current and future generations. Alternatives for every viable remediation are estimated with evidence based on sustainability valuations of the impacts on these spheres. The selected remedial material should have greater sustainability advantages and surpass the detrimental effects on a life-cycle basis. To decrease secondary emissions, energy requirements, the use of resources and ecological impacts, reliable and up-to-date management practices should be utilized. According to the United States

TABLE 14.2
Various Types of NPs/Nanomaterials Used in Nanophytoremediation and the Results Obtained from Their Use

Nanoparticles	Contaminant	Plant species	Results obtained	Reference
nZVI	Endosulfan	*Alpinia calcarata*	Removal rate of endosulfan was enhanced from 81%–100% in the soil	Pillai and Kottekottil (2016)
Fullerene	Trichloroethylene	*Populus deltoides*	Increment in the intake of trichloroethylene provided with 15 mg/l of fullerene NPs was 82%	Ma and Wang (2010)
nZVI	Cadmium	*Boehmeria nivea*	The amount of cadmium increased in leaves from 31%–73%, in the stem from 29%–52% and in the roots from 16–50%	Gong et al. (2017)
Silicon NPs	Chromium	*Pisum sativum*	Chromium amount in the root diminished from 1,472–516.6 mg/kg DW and in the stem from 62.5–32.2 mg/kg DW	Tripathi et al. (2015)
Nano-hydroxyapatite	Lead	*Lolium perenne*	In the shoot and root concentrations, Pb diminished from 13.19–20.3% and from 2.8–21%, respectively	Liang et al. (2017)
Salicylic acid	Arsenic	*Isatis cappadocica*	Arsenic-accumulated concentration in the shoots and roots reached from 705–1,185 mg/Kg DW	Souri et al. (2017)
nZVI	Arsenic, cadmium, lead and zinc	*Helianthus annus* and *Lolium perenne*	Phytostabilization done by nZVI; arsenic, cadmium, lead and zinc concentrations diminished by 50–60% in the shoots and roots compared to the control	Vítková et al. (2018)
Ag NPs	Cadmium, lead and nickel	*Zea mays*	Amount of cadmium, lead and nickel in the shoots enhanced from 0.65–0.73 mg/kg DW	Khan and Bano (2016)
Nano Silica	Cadmium and lead	*Secale montanum*	Highest accumulating concentration for Pb was 533 mg/Kg DW and for Cd was 208mg/kg DW in roots because of nano silica	Moameri and Abbasi Khalaki (2019)
TiO_2 NPs	Cadmium	*Glycine max*	Intake enhanced from 128 µg to 507µg	Singh and Lee (2016)
Nano hydroxyapatite and nano carbon black	Lead	*Lolium perenne*	Pb content in roots and shoots diminished from 2.86–21% and 13.20–20%	Jin et al. (2016)
nZVI	Trinitrotoluene	*Panicum maximum*	nZVI increased the elimination from 85%–100% after a time gap of 120 days	Jiamjitrpanich et al. (2012)

Environmental Protection Agency (USEPA), green remediation employs various approaches to have minimalistic detrimental effects on the environment while eradicating pollutants from their source, reducing waste generation and integrating options that will ultimately enhance the overall benefits of environmental remediation actions (USEPA 2008).

Sustainable remediation takes into consideration the ideal outcome for both present and future generations. Since the environment has a dynamic nature, it must be remembered that sustainable remediation methods should be robust to endure changes in the environment such as the depletion of groundwater, a rise in sea levels and enhanced levels of precipitation. Sustainable remediation should possess certain features to consider these changes including the potential to cope with evolving human health and standards in the environment, a resilience to the dynamic geophysical conditions (climate change) and flexibility in various choices for future site development (Hou and Al-Tabbaa 2014). According to complexity theories, the sustainability of coupled human nature systems requires both change and persistence (Holling 2001). From a remediation perspective, persistence can be summarized as the effectiveness of the remediation platform and an assurance to clean up contaminants. Remediation change involves adaptability, the acceptance of novel ideas from remediation practitioners and the tractability of the remediation system to integrate new technologies. Activities encouraged and supported by nature fall under nature-based solutions (van den Bosch and Sang 2017). Implementation of nature-based solutions to remediate contaminated sites provides us with different environmental, economic and social welfares (Song et al. 2019) such as energy and enhanced material efficiency for promoting resilience to combat environmental change globally (Chi et al. 2017; Liang and Wang 2017). Different substitutes that have been actually utilized in polluted sites are re-vegetating, urban planning, developing green industrial heritage parks and creating nature reserves.

There are basically four primary objectives for sustainable development, specifically, progress at a social level, which takes care of everyone's needs, environmental protection, a judicious use of natural resources and the maintenance of high and stable levels of economic growth and employment.

14.12 PROCESSES OF SUSTAINABLE REMEDIATION

Sustainable remediation assesses the alternatives not only for remediation but also decision making. A pivotal feature is the employment of green and clean processes in environmental remediation. The overall sustainability has a strong interconnection with the alternative of either ex-situ or in-situ remediation. It has been found that remediating groundwater with the in-situ method that utilizes permeable reactive barriers has very low life cycle impingement compared to the ex-situ method with a pump and treat system (Higgins and Olson 2009). Likewise, Lemming et al. (2010) investigated in-situ bio-remediation and in-situ thermal desorption and found that they have minimal life cycle impacts compared to transporting the contaminated medium for off-site treatment and disposal. Several studies have been conducted based on a life cycle assessment, and the results indicated that in-situ remediation is more sustainable than ex-situ handling (Blanc et al. 2004: Harbottle et al. 2008). Quantitatively assessing remediation methods and ground-breaking, pioneering technology for remediation is currently needed to lower the primary, secondary and tertiary environmental impacts and maximize the sustainability benefits.

14.12.1 PHYTOREMEDIATION

Generally, the term phytoremediation indicates the use of plants to eliminate or diminish contaminant levels. It involves various processes such as phytodegradation, phytovolatilization, phytoextraction, phytostabilization and rhizofiltration.

14.12.2 In-Situ Bioremediation

In-situ bioremediation involves treating polluted soil and groundwater biologically while neither unearthing the soil nor driving water for treatment above the soil. Basically, in-situ bioremediation relies upon the use of microbes in plants to degrade or restrain pollutants in situ. Furthermore, this technology provides the option of equilibrium between cost and effectiveness while upholding site availability. If there is a lack in characterising the site and a poor design for application, there may be some problems. In-situ bioremediation can be employed in geochemical circumstances that have an anoxic feature and that rely on pollutants and their mechanism for degradation. For chlorinated solvents, anaerobic in-situ bioremediation is a better choice, with a nominal disruption of the ground surface. However, this technique may sometimes have serious costs (Griffiths 2019), such as effects on groundwater quality including color, smell, turbidity and a changed pH.

Organic contaminants such as coal tar and crude oil from soil can be remediated using a novel technology viz, active remediation for self-sustaining treatment; in this technology, air is used for smoldering, which is basically flameless low-temperature combustion. Through contaminated soil, this self-sustaining reaction moves from the ignition point while destroying pollutants along the way (Gerhard et al. 2019). It involves both in-situ and ex-situ methods. Some studies have described that this novel technique rendered an order-of-magnitude lower environmental footprint compared with thermal desorption and dig-and-haul technologies.

14.12.3 In-Situ Solidification/Stabilization

Stabilization/solidification is a remediation technique that has various advantages such as rapid implementation, swift site development and diminished disposal off-site and utilizes advanced equipment and novel techniques that, in turn, result in reduced risk for workers. Stabilization/solidification remediation treatment bounds the load of the pollutants released from solid harmful wastes. As the name suggests, solidification includes the encapsulation of particular solids, and stabilization hints at constraining pollutants' mobility and converting them to a less-toxic form. The reaction for stabilization involves adsorption, precipitation, isomorphous substitution and lattice incorporation. However, one of the greatest concerns related to this technique is contaminant retention in the ground. The effectiveness of solidification/stabilization has been considered doubtful in the long run as there are various environmental stresses that damage solidification/stabilization materials. On-site solidification/stabilization under natural conditions has been observed to be effective (Jin 2019). Durability and long-term effectiveness are elements in the assessment of the sustainability of contaminated land remediation techniques. However, other aspects of sustainability, for example, energy use and the use of natural resources, are soon likely to become as much of a concern as durability with the increased pressures to address the technical, environmental and social elements of sustainability on contaminated land remediation projects (Harbottle et al. 2005).

14.12.4 Remedial Process Optimization

To assess the performance and efficacy of site remediation and recognize the different strategies for cost effectiveness, one systemic approach is remedial process optimization. This applies the principles of value engineering to the implementation components of remediation (Leu and Hou 2019). From a sustainability point of view, it is recommended that the remediation industry consider all environmental, economic and social benefits. In general, remedial process optimization is an operational way to diminish the life cycle environmental footprints of remedial processes, thereby making them greener and more sustainable

14.13 CRITERIA FOR SUSTAINABILITY REMEDIATION (M.J. HARBOTTLE & A. AL-TABBAA)

14.13.1 First Criterion

The benefits of remediation to future and current generations must outweigh the costs. A quantification of cost and benefits (non-financial) can be performed by including terms such as the quality of the water, groundwater, air and soil, the overall risks to site users and the general public, non-recyclable contaminants and future land use. In addition, financial considerations include the effect on surrounding areas, costs such as capital, operation and maintenance, labor, site investigation, monitoring/post-closure maintenance and off-site contaminant disposal.

14.13.2 Second Criterion

While employing remediation processes, the effect on the environment should be less than the impact of untreated land. Different factors can be used to measure and compare the impact of both remediation on the environment and no action taken on contaminated land such as the imminent risks to human health, effects on water flow, air pollution, the pollutants' fate, impacts on flora and fauna, constraints on the future use of land and influences on the landscape/other sites.

14.13.3 Third Criterion

The environmental impact of the remediation process is minimal and measurable. This requires the environmental impact from the remediation process itself to be minimal. This includes emissions into the air, energy use, the use of secondary materials, waste, the direct use of natural resources and the impact of the materials used in the remediation process. All impacts must be measurable.

14.13.4 Fourth Criterion

The timescale over which the environmental effects arise and, thus, the intergenerational risk are part of the decision-making mechanism. Parameters include long-term surveillance, post-closure maintenance, durability, future underground operations, land use problems, long-term depletion of the pollutants and sustainable use of the soil.

14.13.5 Fifth Criterion

The decision-making process includes an appropriate level of engagement of all stakeholders. However, a discussion of the potential consequences of these technologies for stakeholders should be held.

14.14 NANOPHYTOREMEDIATION CHALLENGES

Although there are many known advantages of nanomaterials, research related to nanophytoremediation has not explored very much. Thus, in the future, diverse studies are required to ensure a proper understanding at the field level, to check the concrete effects of NPs in the nanophytoremediation process and to induce changes in the soil fertility status. NP aggregation is one of the main issues that gives rise to their limited mobility, so NP fabrication is a win-win situation to increase their bioavailability. NPs' influence and safety in contaminated soil have to be estimated and evaluated. Meteorological situations are the main factors on which sustainable nanophytoremediation depends; thus, NPs stable in the environment identification are important. Nanophytoremediation is a suitable method for places that have a moderate amount of contaminants due to the unsuitable plant growth in extremely polluted environments. Pollutant uptake in plants from the soil needs better understanding and will also aid agro-mining, which can be utilized for contaminant extraction from plant biomass even before harvesting.

14.15 CONCLUSION AND FUTURE PERSPECTIVES OF NANOPHYTOREMEDIATION

It is a well-known fact that nanophytoremediation for environmental remediation is a new technology, and studies related to nanophytoremediation are usually limited to labs and culture pots, that is, full-scale research has not yet been conducted. Based on their research, Ji et al. (2011) assessed that performing experiments in the field actually provides practical information towards nanophytoremediation technology development that cannot be delivered by tests in the lab. This difference is the result of culture experiments in the lab and field due to factors such as temperature, pH, soil type, nutrients and the moisture content of the soil, which play roles in the actual field environment (Vangronsveld et al.2009). Environmental remediation through nanotechnology can provide eco-friendly and better substitutes and management without harming nature. Some hyperaccumulators have been identified including plants, fungi and bacteria, which have the potential to accumulate a considerable concentration of metals. These hyperaccumulators serve an important function in bioremediation in heavy metal-polluted areas. Different forms of nanomaterials can be employed for environmental clean up, and there is a basic need to understand the process behind the mobility of NPs in the environment to ameliorate any detrimental effects posed to plants or the environment overall. Accordingly, the selection of suitable NPs and plant species for the environment should be the priority, along with agronomic optimization management for high-end remediation processes. With its innovative potential, multi-disciplinary nature, foreseen opportunities and influences from its probable benefits, nanotechnology has been acknowledged as a field of the utmost importance. To find solutions for hunger and increase production, the hope lies in nanoproducts utilized in an eco-friendly manner.

REFERENCES

Adesina, A.A. 2004. Industrial exploitation of photocatalysis: Progress, perspectives and prospects. *Catalysis Surveys from Asia*, 8:265–273.

Ali, H., Khan, E. and Sajad, M.A. 2013. Phytoremediation of heavy metals—Concepts and applications. *Chemosphere*, 91(7):869–881.

Alizadeh Fard, M., Aminzadeh, B. and Vahidi, H. 2013. Degradation of petroleum aromatic hydrocarbons using TiO2 nanopowder film. *Environmental Technology*, 34:1183–1190.

An, X. and Jimmy, C.Y. 2011. Graphene-based photocatalytic composites. *RSC Advances*, 1:1426–1434.

Anjum, N.A., Singh, N., Singh, M.K., Sayeed, I., Duarte, A.C., Pereira, E. and Ahmad, I. 2014. Single-bilayer graphene oxide sheet impacts and underlying potential mechanism assessment in germinating faba bean (*Vicia faba* L.). *Science of the Total Environment*, 472:834–841.

Araujo, J.A. and Nel, A.E. 2009. Particulate matter and atherosclerosis: Role of particle size, composition and oxidative stress. *Particle and Fibre Toxicology*, 6:1–19.

Arencibia, A., Aguado, J. and Arsuaga, J.M. 2010. Regeneration of thiol-functionalized mesostructured silica adsorbents of mercury. *Applied Surface Science*, 256:5453–5457.

Asante-Badu, B., Kgorutla, L.E., Li, S.S., Danso, P.O., Xue, Z. and Qiang, G. 2020. Phytoremediation of organic and inorganic compounds in a natural and an agricultural environment: A review. *Applied Ecology and Environmental Research*, 18(5):6875–6904.

Bao, Q., Zhang, D. and Qi, P. 2011. Synthesis and characterization of silver nanoparticle and graphene oxide nanosheet composites as a bactericidal agent for water disinfection. *Journal of Colloid and Interface Science*, 360:463–470.

Binderup, M.L., Bredsdorff, L., Beltoft, V.M., Mortensen, A., Löschner, K., Larsen, E.H. and Eriksen, F.D. 2013. *Systemic Absorption of Nanomaterials by Oral Exposure: Part of the "Better Control of Nano" Initiative, 2012–2015*. Danish Environmental Protection Agency. Available from: http://www2.mst.dk/Udgiv/publications/2013/09/978-87-93026-51-3.pdf

Blanc, A., Métivier-Pignon, H., Gourdon, R. and Rousseaux, P. 2004. Life cycle assessment as a tool for controlling the development of technical activities: Application to the remediation of a site contaminated by sulfur. *Advances in Environmental Research*, 8:613–627.

Boente, C., Sierra, C., Martínez-Blanco, D., Menéndez-Aguado, J.M. and Gallego, J.R. 2018. Nanoscale zero-valent iron-assisted soil washing for the removal of potentially toxic elements. *Journal of Hazardous Materials*, 350:55–65.

Bollini, P., Didas, S.A. and Jones, C.W. 2011. Amine-oxide hybrid materials for acid gas separations. *Journal of Materials Chemistry*, 21:15100–15120.

Bosetti, M., Massè, A., Tobin, E. and Cannas, M. 2002. Silver coated materials for external fixation devices: In vitro biocompatibility and genotoxicity. *Biomaterials*, 23:887–892.

Bouwmeester, H., Hollman, P.C. and Peters, R.J. 2015. Potential health impact of environmentally released micro- and nanoplastics in the human food production chain: Experiences from nanotoxicology. *Environmental Science & Technology*, 49:8932–8947.

Brigante, M., Pecini, E. and Avena, M. 2016. Magnetic mesoporous silica for water remediation: Synthesis, characterization and application as adsorbent of molecules and ions of environmental concern. *Microporous and Mesoporous Materials*, 230:1–10.

Burken, J.G. and Schnoor, J.L. 1996. Phytoremediation: Plant uptake of atrazine and role of root exudates. *Journal of Environmental Engineering*, 122:958–963.

Chao, C., Xiangyu, W., Chang, Y. and Huiling, L. 2008. Dechlorination of disinfection by-product monochloroacetic acid in drinking water by nanoscale palladized iron bimetallic particle. *Journal of Environmental Sciences*, 20:945–951.

Chen, F., Huber, C., May, R. and Schröder, P. 2016. Metabolism of oxybenzone in a hairy root culture: Perspectives for phytoremediation of a widely used sunscreen agent. *Journal of Hazardous Materials*, 306:230–236.

Chen, X., Cen, C., Tang, Z., Zeng, W., Chen, D., Fang, P. and Chen, Z. 2013. The key role of pH value in the synthesis of titanate nanotubes-loaded manganese oxides as a superior catalyst for the selective catalytic reduction of NO with NH3. *Journal of Nanomaterials*, 2013.

Chi, T., Zuo, J. and Liu, F. 2017. Performance and mechanism for cadmium and lead adsorption from water and soil by corn straw biochar. *Frontiers of Environmental Science & Engineering*, 11(2).

Cho, M., Chung, H., Choi, W. and Yoon, J. 2005. Different inactivation behaviors of MS-2 phage and Escherichia coli in TiO2 photocatalytic disinfection. *Applied and Environmental Microbiology*, 71:270–275.

Chou, K.S., Lu, Y.C. and Lee, H.H. 2005. Effect of alkaline ion on the mechanism and kinetics of chemical reduction of silver. *Materials Chemistry and Physics*, 94:429–433.

Cule, N., Vilotic, D., Nesic, M., Veselinovic, N., Drazic, D. and Mitrovic, S. 2016. Phytoremediation potential of *Canna indica* L. in water contaminated with lead. *Fresenius Environmental Bulletin*, 25(9):3728–3733.

Da Silva, M.B., Abrantes, N., Nogueira, V., Goncalves, F. and Pereira, R. 2016. TiO2 nanoparticles for the remediation of eutrophic shallow freshwater systems: Efficiency and impacts on aquatic biota under a microcosm experiment. *Aquatic Toxicology*, 178:58–71.

Das, R., Hamid, S.B.A., Ali, M.E., Ramakrishna, S. and Yongzhi, W. 2015. Carbon nanotubes characterization by X-ray powder diffraction—A review. *Current Nanoscience*, 11(1):23–35.

Deng, C.H., Gong, J.L., Zhang, P., Zeng, G.M., Song, B. and Liu, H.Y. 2017. Preparation of melamine sponge decorated with silver nanoparticles-modified graphene for water disinfection. *Journal of Colloid and Interface Science*, 488:26–38.

Diallo, M.S., Christie, S., Swaminathan, P., Johnson, J.H. and Goddard, W.A. 2005. Dendrimer enhanced ultrafiltration. 1. Recovery of Cu (II) from aqueous solutions using PAMAM dendrimers with ethylene diamine core and terminal NH2 groups. *Environmental Science & Technology*, 39(5):1366–1377.

Diallo, M.S., Falconer, K., Johnson, J.H. and Goddard III, W.A. 2007. Dendritic anion hosts: Perchlorate uptake by G5-NH2 poly (propyleneimine) dendrimer in water and model electrolyte solutions. *Environmental Science & Technology*, 41:6521–6527.

Dimkpa, C.O., McLean, J.E., Latta, D.E., Manangón, E., Britt, D.W., Johnson, W.P., Boyanov, M.I. and Anderson, A.J. 2012. CuO and ZnO nanoparticles: Phytotoxicity, metal speciation, and induction of oxidative stress in sand-grown wheat. *Journal of Nanoparticle Research*, 14:1–15.

Di Paola, A., García-López, E., Marcì, G. and Palmisano, L. 2012. A survey of photocatalytic materials for environmental remediation. *Journal of Hazardous Materials*, 211:3–29.

Drese, J.H., Talley, A.D. and Jones, C.W. 2011. Aminosilica materials as adsorbents for the selective removal of aldehydes and ketones from simulated bio-oil. *ChemSusChem*, 4:379–385.

Duan, L., Lin, Q., Peng, H., Lu, C., Shao, C., Wang, D., Rao, S., Cao, H. and Lv, W. 2023. The catalytic reduction mechanisms of metal-doped TiO2 for nitrate produced from non-thermal discharge plasma: The interfacial photogenerated electron transfer and reduction process. *Applied Catalysis A: General*, 650:118995.

Ebrahim, S.E., Sulaymon, A.H. and Saad Alhares, H. 2016. Competitive removal of Cu2+, Cd2+, Zn2+, and Ni2+ ions onto iron oxide nanoparticles from wastewater. *Desalination and Water Treatment*, 57:20915–20929.

Echereme, C.B., Igboabuchi, N.A. and Izundu, A.I. 2018. Phytoremediation of heavy metals and persistent organic pollutants (POPs): A review. *IJSRM Human*, 10(4):107–125.

El-Temsah, Y.S. and Joner, E.J. 2013. Effects of nano-sized zero-valent iron (nZVI) on DDT degradation in soil and its toxicity to collembola and ostracods. *Chemosphere*, 92(1):131–137.

Eskander, S. and Saleh, H. 2017. Phytoremediation: An overview. In *Environmental Science and Engineering, Soil Pollution and Phytoremediation* (Vol. 11, 1st edn., pp. 124–161). Studium Press LLC, Houston, TX.

Fan, D., Gilbert, E.J. and Fox, T. 2017. Current state of in situ subsurface remediation by activated carbon-based amendments. *Journal of Environmental Management*, 204(2):793–803.

Gao, M., Zhang, Z., Lv, M., Song, W. and Lv, Y. 2018. Toxic effects of nanomaterial-adsorbed cadmium on Daphnia magna. *Ecotoxicology and Environmental Safety*, 148:261–268.

Gei, F., Li, M.M., Ye, H. and Zhao, B.X. 2012. Effective removal of heavy metal ions Cd2+, Zn2+, Pb2+, Cu2+ from aqueous solution by polymer-modified magnetic nanoparticles. *Journal of Hazardous Materials*, 211:366–372.

Gerhard, J.I., Grant, G.P. and Torero, J.L. 2019. Star: A uniquely sustainable in situ and ex situ remediation process. In *Sustainable Remediation of Contaminated Soil and Groundwater: Materials, Processes, and Assessment* (pp. 221–246). Butterworth-Heinemann, Woburn.

Ghenaatian, H.R., Baei, M.T. and Hashemian, S. 2013. Zn12O12 nano-cage as a promising adsorbent for CS2 capture. *Superlattices and Microstructures*, 58:198–204.

Godt, J., Scheidig, F., Grosse-Siestrup, C., Esche, V., Brandenburg, P., Reich, A. and Groneberg, D.A. 2006. The toxicity of cadmium and resulting hazards for human health. *Journal of Occupational Medicine and Toxicology*, 1(1): 22–27.

Gogoi, S.K., Gopinath, P., Paul, A., Ramesh, A., Ghosh, S.S. and Chattopadhyay, A. 2006. Green fluorescent protein-expressing escherichia coli as a model system for investigating the antimicrobial activities of silver nanoparticles. *Langmuir*, 22:9322–9328.

Gong, X., Huang, D., Liu, Y., Zeng, G., Wang, R., Wan, J., Zhang, C., Cheng, M., Qin, X. and Xue, W. 2017. Stabilized nanoscale zerovalent iron mediated cadmium accumulation and oxidative damage of *Boehmeria nivea* (L.) Gaudich cultivated in cadmium contaminated sediments. *Environmental Science & Technology*, 51(19):11308–11316.

Grieger, K.D., Fjordbøge, A., Hartmann, N.B., Eriksson, E., Bjerg, P.L. and Baun, A. 2010. Environmental benefits and risks of zero-valent iron nanoparticles (nZVI) for in situ remediation: Risk mitigation or trade-off?. *Journal of Contaminant Hydrology*, 118:165–183.

Griffiths, D.R. 2019. Controlling secondary pollution impacts during enhanced in-situ anaerobic bioremediation. In *Sustainable Remediation of Contaminated Soil and Groundwater: Materials, Processes, and Assessment* (pp. 201–220). Butterworth-Heinemann, Woburn.

Gu, J., Dong, D., Kong, L., Zheng, Y. and Li, X. 2012. Photocatalytic degradation of phenanthrene on soil surfaces in the presence of nanometer anatase TiO2 under UV-light. *Journal of Environmental Sciences*, 24(12):2122–2126.

Guerra, F.D., Attia, M.F., Whitehead, D.C. and Alexis, F., 2018a. Nanotechnology for environmental remediation: Materials and applications. *Molecules*, 23(7):1760.

Guerra, F.D., Campbell, M.L., Attia, M.F., Whitehead, D.C. and Alexis, F. 2018b. Capture of aldehyde VOCs using a series of amine-functionalized cellulose nanocrystals. *ChemistrySelect*, 3:5495–5501.

Guo, M., Weng, X., Wang, T. and Chen, Z. 2017. Biosynthesized iron-based nanoparticles used as a heterogeneous catalyst for the removal of 2, 4-dichlorophenol. *Separation and Purification Technology*, 175:222–228.

Gupta, A. and Silver, S. 1998. Molecular genetics: Silver as a biocide: Will resistance become a problem? *Nature Biotechnology*, 16:888–888.

Han, Y. and Yan, W. 2016. Reductive dechlorination of trichloroethene by zero-valent iron nanoparticles: Reactivity enhancement through sulfidation treatment. *Environmental Science & Technology*, 50:12992–13001.

Harbottle, M.J., Al-Tabbaa, A. and Evans, C.W. 2005. *The Technical Sustainability of In-Situ Stabilisation/ Solidification Department of Engineering*. University of Cambridge, Cambridge.

Harbottle, M.J., Al-Tabbaa, A. and Evans, C.W. 2008. Sustainability of land remediation: Part I: Overall analysis. *Proceedings of the Institution of Civil Engineers-Geotechnical Engineering*, 161:75–92.

Hernández-Morales, V., Nava, R., Acosta-Silva, Y.J., Macías-Sánchez, S.A., Pérez-Bueno, J.J. and Pawelec, B. 2012. Adsorption of lead (II) on SBA-15 mesoporous molecular sieve functionalized with—NH2 groups. *Microporous and Mesoporous Materials*, 160:133–142.

Higgins, M.R. and Olson, T.M. 2009. Life-cycle case study comparison of permeable reactive barrier versus pump-and-treat remediation. *Environmental Science & Technology*, 43:9432–9438.

Hochella Jr, M.F., Moore, J.N., Putnis, C.V., Putnis, A., Kasama, T. and Eberl, D.D. 2005. Direct observation of heavy metal-mineral association from the Clark Fork River Superfund Complex: Implications for metal transport and bioavailability. *Geochimica et Cosmochimica Acta*, 69:1651–1663.

Holling, C.S. 2001. Understanding the complexity of economic, ecological, and social systems. *Ecosystems*, 4:390–405.

Hooshyar, Z., Rezanejade Bardajee, G. and Ghayeb, Y. 2013. Sonication enhanced removal of nickel and cobalt ions from polluted water using an iron based sorbent. *Journal of Chemistry*, 2013.

Hou, D. and Al-Tabbaa, A. 2014. Sustainability: A new imperative in contaminated land remediation. *Environmental Science & Policy*, 39:25–34.

Huang, D., Qin, X., Peng, Z., Liu, Y., Gong, X., Zeng, G. and Hu, Z. 2018. Nanoscale zero-valent iron assisted phytoremediation of Pb in sediment: Impacts on metal accumulation and antioxidative system of *Lolium perenne*. *Ecotoxicology and Environmental Safety*, 153:229–237.

Huang, H.Y., Yang, R.T., Chinn, D. and Munson, C.L. 2003. Amine-grafted MCM-48 and silica xerogel as superior sorbents for acidic gas removal from natural gas. *Industrial & Engineering Chemistry Research*, 42:2427–2433.

Ifang, S., Gallus, M., Liedtke, S., Kurtenbach, R., Wiesen, P. and Kleffmann, J. 2014. Standardization methods for testing photo-catalytic air remediation materials: Problems and solution. *Atmospheric Environment*, 91:154–161.

Ji, P., Sun, T., Song, Y., Ackland, M.L. and Liu, Y. 2011. Strategies for enhancing the phytoremediation of cadmium-contaminated agricultural soils by Solanum nigrum L. *Environmental Pollution*, 159:762–768.

Jiamjitrpanich, W., Parkpian, P., Polprasert, C. and Kosanlavit, R. 2012, June. Enhanced phytoremediation efficiency of TNT-contaminated soil by nanoscale zero valent iron. In *2nd International Conference on Environment and Industrial Innovation IPCBEE* (pp. 82–86). IACSIT Press, Singapore. Available from: https://www.researchgate.net/publication/229090834_Enhanced_Phytoremediation_Efficiency_of_TNT-Contaminated_Soil_by_Nanoscale_Zero_Valent_Iron

Jin, F. 2019. Long-term effectiveness of in situ solidification/stabilization. In *Sustainable Remediation of Contaminated Soil and Groundwater: Materials, Processes, and Assessment* (pp. 247–278). Butterworth-Heinemann, Woburn.

Jin, Y., Liu, W., Li, X.-L., Shen, S.-G., Liang, S.-X., Liu, C., et al. 2016. Nano-hydroxyapatite immobilized lead and enhanced plant growth of ryegrass in a contaminated soil. *Ecological Engineering*, 95:25–29.

Kamat, P.V. and Meisel, D. 2003. Nanoscience opportunities in environmental remediation. *Comptes Rendus Chimie*, 6:999–1007.

Karn, B., Kuiken, T. and Otto, M. 2009. Nanotechnology and in situ remediation: A review of the benefits and potential risks. *Environmental Health Perspectives*, 117:1823–1831.

Khalid, S., Shahid, M., Niazi, N.K., Murtaza, B., Bibi, I. and Dumat, C. 2017. A comparison of technologies for remediation of heavy metal contaminated soils. *Journal of Geochemical Exploration*, 182:247–268.

Khan, A. 2020. Promises and potential of in situ nano-phytoremediation strategy to mycorrhizoremediatie heavy metal contaminated soils using non-food bioenergy crops (*Vetiver zizinoides & Cannabis sativa*). *International Journal of Phytoremediation*, 22(12):900–915.

Khan, N. and Bano, A. 2016. Role of plant growth promoting rhizobacteria and Ag-nano particle in the bioremediation of heavy metals and maize growth under municipal wastewater irrigation. *International Journal of Phytoremediation*, 18:211–221.

Khare, P., Yadav, A., Ramkumar, J. and Verma, N. 2016. Microchannel-embedded metal—carbon—polymer nanocomposite as a novel support for chitosan for efficient removal of hexavalent chromium from water under dynamic conditions. *Chemical Engineering Journal*, 293:44–54.

Kharisov, B.I., Dias, H.R. and Kharissova, O.V. 2014. Nanotechnology-based remediation of petroleum impurities from water. *Journal of Petroleum Science and Engineering*, 122:705–718.

Khodakovskaya, M.V., De Silva, K., Biris, A.S., Dervishi, E. and Villagarcia, H. 2012. Carbon nanotubes induce growth enhancement of tobacco cells. *ACS Nano*, 6:2128–2135.

Kirkham, M.B. 2006. Cadmium in plants on polluted soils: Effects of soil factors, hyperaccumulation, and amendments. *Geoderma*, 137:19–32.

Kouketsu, T., Duan, S., Kai, T., Kazama, S. and Yamada, K. 2007. PAMAM dendrimer composite membrane for CO2 separation: Formation of a chitosan gutter layer. *Journal of Membrane Science*, 287:51–59.

Kumar, A., Chaturvedi, A.K., Yadav, K., Arunkumar, K.P., Malyan, S.K., Raja, P., Kumar, R., Khan, S.A., Yadav, K.K., Rana, K.L. and Kour, D. 2019. Fungal phytoremediation of heavy metal-contaminated resources: Current scenario and future prospects. In *Recent Advancement in White Biotechnology through Fungi: Volume 3: Perspective for Sustainable Environments* (pp. 437–461). Springer, Cham.

Kumar, V., Shahi, K. and Singh, S. 2018. Bioremediation: An eco-sustainable approach for restoration of contaminated sites. In *Microbial Bioprospecting for Sustainable Development* (pp. 115–136). Springer, Singapore.

Lebrun, M., Miard, F., Nandillon, R., Léger, J.C., Hattab-Hambli, N., Scippa, G.S., Bourgerie, S. and Morabito, D. 2018. Assisted phytostabilization of a multicontaminated mine technosol using biochar amendment: Early stage evaluation of biochar feedstock and particle size effects on As and Pb accumulation of two Salicaceae species (*Salix viminalis* and *Populus euramericana*). *Chemosphere*, 194:316–326.

Lemming, G., Hauschild, M.Z., Chambon, J., Binning, P.J., Bulle, C., Margni, M. and Bjerg, P.L. 2010. Environmental impacts of remediation of a trichloroethene-contaminated site: Life cycle assessment of remediation alternatives. *Environmental Science & Technology*, 44(23):9163–9169.

Leu, J. and Hou, D. 2019. Remedial process optimization and sustainability benefits. In *Sustainable Remediation of Contaminated Soil and Groundwater Materials, Processes, and Assessment* (pp. 279–30). Butterworth-Heinemann, Woburn.

Li, D., Cui, F., Zhao, Z., Liu, D., Xu, Y., Li, H. and Yang, X. 2014. The impact of titanium dioxide nanoparticles on biological nitrogen removal from wastewater and bacterial community shifts in activated sludge. *Biodegradation*, 25:167–177.

Li, X.Q. and Zhang, W.X. 2006. Iron nanoparticles: The core– shell structure and unique properties for Ni (II) sequestration. *Langmuir*, 22:4638–4642.

Li, Y., Zhang, P., Du, Q., Peng, X., Liu, T., Wang, Z., Xia, Y., Zhang, W., Wang, K., Zhu, H., et al. 2011. Adsorption of fluoride from aqueous solution by graphene. *Journal of Colloid and Interface Science*, 363:348–354.

Li, Y.H., Wang, S., Cao, A., Zhao, D., Zhang, X., Xu, C., Luan, Z., Ruan, D., Liang, J., Wu, D. and Wei, B. 2001. Adsorption of fluoride from water by amorphous alumina supported on carbon nanotubes. *Chemical Physics Letters*, 350:412–416.

Liang, D. and Wang, S. 2017. Development and characterization of an anaerobic microcosm for reductive dechlorination of PCBs. *Frontiers of Environmental Science & Engineering*, 11(6).

Liang, S.X., Jin, Y., Liu, W., Li, X., Shen, S.G. and Ding, L. 2017. Feasibility of Pb phytoextraction using nano-materials assisted ryegrass: Results of a one-year field-scale experiment. *Journal of Environmental Management*, 190:170–175.

Lien, H.L. and Zhang, W.X. 2005. Hydrodechlorination of chlorinated ethanes by nanoscale Pd/Fe bimetallic particles. *Journal of Environmental Engineering*, 131:4–10.

Limmer, M. and Burken, J. 2016. Phytovolatilization of organic contaminants. *Environmental Science and Technology*, 50(13):6632–6643.

Liou, S.H., Tsou, T.C., Wang, S.L., Li, L.A., Chiang, H.C., Li, W.F., Lin, P.P., Lai, C.H., Lee, H.L., Lin, M.H. and Hsu, J.H. 2012. Epidemiological study of health hazards among workers handling engineered nanomaterials. *Journal of Nanoparticle Research*, 14:1–15.

Liu, X., Pan, L., Lv, T., Lu, T., Zhu, G., Sun, Z. and Sun, C. 2011. Microwave-assisted synthesis of ZnO—graphene composite for photocatalytic reduction of Cr (VI). *Catalysis Science & Technology*, 1:1189–1193.

Lu, C. and Chiu, H. 2006. Adsorption of zinc (II) from water with purified carbon nanotubes. *Chemical Engineering Science*, 61:1138–1145.

Ma, G. and Wang, L. 2018. A critical review of preparation design and workability measurement of concrete material for largescale 3D printing. *Frontiers of Structural and Civil Engineering*, 12:382–400.

Ma, J., Mercer, R.R., Barger, M., Schwegler-Berry, D., Cohen, J.M., Demokritou, P. and Castranova, V. 2015. Effects of amorphous silica coating on cerium oxide nanoparticles induced pulmonary responses. *Toxicology and Applied Pharmacology*, 288:63–73.

Ma, X., Geiser-Lee, J., Deng, Y. and Kolmakov, A. 2010. Interactions between engineered nanoparticles (ENPs) and plants: Phytotoxicity, uptake and accumulation. *Science of the Total Environment*, 408:3053–3061.

Ma, X. and Wang, C. 2010. Fullerene nanoparticles affect the fate and uptake of trichloroethylene in phytoremediation systems. *Environmental Engineering Science*, 27:989–992.

Mahabadi, A.A., Hajabbasi, M.A., Khademi, H. and Kazemian, H. 2007. Soil cadmium stabilization using an Iranian natural zeolite. *Geoderma*, 137:388–393.

Mani, D. and Kumar, C. 2014. Biotechnological advances in bioremediation of heavy metals contaminated ecosystems: An overview with special reference to phytoremediation. *International Journal of Environmental Science and Technology*, 11(3):843–872.

Mauter, M.S. and Elimelech, M. 2008. Environmental applications of carbon-based nanomaterials. *Environmental Science & Technology*, 42:5843–5859.

Meldrum, K., Guo, C., Marczylo, E.L., Gant, T.W., Smith, R. and Leonard, M.O. 2017. Mechanistic insight into the impact of nanomaterials on asthma and allergic airway disease. *Particle and Fibre Toxicology*, 14:1–35.

Misra, S. and Misra, K.G. 2019. Phytoremediation: An alternative tool towards clean and green environment. In *Sustainable Green Technologies for Environmental Management* (pp. 87–109). Springer, Amsterdam.

Mittal, H., Maity, A. and Ray, S.S. 2015. Synthesis of co-polymer-grafted gum karaya and silica hybrid organic—inorganic hydrogel nanocomposite for the highly effective removal of methylene blue. *Chemical Engineering Journal*, 279:166–179.

Moameri, M. and Abbasi Khalaki, M. 2019. Capability of *Secale montanum* trusted for phytoremediation of lead and cadmium in soils amended with nano-silica and municipal solid waste compost. *Environmental Science and Pollution Research International*, 26(24):24315–24322.

Mueller, N.C. and Nowack, B. 2010. Nanoparticles for remediation: Solving big problems with little particles. *Elements*, 6:395–400.

Nagpal, V., Bokare, A.D., Chikate, R.C., Rode, C.V. and Paknikar, K.M. 2010. Reductive dechlorination of gamma-hexachlorocyclohexane using Fe-Pd bimetallic nanoparticles. *Journal of Hazardous Materials*, 175(1–3):680–687.

Nakanishi, K., Tomita, M. and Kato, K. 2015. Synthesis of amino-functionalized mesoporous silica sheets and their application for metal ion capture. *Journal of Asian Ceramic Societies*, 3:70–76.

Nikolić, M. and Stevović, S. 2015. Family Asteraceae as a sustainable planning tool in phytoremediation and its relevance in urban areas. *Urban Forestry & Urban Greening*, 14:782–789.

Nomura, A. and Jones, C.W. 2013. Amine-functionalized porous silicas as adsorbents for aldehyde abatement. *ACS Applied Materials & Interfaces*, 5:5569–5577.

Nomura, A. and Jones, C.W. 2014. Enhanced formaldehyde-vapor adsorption capacity of polymeric amine-incorporated aminosilicas. *Chemistry—A European Journal*, 20:6381–6390.

Otto, M., Floyd, M. and Bajpai, S. 2008. Nanotechnology for site remediation. *Remediation Journal: The Journal of Environmental Cleanup Costs, Technologies & Techniques*, 19:99–108.

Pal, S., Tak, Y.K. and Song, J.M. 2007. Does the antibacterial activity of silver nanoparticles depend on the shape of the nanoparticle? A study of the gram-negative bacterium Escherichia coli. *Applied and Environmental Microbiology*, 73:1712–1720.

Park, J.Y. and Lee, I.H. 2014. Photocatalytic degradation of 2-chlorophenol using Ag-doped TiO2 nanofibers and a near-UV light-emitting diode system. *Journal of Nanomaterials*, 2014.

Parmar, S. and Singh, V. 2015. Phytoremediation approaches for heavy metal pollution: A review. *Journal of Plant Science & Research*, 2(2):135–147.

Pillai, H.P. and Kottekottil, J. 2016. Nano-phytotechnological remediation of endosulfan using zero valent iron nanoparticles. *Journal of Environmental Protection*, 7:734.

Pino, F., Mayorga-Martinez, C.C. and Merkoçi, A. 2017. Nanomaterials-based platforms for environmental monitoring. *Comprehensive Analytical Chemistry*, 77:207–236.

Poguberović, S.S., Krčmar, D.M., Maletić, S.P., Kónya, Z., Pilipović, D.D.T., Kerkez, D.V. and Rončević, S.D. 2016. Removal of As (III) and Cr (VI) from aqueous solutions using "green" zero-valent iron nanoparticles produced by oak, mulberry and cherry leaf extracts. *Ecological Engineering*, 90:42–49.

Pradhan, S., Patra, P., Das, S., Chandra, S., Mitra, S., Dey, K.K., Akbar, S., Palit, P. and Goswami, A. 2013. Photochemical modulation of biosafe manganese nanoparticles on Vigna radiata: A detailed molecular, biochemical, and biophysical study. *Environmental Science & Technology*, 47:13122–13131.

Pradhan, S., Patra, P., Mitra, S., Dey, K.K., Jain, S., Sarkar, S., Roy, S., Palit, P. and Goswami, A. 2014. Manganese nanoparticles: Impact on non-nodulated plant as a potent enhancer in nitrogen metabolism and toxicity study both in vivo and in vitro. *Journal of Agricultural and Food Chemistry*, 62:8777–8785.

Prata, J.C. 2018. Airborne microplastics: Consequences to human health? *Environmental Pollution*, 234:115–126.

Qi, G., Wang, Y., Estevez, L., Duan, X., Anako, N., Park, A.H.A., Li, W., Jones, C.W. and Giannelis, E.P. 2011. High efficiency nanocomposite sorbents for CO2 capture based on amine-functionalized mesoporous capsules. *Energy & Environmental Science*, 4:444–452.

Qu, X., Alvarez, P.J. and Li, Q. 2013. Applications of nanotechnology in water and wastewater treatment. *Water Research*, 47:3931–3946.

Quan, C., Sun, Q., Lippmann, M. and Chen, L.C. 2010. Comparative effects of inhaled diesel exhaust and ambient fine particles on inflammation, atherosclerosis, and vascular dysfunction. *Inhalation Toxicology*, 22:738–753.

Rasalingam, S., Peng, R. and Koodali, R.T. 2014. Removal of hazardous pollutants from wastewaters: Applications of TiO2-SiO2 mixed oxide materials. *Journal of Nanomaterials*, 2014.

Reeves, R.D., Baker, A.J.M., Jaffré, T., Erskine, P.D., Echevarria, G. and van der Ent, A. 2018. A global database for plants that hyperaccumulate metal and metalloid trace elements. *New Phytologist*, 218(2):407–411.

Ren, X., Chen, C., Nagatsu, M. and Wang, X. 2011. Carbon nanotubes as adsorbents in environmental pollution management: A review. *Chemical Engineering Journal*, 170:395–410.

Rengaraj, S. and Li, X.Z. 2006. Enhanced photocatalytic activity of TiO2 by doping with Ag for degradation of 2, 4, 6-trichlorophenol in aqueous suspension. *Journal of Molecular Catalysis A: Chemical*, 243:60–67.

Repo, E., Warchoł, J.K., Bhatnagar, A., Mudhoo, A. and Sillanpää, M. 2013. Aminopolycarboxylic acid functionalized adsorbents for heavy metals removal from water. *Water Research*, 47:4812–4832.

Roberts, A. 2003. Plasmodesmata and the control of symplastic transport. *Plant, Cell and Environment*, 26:103–124.

Rogers, K.R., 2006. Recent advances in biosensor techniques for environmental monitoring. *Analytica Chimica Acta*, 568(1–2):222–231.

Rostamian, R., Najafi, M. and Rafati, A.A. 2011. Synthesis and characterization of thiol-functionalized silica nano hollow sphere as a novel adsorbent for removal of poisonous heavy metal ions from water: Kinetics, isotherms and error analysis. *Chemical Engineering Journal*, 171:1004–1011.

Saha, P., Shinde, O. and Sarkar, S. 2017. Phytoremediation of industrial mines wastewater using water hyacinth. *International Journal of Phytoremediation*, 19:87–96.

Sajid, M., Ilyas, M., Basheer, C., Tariq, M., Daud, M., Baig, N. and Shehzad, F. 2015. Impact of nanoparticles on human and environment: Review of toxicity factors, exposures, control strategies, and future prospects. *Environmental Science and Pollution Research*, 22:4122–4143.

Santhosh, C., Velmurugan, V., Jacob, G., Jeong, S.K., Grace, A.N. and Bhatnagar, A. 2016. Role of nanomaterials in water treatment applications: A review. *Chemical Engineering Journal*, 306:1116–1137.

Sarwar, N., Imran, M., Shaheen, M.R., Ishaque, W., Kamran, M.A., Matloob, A., Rehim, A. and Hussain, S. 2017. Phytoremediation strategies for soils contaminated with heavy metals: Modifications and future perspectives. *Chemosphere*, 171:710–721.

Sattelmacher, B. 2001. The apoplast and its significance for plant mineral nutrition. *New Phytologist*, 149:167–192.

Schwab, F., Zhai, G., Kern, M., Turner, A., Schnoor, J.L. and Wiesner, M.R. 2016. Barriers, pathways and processes for uptake, translocation and accumulation of nanomaterials in plants—Critical review. *Nanotoxicology*, 10:257–278.

Seredych, M. and Bandosz, T.J. 2007. Removal of ammonia by graphite oxide via its intercalation and reactive adsorption. *Carbon (New York, NY)*, 45:2130–2132.

Singh, B. and Singh, K. 2016. Microbial degradation of herbicides. *Critical Reviews in Microbiology*, 42(2):245–261.

Singh, H.P., Kaur, S., Batish, D.R., Sharma, V.P., Sharma, N. and Kohli, R.K. 2009. Nitric oxide alleviates arsenic toxicity by reducing oxidative damage in the roots of *Oryza sativa* (rice). *Nitric Oxide*, 20:289–297.

Singh, J. and Lee, B.K. 2016. Influence of nano-TiO2 particles on the bioaccumulation of Cd in soybean plants (Glycine max): A possible mechanism for the removal of Cd from the contaminated soil. *Journal of Environmental Management*, 170:88–96.

Singh, R., Singh, S., Parihar, P., Singh, V.P. and Prasad, S.M. 2015. Arsenic contamination, consequences and remediation techniques: A review. *Ecotoxicology and Environmental Safety*, 112:247–270.

Sly, P.D. and Schüepp, K. 2011. Nanoparticles and children's lungs: Is there a need for caution?. *Paediatric Respiratory Reviews*, 13:71–72.

Soler, L. and Sánchez, S. 2014. Catalytic nanomotors for environmental monitoring and water remediation. *Nanoscale*, 6:7175–7182.

Song, B., Xu, P., Chen, M., Tang, W., Zeng, G., Gong, J., . . . and Ye, S. 2019. Using nanomaterials to facilitate the phytoremediation of contaminated soil. *Critical Reviews in Environmental Science and Technology*, 49(9):791–824.

Souri, Z., Karimi, N., Sarmadi, M. and Rostami, E. 2017. Salicylic acid nanoparticles (SANPs) improve growth and phytoremediation efficiency of Isatis cappadocica Desv., under As stress. *IET Nanobiotechnology*, 11:650–655.

Srisitthiratkul, C., Pongsorrarith, V. and Intasanta, N. 2011. The potential use of nanosilver-decorated titanium dioxide nanofibers for toxin decomposition with antimicrobial and self-cleaning properties. *Applied Surface Science*, 257:8850–8856.

Subramanian, K.S., Manikandan, A., Thirunavukkarasu, M. and Rahale, C.S. 2015. Nano-fertilizers for balanced crop nutrition. In *Nanotechnologies in Food and Agriculture* (pp. 69–80). Springer, Cham.

Sugunan, A., Thanachayanont, C., Dutta, J. and Hilborn, J.G. 2005. Heavy-metal ion sensors using chitosan-capped gold nanoparticles. *Science and Technology of Advanced Materials*, 6:335–340.

Tee, Y.H., Grulke, E. and Bhattacharyya, D. 2005. Role of Ni/Fe nanoparticle composition on the degradation of trichloroethylene from water. *Industrial & Engineering Chemistry Research*, 44:7062–7070.

Theron, J., Walker, J.A. and Cloete, T.E. 2008. Nanotechnology and water treatment: Applications and emerging opportunities. *Critical Reviews in Microbiology*, 34:43–69.

Thines, R.K., Mubarak, N.M., Nizamuddin, S., Sahu, J.N., Abdullah, E.C. and Ganesan, P. 2017. Application potential of carbon nanomaterials in water and wastewater treatment: A review. *Journal of the Taiwan Institute of Chemical Engineers*, 72:116–133.

Tratnyek, P.G. and Johnson, R.L. 2006. Nanotechnologies for environmental cleanup. *Nano Today*, 1:44–48.

Tripathi, D.K., Singh, V.P., Prasad, S.M., Chauhan, D.K. and Dubey, N.K. 2015. Silicon nanoparticles (SiNp) alleviate chromium (VI) phytotoxicity in Pisum sativum (L.) seedlings. *Plant Physiology and Biochemistry*, 96:189–198.

Tripathi, S., Srivastava, P., Devi, R.S. and Bhadouria, R. 2020. Influence of synthetic fertilizers and pesticides on soil health and soil microbiology. In *Agrochemicals Detection, Treatment and Remediation: Pesticides and Chemical Fertilizers* (pp. 25–54). Butterworth-Heinemann, Woburn.

Tsai, C.H., Chang, W.C., Saikia, D., Wu, C.E. and Kao, H.M. 2016. Functionalization of cubic mesoporous silica SBA-16 with carboxylic acid via one-pot synthesis route for effective removal of cationic dyes. *Journal of Hazardous Materials*, 309:236–248.

Tungittiplakorn, W., Lion, L.W., Cohen, C. and Kim, J.Y., 2004. Engineered polymeric nanoparticles for soil remediation. *Environmental Science & Technology*, 38:1605–1610.

USEPA. 2008. *Green Remediation: Incorporating Sustainable Environmental Practices into Remediation of Contaminated Sites*. EPA 542-R-08-002. United States Environmental Protection Agency (USEPA), Washington, DC.

van den Bosch, M. and Sang, A.O. 2017. Urban natural environments as nature-based solutions for improved public health—A systematic review of reviews. *Environmental Research*, 158:373–384.

van der Ent, A., Baker, A.J.M., Reeves, R.D., Pollard, A.J. and Schat, H. 2013. Hyperaccumulators of metal and metalloid trace elements: Facts and fiction. *Plant and Soil*, 362:319–334.

Vangronsveld, J., Herzig, R., Weyens, N., Boulet, J., Adriaensen, K., Ruttens, A., Thewys, T., Vassilev, A., Meers, E., Nehnevajova, E. and van der Lelie, D. 2009. Phytoremediation of contaminated soils and groundwater: Lessons from the field. *Environmental Science and Pollution Research*, 16:765–794.

Varma, A. and Khanuja, M. 2017. Role of nanoparticles on plant growth with special emphasis on Piriformospora indica: A review. *Nanoscience and Plant—Soil Systems*, 387–403.

Vijaya, J.J., Adinaveen, T. and Bououdina, M., 2018. Economic aspects of functionalized nanomaterials for environment. In *Nanotechnology in Environmental Science*, ed. Hussain, C.M. and Mishra, A.K. Wiley-VCH Verlag GmbH & Co. KGaA, Weinheim.

Villalobos, M., Toner, B., Bargar, J. and Sposito, G. 2003. Characterization of the manganese oxide produced by Pseudomonas putida strain MnB1. *Geochimica et Cosmochimica Acta*, 67:2649–2662.

Vithanage, M., Dabrowska, B.B., Mukherjee, A.B., Sandhi, A. and Bhattacharya, P. 2012. Arsenic uptake by plants and possible phytoremediation applications: A brief overview. *Environmental Chemistry Letters*, 10:217–224.

Vítková, M., Puschenreiter, M. and Komárek, M. 2018. Effect of nano zero-valent iron application on As, Cd, Pb, and Zn availability in the rhizosphere of metal (loid) contaminated soils. *Chemosphere*, 200:217–226.

Wang, D., Pillai, S.C., Ho, S.H., Zeng, J., Li, Y. and Dionysiou, D.D. 2018. Plasmonic-based nanomaterials for environmental remediation. *Applied Catalysis B: Environmental*, 237:721–741.

Wang, S., Sun, H., Ang, H.M. and Tadé, M.O. 2013. Adsorptive remediation of environmental pollutants using novel graphene-based nanomaterials. *Chemical Engineering Journal*, 226:336–347.

Wang, S., Wang, K., Dai, C., Shi, H. and Li, J. 2015. Adsorption of Pb2+ on amino-functionalized core—shell magnetic mesoporous SBA-15 silica composite. *Chemical Engineering Journal*, 262:897–903.

Werkneh, A.A. and Rene, E.R. 2019. Applications of nanotechnology and biotechnology for sustainable water and wastewater treatment. In *Water and Wastewater Treatment Technologies*, ed. Bui, X.T., Chiemchaisri, C., Fujioka, T. and Varjani, S. (pp. 405–430). Springer, Amsterdam.

Xiao, W., Li, D., Ye, X., Xu, H., Yao, G., Wang, J., Zhang, Q., Hu, J. and Gao, N. 2017. Enhancement of Cd phytoextraction by hyperaccumulator *Sedum alfredii* using electrical field and organic amendments. *Environmental Science and Pollution Research*, 24:5060–5067.

Xie, Y., Fang, Z., Cheng, W., Tsang, P.E. and Zhao, D. 2014. Remediation of polybrominated diphenyl ethers in soil using Ni/Fe bimetallic nanoparticles: Influencing factors, kinetics and mechanism. *Science of the Total Environment*, 485:363–370.

Xiong, Z., Ma, J., Ng, W.J., Waite, T.D. and Zhao, X.S. 2011. Silver-modified mesoporous TiO2 photocatalyst for water purification. *Water Research*, 45:2095–2103.

Yadav, K.K., Gupta, N., Kumar, A., Reece, L.M., Singh, N., Rezania, S. and Khan, S.A. 2018. Mechanistic understanding and holistic approach of phytoremediation: A review on application and future prospects. *Ecological Engineering*, 120:274–298.

Yadav, K.K., Singh, J.K., Gupta, N. and Kumar, V. 2017. A review of nanobioremediation technologies for environmental cleanup: A novel biological approach. *Journal of Materials and Environmental Sciences*, 8(2):740–757.

Yan, Z., Tao, S., Yin, J. and Li, G. 2006. Mesoporous silicas functionalized with a high density of carboxylate groups as efficient absorbents for the removal of basic dyestuffs. *Journal of Materials Chemistry*, 16:2347–2353.

Yang, J., Cao, W. and Rui, Y. 2017. Interactions between nanoparticles and plants: Phytotoxicity and defense mechanisms. *Journal of Plant Interactions*, 12:158–169.

Yang, M.Q., Zhang, N. and Xu, Y.J. 2013a. Synthesis of fullerene—, carbon nanotube—, and graphene—TiO2 nanocomposite photocatalysts for selective oxidation: A comparative study. *ACS Applied Materials & Interfaces*, 5:1156–1164.

Yew, S.P., Tang, H.Y. and Sudesh, K. 2006. Photocatalytic activity and biodegradation of polyhydroxybutyrate films containing titanium dioxide. *Polymer Degradation and Stability*, 91:1800–1807.

Yildiz, O. and Bradford, P.D. 2013. Carbon aligned carbon nanotube sheet high efficiency particulate air filters. *Carbon*, 64:295–304.

Younas, H., Qazi, I.A., Hashmi, I., Awan, M.A., Mahmood, A. and Qayyum, H.A. 2014. Visible light photocatalytic water disinfection and its kinetics using Ag-doped titania nanoparticles. *Environmental Science and Pollution Research*, 21:740–752.

Yunus, I.S., Harwin, Kurniawan, A., Adityawarman, D. and Indarto, A., 2012. Nanotechnologies in water and air pollution treatment. *Environmental Technology Reviews*, 1:136–148.

Zhang, N., Yang, M.Q., Tang, Z.R. and Xu, Y.J. 2013. CdS—graphene nanocomposites as visible light photocatalyst for redox reactions in water: A green route for selective transformation and environmental remediation. *Journal of Catalysis*, 303:60–69.

Zhang, Y., Tang, Z.R., Fu, X. and Xu, Y.J. 2010a. TiO2– graphene nanocomposites for gas-phase photocatalytic degradation of volatile aromatic pollutant: Is TiO2– graphene truly different from other TiO2– carbon composite materials? *ACS Nano*, 4:7303–7314.

Zhu, N., Luan, H., Yuan, S., Chen, J., Wu, X. and Wang, L. 2010. Effective dechlorination of HCB by nanoscale Cu/Fe particles. *Journal of Hazardous Materials*, 176:1101–1105.

Züttel, A., Sudan, P., Mauron, P., Kiyobayashi, T., Emmenegger, C. and Schlapbach, L. 2002. Hydrogen storage in carbon nanostructures. *International Journal of Hydrogen Energy*, 27:203–212.

15 Emerging Techniques for the Treatment of Wastewater

Neha Naaz, Zubair Ahmad, Adil Shafi, and Sana Choudhary

15.1 INTRODUCTION

Over recent decades, drugs, personal care products, xenobiotics, steroid hormones and volatile organic compounds have been described as emerging contaminants (ECs) because of their probable inimical effects on aquatic environments and their anthropological strength, and they have been extensively reported to be in different water matrices such as surface water, drinking water treatment plants, wastewater and wastewater treatment plant effluents (Anawar and Ahmed, 2019; Sgroi et al., 2017; Tran et al., 2018). Thus, the elimination of ECs is imperative to the generation of safe drinking water, and water supply and treatment has frequently taken precedence over wastewater collection and treatment in the past (Grassi et al., 2012). Some of the main sources of ECs are untreated urban wastewater and wastewater treatment plant (WWTP) effluents that cause serious soil and water pollution (Petrović et al., 2003). Conventional techniques cannot eradicate emerging pollutants and their metabolites from the water. Consequently, ECs end up threatening all living organisms through the feminization of aquatic organisms, endocrine disruption, bacterial resistance and cancer (Mohapatra and Kirpalani, 2019). Hence, the development of advanced wastewater treatment technologies is necessary to recover the final status or biodegradability of the water. Emerging technologies include innovations that correspond to an advanced progression within a domain of competitive worth.

15.2 EMERGING TECHNIQUES FOR THE TREATMENT OF WASTEWATER

15.2.1 Advanced Oxidation Processes

Glaze (1987) interpreted advanced oxidation processes (AOPs) as near-ambient temperature treatment procedures attributed to extremely reactive radicals; specifically, the hydroxyl radical (·OH) is the primary oxidizing agent. The hydroxyl radical is a potent, unselective reactive species that oxidizes several reduced compounds and starts a sequence of oxidative degeneration reactions (Goi, 2005; Bajpai, 2017). Table 15.1 illustrates the redox potential of several chemical oxidizing agents that are generally used in water and wastewater treatment. Various strong oxidants applied in AOPs are shown in Table 15.2. The OH radicals are usually generated by integrating ozone (O_3), titanium dioxide (TiO_2), hydrogen peroxide (H_2O_2), ultraviolet (UV) irradiation, high electron beam irradiation and heterogeneous photo-catalysis. Of these, O_3/H_2O_2, O_3/UV and H_2O_2/UV facilitate the detoxification of water and wastewater (Zhou and Smith, 2002).

Usually, extremely reactive hydroxyl radicals recruit a sequence of oxidative deterioration reactions of organic compounds existing in wastewater through subsequent methods (Huang et al., 1993):

1. Hydroxyl radical addition reaction:
   $OH\cdot + R \rightarrow ROH\cdot \rightarrow$ hydroxylated products $K \approx 10^7\text{–}10^{10}\ M^{-1}s^{-1}$

2. H_2 abstraction:
   $OH\cdot + R \rightarrow R\cdot + H_2O \rightarrow$ oxidized products $K \approx 10^7\text{–}10^{10}\ M^{-1}s^{-1}$

 DOI: 10.1201/9781003186298-15

However, the hydroxyl radical causes the partial oxidation of non-decomposable organic compounds. Accordingly, the biodegradability of effluent will intensify and could be decisive for the proper elimination of chemical oxygen demand (COD) in a post-biological treatment method (Chang et al., 2004; Balcioğlu et al., 2007; Bijan and Mohseni, 2008; Simstich and Oeller, 2010).

AOPs are employed as pre- and/or post-procedures of biological treatment (Figure 15.1). AOPs endeavor to improve the quality of wastewater and facilitate their process of treatment through ordinary microorganisms in the pre-procedure (Karimi et al., 1997; Karrer et al., 1997; Yu et al., 1998; Bila et al., 2005; Cesaro et al., 2013). In the post-procedure, the process of oxidation excludes the pollutants, but they are not entirely destroyed throughout the biological treatment (Poole, 2004; Cesaro et al., 2013). High concentrations of aliphatic carboxylic acids or hydroxyl radical scavengers in the effluents are difficult to oxidize, resulting in reduced degradation rates (Bajpai, 2017).

15.3 OZONATION

Ozonation reveals a strong oxidizing power with brief reaction periods, although it is an energy-intensive process described by high operating costs while allowing the treatment of a considerable

TABLE 15.1
Redox Potential of Several Chemical Oxidants (Zhou and Smith, 2002)

Oxidants	Redox potential (V)
Hydroxyl radical	2.80
Oxygen (atomic)	2.42
Ozone (O_3)	2.08
Hydrogen peroxide (H_2O_2)	1.78
Permanganate	1.69
Chlorine dioxide	1.56
Chlorine	1.36
Oxygen (molecular)	1.23

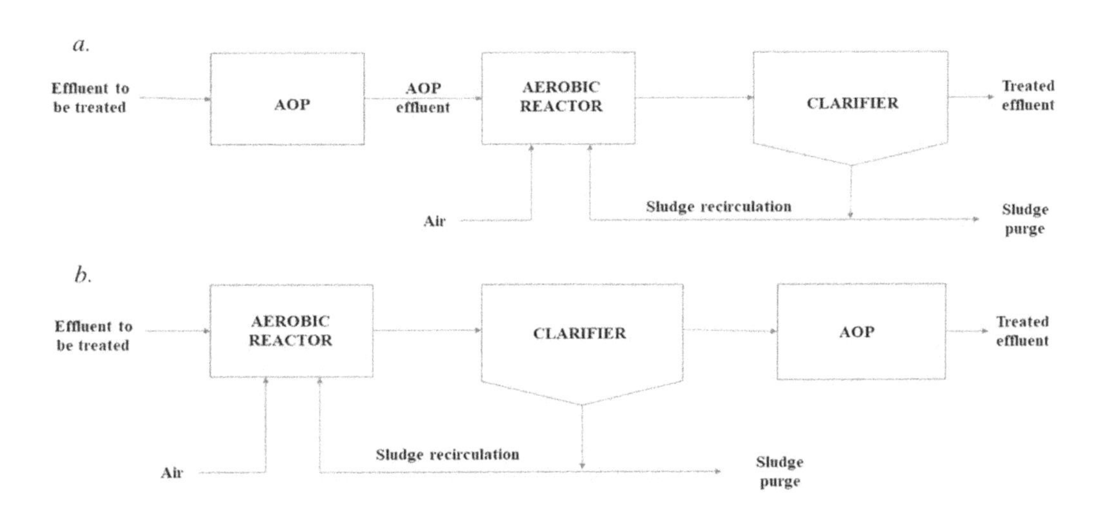

FIGURE 15.1 AOPs as a (a.) pretreatment and (b.) posttreatment of the biological system.

TABLE 15.2
Strong Oxidants Used in Advanced Oxidation Processes (Ozone)

Ozone/H_2O_2 (Peroxone)
Ozone/UV
Ozone/UV/H_2O_2
Ozone/TiO_2/H_2O_2
Ozone/TiO_2/Electron beam irradiation
Ozone/Electron beam irradiation
Ozone/Ultrasonics
H_2O_2/UV
H_2O_2/UV/Fe_2^+ (photo assisted Fenton)

amount of wastewater (Cesaro et al., 2013). Ozone (O_3) is a strong oxidizing agent for water and wastewater with a redox potential of 2.07 V. Once O_3 is dissolved in water, it undergoes a complex series of reactions in two dissimilar ways: through direct oxidation or by the creation of OH radicals and superoxide radicals (Al-Kdasi et al., 2004).

$$O_3 + \text{OH-} \rightarrow O_2^{*-} + HO_2^*$$

$$\text{O}_3 + O_2^{*-} \rightarrow O_3^{*-} + \text{O}_2$$

$$O_3^{*-} \rightarrow HO_3^{*-}$$

$$HO_3^* \rightarrow \text{OH}* + \text{O}_2$$

Ozonation is feasibly carried out at diverse pH values. Of these, the best results are shown at high pH values because of the reaction of organic and inorganic complexes with the ozone and radicals. Another significant parameter is the dose of ozone, which affects the extent of oxidation. A ten-fold increase of ozone dosage enhances the rate of the discoloration of treated water (Bajpai, 2017).

Various devices can be operated to transfer the generated ozone into water including jet reactors, countercurrent bubble columns, static mixers, agitated vessels and packed and plate columns (Bowers et al., 1973; Glaze et al., 1987; Munter et al., 1993). Figure 15.2 depicts the equipment generally employed for ozonation (Goi, 2005).

Presently, there are more than 4,000 ozonation plants functioning globally. Moreover, ozone applications can be categorized as a potent disinfectant and a strong oxidizing agent (Paillard et al., 1989). For disinfection, ozone effectively inactivates the bacterial species including the coliform bacteria and other resistant pathogenic microbes, specifically, *Cryptosporidium* spp. and *Giardia* spp., from municipal waters through water treatment. As an oxidant, ozone will become more potent after disintegration and turn into extremely reactive radicals to oxidize the chemical contaminants (Zhou and Smith, 2002). Sangave et al. (2007a) considered what ozonation meant for the biological processing of distillery wastewater alone as pre- and post-treatment, which achieved a 79% reduction of pollutants; moreover, a 35% COD reduction occurred over the non-ozonated sample.

Ozone should be united with other oxidizing agents, particularly UV irradiation or hydrogen peroxide, thereby increasing the formation of radicals and thus degrading the contaminants at higher rates (Gogate and Pandit, 2004).

On combining O_3 with UV or peroxide, AOPs are generated, which reduce the highly persistent substances and assist in rendering nondegradable water contaminants to be non-toxic (Bajpai, 2017).

The UV/ozone system is an efficient technique intended for the oxidation and demolition of organic compounds existing in water (Rodríguez, 2003). UV photons stimulate the ozone molecules,

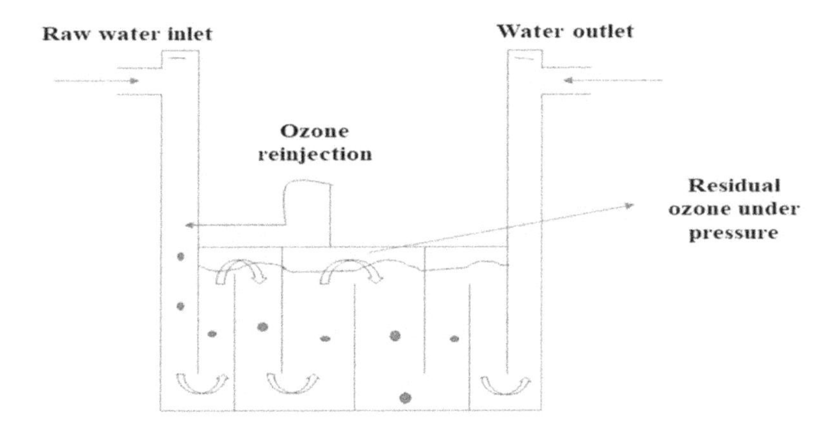

FIGURE 15.2 Torricelli apparatus used for ozonation.

thus enabling the production of hydroxyl radicals (Peyton and Glaze, 1988). Hydrogen peroxide is generated through the photolysis of O_3 in water and is then photolyzed into hydroxyl radicals (García et al., 2017). Several researchers have examined the efficacy of this process with distinct aromatic compounds. It was observed that the ozone/UV system is more efficient, except for the UV/H_2O_2 system, for the deprivation of p-chloronitrobenzene (Guittonneau et al., 1990). However, it was stated that ozone alone was not more economical than the O_3/UV system (Rodríguez et al., 1999). The UV/O_3 process improves the aerobic deprivation of distillery wastewater and reduces the COD up to around 45% (Sangave et al., 2007b).

O_3/H_2O_2 (Peroxone): The incorporation of H_2O_2 into ozone in water accelerates the degradation of ozone and leads to the production of hydroxyl radicals, where H_2O_2 is partially segregated in the aqueous solution as hydroperoxide anion (HO_2^-), which reacts with ozone and initiates a series of chain reactions (Rodríguez, 2003).

$$H_2O_2 + H_2O \rightarrow HO_2^- + H_3O^+$$

$$HO_2^- + O_3 \rightarrow \cdot OH + O_2^- + O_2$$

At lower concentrations, the hydroperoxide (HO_2^-) ion is efficient in commencing ozone degradation and facilitates the OH radical's creation. The degradation of ozone increases with an increase in the pH level.

Alnaizy and Akgerman (2000) reported that the H_2O_2/UV combination improved the derogation of phenol compared to direct photolysis.

In the H_2O_2/UV system, hydroxyl radicals are formed through the photolysis of H_2O_2. Then, OH radicals react with organic impurities or undergo an H_2O_2 degeneration creation cycle (Crittenden et al., 1999).

$$H_2O_2 + h\nu \rightarrow 2 \cdot OH$$

$$H_2O_2 + \cdot OH \rightarrow H_2O + HO_2$$

$$HO_2 + HO_2 \rightarrow H_2O_2 + O_2$$

Alaton and Balciog (2002) demonstrate the efficacy of the H_2O_2/UV approach even as a pretreatment method or by integrating it with additional AOPs of textile wastewater.

15.4 FENTON PROCESS

This process was reported for maleic acid oxidation by Fenton in 1884. Amongst all AOPs, The Fenton process has been demonstrated to be easier to operate and is found to be more efficient with reference to the removal rate. Fenton's reagent targets organic pollutants such as polyphenols, soluble substances and aromatic nucleophilic compounds (Bajpai, 2017). According to Haber and Weiss (1934), the Fenton reaction is built on the electron transfer between H_2O_2 and Fe^{2+} and acts as a catalyst to produce an OH radical, which, in turn, degrades the organic compounds:

$$Fe^{2+} + H_2O_2 \rightarrow Fe^{3+} + OH- + OH$$

The factors affecting the extent of Fenton processes are the concentration of H_2O_2, pH of the solution, amount of Fe^{2+} ions, initial concentration of contaminants and the occurrence of other ions (Gogate and Pandit, 2004). Furthermore, Fenton reagent action can be enhanced once it is subjected to UV radiation (Arana et al., 2001). Pirkanniemi et al. (2007) analyzed the meaning of the Fenton procedure for the elimination of EDTA from ECF bleaching effluent and concluded that it can be utilized as an efficient pretreatment that precedes biological treatment. Fenton processes employed to industrial wastewaters, such as tannery effluents with low pH and the presence of aromatic compounds at high temperature (Cesaro et al., 2013). Pera-Titus et al. (2004), proved to be beneficial in terms of the removal of insecticides and contaminants from the refinery, rubber and plastic industries. By considering some disadvantages of Fenton's process, Fenton's method is applied in combination with the cavitation process. Here, ultrasound waves are used to produce hydroxyl radicals to enhance the conversion rate and degeneration of organic compounds present in wastewater. Moreover, cavitation, along with the ultrasound waves, affect the organic contaminants through pyrolysis, as many of them are not influenced by OH radicals (Sharma et al., 2019).

15.5 PHOTO-FENTON PROCESS

Fenton's elemental incorporation with UV light reduces the ferric ions and thus produces ferrous ions along with hydroxyl radical content through photolysis (Hermosilla et al., 2015):

$$Fe\ (III)\ OH^{2+} + h\nu \rightarrow Fe^{2+} + OH$$

Photo-Fenton process proved to be better at eliminating organic compounds and appears in the concentrate from reverse osmosis (RO) (Hermosilla et al., 2012). This method is more economical in the deprivation of effluent elements compared to its dark version and thus enhances both the amount and the rate of total organic carbon (TOC) decline in contrast with the UV/H_2O_2approach (Catalkaya and Kargi, 2007; Karimi et al., 2010; Hermosilla et al., 2009, 2012).

15.6 HETEROGENEOUS PHOTOCATALYSIS

Heterogeneous photocatalysis processes enhanced via the activity of a semiconductor catalyst are broadly utilized, and the most efficient is TiO_2 (Vogelpohl, 2007). The anatase form of TiO_2 possesses the remarkable attributes of being economical and highly robust and shows good performance (Andreozzi et al., 1999). The mechanism involves the transition of electrons from the valence to the conduction band as a product of the light irradiation of the catalyst. Thus, drifting electrons, along with the holes formed in the valence band, can contribute to redox reactions with compounds immersed on the photocatalyst (Rizzo, 2011). TiO_2/UV and TiO_2/H_2O_2/UV are the most significant heterogeneous photocatalytic processes.

15.7 AOPS COMBINED WITH BIOLOGICAL TREATMENTS

Biological analyses of groundwater, hazardous waste and wastewater are considered the most efficient alternatives compared with other analyses (Rodríguez, 2003). The capability of a compound to experience biological deprivation is influenced by various aspects (Bajpai, 2017):

- Chemical structure,
- Substituents of the target compound,
- Concentration,
- pH, and
- Occurrence of inhibitory compounds.

According to Scott and Ollis (1995), AOPs combined with biological processes can successfully treat industrial wastewater containing bio-resistant and recalcitrant compounds along with wastewaters holding toxins that cause noxiousness to microorganisms. Several researchers have exhibited the efficiency of a coupled chemical and biological treatment leading to the mineralization of biorecalcitrant industrial impurities (Scott and Ollis, 1995; Pulgarin et al., 1999; Parra et al., 2000; Sarria et al., 2001). According to Alaton and Balciog (2002), ozonation, in addition to H_2O_2/UV, for the pretreatment of textile wastewater improves the efficacy of a subsequent biological treatment.

15.8 MEMBRANE FILTRATION TECHNOLOGIES

Membrane filtration processes offer several advantages over conventional water treatment procedures and reduce the environmental effect of effluents while offering reduced land obligations and the capability for mobile analysis units (Owen et al., 1995). Membrane filtration is an innovative technique for the exclusion of tint, COD and salinity through the treatment of wastewater (Zheng et al., 2013). The membranes used in filtration techniques may be polymeric, organo-mineral, metallic or ceramic (Van Der Bruggen et al., 2003). Membrane filtration is classified with respect to the pore sizes of the membranes. However, the membrane's potential is estimated over the permeates flux and the rejection mechanism (Ahmad et al., 2015).

15.8.1 ULTRAFILTRATION

Ultrafiltration (UF) membranes possess pore sizes ranging from 0.1 to 0.001μm. However, lower rejection can be achieved through a larger pore size. This filtration process requires lower pressure to be more economical than nanofiltration and reverse osmosis. UF membranes are prepared

TABLE 15.3
Main Features of Membrane Filtration Processes

Membranes	Pore size (μm)	Permeate flux	Transmembrane pressure (MPa)	Applications
Reverse osmosis (RO)	<0.001	Low	5–8	Desalination
Nanofiltration (NO)	0.001–0.008	Medium	0.5–1.5	Removal of small organics and multivalent ions
Ultrafiltration (UF)	0.001–0.1	High	0.05–0.5	Removal of macromolecules (natural organic material)
Microfiltration (MF)	>0.05	High	0.03–0.3	Disinfection barrier

with polysulphone/poly(ether sulfone)/sulphonated polysulfone, polyacrylonitrile and related block copolymers, polyvinylidene fluoride, cellulosic, etc. by phase inversion. Polymer blends with polyvinylpyrrolidone (PVP) are frequently used to increase the membrane's hydrophilicity (Van Der Bruggen et al., 2003). Alventosa-deLara et al. (2012) used the ceramic membrane of 150kDa as an ultrafiltration membrane to eradicate Reactive Black 5. More than a 79.8% rejection was achieved through RSM optimization, and the rejection persisted over 70% even at a high dye application (500 mg/L). Mondal et al. (2012) studied polyelectrolyte enhanced ultrafiltration (PEUF). Here, a polymer such as poly (ammonium acrylate), poly (acrylic acid) and cellulose membrane of 10kDa is used. The polymer molecules are complexified with solutes to produce macromolecules that could be preserved by the ultrafiltration membrane. Consequently, there was an increase in rejection with the polymer concentration. Dye-polymer interactions are pH-dependent. Dye removal will be higher at a 2-bar transmembrane pressure. Micellar enhanced ultrafiltration (MEUF) augments UF by adding surfactants to the dye solution and forming micelles with dye molecules trapped in it. Ngang et al. (2012) studied MEUF including SDS as the surfactant by utilizing a polysulfone membrane, and the polyvinylidene fluoride-titanium dioxide (PVDF-TiO_2) mixed membranes showed 99.3% rejection at 300kPa and 99% rejection at 0.5 bar, respectively. Therefore, PVDF-TiO_2 has considerable UV disinfecting activity, which simplifies its management.

15.9 NANOFILTRATION

Nanofiltration (NF) is a recently established membrane technique for the treatment and decontamination of wastewater. NF membranes have pore sizes ranging from 1 to 10 Å and are designed for the elimination of organic pollutants. NF membranes offer a higher permeate flux, lower osmotic potential and higher retention of multivalent salts over UF and RO membranes, but they are not efficient for the rejection of monovalent ions, especially chloride (Mänttäri and Nyström, 2007). The membranes are fabricated of polysulfone/poly(ether sulfone)/sulfonated polysulfone, aromatic polyamide, poly(piperazine amide) or cellulose acetate. NF membranes can reduce the hardness, total dissolved solids (TDS), tint and odor and can eliminate heavy metals from groundwater (Sharma and Sharma, 2012). NF has the ability to eliminate 98% of the color from industrial effluents, while UF eliminates only 90% (Aouni et al., 2012). Interfacial polymerization has led to the development of innovative NF composite membranes, which are able to eliminate over 90% of Safranin O and Aniline Blue dyes at pH 11 (Shao et al., 2013).

15.10 NANOTECHNOLOGY

Recent innovations in environmental techniques have revealed nanoparticles as efficient solutions to various current concerns involving energy utilization, pollution and waste control. The nano dimensional particles have accelerated innovative and upgraded technologies in several facets of environmental preservation and remediation (Dervin et al., 2016). The approach of nanotechnology has offered innumerable prospects to decontaminate water even at the ionic state (Das et al., 2014).

15.11 NANOPARTICLES

A nanoparticle is very minute, possesses a high surface area-to-volume ratio and is employed to distinguish subtle contaminants (Lu and Zhao, 2004). Nanomaterials have advantages of a high surface area, advanced reactivity and better discarding ability (Yunus et al., 2012). Nanomaterials have displayed effective antimicrobial activities through diverse mechanisms, including the photocatalytic creation of reactive oxygen species (ROS) that disfigure cell constituents and viruses, and they comprise a bacterial cell envelope such as carboxyfullerene, carbon nanotubes (CNTs), ZnO and silver nanoparticles (nAg). Nanomembranes and nanosensors have also been utilized effectively to generate potable water and to detect single cells, respectively (Ahmed et al., 2014).

Several researchers have reported that on applying an external magnetic field, magnetic nanoparticles have the potential to remove dyes from wastewater (Kong et al., 2012; Xu et al., 2013b). Acid leaching-precipitation and co-precipitation are the procedures to produce magnetite nanoparticles. Recently, manganese ferrite nanoparticles have been synthesized via manganese nitrate and iron nitrate for the exclusion of dyes (Direct green 6, Acid red 18 and Direct red 31) from the binary system (Mehmoodi, 2014). Additionally, non-magnetic nanoparticles were tested for the exclusion of dyes from wastewater. Assefi et al. (2014) synthesized a cobalt (III) oxide (Co_2O_3) nanoparticle stacked on an adsorbent (activated carbon) and detected it as an excellent adsorbent for the exclusion of Eosin Y, a perilous dye, from the solution. As a result, it was determined that Co_2O_3-NP-AC is considered an efficacious, green and economical adsorbent with high adsorption ability for removing dyes from solutions.

15.12 SILVER NANOPARTICLES

Silver nanoparticles can be used against bacteria, fungi and viruses due to their biocidal efficacy in the exclusion of small impurities, water filtration and water quality assessment (Ahmed et al., 2014). Some researchers have evaluated the biocidal action of Ag nanoparticles (Panáček et al., 2006; Kim et al., 2007; Shrivastava et al., 2007). Silvestry-Rodriguez et al. (2007) reported that silver diminishes aquatic pathogens such as *Pseudomonas aeruginosa* and *Aeromonas hydrophila* by immobilizing them. Silver nanoparticles exhibit a prominent antibacterial/antifungal impact (Khaydarov et al., 2009). Another finding revealed that the bactericidal impact of silver (Ag) nanoparticles on gram-negative bacteria is size-reliant (Morones et al., 2005).

The usage of Ag nanoparticles along with ultrasonic emissions for a brief time in the case of coliform bacteria improves the antibacterial outcome (Tiwari et al., 2008). It was reported that polypropylene filters coated with Ag nanoparticles of size 45 nm were used in water purification. After 6.5 hours, no bacteria were identified in the solution when the water had a bacterial load of 103 CFU/ml. Furthermore, nano silver-coated filters are absolutely active in the case of E. coli, and the leaching of Ag nanoparticles examined through the inductively coupled plasma/mass spectrometry of a filtered water sample was found to be nil (Heidarpour et al., 2011). Water filters fabricated of Ag nanoparticles coated on polyurethane foam were utilized, and leaching was not seen (Jain and Pradeep, 2005).

15.13 NANO ZERO-VALENT IRON

Nano zero-valent iron (nZVI) is a promising treatment for polluted soil and groundwater. Until 1998, a common system was used to rectify water, namely, the 'pump and treat' system. Currently, another approach to rectify water is a permeable reactive barrier (PRB), which purifies groundwater. This system is utilized to clean up contaminants including aromatic nitro compounds, chlorinated hydrocarbons, pesticides, polychlorinated biphenyls (PCBs) and chromate compounds (Yunus et al., 2012). Uyttebroek et al. (2010) observed that zero-valent metals, particularly iron (ZVI), which is recommended as a filter material of PRB, reduces hazardous pollutants from the water in bulk amounts. The nano-iron is efficient in its multifunctional utilization for remediation purposes. It was recognized that nZVI possesses high reactivity for a substantial quantity of chemicals, involving chlorinated hydrocarbons, Cu^{2+}, Eqn0008.eps and nitrate (Yunus et al., 2012). Ferritin is an iron-encompassing protein and has the potential ability to rectify toxic metals and chlorocarbon in solar radiation or visible light (Moretz, 2004). Ferritin is more stable and does not react under photoreduction, which is an advantage over ordinary iron.

15.14 NANO-ADSORBENTS

Adsorption is a highly technical and economically sustainable approach for removing organic and inorganic pollutants by treating water and wastewater. Nano-adsorbents offer a large surface area, short intraparticle diffusion gap, associated adsorption sites, surface chemistry and tunable pore dimension (Qu et al., 2013). Activated carbon, aluminosilicate, zeolites, clay and peat kaolin are the different materials used for removing metals from solutions through wastewater treatment. Currently, carbon nanomaterials (CNMs) acquire an arrangement of carbon nanofibers (CNFs), and carbon nanotubes (CNTs) are novel adsorbents with exceptional performance because of their elevated aspect ratio and high specific surface area (Sharma and Sharma, 2012). Based on several findings, CNTs are found to be superior adsorbents in contrast to activated carbon in terms of heavy metals (e.g., Cu^{2+}, Pb^{2+}, Cd^{2+} and Zn^{2+}), and because of the short intraparticle diffusion gap and highly exposed adsorption sites, the adsorption kinetics are rapid on CNTs (Qu et al., 2013). Gao et al. (2011) stated that sand pellets encrusted by graphite oxide were effective in eliminating Hg_2^+ and a large dye particle (Rhodamine B) in contrast to commercially activated carbon. Metal oxide nanoparticles such as TiO_2, iron oxide and alumina have been investigated to exclude certain types of heavy metals, particularly mercury, cadmium, arsenic, chromium, lead, copper and nickel, and have a significant ability to defeat activated carbon (Qu et al., 2013). According to Li et al. (2003b), multiwall carbon nanotubes (MWCNTs) have a metal-ion adsorption capacity 3–4 times greater than granular activated carbon (GAC) and powder.

15.15 NANOFIBERS AND NANOBIOCIDES

Nanofiber technology, combined with biological treatment, is an advanced method for the exclusion of noxious xenobiotics in industrial wastewater treatment procedures. Nanobiocides (metal nanoparticles and engineered nanoparticles) are effectively merged among nanofibers, and they exhibit both antimicrobial action and stability in water. The quality of water is reduced by membrane fouling caused by bacteria, and these bacteria are immobilized by triggering surface-modified nanofibers. On the basis of these outcomes, polyvinyl alcohol (PVA) and polyacrylonitrile (PAN) nanofibers, including Ag nanoparticles, have antimicrobial action, with a 91–99% reduction of bacteria in a polluted sample by PVA nanofibers and 100% by PAN nanofibers (Yunus et al., 2012). Regarding the polymer nature, nanofibers are moldable, durable and chemically resistant. An important benefit of nanofiber is its contrasting nature with the surface morphology, biocompatibility and dimension of microorganisms, which allows surface colonization by the microorganisms (Sharma and Sharma, 2012).

15.16 NANOSENSORS

New sensor techniques have been established through merging fabrication approaches at the microscale and nanoscale to generate tiny, compact and extremely specific sensors that identify the minute quantity of biochemical and chemical material in water (Hillie et al., 2006). The analysis of water quality can be conducted via nanosensors as these are utilized to detect E. coli and other bacteria. They can distinguish bacteria in a short period of time, as nanosensors flash light on the existence of bacteria in water samples. Nanoshells, quantum dots and CNTs are the nanomaterials employed within nanosensors to analyze water status (Ahmed et al., 2014). Schoen et al. (2010) stated that a gravity-fed device functioning at 100,000 l/(hm^2) inactivates >98% of microorganisms at minute flashes of the incubation period through silver nanowires, cotton and CNTs. The recognition of microbes through gold nanoshells associated with antibodies is one nanotechnology-based method (Nostrand, 2007). Antibodies identify the antigens and apply infrared (IR) radiation for 5 min at 808 nm at 1 W cm^2; the transmission of energy leads to bacterial cell lysis.

15.17 PHOTOCATALYSIS

Photocatalysis utilizes nanostructures of a metal oxide semiconductor, particularly titania (TiO), ZnO, zinc stannate (Zn_2SnO) and tungsten oxide (WO_3). Photocatalysis is a mode of water decontamination, as it removes chemical and biological pollutants (Ahmed et al., 2014). TiO_2 has been extensively exploited for water or wastewater treatment as it is a semiconductor photo-catalyst due to its chemical firmness, economical attribute, low toxicity and profusion as a raw material. It creates an electron/hole (e^-/h^+) pair upon imbibing a UV photon, forming ROS or undergoing an undesired recombination. In optimizing the particle size and shape, the photo-activity of nano-TiO_2 can be improved by diminishing e^-/h^+ recombination via noble metal doping, amplifying reactive facets and enhancing surface treatment pollutant adsorption. Nano-TiO_2-assisted solar disinfection (SODIS) has generally been tested and exists as a viable alternative to generate safe drinking water in remote zones of developing countries. The conventional technique of photocatalysis applied to water decontamination can be operative by using ZnO and TiO_2 nanostructures. Through this method, E. coli concentration in water reduces in just 1 hour to a safe level (Li et al., 2010). Accordingly, ZnO and TiO_2 photocatalysis has emerged as an alternative technique designed for the decontamination of industrial run-off and drinking water (Grieken et al., 2009; Baruah and Dutta, 2009; Zhang et al., 2012).

15.18 NANOCELLULOSE

Because of its high strength and significant surface area, nanocellulose is utilized in eliminating pollutants and disinfecting industrial and drinking water systems (Wang, 2019). Cellulosic materials in mechanical or chemical processes can be transformed within cellulose nanofibers (CNFs), nanocrystals (CNCs) and nanowhiskers (CNWs), which have excellent capability for water purification. The pioneering effort by Saito and Isogai (2005) directed the practice of TEMPO-oxidized cellulose nanofibers (TOCNFs) for the exclusion of different metals including Pb^{2+}, La^{3+} and Ag^{1+} from water. TEMPO oxidation of cellulose pulp results in the formation of 2,3,6-tricarboxy cellulose nanofibers (TPC-CNFs), followed by periodate-chlorite oxidation (Abou-Zeid et al., 2018). TPC-CNFs assist in the effective adsorption of Cu^{2+}, Ca^{2+} and Pb^{2+} (heavy metal ions) from samples.

Nanocellulose with succinylation leads to stable, effective and biodegradable adsorbents. Furthermore, Yu et al. (2013) reported that the esterification of OH groups on CNC associated with succinic anhydride results in the maximal adsorption of Pb^{2+} and Cd^{2+} at pH 5.5 and 6.5, respectively. Srivastava et al. (2012) demonstrated the ability of functionalized CNFs attained by succinic anhydride-modified mercerization to adsorb Cr^{3+}, Ni^{2+}, Pb^{2+} and Cd^{2+} with superior efficacies compared to unfunctionalized CNFs. The modification in succinylated CNFs results in an expanded surface area and the approachability of a large quantity of hydroxyl moieties for succinylation.

Mercury ions are believed to be a major health issue because of their noxiousness and bioaccumulation. It was suggested that the feasibility of spherical nanocellulose (SNC) be assessed as adsorbents for metallic ions from water (Peng et al., 2014; Luo et al., 2015). Other efforts stated that the production of novel thiolated-SNCs through acid hydrolysis and lipase-catalyzed esterification with 3-mercaptopropionic acid (3-MPA) exhibits prompt adsorption of 98% Hg^{2+} ions within 20 mins from a 100 ppm solution (Ram and Chauhan, 2018). The carboxylated CNCs are produced via the esterification of surface hydroxyl moeities with maleic anhydride, which shows a higher uptake efficiency of cationic dyes, including methylene blue, crystal violet and malachite green (Qiao et al., 2015). Moreover, phenol uptake in the water was examined for a composite film comprising CNF, along with in-situ fabricated hydroxyapatite particles by wet chemical precipitation (Narwade et al., 2017). Jackson et al. (2011) confirmed that sulfated CNCs could imbibe drugs such as tetracycline hydrochloride (TC) and doxorubicin hydrochloride (DOX) from the water. Additionally, deposits of imipramine hydrochloride and procaine hydrochloride were adsorbed by biodegradable b-cyclodextrin-modified CNC/Fe_3O_4/SiO_2 superparamagnetic nanorods from wastewater (Chen et al., 2014). The nanocellulose approach is believed to be a favorable mode to combat the evolving water pollution crisis worldwide. However, functionalized nanocellulose exhibits adsorption performance, which is better for conventional adsorbents meant for the elimination of organic contaminants and heavy metals.

15.19 CARBON NANOTUBE MEMBRANES

CNTs are innovative membrane techniques that offer minimal energy solutions for water detoxification. CNTs allow an ultraefficient route to water due to their high aspect ratio, smooth hydrophobic walls and inner pore diameter. Thomas and Mcgaughey (2008) stated that water does not follow a simple fluidics mechanism. Ahn et al. (2012) explained "Nanofluidics" to simulate water movement through a nanotube structure. It was observed that the frictionless passage of water had high-level of velocities of 9.5 to 43.0 cm s^{-1}/bar over a 7-nm diameter of membrane pore. It was 4–5-fold more rapid than standard fluid movement (Majumder et al., 2005). CNT membranes are fabricated with various nanoparticles for example, Cu, Pt, TiO_2, Ag, Pd, Au, biomolecules (DNA, proteins and pollutant degradative enzymes) and polymers that have effectual membrane properties, which consequently widened the use of CNTs in water salination (Hooijdonk et al., 2013). CNTs with nano-porous surfaces are suitable for declining micropollutants and ions in the sample. The cytotoxic effects reduce biofouling and enhance membrane survival by killing and eliminating pathogens (Das et al., 2014).

The functionalization of CNT membranes is a prerequisite for CNT-based water desalination. CNTs are usually tainted with impurities, metal catalysts and physical heterogeneities (Mauter and Elimelech, 2008). In addition, CNTs are capped into the hemisphere like fullerene-type curvatures throughout production and decontamination (Li et al., 2003a, 2003b). Functionalization can attach negative (-COO-, sulfonic groups), positive ($-NH^{3+}$) and hydrophobic (aromatic rings) moeities on CNT surfaces (Goh et al., 2013; Guillen and Hoek, 2009). This makes CNT membranes selective for specific pollutant retention and improves water influx over the nanotube hole. CNTs show mechanical and thermal strength, good water permeability, fouling resistance, self-cleaning functions and pollutant degradation (Qu et al., 2013). Chan et al. (2013) attained 100% ion rejection with simulated and manufactured CNTs that possessed two zwitter ions at their apex while having 0% in the non-functionalized membrane. Corry (2011) successfully eliminated 100% Na^{2+} and Cl- through functionalized CNTs holding -COOH, $-NH_3$ and -OH moieties. Yang et al. (2013) investigated modified CNTs through plasma treatment, and the ability of salt uptake exceeded 400% by weight, which is higher than the activated carbon-based material by two orders of magnitude, for water desalination. CNTs with an elevated water flux and selective salt rejection have elevated their potential for water detoxification.

15.20 2D NANOSTRUCTURES

2D nanostructures, because of their large specific surface area, atomic thickness, mechanical potency and extreme durability, are considered to be suitable as a detoxification and water decontamination membrane substance (Dervin et al., 2016). David Cohen-Tanugi and Jeffrey Grossman (2012) observed the desalination ability of graphene membranes that reveals synthetic nanopores that essentially filter sodium chloride salt from water, which have a permeability larger than traditional RO membranes. The outcomes indicated that graphene membranes were efficient in the 100% rejection of salt ions, thus allowing water transport at speeds ranging from 10–100 L/cm^2/day/MPa (Nicolaï et al., 2014). Graphene oxide framework (GOF) materials are nanoporous materials that include loaded sheets of graphene oxide (GO) covalently bonded with one another via a linear boronic acid backbone, which are called linkers. Thus, the hydrophobicity of graphene and linker concentration offers tremendous water permeation and filtration, respectively (Srinivas et al., 2011). Nicolaï et al. (2014) highlighted that the water permeability of GOF membrane at levels ranging from~5 to 400 L/cm^2/day/MPa with a linker concentration of $n \leq 32$ exhibits a 100% rejection. GO combined with polymer enhances water flux and antimicrobial and mechanical properties (Hu and Mi, 2013). A 1,3,5-benzene tricarbonyl trichloride associated GO membrane resulted in a 46–66% rejection of methylene blue and 93–95% rejection of Rhodamine-WT at flow rates ranging from 80–276 L/M/H/MPa. Ultimately, polymer cross-link GO membranes are model water distillation, pharmaceutical and fuel separation membranes (Ma et al., 2017). Xu et al. (2013a) determined that GO-TiO_2 composite membranes successfully eliminate 100% of methyl orange and Rhodamine B from water flux at a 7 L/m^2/h flow rate. The GO-TiO_2-decorated polysulfone membrane rejects 90% of methylene blue at an increased water flux of 45 L/m^2/h (Gao et al., 2014). The aPES/GO/aGO functionalized membrane showed a water flux of 28 L/m^2/h and a 98% salt rejection (Kim et al., 2013). The SWCNT intercalated GO ultrathin film retained 97.4% to 98.75 of contaminants comprising Bovine Serum Albumin (BSA), Coomassie Brilliant Blue, cytochrome c and Rhodamine B, which signifies ten times more water flux (660–720 L/m^2/h/bar) than traditional NF membranes (Gao et al., 2015). MWCNT intercalated rGO membranes improved the water flux on rising interlayer spacing. They reject >96% of organic dyes including methyl orange and Direct yellow, along with 51.4% of NaCl ions and 83.5% of Na_2SO_4 (Han et al., 2015). Sun et al. (2013) depicted the rejection of 89% and 98% of Evans blue and cytochrome C when a flexible laminar partition membrane made from MoS_2 sheets has 3–5 times more water flux than GO at 245 L/m^2/h/bar. Mxenes, which are 2D materials with a few atoms and dense sheets of transition metal carbonitrides and carbides, were utilized as filtration membranes meant for the segregation of ions and molecules in water. Ti_3C_2Tx membranes are flexible, robust and hydrophilic, and they exhibit a water flux of 37.4 L/m2/h/bar and an increased dissociation of higher charge cations such as K^+, Mg^{2+}, Li^+, Al^{3+}, Na^+, Ca^{2+}, Ni^{2+}, methylthioninium+ and MB dye cations compared to GO (Ren et al., 2015). Hence, 2D nanostructured membranes can offer a resolution to the potential water crisis by enhancing the execution and commercial feasibility of existing water analysis performances.

15.21 MAGNETIC SEPARATION

Magnetic separation is a physical property that assists water decontamination; along with other procedures, it facilitates an improved purification technique. High-gradient magnetic separation (HGMS) is a widely exploited method in magnetic separations (Ambashta and Sillanpää, 2010). This method operates by the principle of coagulation and water purification by applying a magnetic field. Despite this, the steel and power industries utilize HGMS for coolant pipeline maintenance. Magnetically aided water purification can be categorized by relying on the variation in the adoption of the physical method:

- Direct decontamination,
- Seeding and parting of magnetic flocculent,
- Magnetic adsorbent use in organic and inorganic pollutants, and
- Combination activities aided by magnetic assistance.

In direct decontamination, the most common method is anti-scaling. The common elements of scale are $CaSO_4.2H_2O$, $CaCO_3$ and silica, whereas additional possible scale elements are $SrSO_4$, $BaSO_4$, $Ca_3(PO_4)^2$ and aluminum and ferric hydroxides. The salt concentration, intensity of the magnetic field, anion present in salt, physicochemical properties of water, pipe material and extent of magnetic field usage are only some parameters that alter the de-scaling mechanism. Magnetic pulse compressors are an emerging technique for the disinfection and decontamination of water (Narsetti et al., 2006). A brief period of electrical pulses from the pulse compressor is applied for immobilizing bacteria, spores and viruses in drinking water.

The development of a continuous low magnetic field filtration system is used together with specifications for seeding and flocculating phytoplankton to disinfect lake water (Saho et al., 1999). The exclusion of algae, total nitrogen (TN), total phosphorus and biochemical oxygen demand (BOD) was simultaneously detected. About 93% of the phytoplankton can be flocculated out from the lake water by using a superconducting magnetic filter at 400 m^3/day flow rates. Particular elements such as arsenic and phosphate pollutants have been eliminated by the seeding process for water decontamination (Gokon et al., 2002; Okada et al., 2004). Through 15 mg/L of $FeCl_3$ solution, phosphate ion recovery improved from 20% to 80% without the addition of magnetite but with a rising magnetic application from 0.8 to 8 T.

In 1995, magnetic ion exchange resins (MIEX) were established to exclude natural organic matter (NOM) from water, which was advanced to the coagulation process. Coagulation eliminates 60% of dissolved organic carbon (DOC) linked by the 1–10k fraction, although it has a slight influence on the DOC concentration of the <1k fraction, while the MIEX eliminates 80% of DOC linked by the 1–10k fraction and 60% of the DOC related to the <10k fraction (Ambashta and Sillanpää, 2010). Magnetic chitosan gel particles, magnetic alginates and magnetic charcoal assist in the exclusion of malachite green, polycyclic dyes, crystal white and additional organic dyes from water solutions and suspensions (Rocher et al., 2008). The high-level efficacy of magnetic separations has been directed to a another arena of investigation, namely, the range of magnetically aided chemical separation of radionuclides (Ngomsik et al., 2005). Dendrimers associated with the terminal amino groups linked to the magnetic particle surface displayed a 50–400-fold increment in the distribution coefficients for americium and europium, which is comparable to those devoid of dendrimers (Grüttner et al., 2005). Furthermore, 10 ppb uranium ions in the sample can be lowered up to or less than a 1-ppb level on using a 10 T superconducting high-gradient magnetic separator with a 100-ppm magnetic adsorbent (Nishimura et al., 2009). To exclude polluted oil from radioactive liquid waste through HGMS, the seeding process principle was followed effectively. Subsequently, nanoscale iron materials are very efficient when they are intended for detoxification and the transformation of environmental chemicals, particularly organochlorine pesticides, phenols, trinitrotoluenes, chlorinated organic solvents, amino carboxylic acids, herbicide molinate and p-hydroxybenzoic acid. It was proved that in addition to organic chemicals, inorganic anionic chemicals (NO^{3-}, $Cr^{2}O_7^{2-}$) could be deteriorated by iron nanoparticles (Cundy et al., 2008).

15.22 CONCLUSION

Accordingly, wastewater treatment using emerging treatment methods such as AOPs, nanomembranes and nanoadsorbents prove to be efficient remediation techniques for wastewater treatment. In addition, the nanoremediation techniques can treat persistent pollutants by both chemical reduction and catalytic processes and can be beneficial in reducing the clean-up time and avoiding degradation intermediates.

The conventional techniques are not able to eliminate ECs and their metabolites from water; thus, they are considered less efficient compared to modern and emerging treatment methods. The emerging treatment methods discussed in this chapter require less manpower and are cost-effective and facile.

REFERENCES

Abou-Zeid, R. E., Dacrory, S., Ali, K. A., & Kamel, S. (2018). Novel method of preparation of tricarboxylic cellulose nanofiber for efficient removal of heavy metal ions from aqueous solution. *International Journal of Biological Macromolecules*, *119*, 207–214.

Ahmad, A., Mohd-Setapar, S. H., Chuong, C. S., Khatoon, A., Wani, W. A., Kumar, R., & Rafatullah, M. (2015). Recent advances in new generation dye removal technologies: Novel search for approaches to reprocess wastewater. *RSC Advances*, *5*(39), 30801–30818.

Ahmed, T., Imdad, S., Yaldram, K., Butt, N. M., & Pervez, A. (2014). Emerging nanotechnology-based methods for water purification: A review. *Desalination and Water Treatment*, *52*(22–24), 4089–4101.

Ahn, C. H., Baek, Y., Lee, C., Kim, S. O., Kim, S., Lee, S., . . . & Yoon, J. (2012). Carbon nanotube-based membranes: Fabrication and application to desalination. *Journal of Industrial and Engineering Chemistry*, *18*(5), 1551–1559.

Al-Kdasi, A., Idris, A., Saed, K., & Guan, C. T. (2004). Treatment of textile wastewater by advanced oxidation processes—a review. *Global NEST: The International Journal*, *6*(3), 222–230.

Alnaizy, R., & Akgerman, A. (2000). Advanced oxidation of phenolic compounds. *Advances in Environmental Research*, *4*(3), 233–244.

Alventosa-deLara, E., Barredo-Damas, S., Alcaina-Miranda, M. I., & Iborra-Clar, M. I. (2012). Ultrafiltration technology with a ceramic membrane for reactive dye removal: Optimization of membrane performance. *Journal of Hazardous Materials*, *209*, 492–500.

Ambashta, R. D., & Sillanpää, M. (2010). Water purification using magnetic assistance: A review. *Journal of Hazardous Materials*, *180*(1–3), 38–49.

Anawar, H. M., & Ahmed, G. (2019). Combined electrochemical-advanced oxidation and enzymatic process for treatment of wastewater containing emerging organic contaminants. In *Emerging and Nanomaterial Contaminants in Wastewater* (pp. 277–307). Elsevier.

Andreozzi, R., Caprio, V., Insola, A., & Marotta, R. (1999). Advanced oxidation processes (AOP) for water purification and recovery. *Catalysis Today*, *53*(1), 51–59.

Aouni, A., Fersi, C., Cuartas-Uribe, B., Bes-Pía, A., Alcaina-Miranda, M. I., & Dhahbi, M. (2012). Reactive dyes rejection and textile effluent treatment study using ultrafiltration and nanofiltration processes. *Desalination*, *297*, 87–96.

Arana, J., Rendón, E. T., Rodríguez, J. D., Melián, J. H., Díaz, O. G., & Peña, J. P. (2001). Highly concentrated phenolic wastewater treatment by the Photo-Fenton reaction, mechanism study by FTIR-ATR. *Chemosphere*, *44*(5), 1017–1023.

Arslan Alaton, I., & Balcıog, I. A. (2002). The effect of pre-ozonation on the H2O2/UV-C treatment of raw and biologically pre-treated textile industry wastewater. *Water Science and Technology*, *45*(12), 297–304.

Assefi, P., Ghaedi, M., Ansari, A., Habibi, M. H., & Momeni, M. S. (2014). Artificial neural network optimization for removal of hazardous dye Eosin Y from aqueous solution using Co2O3-NP-AC: Isotherm and kinetics study. *Journal of Industrial and Engineering Chemistry*, *20*(5), 2905–2913.

Bajpai, P. (2017). Chapter 7-emerging technologies for wastewater treatment. *Pulp and Paper Industry*, 93–179.

Balcıoğlu, I. A., Tarlan, E., Kıvılcımdan, C., & Saçan, M. T. (2007). Merits of ozonation and catalytic ozonation pre-treatment in the algal treatment of pulp and paper mill effluents. *Journal of Environmental Management*, *85*(4), 918–926.

Baruah, S., & Dutta, J. (2009). Hydrothermal growth of ZnO nanostructures. *Science and Technology of Advanced Materials*, *10*(1), 013001.

Bijan, L., & Mohseni, M. (2008). Novel membrane pretreatment to increase the efficiency of ozonation-biooxidation. *Environmental Engineering Science*, *25*(2), 229–238.

Bila, D. M., Montalvao, A. F., Silva, A. C., & Dezotti, M. (2005). Ozonation of a landfill leachate: Evaluation of toxicity removal and biodegradability improvement. *Journal of Hazardous Materials*, *117*(2–3), 235–242.

Bowers, A., Netzer, A., & Norman, J. D. (1973). Ozonation of wastewater—some technical and economic aspects. *The Canadian Journal of Chemical Engineering*, *51*(3), 332–336.

Catalkaya, E. C., & Kargi, F. (2007). Color, TOC and AOX removals from pulp mill effluent by advanced oxidation processes: A comparative study. *Journal of Hazardous Materials*, *139*(2), 244–253.

Cesaro, A., Naddeo, V., & Belgiorno, V. (2013). Wastewater treatment by combination of advanced oxidation processes and conventional biological systems. *Journal of Bioremediation and Biodegradation*, *4*(8), 1–8.

Chan, W. F., Chen, H. Y., Surapathi, A., Taylor, M. G., Shao, X., Marand, E., & Johnson, J. K. (2013). Zwitterion functionalized carbon nanotube/polyamide nanocomposite membranes for water desalination. *ACS Nano*, *7*(6), 5308–5319.

Chang, C. N., Ma, Y. S., Fang, G. C., Chao, A. C., Tsai, M. C., & Sung, H. F. (2004). Decolorizing of lignin wastewater using the photochemical UV/TiO2 process. *Chemosphere*, *56*(10), 1011–1017.

Chen, L., Berry, R. M., & Tam, K. C. (2014). Synthesis of β-cyclodextrin-modified cellulose nanocrystals (CNCs)@ Fe3O4@ SiO2 superparamagnetic nanorods. *ACS Sustainable Chemistry & Engineering*, *2*(4), 951–958.

Cohen-Tanugi, D., & Grossman, J. C. (2012). Water desalination across nanoporous graphene. *Nano Letters*, *12*(7), 3602–3608.

Corry, B. (2011). Water and ion transport through functionalised carbon nanotubes: Implications for desalination technology. *Energy & Environmental Science*, *4*(3), 751–759.

Crittenden, J. C., Hu, S., Hand, D. W., & Green, S. A. (1999). A kinetic model for H2O2/UV process in a completely mixed batch reactor. *Water Research*, *33*(10), 2315–2328.

Cundy, A. B., Hopkinson, L., & Whitby, R. L. (2008). Use of iron-based technologies in contaminated land and groundwater remediation: A review. *Science of the Total Environment*, *400*(1–3), 42–51.

Das, R., Ali, M. E., Hamid, S. B. A., Ramakrishna, S., & Chowdhury, Z. Z. (2014). Carbon nanotube membranes for water purification: A bright future in water desalination. *Desalination*, *336*, 97–109.

Dervin, S., Dionysiou, D. D., & Pillai, S. C. (2016). 2D nanostructures for water purification: Graphene and beyond. *Nanoscale, 8*(33), 15115–15131.

Gao, S. J., Qin, H., Liu, P., & Jin, J. (2015). SWCNT-intercalated GO ultrathin films for ultrafast separation of molecules. *Journal of Materials Chemistry A, 3*(12), 6649–6654.

Gao, W., Majumder, M., Alemany, L. B., Narayanan, T. N., Ibarra, M. A., Pradhan, B. K., & Ajayan, P. M. (2011). Engineered graphite oxide materials for application in water purification. *ACS Applied Materials & Interfaces, 3*(6), 1821–1826.

Gao, Y., Hu, M., & Mi, B. (2014). Membrane surface modification with TiO2—graphene oxide for enhanced photocatalytic performance. *Journal of Membrane Science, 455*, 349–356.

García, A. M., Torres-Palma, R. A., Galeano, L. A., Vicente, M. Á., & Gil, A. (2017). Separation and characterization of NOM intermediates along AOP oxidation. In *Applications of Advanced Oxidation Processes (AOPs) in Drinking Water Treatment* (pp. 99–132). Springer.

Glaze, W. H. (1987). Drinking-water treatment with ozone. *Environmental Science & Technology, 21*(3), 224–230.

Glaze, W. H., Kang, J. W., & Chapin, D. H. (1987). The chemistry of water treatment processes involving ozone, hydrogen peroxide and ultraviolet radiation. *The Journal of the International Ozone Association, 9*(4), 335–352.

Gogate, P. R., & Pandit, A. B. (2004). A review of imperative technologies for wastewater treatment I: Oxidation technologies at ambient conditions. *Advances in Environmental Research, 8*(3–4), 501–551.

Goh, P. S., Ismail, A. F., & Ng, B. C. (2013). Carbon nanotubes for desalination: Performance evaluation and current hurdles. *Desalination, 308*, 2–14.

Goi, A. (2005). *Advanced Oxidation Processes for Water Purification and Soil Remediation*. Tallinn University of Technology Press.

Gokon, N., Shimada, A., Hasegawa, N., Kaneko, H., Kitamura, M., & Tamaura, Y. (2002). Ferrimagnetic coagulation process for phosphate ion removal using high-gradient magnetic separation. *Separation Science and Technology, 37*(16), 3781–3791.

Grassi, M., Kaykioglu, G., Belgiorno, V., & Lofrano, G. (2012). Removal of emerging contaminants from water and wastewater by adsorption process. In *Emerging Compounds Removal from Wastewater* (pp. 15–37). Springer.

Grüttner, C., Böhmer, V., Casnati, A., Dozol, J. F., Reinhoudt, D. N., Reinoso-Garcia, M. M., . . . & Wang, P. (2005). Dendrimer-coated magnetic particles for radionuclide separation. *Journal of Magnetism and Magnetic Materials, 293*(1), 559–566.

Guillen, G., & Hoek, E. M. (2009). Modeling the impacts of feed spacer geometry on reverse osmosis and nanofiltration processes. *Chemical Engineering Journal, 149*(1–3), 221–231.

Guittonneau, S., De Laat, J., Duguet, J. P., Bonnel, C., & Dore, M. (1990). Oxidation of parachloronitrobenzene in dilute aqueous solution by O3+ UV and H2O2+ UV: A comparative study. *The Journal of the International Ozone Association, 12*(1), 73–94.

Haber, F., & Weiss, J. (1934). The catalytic decomposition of hydrogen peroxide by iron salts. *Proceedings of the Royal Society of London. Series A-Mathematical and Physical Sciences, 147*(861), 332–351.

Han, Y., Jiang, Y., & Gao, C. (2015). High-flux graphene oxide nanofiltration membrane intercalated by carbon nanotubes. *ACS Applied Materials & Interfaces, 7*(15), 8147–8155.

Heidarpour, F., Ghani, W. A. W. A. K., Fakhru'l-Razi, A., Sobri, S., Torabian, A., Heydarpour, V., & Zargar, M. (2011). New trends on microbiological water treatment. *Digest Journal of Nanomaterials & Biostructures (DJNB), 6*(2).

Hermosilla, D., Cortijo, M., & Huang, C. P. (2009). The role of iron on the degradation and mineralization of organic compounds using conventional Fenton and photo-Fenton processes. *Chemical Engineering Journal, 155*(3), 637–646.

Hermosilla, D., Merayo, N., Gascó, A., & Blanco, Á. (2015). The application of advanced oxidation technologies to the treatment of effluents from the pulp and paper industry: A review. *Environmental Science and Pollution Research, 22*(1), 168–191.

Hermosilla, D., Merayo, N., Ordóñez, R., & Blanco, Á. (2012). Optimization of conventional Fenton and ultraviolet-assisted oxidation processes for the treatment of reverse osmosis retentate from a paper mill. *Waste Management, 32*(6), 1236–1243.

Hillie, T., Munasinghe, M., Hlope, M., & Deraniyagala, Y. (2006). *Nanotechnology, Water and Development*. Meridian Institute. Available from: https://eldis.org/document/A21102

Hu, M., & Mi, B. (2013). Enabling graphene oxide nanosheets as water separation membranes. *Environmental Science & Technology, 47*(8), 3715–3723.

Huang, C. P., Dong, C., & Tang, Z. (1993). Advanced chemical oxidation: Its present role and potential future in hazardous waste treatment. *Waste Management*, *13*(5–7), 361–377.

Jackson, J. K., Letchford, K., Wasserman, B. Z., Ye, L., Hamad, W. Y., & Burt, H. M. (2011). The use of nanocrystalline cellulose for the binding and controlled release of drugs. *International Journal of Nanomedicine*, *6*, 321.

Jain, P., & Pradeep, T. (2005). Potential of silver nanoparticle-coated polyurethane foam as an antibacterial water filter. *Biotechnology and Bioengineering*, *90*(1), 59–63.

Karimi, A. A., Redman, J. A., Glaze, W. H., & Stolarik, G. F. (1997). Evaluating an AOP for TCE and PCE removal. *Journal-American Water Works Association*, *89*(8), 41–53.

Karimi, S., Abdulkhani, A., Karimi, A., Ghazali, A. H. B., & Ahmadun, F. L. R. (2010). The effect of combination enzymatic and advanced oxidation process treatments on the colour of pulp and paper mill effluent. *Environmental Technology*, *31*(4), 347–356.

Karrer, N. J., Ryhiner, G., & Heinzle, E. (1997). Applicability test for combined biological-chemical treatment of wastewaters containing biorefractory compounds. *Water Research*, *31*(5), 1013–1020.

Khaydarov, R. R., Khaydarov, R. A., Gapurova, O., Estrin, Y., Evgrafova, S., Scheper, T., & Cho, S. Y. (2009). Antimicrobial effects of silver nanoparticles synthesized by an electrochemical method. In *Nanostructured Materials for Advanced Technological Applications* (pp. 215–218). Springer.

Kim, J. S., Kuk, E., Yu, K. N., Kim, J. H., Park, S. J., Lee, H. J., . . . & Kim, Y. K. (2007). Antimicrobial effects of silver nanoparticles. *Nanomedicine: Nanotechnology, Biology and Medicine*, *3*(1), 95–101.

Kim, S. G., Hyeon, D. H., Chun, J. H., Chun, B. H., & Kim, S. H. (2013). Novel thin nanocomposite RO membranes for chlorine resistance. *Desalination and Water Treatment*, *51*(31–33), 6338–6345.

Kong, L., Gan, X., bin Ahmad, A. L., Hamed, B. H., Evarts, E. R., Ooi, B., & Lim, J. (2012). Design and synthesis of magnetic nanoparticles augmented microcapsule with catalytic and magnetic bifunctionalities for dye removal. *Chemical Engineering Journal*, *197*, 350–358.

Li, J., Ng, H. T., Cassell, A., Fan, W., Chen, H., Ye, Q., . . . & Meyyappan, M. (2003a). Carbon nanotube nanoelectrode array for ultrasensitive DNA detection. *Nano Letters*, *3*(5), 597–602.

Li, Q., Li, Y. W., Liu, Z., Xie, R., & Shang, J. K. (2010). Memory antibacterial effect from photoelectron transfer between nanoparticles and visible light photocatalyst. *Journal of Materials Chemistry*, *20*(6), 1068–1072.

Li, Y. H., Ding, J., Luan, Z., Di, Z., Zhu, Y., Xu, C., . . . & Wei, B. (2003b). Competitive adsorption of Pb2+, Cu2+ and Cd2+ ions from aqueous solutions by multiwalled carbon nanotubes. *Carbon*, *41*(14), 2787–2792.

Lu, G. Q., & Zhao, X. S. (2004). Nanoporous materials—science and engineering. In *Nanoporous Materials-an Overview*. World Scientific.

Luo, X., Zeng, J., Liu, S., & Zhang, L. (2015). An effective and recyclable adsorbent for the removal of heavy metal ions from aqueous system: Magnetic chitosan/cellulose microspheres. *Bioresource Technology*, *194*, 403–406.

Ma, J., Ping, D., & Dong, X. (2017). Recent developments of graphene oxide-based membranes: A review. *Membranes*, *7*(3), 52.

Mahmoodi, N. M. (2014). Synthesis of core—shell magnetic adsorbent nanoparticle and selectivity analysis for binary system dye removal. *Journal of Industrial and Engineering Chemistry*, *20*(4), 2050–2058.

Majumder, M., Chopra, N., Andrews, R., & Hinds, B. J. (2005). Enhanced flow in carbon nanotubes. *Nature*, *438*(7064), 44–44.

Mänttäri, M., & Nyström, M. (2007). Membrane filtration for tertiary treatment of biologically treated effluents from the pulp and paper industry. *Water Science and Technology*, *55*(6), 99–107.

Mauter, M. S., & Elimelech, M. (2008). Environmental applications of carbon-based nanomaterials. *Environmental Science & Technology*, *42*(16), 5843–5859.

Mohapatra, D. P., & Kirpalani, D. M. (2019). Advancement in treatment of wastewater: Fate of emerging contaminants. *The Canadian Journal of Chemical Engineering*, *97*(10), 2621–2631.

Mondal, S., Ouni, H., Dhahbi, M., & De, S. (2012). Kinetic modeling for dye removal using polyelectrolyte enhanced ultrafiltration. *Journal of Hazardous Materials*, *229*, 381–389.

Moretz, P. (2004). Nanoparticles developed that could clean environment. *Temple Times*. Available from: https://www.temple.edu/temple_times/9-9-04/nanoparticles.html

Morones, J. R., Elechiguerra, J. L., Camacho, A., Holt, K., Kouri, J. B., Ramírez, J. T., & Yacaman, M. J. (2005). The bactericidal effect of silver nanoparticles. *Nanotechnology*, *16*(10), 2346.

Munter, R., Preis, S., Kamenev, S., & Siirde, E. (1993). Methodology of ozone introduction into water and wastewater treatment. *The Journal of the International Ozone Association*, *15*(2), 149–165.

Narsetti, R., Curry, R. D., McDonald, K. F., Clevenger, T. E., & Nichols, L. M. (2006). Microbial inactivation in water using pulsed electric fields and magnetic pulse compressor technology. *IEEE Transactions on Plasma Science*, *34*(4), 1386–1393.

Narwade, V. N., Khairnar, R. S., & Kokol, V. (2017). In-situ synthesised hydroxyapatite-loaded films based on cellulose nanofibrils for phenol removal from wastewater. *Cellulose*, *24*(11), 4911–4925.

Ngang, H. P., Ooi, B. S., Ahmad, A. L., & Lai, S. O. (2012). Preparation of PVDF—TiO2 mixed-matrix membrane and its evaluation on dye adsorption and UV-cleaning properties. *Chemical Engineering Journal*, *197*, 359–367.

Ngomsik, A. F., Bee, A., Draye, M., Cote, G., & Cabuil, V. (2005). Magnetic nano-and microparticles for metal removal and environmental applications: A review. *Comptes Rendus Chimie*, *8*(6–7), 963–970.

Nicolaï, A., Sumpter, B. G., & Meunier, V. (2014). Tunable water desalination across graphene oxide framework membranes. *Physical Chemistry Chemical Physics*, *16*(18), 8646–8654.

Nishimura, K., Miura, O., Ito, D., Tsunasima, Y., & Wada, Y. (2009). Removal of radioactive heavy metal ions from solution by superconducting high-gradient magnetic separation with schwertmannite and zirconium-ferrite adsorbents. *IEEE Transactions on Applied Superconductivity*, *19*(3), 2162–2164.

Okada, H., Kudo, Y., Nakazawa, H., Chiba, A., Mitsuhashi, K., Ohara, T., & Wada, H. (2004). Removal system of arsenic from geothermal water by high gradient magnetic separation-HGMS reciprocal filter. *IEEE Transactions on Applied Superconductivity*, *14*(2), 1576–1579.

Owen, G., Bandi, M., Howell, J. A., & Churchouse, S. J. (1995). Economic assessment of membrane processes for water and waste water treatment. *Journal of Membrane Science*, *102*, 77–91.

Paillard, H., Legube, B., Bourbigot, M. M., & Lefebvre, E. (1989). Iron and manganese removal with ozonation in the presence of humic substances. *Ozone: Science & Engineering*, *11*(1), 93–113.

Panáček, A., Kvitek, L., Prucek, R., Kolář, M., Večeřová, R., Pizúrová, N., . . . & Zbořil, R. (2006). Silver colloid nanoparticles: Synthesis, characterization, and their antibacterial activity. *The Journal of Physical Chemistry B*, *110*(33), 16248–16253.

Parra, S., Sarria, V., Malato, S., Péringer, P., & Pulgarin, C. (2000). Photochemical versus coupled photochemical—biological flow system for the treatment of two biorecalcitrant herbicides: Metobromuron and isoproturon. *Applied Catalysis B: Environmental*, *27*(3), 153–168.

Peng, S., Meng, H., Ouyang, Y., & Chang, J. (2014). Nanoporous magnetic cellulose—chitosan composite microspheres: Preparation, characterization, and application for Cu (II) adsorption. *Industrial & Engineering Chemistry Research*, *53*(6), 2106–2113.

Pera-Titus, M., García-Molina, V., Baños, M. A., Giménez, J., & Esplugas, S. (2004). Degradation of chlorophenols by means of advanced oxidation processes: A general review. *Applied Catalysis B: Environmental*, *47*(4), 219–256.

Petrović, M., Gonzalez, S., & Barceló, D. (2003). Analysis and removal of emerging contaminants in wastewater and drinking water. *TrAC Trends in Analytical Chemistry*, *22*(10), 685–696.

Peyton, G. R., & Glaze, W. H. (1988). Destruction of pollutants in water with ozone in combination with ultraviolet radiation. 3. Photolysis of aqueous ozone. *Environmental Science & Technology*, *22*(7), 761–767.

Pirkanniemi, K., Metsärinne, S., & Sillanpää, M. (2007). Degradation of EDTA and novel complexing agents in pulp and paper mill process and waste waters by Fenton's reagent. *Journal of Hazardous Materials*, *147*(1–2), 556–561.

Poole, A. J. (2004). Treatment of biorefractory organic compounds in wool scour effluent by hydroxyl radical oxidation. *Water Research*, *38*(14–15), 3458–3464.

Pulgarin, C., Invernizzi, M., Parra, S., Sarria, V., Polania, R., & Péringer, P. (1999). Strategy for the coupling of photochemical and biological flow reactors useful in mineralization of biorecalcitrant industrial pollutants. *Catalysis Today*, *54*(2–3), 341–352.

Qiao, H., Zhou, Y., Yu, F., Wang, E., Min, Y., Huang, Q., . . . & Ma, T. (2015). Effective removal of cationic dyes using carboxylate-functionalized cellulose nanocrystals. *Chemosphere*, *141*, 297–303.

Qu, X., Alvarez, P. J., & Li, Q. (2013). Applications of nanotechnology in water and wastewater treatment. *Water Research*, *47*(12), 3931–3946.

Ram, B., & Chauhan, G. S. (2018). New spherical nanocellulose and thiol-based adsorbent for rapid and selective removal of mercuric ions. *Chemical Engineering Journal*, *331*, 587–596.

Ren, C. E., Hatzell, K. B., Alhabeb, M., Ling, Z., Mahmoud, K. A., & Gogotsi, Y. (2015). Charge-and size-selective ion sieving through Ti3C2T x MXene membranes. *The Journal of Physical Chemistry Letters*, *6*(20), 4026–4031.

Rizzo, L. (2011). Wastewater treatment by solar driven advanced oxidation processes. *Water, Wastewater and Soil Treatment by Advanced Oxidation Processes (AOPs)*, *183*.

Rocher, V., Siaugue, J. M., Cabuil, V., & Bee, A. (2008). Removal of organic dyes by magnetic alginate beads. *Water Research*, *42*(4–5), 1290–1298.

Rodríguez, J., Contreras, D., Parra, C., Freer, J., Baeza, J., & Durán, N. (1999). Pulp mill effluent treatment by Fenton-type reactions catalyzed by iron complexes. *Water Science and Technology*, *40*(11–12), 351–355.

Rodríguez, M. (2003). *Fenton and UV-vis Based Advanced Oxidation Processes in Wastewater Treatment: Degradation, Mineralization and Biodegradability Enhancement*. Universitat de Barcelona.

Saho, N., Isogami, H., Takagi, T., & Morita, M. (1999). Continuous superconducting-magnet filtration system. *IEEE Transactions on Applied Superconductivity*, *9*(2), 398–401.

Saito, T., & Isogai, A. (2005). Ion-exchange behavior of carboxylate groups in fibrous cellulose oxidized by the TEMPO-mediated system. *Carbohydrate Polymers*, *61*(2), 183–190.

Sangave, P. C., Gogate, P. R., & Pandit, A. B. (2007a). Combination of ozonation with conventional aerobic oxidation for distillery wastewater treatment. *Chemosphere*, *68*(1), 32–41.

Sangave, P. C., Gogate, P. R., & Pandit, A. B. (2007b). Ultrasound and ozone assisted biological degradation of thermally pretreated and anaerobically pretreated distillery wastewater. *Chemosphere*, *68*(1), 42–50.

Sarria, V., Parra, S., Invernizzi, M., Péringer, P., & Pulgarin, C. (2001). Photochemical-biological treatment of a real industrial biorecalcitrant wastewater containing 5-amino-6-methyl-2-benzimidazolone. *Water Science and Technology*, *44*(5), 93–101.

Schoen, D. T., Schoen, A. P., Hu, L., Kim, H. S., Heilshorn, S. C., & Cui, Y. (2010). High speed water sterilization using one-dimensional nanostructures. *Nano Letters*, *10*(9), 3628–3632.

Scott, J. P., & Ollis, D. F. (1995). Integration of chemical and biological oxidation processes for water treatment: Review and recommendations. *Environmental Progress*, *14*(2), 88–103.

Sgroi, M., Roccaro, P., Korshin, G. V., Greco, V., Sciuto, S., Anumol, T., . . . & Vagliasindi, F. G. (2017). Use of fluorescence EEM to monitor the removal of emerging contaminants in full scale wastewater treatment plants. *Journal of Hazardous Materials*, *323*, 367–376.

Shao, L., Cheng, X. Q., Liu, Y., Quan, S., Ma, J., Zhao, S. Z., & Wang, K. Y. (2013). Newly developed nanofiltration (NF) composite membranes by interfacial polymerization for Safranin O and Aniline blue removal. *Journal of Membrane Science*, *430*, 96–105.

Sharma, N., Singh, A., & Batra, N. (2019). Modern and emerging methods of wastewater treatment. In *Ecological Wisdom Inspired Restoration Engineering* (pp. 223–247). Springer.

Sharma, V., & Sharma, A. (2012). Nanotechnology: An emerging future trend in wastewater treatment with its innovative products and processes. *Nanotechnology*, *1*(2).

Shrivastava, S., Bera, T., Roy, A., Singh, G., Ramachandrarao, P., & Dash, D. (2007). Characterization of enhanced antibacterial effects of novel silver nanoparticles. *Nanotechnology*, *18*(22), 225103.

Silvestry-Rodriguez, N., Bright, K. R., Uhlmann, D. R., Slack, D. C., & Gerba, C. P. (2007). Inactivation of Pseudomonas aeruginosa and Aeromonas hydrophila by silver in tap water. *Journal of Environmental Science and Health, Part A*, *42*(11), 1579–1584.

Simstich, B., & Oeller, H. J. (2010). Membrane technology for the future treatment of paper mill effluents: Chances and challenges of further system closure. *Water Science and Technology*, *62*(9), 2190–2197.

Srinivas, G., Burress, J. W., Ford, J., & Yildirim, T. (2011). Porous graphene oxide frameworks: Synthesis and gas sorption properties. *Journal of Materials Chemistry*, *21*(30), 11323–11329.

Srivastava, S., Kardam, A., & Raj, K. R. (2012). Nanotech reinforcement onto cellulosic fibers: Green remediation of toxic metals. *International Journal of Green Nanotechnology*, *4*(1), 46–53.

Sun, L., Huang, H., & Peng, X. (2013). Laminar MoS 2 membranes for molecule separation. *Chemical Communications*, *49*(91), 10718–10720.

Thomas, J. A., & McGaughey, A. J. (2008). Reassessing fast water transport through carbon nanotubes. *Nano Letters*, *8*(9), 2788–2793.

Tiwari, D. K., Behari, J., & Sen, P. (2008). Application of nanoparticles in waste water treatment. *World Applied Sciences Journal*, *3*(3).

Tran, N. H., Reinhard, M., & Gin, K. Y. H. (2018). Occurrence and fate of emerging contaminants in municipal wastewater treatment plants from different geographical regions-a review. *Water Research*, *133*, 182–207.

Uyttebroek, M., Baillieul, H., Vermeiren, N., Scholiers, R., Devleeschauwer, P., Gemoets, J., & Bastiaens, L. (2010). In situ remediation of a chlorinated ethene contaminated source zone by injection of zero-valent iron: From lab to field scale. *Permeable Reactive Barriers & Reactive Zones, 20.*

Van der Bruggen, B., Vandecasteele, C., Van Gestel, T., Doyen, W., & Leysen, R. (2003). A review of pressure-driven membrane processes in wastewater treatment and drinking water production. *Environmental Progress*, *22*(1), 46–56.

van Grieken, R., Marugán, J., Sordo, C., Martínez, P., & Pablos, C. (2009). Photocatalytic inactivation of bacteria in water using suspended and immobilized silver-TiO2. *Applied Catalysis B: Environmental*, *93*(1–2), 112–118.

Van Hooijdonk, E., Bittencourt, C., Snyders, R., & Colomer, J. F. (2013). Functionalization of vertically aligned carbon nanotubes. *Beilstein Journal of Nanotechnology*, *4*(1), 129–152.

Van Nostrand, J. E. (2007). *Detection and Destruction of Escherichia Coli Bacteria and Bacteriophage Using Biofunctionalized Nanoshells* (Doctoral dissertation, Wright State University).

Vogelpohl, A. (2007). Applications of AOPs in wastewater treatment. *Water Science and Technology*, *55*(12), 207–211.

Wang, D. (2019). A critical review of cellulose-based nanomaterials for water purification in industrial processes. *Cellulose*, *26*(2), 687–701.

Xu, C., Cui, A., Xu, Y., & Fu, X. (2013a). Graphene oxide—TiO2 composite filtration membranes and their potential application for water purification. *Carbon*, *62*, 465–471.

Xu, H., Zhang, Y., Jiang, Q., Reddy, N., & Yang, Y. (2013b). Biodegradable hollow zein nanoparticles for removal of reactive dyes from wastewater. *Journal of Environmental Management*, *125*, 33–40.

Yang, H. Y., Han, Z. J., Yu, S. F., Pey, K. L., Ostrikov, K., & Karnik, R. (2013). Carbon nanotube membranes with ultrahigh specific adsorption capacity for water desalination and purification. *Nature Communications*, *4*(1), 1–8.

Yu, G., Zhu, W., & Yang, Z. (1998). Pretreatment and biodegradability enhancement of DSD acid manufacturing wastewater. *Chemosphere*, *37*(3), 487–494.

Yu, X., Tong, S., Ge, M., Wu, L., Zuo, J., Cao, C., & Song, W. (2013). Adsorption of heavy metal ions from aqueous solution by carboxylated cellulose nanocrystals. *Journal of Environmental Sciences*, *25*(5), 933–943.

Yunus, I. S., Harwin, Kurniawan, A., Adityawarman, D., & Indarto, A. (2012). Nanotechnologies in water and air pollution treatment. *Environmental Technology Reviews*, *1*(1), 136–148.

Zhang, L., Yan, J., Zhou, M., & Liu, Y. (2012). Photocatalytic inactivation of bacteria by TiO2-based compounds under simulated sunlight irradiation. *International Journal of Material Science*, *2*(2), 43–46.

Zheng, Y., Yao, G., Cheng, Q., Yu, S., Liu, M., & Gao, C. (2013). Positively charged thin-film composite hollow fiber nanofiltration membrane for the removal of cationic dyes through submerged filtration. *Desalination*, *328*, 42–50.

Zhou, H., & Smith, D. W. (2002). Advanced technologies in water and wastewater treatment. *Journal of Environmental Engineering and Science*, *1*(4), 247–264.

Index

For Product Safety Concerns and Information please contact our EU
representative GPSR@taylorandfrancis.com
Taylor & Francis Verlag GmbH, Kaufingerstraße 24, 80331 München, Germany

www.ingramcontent.com/pod-product-compliance
Ingram Content Group UK Ltd.
Pitfield, Milton Keynes, MK11 3LW, UK
UKHW011452240425
457818UK00008B/293